INTRODUCTION TO NANOMEDICINE AND NANOBIOENGINEERING

WILEY SERIES IN BIOMEDICAL ENGINEERING AND MULTIDISCIPLINARY INTEGRATED SYSTEMS

KAI CHANG, SERIES EDITOR

Advances in Optical Imaging for Clinical Medicine
Nicusor Iftimia, William R. Brugge, and Daniel X. Hammer (Editors)

Antigen Retrieval Immunohistochemistry Based Research and Diagnostics
Shan-Rong Shi and Clive R. Taylor

Introduction to Nanomedicine and Nanobioengineering
Paras N. Prasad

INTRODUCTION TO NANOMEDICINE AND NANOBIOENGINEERING

PARAS N. PRASAD

WILEY

A JOHN WILEY & SONS, INC., PUBLICATION

Published by John Wiley & Sons, Inc., Hoboken, New Jersey
Published simultaneously in Canada

Library of Congress Cataloging-in-Publication Data:

Prasad, Paras N.
Introduction to nanomedicine and nanobioengineering / Paras N. Prasad.
p. ; cm.
Transforming healthcare with nanotechnology
Includes bibliographical references.
ISBN 978-1-118-09343-6
I. Title. II. Title: Transforming healthcare with nanotechnology.
[DNLM: 1. Nanomedicine–methods. QT 36.5]
610.28'4–dc23

2012002564

Printed in the United States of America

10 9 8 7 6 5 4 3 2 1

CONTENTS

PREFACE

Nanomedicine and nanobioengineering, defined respectively by the fusion of nanotechnology with medicine and bioengineering, are emerging new frontiers, providing challenges for fundamental research and opportunities for new biotechnologies. These fields promise to make a major impact on healthcare worldwide. They are multidisciplinary fields, creating opportunities in physics, chemistry, applied sciences, engineering, and biology, as well as in biomedical technology and drug development.

A good number of books and reviews discuss selective aspects of nanomedicine and nanobioengineering. However, there is a need for a comprehensive monograph that provides a unified synthesis of these two fields and their joint impact on global healthcare. This book fills this need, by providing an integrated description of nanomedicine and nanobioengineering for new-generation diagnostics and therapy. The comprehensive, multidisciplinary program at our Institute in this integrated field of nanomedicine received early, generous funding from the John R. Oishei Foundation in Buffalo, New York; and the resulting work provided the inspiration and much of the impetus to undertake this writing.

The objective of this book is to provide a basic knowledge of a broad range of topics in an integrated manner so that individuals in all disciplines can rapidly acquire the minimal necessary background for research and development in this field. The author intends this monograph to serve both as a textbook for education and training and as a reference book that aids research and development in areas integrating nanotechnology with medicine and bioengineering. Another aim of the book is to stimulate the interest of researchers, industries, and businesses, as well as to foster collaboration through multidisciplinary programs in these emerging frontiers of bioscience and biotechnology to yield a new dimension for healthcare.

This book encompasses the fundamentals of nanomaterials design, bioengineering, nanodiagnostics, and nanotherapy. Each chapter begins with an introduction describing what a reader will find in that chapter. Each chapter ends with highlights that are basically the "take-home message" and may serve as a review of the materials presented. In addition, exercises are provided to assist in teaching. Throughout the book, examples are given from our research work wherever possible, merely because of convenience; this should be considered supplemental to the excellent work being conducted at various institutions worldwide.

In writing this book, which covers a very broad range of topics, I received help from a large number of individuals at the Institute of Lasers, Photonics, and Biophotonics at the University at Buffalo, State University of New York, as well as from elsewhere. This help has consisted of furnishing technical information, creating illustrations, providing critiques, and preparing the manuscript. A separate Acknowledgments section recognizes these individuals.

Here I would like to acknowledge the individuals whose broad-based support has been of paramount value in completing the book. I express my sincere appreciation to my colleague Professor Paul Knight, M.D., for his endless help and encouragement. I thank Dr. Indrajit Roy, Dr. Andrey Kuzmin, Dr. Wing Cheung Law, and Dr. Artem Pliss for their valuable support and technical help throughout the book. I owe thanks to my administrative assistant, Ms. Margie Weber, for her encouragement and for assuming responsibility of many of the noncritical administrative issues at the Institute, in order to free up my time for writing this monograph. I thank Ms. Barbara Raff, whose clerical help in manuscript preparation was invaluable.

Finally, I am indebted to my daughters, my princesses, Natasha and Melanie, for showing their love, understanding, and encouragement.

Buffalo, New York PARAS N. PRASAD

ACKNOWLEDGMENTS

Technical Contents

Professor Stephen Arnold, Dr. Ana Karla Braz, Professor Edward Furlani, Professor Indrajit Roy, Professor Paul Knight, Dr. Rajiv Kumar, Dr. Andrey Kuzmin, Dr. Wing Cheung Law, Dr. Supriya Mahajan, Professor Gene Morse, Dr. Tymish Ohulchanskyy, Dr. Artem Pliss, Dr. Haridas Pudavar, Dr. Yudhisthira Sahoo, Professor Mark Swihart.

Technical Illustrations and References

Dr. Adela Bonoui, Dr. Ana Karla Braz, Dr. Folarin Erogbogbo, Professor Edward Furlani, Professor Indrajit Roy, Dr. Rajiv Kumar, Dr. Andrey Kuzmin, Dr. Wing Cheung Law, Dr. Supriya Mahajan, Dr. Tymish Ohulchanskyy, Dr. Artem Pliss.

Chapter Critiques

Professor Stelios Andreadis, Professor Robert Baier, Ms. Cathy Carfagna, Professor Heather Clark, Professor Howard Gendelman, Dr. Piotr Grodzinski, Professor Jiang Feng, Dr. George Hinkal, Dr. Aliaksandr Kachynski, Ms. Dana Knight, Professor Raoul Kopelman, Dr. Ewa Anna Kucz, Dr. Howard Lippes, Professor Mona Marei, Dr. Anil Patri, Ms. Melanie Prasad, Professor Marek Samoc, Dr. Stanley Schwartz, Dr. Hulda Swai, Professor Kenneth Tramposch, Dr. Paul Wallace, Ms. Hendriette Van der Walt.

Manuscript Preparation

Dr. Andrey Kuzmin, Ms. Barbara Raff, Ms. Margie Weber.

CHAPTER 1

INTRODUCTION

1.1. NANOMEDICINE: A GLOBAL VISION

This monograph describes emerging interdisciplinary frontiers created by the fusion of nanotechnology, engineering, and medicine that provide a global vision to produce breakthrough approaches for meeting our current and future healthcare challenges. Traditionally, nanomedicine is defined as the application of nanotechnology to medicine; nanobioengineering is often used to describe nanotechnology applied to bioengineering, which includes imaging, sensing, diagnostics, blood fluids, and tissue engineering. This book brings a much-needed integration of nanomedicine and nanobioengineering to produce a broadened nanomedicine platform that utilizes nanotechnology to generate exciting new approaches for diagnostics, bioengineering, and targeted therapy. Such an integration could lead to multifunctional nanomedicines that can, as a single formulation, be used to diagnose, treat, and evaluate treatment effectiveness in real time. Collectively, these agents are termed nanotheranostics.

We live in a complex world where our health is determined by an interplay of our genetic inheritance, the environment we live in, and the lifestyle we choose. As the barriers between social, ethnic, religious, regional, and national divides come down and the world becomes a melting pot for the human race, healthcare issues (whether genetic, environmental, or lifestyle originated) do not remain localized. Today these issues are not the problems of a specific

Introduction to Nanomedicine and Nanobioengineering, First Edition. Paras N. Prasad.

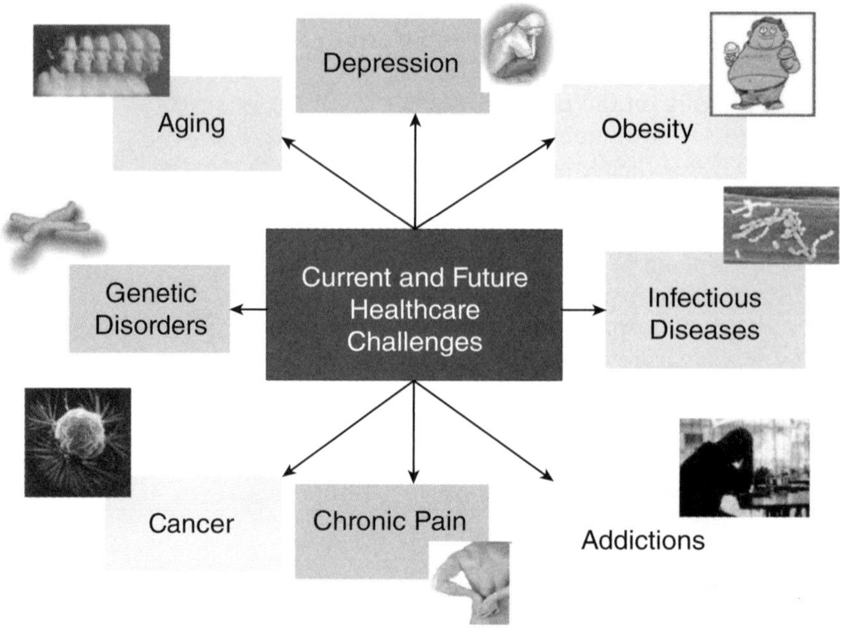

Figure 1.1. Current and future healthcare challenges.

society or a specific nation, but a global concern and global priority. As technological advances facilitate rapid travel through geographic variation, different time zones, and diverse climates, infections are no longer confined to their old boundaries and instead travel all over the globe, spreading like wildfire. Thus, as new healthcare challenges emerge in the future, we must face them together as a single global community and find effective solutions for them collectively.

The healthcare challenges that we now face and can anticipate for the future are many, and they pose an almost insurmountable task for us. Figure 1.1 lists some major challenges that our global community faces. Despite tremendous progress in winning some cancer battles, cancer remains a major healthcare challenge. Take, for example, pancreatic cancer, the survival rate for which beyond five years after detection using the current diagnostics is only 4%. Consequently, there is a need for early diagnosis, preferably at a precancerous stage when many options to treat may be available, as well as for a more effective treatment. Some treatments for cancer can be very harsh, where the patient's quality of life is seriously compromised. A more effective treatment or an alternative gentler therapy would be of significant value to such patients. New strains of infectious diseases, as well as existing ones, are another major challenge we face. New infections such as the bird flu or swine flu may originate in one small region, but it does not take long for them to spread around the world and become a pandemic. Infections such as tuberculosis and malaria, often referred to as poverty-related diseases (PRDs), are on the rise and spreading worldwide. Diseases that are manifestations of genetic disorder are again on the rise worldwide, as a result of a complex interplay of our genes, the environment, and our dietary intake. Depression and chronic pain are other healthcare problems what are highly detrimental to the quality of life that we wish to have.

Then there are healthcare issues that we create by the lifestyle we choose. The examples given in Figure 1.1 are obesity and addiction. Obesity rates are rapidly rising, and of particular concern is child obesity. Obesity creates not only physical handicap, but also a cascade of other disease manifestations, such as diabetes and cardiovascular diseases. While obesity may start in many cases from a lifestyle of eating unhealthy and fattening food, it soon becomes a biochemical addiction in which overactive bad genes in the brain create a constant need to eat. Similarly, addictions to medication, drugs, and alcohol are biochemical in nature, generating specific biochemical signatures in the brain—again, produced by the lifestyle we choose. Such addictions have now become major health issues worldwide.

Aging is not a disease, but it does affect quality of life and increases an individual's vulnerability to various diseases and infections. The world's population is aging. Accordingly, more people are suffering from neurological disorders such as Alzheimer's disease, impairment of body functions, chronic pain in joints, loss of hearing, and a reduction in eyesight. While we cannot permanently reverse aging (for which we must wait for rejuvenation therapy, discussed in Chapter 16, to develop), we can certainly use new medical advances in utilizing stem cells (discussed in Chapter 17) to replenish nonfunctioning cells, and tissue engineering (presented in Chapter 18) to replace nonfunctioning organs. We can also explore the promise of gene therapy (covered in Chapter 14) and stem cell therapy to treat neurological diseases, as well as to effectively boost the immunity to fight infections (discussed in Chapter 15). This book will address how an integrated nanomedicine platform provides new, revolutionary approaches to tackle these major healthcare issues.

1.2. THE NANOTECHNOLOGY REVOLUTION: REALIZATION OF ASIMOV'S FICTION

Nanotechnology is an emerging discipline of science and technology that has captured the imaginations of the world. Many countries have recognized nanotechnology as a national priority and allocated major resources to develop this area. It has a high societal impact, because it provides promising new solutions to numerous technical needs that the world faces (subject of global priorities), some of which are listed in Figure 1.2. In commonly adopted definitions, nanotechnology deals with materials, structures, and devices that are of dimensions in the range of 1–100 nanometers (1 nanometer is one billionth of a meter;

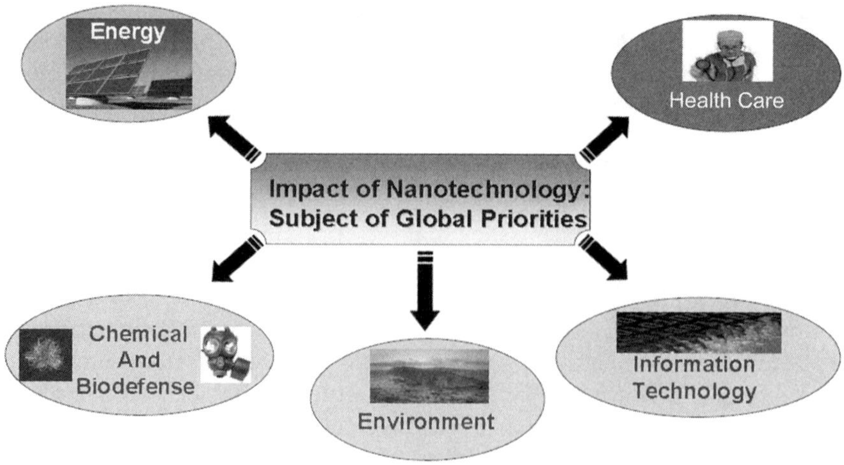

Figure 1.2. Examples of global priorities on which nanotechnology has made an impact.

nanometer abbreviated as nm) and thus are so ultrasmall that they cannot even be seen under a normal optical microscope. In reality, nanotechnology is not so new. Although it is hard to place an exact period when it was introduced, there are plenty of examples of their use in the production of tinted glass widely used in cathedrals and churches from the medieval period onwards. A beautiful example is shown in Figure 1.3, which is from the cathedral of Notre Dame in Paris. It was not until 1860 that Michael Faraday actually recognized that these bright colors in the glass were imparted by metallic gold or silver nanoparticle inclusions formed during glass processing. However, many consider Feynman as the father of modern nanotechnology when he stated in his famous lecture of 1970, "There is plenty of room at the bottom. . . ." This referred to the fact that many, many objects (particles or structures) of nanometer dimensions can be packed even in a small volume.

A major impact area for nanotechnology is healthcare. A nanoscale object (such as a nanoparticle) can provide new approaches to diagnostics and therapy, which constitutes the field of nanomedicine. Such developments demonstrate a realization of the fiction novel *Fantastic Voyage* by Isaac Asimov in 1966, which was later dramatized in a film by Richard Fleischer. These works presented a visionary fiction in which a submarine carrying a crew and a medical team was reduced to microscopic size and injected into the bloodstream of a diplomat. As depicted in Figure 1.4, the submarine navigated to a blood clot, which was then zapped with a laser beam to remove it and thus save the diplomat. In 2002, we used the term "nanoclinics" to describe the modern approach of using nanoparticles as carriers for targeting and circulating agents that can be directed to a desired biological site in a body. These nanoparticles can be armed with various diagnostic probes to provide on-site diagnosis and then treat and/or repair a disease manifestation. (For a more

Figure 1.3. Metallic nanoparticles doped stained glass windows in Notre Dame Cathedral in Paris.

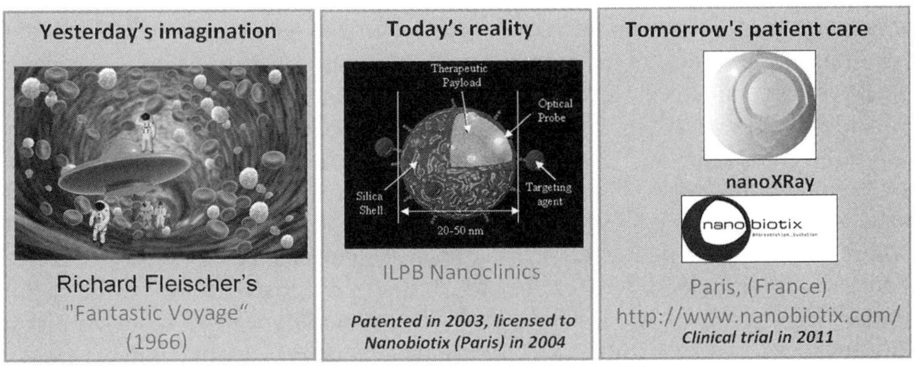

Figure 1.4. The imaginary nanosubmarine in the 1966 science fiction movie *Fantastic Voyage*, shown with the concept of a nanoclinic developed by us in 2002. This technology is currently being used in a clinical trial at Nanobiotix for X-ray nanotherapy.

detailed discussion, please see Chapter 3.) This approach is an excellent example of how the imagination of yesterday can become a reality today, and it can be further refined to produce high societal impact in the future. As shown in Figure 1.4, our nanoclinic concept licensed by the company Nanobiotix has just entered a clinical trial for X-ray nanotherapy.

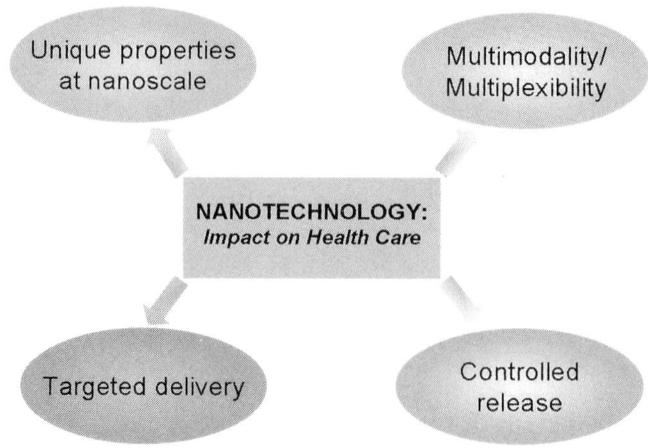

Figure 1.5. Features provided by nanotechnology that impact healthcare.

The possible impacts of nanotechnology on healthcare and society are tremendous. Some of the features offered by nanotechnology for healthcare are illustrated in Figure 1.5. First, materials, when reduced to nanometer size, exhibit physical properties that can be different from their bulk form. Furthermore, this property may become size-dependent on nanoscale. An example discussed in detail in Chapter 5 is the light absorption and emission by nanoparticles of inorganic semiconductors, such as CdSe or Si. Once their size becomes smaller than a certain length, the wavelength of light they absorb and the resulting emission color (the emission wavelength) become size-dependent. These semiconductor nanomaterials are quantum dots and quantum rods, which are presented in Chapter 5. This size dependence can be utilized for multiplexed optical bioimaging using quantum dots or quantum rods of various sizes. Another example is metallic nanoparticles, which on nanoscale are not reflective. As shown in Figure 1.3, the metallic nanoparticles exhibit bright colors derived from new optical absorptions called surface plasmon resonance bands (also discussed in Chapter 5) that do not exist in the bulk metal form. The other feature is building multifunctionality onto a nanostructure/nanoparticle platform. For example, a nanoparticle can be loaded with a number of imaging agents for multimodal medical imaging such as optical bioimaging, magnetic resonance imaging (MRI), and positron emission tomography (PET) which are covered in Chapter 8. Even in optical imaging, one can use different dyes or quantum dots combinations, and thus multiple color staining (labeling) for multiplexed optical imaging can be realized to enhance detection specificity. The nanostructured materials are promising scaffolds for tissue regeneration, an evolving field also known as tissue engineering (Chapter 18). For therapy, one can introduce a combination of therapeutic modalities such as light-induced therapy, magnetic therapy, thermal therapy, radiotherapy, and chemotherapy into a simple nanoparticle. Nanotechnology also holds promise for stem cell biotechnology, which is discussed in Chapter 17 ("Stem Cell Biotechnology").

Targeted delivery is another important feature whereby one can introduce biorecognition (by antibody or other biospecific units) on a nanoparticle to identify a specific biomarker (signature) of a disease and thus target the disease site. One can build multiple targeting ability on the nanoparticle to enhance its specificity and thus increase its targeting ability.

Controlled release of a drug or therapeutic payload carried by a nanostructure or nanoparticle is another important feature offered by the nanotechnology approach. A nanoparticle offers tremendous structural flexibility for inclusion of various payloads and their controlled release. First, the volume of the nanoparticle can itself serve as a diagnostic or therapeutic agent (pure nanoformulation). Second, a diagnostic or therapeutic agent can be attached on the surface or included in the interior of a nanoparticle. The release can be controlled in a number of ways. First, one can introduce external control by using a magnetic, optical, or radio-frequency (rf) field to break a nanoparticle or cleave a labile chemical linkage in order to release the payload. Second, one can manipulate the pores on the nanoparticle, either by enzymatic activities or by local heating using light or magnetic field, to control the release kinetics. Finally, one can take advantage of the enzymatic activities in the targeted cells to break down the nanoparticle to make the payload active.

While nanotechnology can offer many benefits to healthcare, there is also a growing concern about potential health hazards that may be caused by nanoparticles. The short- and the long-term toxicity of nanoparticles in the body must be thoroughly investigated. With the growing euphoria about the vast potential of nanotechnology in so many industrial sectors, there is also a concern that airborne nanoparticles in a workplace can lead to organ damage and health problems. Thus, nanotoxicity (discussed in Chapter 21) is an integral factor in developing nanomedicine. Therefore, for each nanomedicine application we must weigh the benefits versus the risks.

1.3. NANOMEDICINE: A NEW ERA IN PERSONALIZED MEDICINE

Nanomedicine, inclusive of nanobioengineering in its broad scope, is a nanobiotechnology utilizing a specifically engineered nanoplatform to carry various payloads for new, minimally invasive diagnosis, targeted delivery of therapeutics, enhanced efficacy of an existing therapy/treatment, and real-time monitoring of a treatment.

The scope and applications of nanomedicine, together with nanobioengineering, are highlighted in Figure 1.6. First, *in vitro* diagnosis in a laboratory to profile a disease can utilize various body fluids/excretions such as blood,

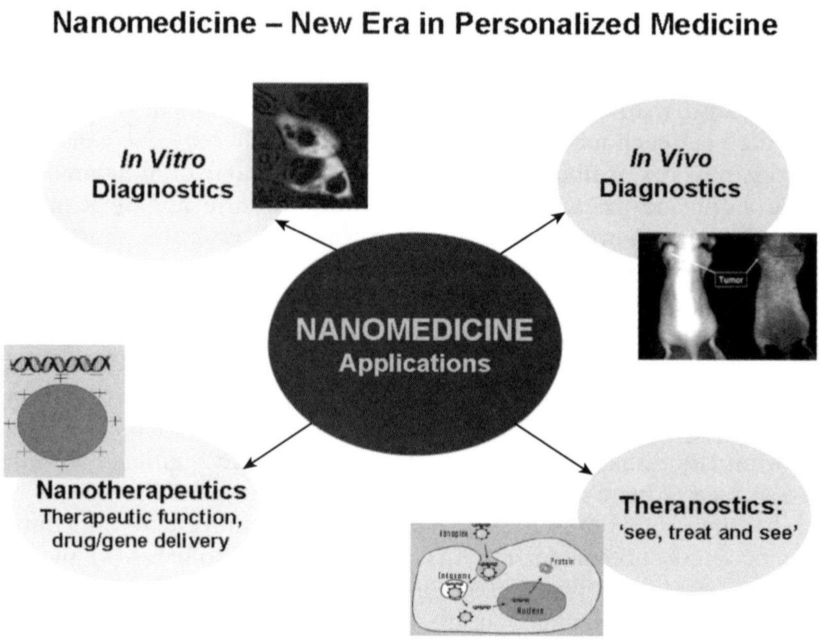

Figure 1.6. The broad scope of applications for nanomedicine.

urine, saliva, sputum, and feces. Some studies even focus on using the exhaled breath for analysis of diseases. These body fluids can interact with specially designed nanoparticles to create biological responses for identifying diseases, even at the molecular and cellular levels. This can lead ultimately to a molecular understanding of disease mechanism and sensitivity of detection at single cell levels, which can be key to early detection and personalized molecular medicine. The *in vitro* diagnosis, using a multipronged detection and quantification enabled by a nanoparticle platform, will be able to elucidate drug intake, its biodistribution, its cellular pathway, and subsequent intracellular interactions. This information can be tremendously effective in drug development and screening of various possible therapies for a given disease. Since the testing is *in vitro*, nanotoxicity is not of concern in such a scenario. For this reason, I envision that a full implementation of nanoparticle-based *in vitro* diagnosis is the first realized application of nanomedicine.

In vivo diagnosis with a nanomedicine approach offers the benefit of combining the various diagnostic modalities in a single nanoplatform (e.g., nanoparticles). For example, one can combine optical imaging and spectrometry with MRI and PET imaging to do a more thorough disease profiling based on molecular, structural, and morphological changes as a result of disease manifestations. Also, packaging them in the small nanovolume of a biocompatible nanoparticle with the ability to localize (due to the presence of targeting group) at the disease site enhances the sensitivity of detection and minimizes the potential for systemic toxicity of the imaging agent. The simultaneous presence of various diagnostic agents in the same nanoformulation also allows a medical facility to use them at the same time, without requiring separate preparation for each modality.

Of course, a major function of nanomedicine is to provide a nanoformulation that opens new modality of therapy or increases the effectiveness of an existing therapy, as well as to create the prospect of using more than one therapeutic approach in tandem. Examples of new approaches include (a) magnetic therapy using magnetic nanoparticles and (b) photothermal therapy using metallic nanoparticles. An example of improving the efficacy of an existing therapy can be demonstrated by (a) enhancing the biodistribution and circulation of a hydrophobic drug by using a nanoparticle carrier with hydrophilic surface and (b) targeting the carrier to localize a large concentration of the drug at the diseased site. Additional merits offered by nanotherapeutics include controlled and sustained release of a drug. One can control the release by manipulation of pores in the nanoparticle or external stimulation using light, magnetic field, heat, or radio-frequency field.

Finally, the biggest payoff of nanomedicine lies in the realization of theranostics, the combined function of therapy and diagnostics. In other words, the functions of targeting, effective biodistribution, multiple diagnostics, and multimodal therapy can be combined in a single nanoformulation. This allows one to follow the process of therapy to see (and monitor) a therapeutic process at work and to assess its effectiveness in real time. Real-time monitoring of therapeutic action will be of tremendous value to a patient, because one does not have to wait post treatment to determine the outcome.

1.4. NANOMEDICINE: A PROMISE OR REALITY?

In any emerging field showing great promise (and often generating considerable hype), expectations generally run ahead of the real progress. Naturally, the question may arise whether the field of nanomedicine is only a promise for the future (which may or may not materialize) or if there is evidence that nanomedicine is already impacting healthcare. This section provides a very brief account of what has been already achieved in nanomedicine.

Within the realm of *in vitro* diagnostics, in which tests are conducted in the laboratory on biological fluids outside of the body, nanotechnology is well poised to make a significant, immediate impact. There are already examples of nanoparticle-based colorimetric detection modalities, such as those used for home pregnancy kits, in which color changes are introduced by aggregation of metallic nanoparticles caused by the biomarker signature (expressions) of pregnancy. The surface plasmon resonance (SPR) biosensors discussed in Chapter 9 are widely used in biological laboratories and biomedical research worldwide.

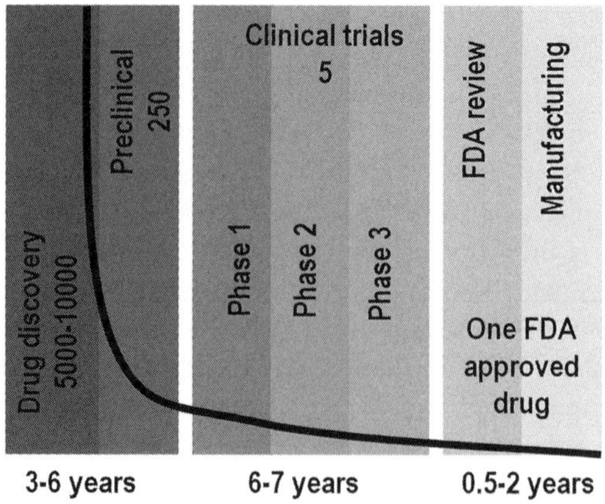

Figure 1.7. The different steps of drug development and clearance by the FDA. Data from http://www.phrma.org/.

In vivo diagnostics, nanocarrier drug delivery, tissue regeneration, and organ replacement require introduction of a foreign nanostructure in the body. The procedure for regulatory clearance [such as by the Food and Drug Administration (FDA) in the United States] is quite complex, as illustrated by Figure 1.7, and requires several steps of clinical trials.

However, several nanoformulations of drugs—such as for cancer therapy—are already FDA approved and are being used. In addition, many nanoformulations are undergoing different stages of clinical trial. Chapter 13 provides examples of nanoformulations of chemotherapy drugs for cancer treatment. Two examples are Doxil® (a nanoparticle formulation of the drug Doxorubicin, FDA approved in 1995) and Abraxane® (a nanoparticle formulation of the drug Paelitaxel, FDA approved in 2005). These are discussed in Chapter 13, along with other nanoformulations that have been approved or are in clinical trials for cancer therapy.

1.5. A NEW FRONTIER: MULTIDISCIPLINARY CHALLENGES AND OPPORTUNITIES

Nanomedicine, in its broad scope (of which nanobioengineering is a major component), is a new frontier that faces multidisciplinary challenges—from a proper formulation of nanoplatform, to bedside implementation of nanotheranostics. It thus requires a close collaboration between biologists, chemists, physicists, engineers, pharmacologists, and clinicians. Some key multidisciplinary challenges (which in turn provide opportunities for a given discipline)

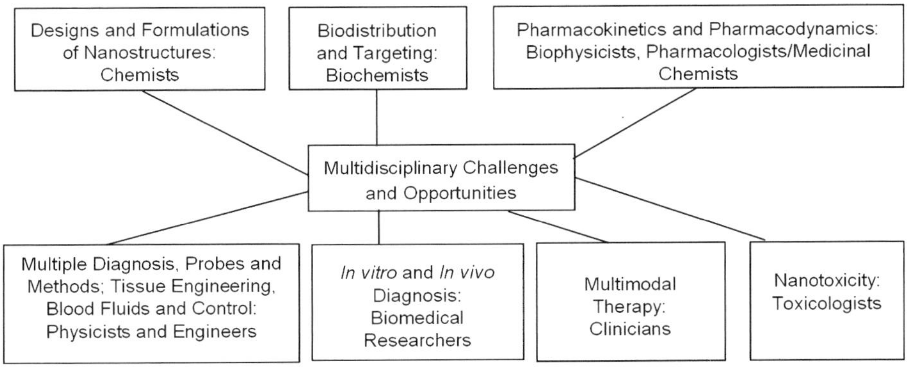

Figure 1.8. Multidisciplinary challenges and opportunities offered by integrating nanomedicine and nanobioengineering.

are summarized in Figure 1.8. For chemists, the challenges and opportunities include producing effective nanoformulations that are chemically and environmentally stable, as well as biocompatible, and that provide appropriate linkage and caging sites to attach and/or encapsulate the following: (a) various diagnostic probes, (b) therapeutic agents, and (c) groups enhancing circulation and producing targeting. Because a nanoparticle has a large surface-to-volume ratio, control of the surface composition and structure (surface chemistry) plays an important role in chemical design and synthesis. Identification of biomarkers and the selection of targeting group is another important aspect of creating a nanosize magic bullet that makes a precise hit of the target (in our case, a diseased site or a tumor). Ensuring effective kinetics of biodistribution, circulation, and selective localization of the nanoparticles at the targeted site is another important challenge that requires a multidisciplinary input from biophysicists, pharmacologists, and medicinal chemists. Multiple diagnoses, using a combination of various probes and methods and utilizing a number of physical and chemical principles, require input from physicists and engineers. A growing discipline worldwide is biomedical engineering, which cross-fertilizes biomedical sciences with engineering. This is a very welcome new discipline, which can play a major role in nanomedicine through the inclusion of nanobioengineering.

In vitro and *in vivo* diagnostics provide a comprehensive approach for early disease diagnosis, as well as for monitoring its progression and drug-induced depression. Active engagement of practicing physicians in clinical trials and subsequent translation to bedside of a patient is of vital importance. We have to engage clinicians from an early stage of nanomedicine, because their feedback is crucial in advancing the frontier of nanomedicine toward real patient care. Finally, a multidisciplinary effort is necessary to evaluate and validate the safety of nanomedicine. Toxicity concerns include chemical toxicity, immunotoxicity, organ injury, and interference in physiological functions. Hence, the continued development of nanomedicine requires thorough study at the cellular, tissue, animal, and human levels.

1.6. SCOPE OF THE BOOK: MULTIDISCIPLINARY EDUCATION, TRAINING, AND RESEARCH

Like any new frontier, advances in nanomedicine will require engagement of various disciplines as described in Section 1.5. One major challenge is that these disciplines do not even use the same set of vocabularies and acronyms. An effective cross-fertilization among these disciplines will require giving them common vocabulary terms, and the introduction of multidisciplinary concepts that can provide collaborators with the ability to understand and communicate with each other on real issues. Although a good number of books and reviews cover selective aspects of nanomedicine and nanobioengineering [e.g.: Jain, 2008; Tibbols, 2011], there is a need for a comprehensive monograph that introduces the integration of unified introductory concepts and provides a broad multidisciplinary exposure of the field to new researchers. This book is intended to fill this void and act as an introduction, providing basic concepts for the benefit of readers from the disciplines of chemistry, physics, biology, biomedical sciences, biomedical engineering, medical school, pharmacy school, and dental school, as well as from the pharmaceutical and cosmetic industries. To serve this purpose, Chapters 2–7 are designed to present basic materials, elucidate concepts, and provide an overview of the current status in meeting specific challenges of the areas covered in these chapters.

For a researcher either entering the field or interested in expanding his/her research scope, for a drug developer in a pharmaceutical industry, for a biomedical engineer interested in developing appropriate engineering tools, for a dentist applying nanotechnology for dental care, for a cosmetic industry person developing nanocosmetics, or for a clinician interested in nanomedicine therapeutic approaches, the subsequent chapters introduce specific applications and needs.

Each chapter begins with a brief outline of what the reader can expect from it, and then it ends with a highlight of the chapter. These highlights succinctly summarize the key points from the chapter, which is a very convenient listing of the take-home message from that chapter. For assisting in the teaching of this subject, each chapter also provides exercises. The chapters are written largely in a self-contained manner, so that it is not necessary to read the chapters in the sequential order as presented here—the reader can skip a chapter to move on to another one, depending on interest and need.

It is my hope that this monograph—with its comprehensive, yet introductory, coverage of the basics, applications, and needs of nanomedicine—will serve as a resource for educating and training a new generation of multidisciplinary researchers while helping to advance this new frontier toward applications needed for real patient care.

REFERENCES

Jain, K. K., *The Handbook of Nanomedicine*, Humana Press, Totowa, NJ (2008).

Tibbols, H. F., *Medical Nanotechnology and Nanomedicine*, CRC Press, Boca Raton, FL (2011).

CHAPTER 2

THE HUMAN BODY

Nanomedicine involves the interaction of engineered nanoparticles and biological systems with intent of generating a beneficial result whether it is diagnostic, prophylactic, or therapeutic. These interactions cover the molecular, subcellular/cellular, tissue/organ, and whole-body levels. In order to grasp the nature of these interactions, it is imperative to understand some of the fundamental biological concepts and processes taking place, from subcellular levels to functioning of organs and organ systems, in the human body. This chapter introduces the basics of such biological processes, along with the necessary terminology and concepts of biology relevant for nanomedicine. The chapter is written primarily for those unfamiliar with biological concepts, or those wishing to refresh their background in this subject. The chapter will also serve as a source of vocabulary of relevant biological terms. A large portion of this chapter is adopted from this author's previous book, *Introduction to Biophotonics* [Prasad, 2003], with the permission of the publisher.

The design of nanoprobes for biomedical applications relies heavily on the concepts of cell biology. Sections 2.1–2.6 provide a detailed description of the cell, looking into the various structural components of the cell and their functions, most of which modify the behavior of a foreign nanoparticle within the cell. The diversity of the cell types present in the body is discussed in Section 2.3, highlighting their unique structural and functional aspects and describing how they can be manipulated for a therapeutic purpose.

Introduction to Nanomedicine and Nanobioengineering, First Edition. Paras N. Prasad.

The next level of hierarchy in a biological system is a tissue, discussed in Sections 2.7 and 2.8, which arises from an assembly of different cells of the same origin. The organization of a tissue in terms of assembling of cells and extracellular matrix is also described. Finally, the various tissues that combine to form an organ and organ system are presented in Section 2.9.

Finally, this chapter concludes with an introduction to malignancies in Section 2.10, as a special case of tissue formation. Further references on cellular and tissue structures and functions are books by Alberts et al. (2002) and by Lodish et al. (2000). For biochemical aspects that deal with chemical building blocks, biosynthesis, enzyme catalysis, and cell energy production, suggested references are books by Horton et al. (2006), Stryer (1995), Lehninger et al. (2008), and Voet et al. (2002). General chemical principles are covered in a book by Solomon (1987).

2.1. INTRODUCTORY CONCEPTS

All living creatures are made up of cells. They exist in a wide variety of forms, from a single cell in free-living organisms (i.e., bacteria, fungi, and protozoa) to those present in complex biological organisms (i.e., eukaryotic cells). Although living systems exhibit a great diversity, they are, amazingly, composed of the same biochemical make up, and they utilize common principles in replication, metabolism, and the ability to hierarchically organize themselves to form higher organisms.

This section is intended to provide the fundamental knowledge required for understanding the structure and function of a living cell and its interaction with its surroundings. The dimensions of individual cells and microorganisms vary widely, ranging from viruses measuring 20–200 nm to eukaryotic cells usually measuring 10–100 μm. Table 2.1 shows the hierarchical structures in the biological evolution of life.

While Table 2.1 lists only animal cells as an example of the eukaryotic cells; other types of cells falling in this category are yeasts/fungi and protozoa, which are also self-replicating unicellular organisms. Cells of plants are also classified as eukaryotic, but they contain a cell wall structure consisting of cellulose. This cell wall is absent in animal cells. An important feature of the eukaryotic cell is its ability to differentiate and produce a variety of cells, each carrying out a specialized function. The complex assembly of these differentiated cells leads to higher organizations, which eventually form a higher organism, as shown in Table 2.2.

Thus, living organisms formed from cells are highly complex and organized and perform a variety of functions [Stryer, 1995; Lodish et al., 2000]:

- They extract and transform energy from their environment (food chain).
- They build and maintain their intricate structures from simple raw materials.

TABLE 2.1. Molecular Hierarchy of the Biological Evolution

Small molecules:
Amino acids
Nucleic acids
Water
Lipids
Ions

↓

Polymerization of nucleic acids to form DNA and RNA, and polymerization of amino acids to form protein

↓

Virus

- Size scale: 20–200 nm
- Structure: Single- or double-stranded RNA or DNA
- Function: Infectious but not self-replicating

Prokaryotic cells: Bacteria

- Size scale: ~1–10 μm
- Structure: Single-cell organism consisting of single closed compartment that lacks a defined nucleus
- Function: Free-living and self-replicating

↓

Eukaryotic cells: Animal/plant cells

- Size scale: ~10–100 μm
- Structure: Complex structure surrounded by a lipid membrane, contains an organized nuclear structure
- Function: Self-replicating and able to assemble to form tissues

TABLE 2.2. Hierarchy of Biological Organization

Cells
↓ Cell differentiation and association
Tissues
↓ Organization to perform a function
Organs
↓ Integration of various functions
Organism

- They carry on mechanical work using muscles.
- They use a highly regulated process of cell replication to reproduce cells identical in mass, shape, and internal structure.
- They use an intricate set of communications to orchestrate multiple biochemical processes in individual cells.

- They use internal defenses to fight diseases and carry out self-repair of damage due to injury.

The following sections describe the structures and functions of cells and tissues that are of particular interest in nanomedicine. The description follows a bottom-up approach, starting from the makeup of cells to their assembly into tissues and subsequently tissues forming functioning organs. This also illustrates the miracle of Nature: how living organisms conduct live functions using inert molecules.

2.2. CELLULAR STRUCTURE

Biological systems are essentially a complex and synergistic assemblies of biomolecules where water, amino acids, carbohydrates (sugars), fatty acids, and ions account for 75–80% of the matter in cells. The remainder of the cell mass is accounted for by macromolecules or biopolymers, which include peptides/proteins (formed from amino acids), polysaccharides (formed from sugars), DNA (deoxyribonucleic acid, formed from nucleotide bases and deoxyribose sugar), RNA (ribonucleic acid, formed from nucleotide bases and ribose sugar), and phospholipids (formed from fatty acids). These biopolymers organize to form living cells, surrounded by a semipermeable membrane (phospholipid bilayer). Within this contained biological microcosm, two types of organized cells exist, as shown in Table 2.1. Prokaryotic cells (bacteria) are cells with little internal structure and no defined nucleus. On the other hand, eukaryotic cells have a significantly more complex internal architecture, including defined, membrane-bound organelles such as cell nucleus, mitochondria, lisosomes, and others. The smallest organized particle is a virus. The smallest self-replicating cells are bacteria. Eukaryotic cells, for the most part, organize to form complex living organisms. Eukaryotic (animal) cell structure is shown in detail in Figure 2.1. From a single pluripotent cell (a cell with the capacity to differentiate into diverse cellular types) arise tissues and organs, and finally a complex living organism, as shown in Table 2.2.

The structure of a cell—specifically, eukaryotic cells, which are of relevance to nanomedicine—can be described in terms of the various subcellular compartments and the constituent chemical species they contain. The main structural components of a cell are:

- Plasma membrane, which defines the outer boundary of a cell. This is present in all cells.
- Cell wall, which exists in the prokaryotic cells as well as in the eukaryotic cells of plants and fungi, but not in animals.
- Cytoplasm, which represents the semi liquid matrix contained within the plasma membrane, holding all the cellular organelles except for the

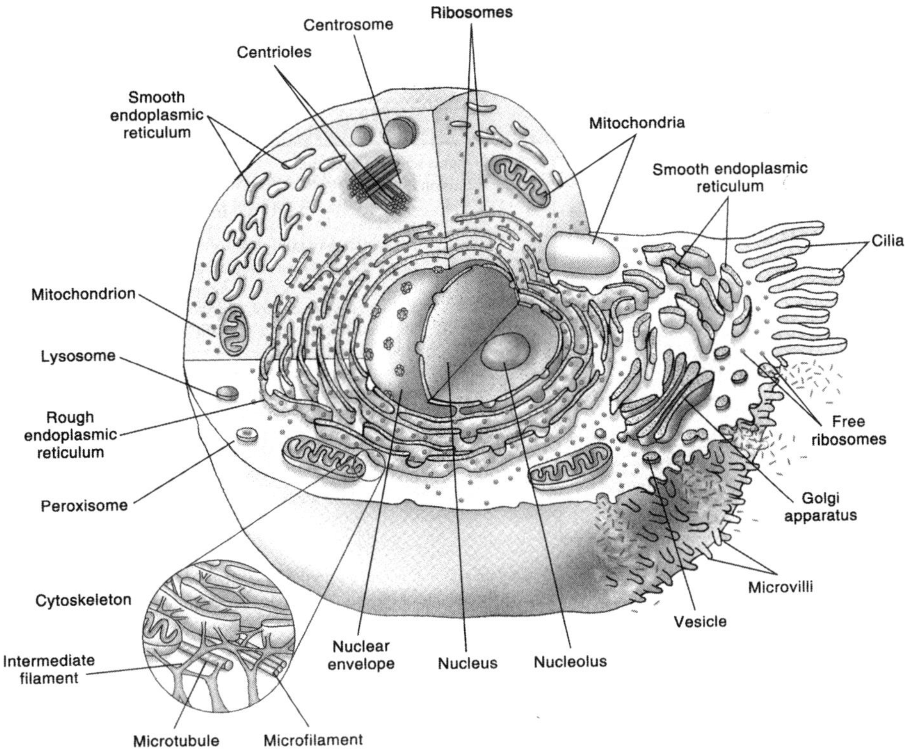

Figure 2.1. Animal cell structure. (Reproduced with permission from Thibodeau and Patton [2003]).

nucleus. The liquid component of the cytoplasm surrounding the organelles is named cytosol.

- Organelles (i.e., mitochondria), which are subcellular structures within the cytoplasm where specific cellular processes take place.
- Nucleus, which is the largest organelle which contains the genome and participates in regulation of essentially all cell functions.

Subcellular organelles are like little organs within a cell that perform specialized cellular functions, just like organs performing various tasks in a living system. These intracellular organelles include the nucleus, mitochondria, Golgi apparatus, endoplasmic reticulum, lysosomes, and peroxisomes. In the case of plant cells, other organelles include plastids, chloroplasts, vacuoles, and the cell wall (already listed above). The following sections describe some structural aspects of various cellular components and functions they perform [Audesirk et al., 2011].

Plasma Membrane. This forms a semipermeable outer boundary of both prokaryotic and eukaryotic cells. The outer membrane, about 4–5 nm thick, is a heterogeneous bilayer of phospholipids in which proteins, glycolipids, and cholesterol are also intercalated within a bilayer matrix. A phospholipid molecule has a long tail of alkyl chain, which is hydrophobic (repels water), and a hydrophilic head (likes water) that carries a charge (and is thus ionic). Phospholipid molecules spontaneously orient (or self-organize) to form a bilayer within which the hydrophobic tails are pointed inwards (shying away from the outer aqueous environment). The hydrophilic, ionic head groups point to the exterior and are thus in contact with the surrounding aqueous environment, both on the outside and inside the cell. The membrane derives its rigidity by inclusion of cholesterol molecules, which are interdispersed in the phospholipid bilayer. Also embedded are membrane proteins (i.e., receptors, pores, and enzymes) that are important for a number of cell activities, including communication between the intracellular and extracellular environments. The plasma membrane controls the transport of food, water, nutrients, and ions such as Na^+, K^+, and Ca^{2+} (through so-called ion channels) to and from the cell, as well as signals (cell signaling) necessary for proper cell function.

Cytoplasm. As indicated above, the **cytoplasm** is the gel-like substance residing within the plasma membrane holding all the cell's internal substructures (i.e., the organelles). It is present in all cells where metabolic reactions occur. It consists mainly of a viscous fluid medium (cytosol) and contains salts, sugars, lipids, vitamins, nucleotides, amino acids, RNA, and various proteins. Many of the functions for cell growth, metabolism, and replication are carried out within the cytoplasm.

Cytoskeleton. The cytoskeleton structure, located just under the membrane, comprises a network of fibers composed of proteins, such as actin microfilaments, microtubules, and intermediate filaments. This complex structure is connected to plasma membrane and various organelles. In animal cells, it is often organized from an area near the nucleus, and it extends to the plasma membrane in every direction. These arrays of protein filaments perform a variety of functions:

- Establish the cell shape.
- Provide mechanical strength to the cell.
- Are responsible for the cellular shortening involved in muscle contraction.
- Control changes in cell shape and may promote cellular migration.
- Provide anchors for chromosome separation in mitosis and meiosis (these processes are discussed below).
- Participate in intracellular transport of organelles and other vesicles along the microtubules.

Nucleus. The nucleus is often called the "control center" of the cell. It is the largest organelle in the cell, usually spherical with a diameter of 4–10 μm, and is separated from the cytoplasm by an envelope (the nuclear envelope) consisting of an inner and an outer membrane. All eukaryotic cells have a nucleus. The nucleus contains DNA distributed among structures called chromosomes, which determine the genetic makeup of the organism. The chromosomal DNA is packaged into chromatin fibers by association with an approximately equal mass of histone proteins. Transport of nutrients and other essential biomolecules between the nucleus and the cytosol takes place through a highly specialized "gate" present in the nuclear envelope, known as the nuclear pore complex (~120 nm across).

Figure 2.2 shows the schematic of a nucleus. The nucleus is the site of replication of genomic DNA and its transcription into RNA. The inside of the nucleus contains a prominent domain, called a nucleolus, exhibiting an ellipsoid or crescent-shaped structure. The major function of nucleolus is to produce ribosomes by synthesizing ribosomal RNA and packaging it with ribosomal proteins. Matured ribosomes are transported to the cytoplasm where they synthesize polypeptide chains from the RNA templates in a process called translation. These processes will be discussed in a later section.

Mitochondria. Mitochondria are large organelles, globular in shape (almost like fat sausages), which are 0.5–1.5 μm wide and 3–10 μm long. They occupy about 20% of the cytoplasmic volume. They contain an outer and an inner

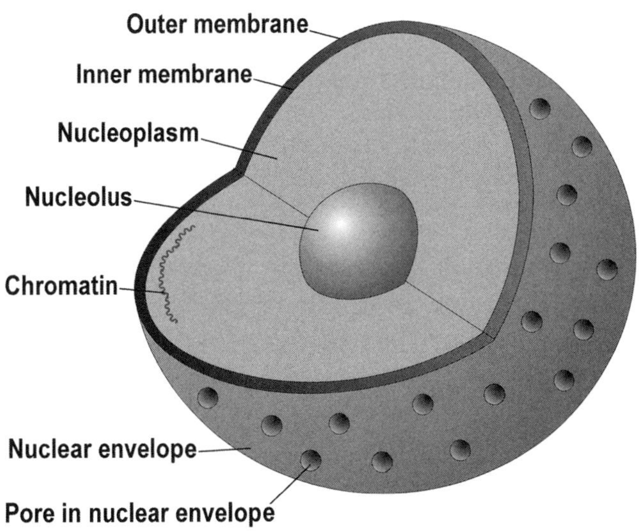

Figure 2.2. Schematic of cell nucleus.

membrane, which differ in lipid composition and in enzymatic activity. The inner membrane, which surrounds the matrix base, has many infoldings, called cristae, which provide a large surface area for attachment of enzymes involved in respiration. The matrix space enclosed by the inner membrane is rich in enzymes and contains the mitochondrial DNA. Mitochondria are self-replicating organelles that serve as the energy generating plant of a eukaryotic cell. They produce, through a process known as oxidative metabolism, most of the cellular adenosine triphosphate (ATP), which is the major source of the energy in the cell. The chemical energy stored in the phosphate bonds of the ATP molecule is used to power multiple types of intracellular reactions. In plants, the counterpart of mitochondria is the chloroplast, which utilizes a different mechanism, photosynthesis, to harness energy for the synthesis of high-energy phosphate bonds.

Endoplasmic Reticulum. The endoplasmic reticulum consists of flattened sheets, sacs, and tubes of membranes that extend throughout the cytoplasm of eukaryotic cells and enclose a large intracellular space called lumen. There is a continuum of the lumen between membranes of the nuclear envelope. The rough endoplasmic reticulum (rough ER) is close to the nucleus and is the site of attachment of the ribosomes. Ribosomes are small and dense structures, 20 nm in diameter, that are present in great numbers in the cell, mostly attached to the surface of rough ER, but can also be free in the cytoplasm. They are assembled from ribosomal RNA and multiple proteins in the nucleolar domain of the nucleus and are then transported to the cytoplasm. Size of the ribosomes is traditionally measured in Svedberg units of sedimentation; eukaryotes have 80S ribosomes, each consisting of a small 40S and large 60S subunit. Ribosomes are the sites of protein synthesis. The process of protein synthesis using a messenger RNA template is described below. The rough ER transitions into a smooth endoplastic reticulum (smooth ER), which is generally more tubular and lacks attached ribosomes. The smooth ER is the primary site of (a) synthesis of lipids and sugars and (b) assembly of lipo- and glycol proteins. The smooth ER also contains degradative enzymes, which detoxify many organic molecules.

Golgi Apparatus. This organelle is named after Italian scientist Camillo Golgi, who described it. It consists of stacked, flattened membrane sacs or vesicles, which function like shipping and receiving departments because they are involved in modifying, sorting, and packaging proteins for delivery to other organelles, or for secretion outside of the cell. There are numerous membrane-bound vesicles (<50 nm) around the Golgi apparatus, which are involved in the processing and transport of lipo- and glycol proteins between this organelle and different compartments of the cell.

Lysosomes. These are bags (technical term: vesicles) of hydrolytic enzymes that are 0.2–0.5 μm in diameter and are single-membrane bound. They have an acidic interior and contain about 40 hydrolytic enzymes involved in intracellular digestions.

Peroxisomes. These are membrane-bound vesicles containing oxidative enzymes that generate and metabolize hydrogen peroxide. They are 0.2–0.5 μm in diameter.

Chloroplasts. This cell organelle generates energy from the sun light. This organelle is found in plants and does not exist in animal systems.

2.3. VARIOUS TYPES OF CELLS

Cells come in many shapes, sizes, and compositions. The human body is made up of over 200 different types of cells. These cell variations are produced by cell differentiation. The different types of cells assemble together to form multicellular tissues or organisms. The human body also consists of nonliving matter such as hairs, fingernails, outermost cornified layer of skin, and hard parts of bone and teeth, which are mostly composed of extracellular fibers, dead cells, or both. Presented here is a small selection representing the most common types of cells in animals.

Epithelial Cells. Epithelial cells form sheets, called epithelia, which line the inner and outer surfaces of the body. Some of the specialized types of cells are (i) absorptive cells, which have numerous hair-like microvilli projecting from their surface to increase the absorption area, (ii) ciliated cells, which move substances such as mucus over the epithelial sheet, (iii) secretory cells, which form exocrine glands that secrete tears, mucus, and gastric juices, (iv) endocrine glands, which secrete hormones into the blood, and (v) mucosal cells, which protect tissues from invasive microorganisms, dirt, and debris.

Blood Cells. These are the cellular elements of the blood. Blood is a heterogeneous fluid consisting of a number of different types of cells. Cells comprise about 45% of the blood volume. The other component of blood is plasma which is a colloidal suspension of proteins dispersed throughout an electrolyte solution (liquid phase) with a high NaCl concentration. The three different types of blood cells are (i) erythrocytes (commonly known as red blood cells; often abbreviated as RBC), (ii) leukocytes (commonly known as white blood cells), and (iii) thrombocytes (also known as platelets). Erythrocytes or red blood cells are very small cells, 7–9 μm in diameter, with a biconcave, disc shape. In humans they usually have no nucleus. One cubic centimeter of blood contains about 5 billion erythrocytes, the actual number depending on a number of factors such as age, gender, and health. They contain an oxygen-binding protein called hemoglobin and, thus, perform the important function of transporting O_2 and CO_2.

Leukocytes or white blood cells provide protection against infection. They exist in a ratio of one white blood cell for about every 1000 red blood cells. They are usually larger than red blood cells. There are a number of different types of leucocytes, including (i) lymphocytes, which are responsible for immune responses (there are many kinds of lymphocytes most notably T cells, which are associated cell-mediated immunity, and B cells, which are associated with antibody production), and (ii) macrophages and neutrophils, which move to sites of infection where they ingest bacteria and debris. The platelets are 2–5 μm in diameter (smaller than red blood cells). Technically, platelets are not whole cells and are devoid of nucleus, because they are derived from fragmentation of precursor megakaryocytes. Platelets participate in blood coagulation process.

Muscle Cells. These specialized cells form muscle tissues such as skeletal muscles to move joints, cardiac muscles to produce heartbeat, and smooth muscle tissues found around the internal organs (i.e., stomach and intestines) and large blood vessels. Muscle cells produce mechanical force by their contraction and relaxation.

Nerve Cells or Neurons. Neurons are cells specializing in communication. The brain and spinal cord, for example, are composed of a network of neurons that extend out from the spinal cord into the body.

Sensory Cells. These cells form the sensory organs such as (a) hair cells of the inner ear, which act as detectors of sound, and (b) retina, where rod cells specialize in responding to light. The rod cells contain a photosensitive region consisting of light-sensitive pigments (chromophores) called rhodopsin.

Germ Cells. Germ cells are responsible for reproduction and serve to produce gametes, which are haploid (cells containing one member or a copy of each pair of chromosome). Gametes are specialized for sexual fusion and include (i) a larger, nonmotile (nonmoving) cell called the egg (or ovum) from a female and (ii) a small, motile cell referred to as sperm (or spermatozoon) from a male. A sperm fuses with an egg to form a new diploid organism (containing both chromosomes). Bacteria are another example of haploid cells.

Stem Cells. Another type of cell that has received considerable attention during recent years is the stem cell. Stem cells can be thought of as blank cells that have yet to become specialized (differentiated), giving them the characteristics of a particular type of cell such as the ones described above. Stem cells thus have the ability to become any type of cell to form any type of tissue (bone, muscle, nerve, etc.). These cells are further discussed in Chapter 17 on Stem Cell Biotechnology.

Embryonic stem cells are classified as pluripotent because they can differentiate into any cell type. Fetal and adult stem cells, on the other hand, are multipotential in that they have already undergone some differentiation. Pluripotent stem cells have greater versatility, but unfortunately they are also more likely to become malignant when used *in vivo*. However, recent research suggests that multipotent adult stem cells can have pluripotent capability. Stem cells can provide a solution in regard to curing diseases caused by cell failure and repairing tissues that do not self-repair by allowing one to produce appropriate cells and grow the needed tissues. Some of the diseases, which are projected to benefit from stem cell research are heart damage, spinal cord injuries, Parkinson's disease, leukemia, and diabetes.

2.4. BIOCHEMICAL MAKEUP OF CELLS

This section describes the chemical makeup of the principal constituents of a cell: (i) nucleic acids in the form of DNA and RNA, (ii) proteins, (iii) saccharides (sugar derivatives), and (iv) lipids [Alberts et al., 2002; Stryer, 1995].

Nucleic Acids. Nucleic acids exist in a cell in two forms: (i) deoxyribonucleic acids (DNA) and (ii) ribonucleic acids (RNA). Both forms consist of three chemical building blocks: (i) nitrogen-containing ring compounds that are either purine or pyrimidine bases, (ii) a sugar (either deoxyribose for DNA or ribose for RNA), and (iii) a phosphate. The four bases constituting DNA are adenine (A), guanine (G), thymine (T), and cytosine (C). In the case of RNA, the base thymine (T) is replaced by another base, uracil (U). The structures of these bases are shown in Figure 2.3. The constituent sugars are ribose for RNA and 2-deoxyribose for DNA. The phosphate group is the tetrahedral PO_4 group.

The following hierarchy of chemical coupling represents the nucleic acid formation. It also provides the nomenclature used to describe the nucleic acid chemistry [Stryer, 1995; Voet et al., 2002]:

The polymerization of the nucleotides to form nucleic acids occurs in a directed order. The convention is that the synthesis occurs from 5′ to 3′. In this manner, the phosphate group located on the 5′ carbon is added to the hydroxyl group on the 3′ carbon of the growing chain. And, by convention, the polynucleotide sequences are also read 5′ to 3′. For example, AUG represents 5′-AUG-3′. The sequence of nucleotides in the DNA defines the genetic makeup of the organism. All nucleotides involve a phosphate group linked by a phosphoester bond to a five-membered sugar molecule, pentose, which in turn is bonded to an organic base.

The DNA chemical structure involves the four nucleotides A, G, T, C as a continually varying sequence where the nucleotide bases are connected by a phosphodiester bond involving condensation of the hydroxyl (OH) group on one, with the acid, –OH, group on the phosphate of the other. The sugar unit is a five-membered ring, deoxyribose. Here, B1 and B2 are the two-nucleotide bases. The OH group on the other sugar unit provides a site for bonding with

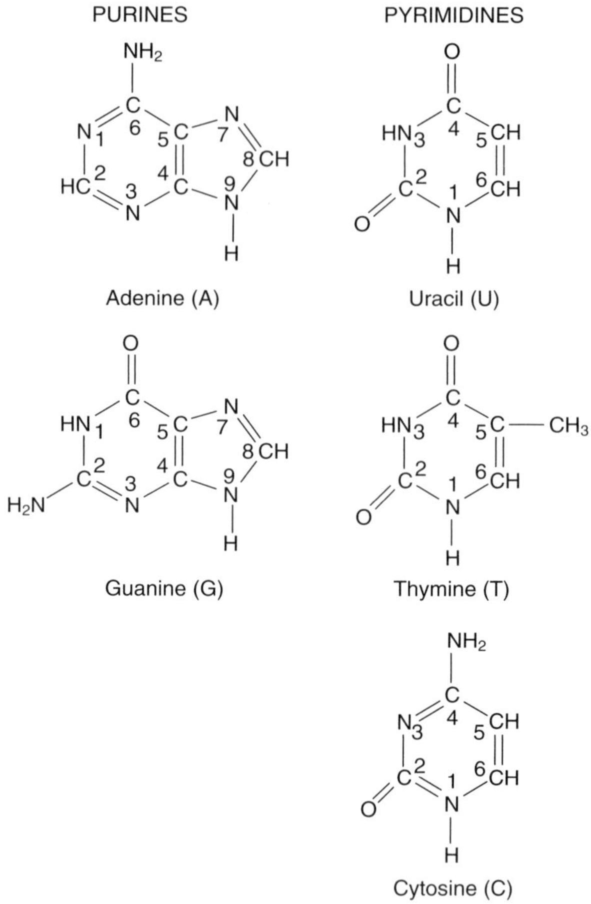

Figure 2.3. Chemical structures of purine or pyrimidine bases.

a phosphate group on another nucleotide to repeat the process, ultimately yielding the polymeric nucleic acid of DNA.

The chemical structure of RNA is similar to DNA, but the sugar component of RNA has an additional hydroxyl group at the 2′ position (the sugar is ribose) and the thymine (T) in DNA is replaced by uracil (U) in RNA. DNA is contained mostly in the nucleus of a cell and is distributed in supercoiled structures called chromosomes. The number of chromosomes is the same in all types of cells of a specific organism, but varies between biological species. A small amount of DNA is also present in mitochondria.

Proteins. Proteins are formed during a polymerization process that links amino acids. Only 20 different amino acids form a vast array of proteins capable of highly diverse tasks. The functional nature of each of these amino

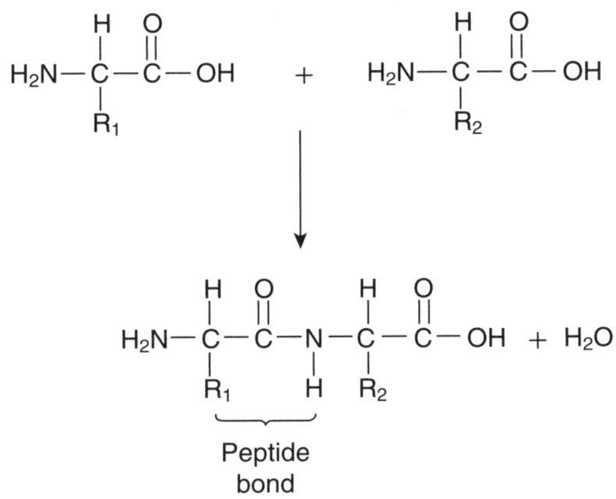

Figure 2.4. Peptide bond formed by condensation of the amino group of one amino acid with the carboxylic group.

acids resides in the side-chain group (R group). The amino acids link together by a peptide bond formed by condensation of the amino group of one amino acid with the carboxylic group of another, as shown in Figure 2.4. This process continues on both ends (involving the NH_2 and COOH groups) to produce the polymeric chemical structures of a protein. R1 and R2 are side groups (or chains) that may simply be an alkyl, acidic, basic, or aromatic group. Proteins are ubiquitous throughout the cell.

Sugars. Sugars provide a major source of energy for a cell. This energy is required for the various cellular functions. The simplest sugars are called monosaccharides and are of general chemical formula $(CH_2O)_n$. Glucose is a six-membered ring and is one of the simplest sugars used as a source of energy for cells. The five-membered ring sugars, ribose and deoxyribose, are utilized by cells only in the formation of RNA and DNA, respectively. Glucose has a number of stereoisomers, galactose and mannose, which differ in the orientation of their hydroxy groups. Also, natural glucose only exists in one optically active D-form, which can have two conformations, a and b. Complex sugars form when multiple monosaccharide units are linked together. Two monosaccharides can link together to form a disaccharide. An oligosaccharide involves the linkage of more than two monosaccharides. A polysaccharide is formed by the linkage of thousands of monosaccharide units, also called residues. Examples of polysaccharides are glycogens in animal cells and starches in plants, which are composed solely of glucose residues. Complex polysaccharides form important extracellular structures and are also found covalently linked to proteins in the form of glycoproteins, as well as to lipids in the form of glycolipids.

Lipids. Lipids which are another significant energy source for cells refer to a group of compounds in living systems that are soluble in nonpolar solvents

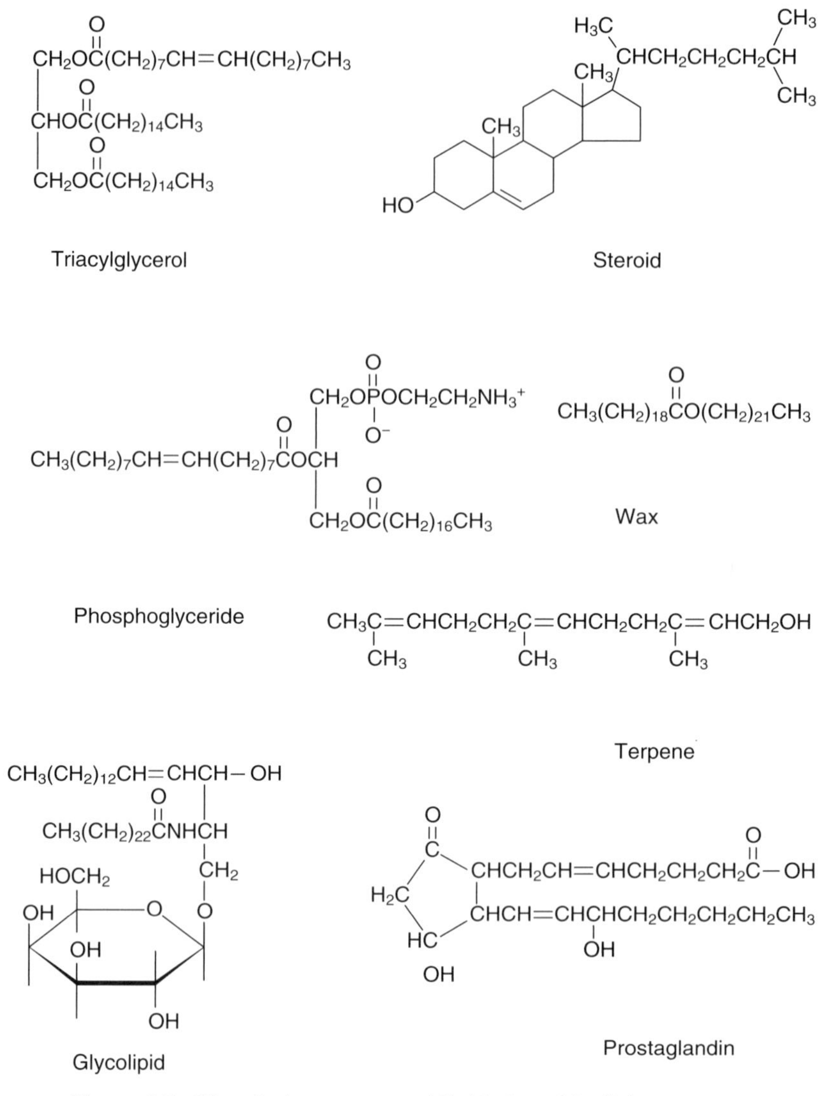

Figure 2.5. Chemical structures of lipids found in living systems.

like hexane (a hydrocarbon, C_6H_{12}). The various kinds of lipids found in living systems are represented in Figure 2.5. Most lipids are derived from fatty acids with a general formula of R-COOH. It is a carboxylic acid in which the R group is a long alkyl chain. The most nonpolar of the lipids is triacylglycerol, classified as a fat, which has no polar group. It is stored in the body as a fat and serves as an energy reservoir that is metabolized when needed. Triacylglycerol is formed when one molecule of glycerol reacts with three fatty acid molecules. Phosphoglycerides, the structural components of the membrane, also known as phospholipids are formed from a combination of two fatty acids, one glycerol, one phosphoric acid, and one other hydrophilic group. These groups include choline (phosphatidylcoline), ethanolamine (phosphatidylethanolamine), serine (phosphatidylserine), and inositol (phosphatidylinositol).

Steroids found in living systems are hydroxy-substituted ring hydrocarbons. The most abundant steroid in living systems is cholesterol, whose structure is shown in Figure 2.5. Prostaglandins are lipids found in seminal fluids, as in prostate glands, as well as in other animal tissues. Glycolipids, also called sugar lipids, are predominantly found in membranes of brain tissues; however, selected types are universal in all cell membranes. Lipid-soluble vitamins form a minor subset of hydrophobic molecules, which are important in nutrition. The precursor molecules for most of these vitamins are terpenes, which are multiples of 2-methyl-1,3-butadiene (isoprene). Wax, another group of lipids present in nature, serves as a protective coating in leaves, skin, feathers, fur, and the outer skeleton of some insects.

2.5. OTHER IMPORTANT CELLULAR COMPONENTS

Other important cellular constituents are AMP (adenosine monophosphate), ADP (adenosine diphosphate), and ATP (adenosine triphosphate). During a chemical process called oxidative/phosphorylation, ATP synthesis in the mitochondria provides the primary source of chemical energy to drive many cellular processes. The structure of ATP is shown in Figure 2.6. It involves the formation of phosphoanhydride bonds to store energy that is released in a reaction with water (hydrolysis) to produce adenosine diphosphate (ADP), which contains two phosphate groups, or adenosine monophosphate (AMP), which contains only one phosphate group. Within the mitochondria, phosphate is added back to AMP and ADP to form ATP (stored chemical energy).

Nicotinamide adenine dinucleotide acts as a coenzyme in enzymatically catalyzed reactions that involve electron/proton transfer. The structure of

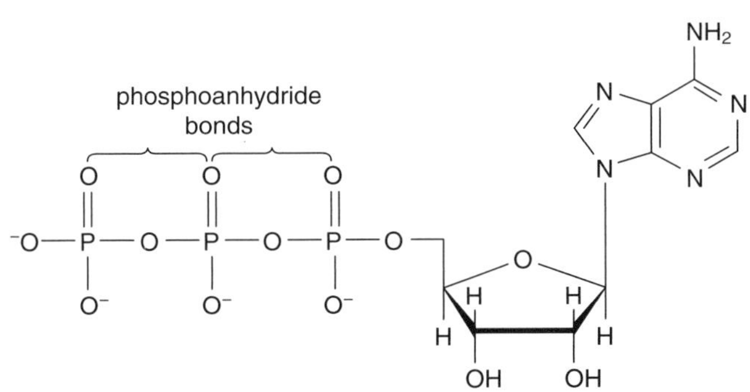

Figure 2.6. Chemical structure of ATP.

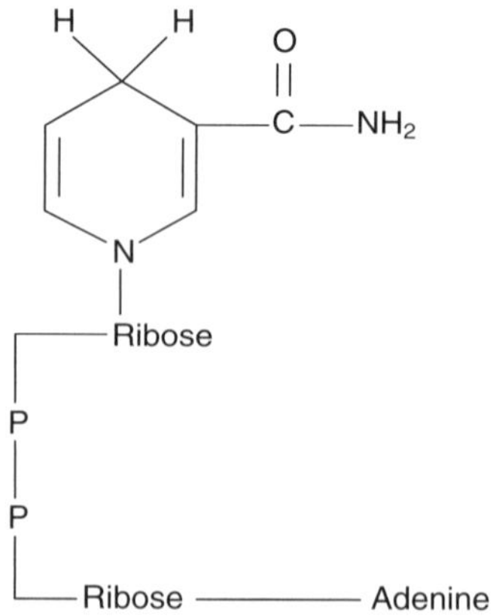

Figure 2.7. Chemical structure of NADH.

NADH is shown in Figure 2.7. As discussed in Section 2.4, ribose is a five-membered ring sugar and adenosine is a nucleotide. A related compound is $NADP^+$ and its protonated form NADPH (nicotinamide adenine dinucleotide phosphate) in which a phosphate group replaces an H atom in the adenosine group.

2.6. CELLULAR PROCESSES

A living cell is a highly dynamic system in all its functions, from replication, to operation, to communication between cells. The processes involve both physical and chemical changes. Furthermore, the chemical changes can be permanent (such as protein synthesis, DNA replication) or cyclic (such as conversion of ATP into ADP and back). The chemical changes occurring are highly complex, often catalyzed (accelerated) by enzymes (reaction-specific proteins) and coenzymes (small molecules such as NADH). The various dynamic processes occurring in a cell are illustrated in Table 2.3. Some classes of cellular processes fall in more than one category. For example, most enzyme-catalyzed reactions are irreversible; however, a few are reversible.

On the basis of their biological function, the cellular processes can also be divided as shown in Table 2.4. The cellular processes listed in Table 2.4 are further discussed here.

Cell Proliferation. The complex cellular metabolic process is necessary not only for the cell to survive but, even more important, to replicate. The primary driving force in life is the ability to proliferate for the continuing survival of the species. In eukaryotic and prokaryotic cells, proliferation produces identical daughter cells. Eukaryotic cells replicate through an orderly series of events constituting the cell cycle [Lodish et al., 2000]. During this cycle, the

TABLE 2.3. Intracellular Metabolic Processes

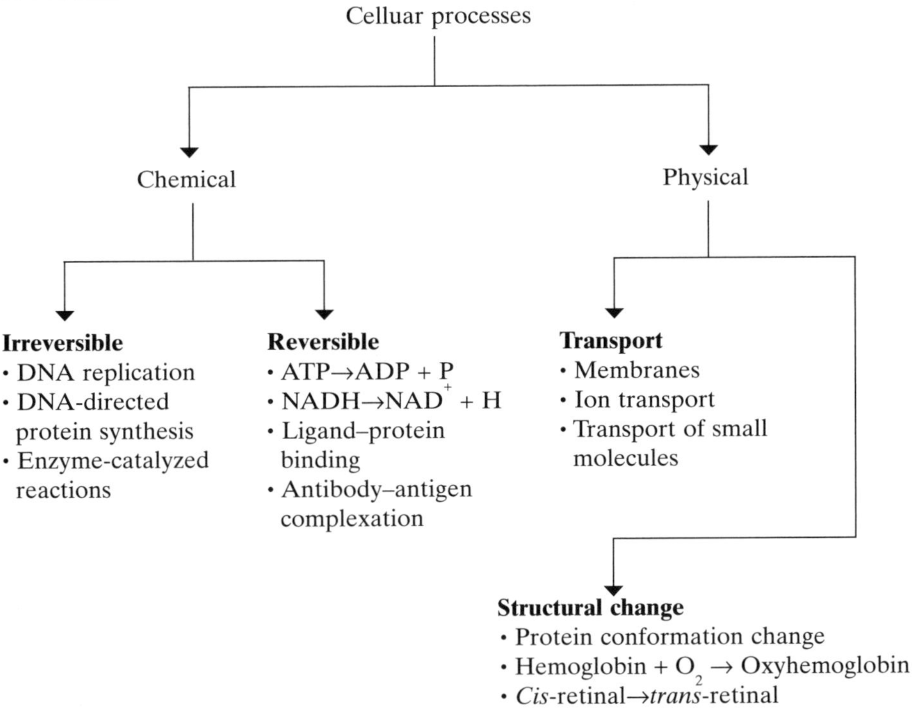

chromosomes are replicated and their copies segregate into each daughter cell. Strict regulation of the cell cycle is required for normal development and maintenance of the organism. This regulation is found in the nucleus of the cell coupled with the replication of the genomic material. Loss of this control ultimately leads to cancer. In eukaryotic cells, the process of cell division is called mitosis.

Most animal cells are diploids, because they contain two copies of each chromosome. Mitotic cell cycle is the process in which a diploid parent cell duplicates its DNA, condenses it into chromosomes, and then splits to form two new cells. Thus, a parent cell produces two genetically equivalent daughter cells, each receiving two copies of each chromosome. Mitosis is the most common type of cell division.

Meiosis is another type of cell division process by which a cell duplicates its DNA, condenses it into chromosomes, but then forms new cells, each containing only one copy of each chromosome—that is, half the number of chromosomes of the parent cell. Therefore, only one step duplicates DNA; the second step does not produce DNA. Meiosis produces sex cells, also known as gametes. The genetically nonequivalent cells formed containing only one copy of each chromosome are also called haploid cells or germ cells.

TABLE 2.4. The Cellular Processes Listed by Biological Function

- **Cell replication:** Process to produce or replenish cells
  Mitosis: Process whereby the nucleus divides into two genetically equivalent daughter cells
  Meiosis: Specialized process whereby the two successive rounds of nuclear and cellular division with a single round of DNA replication yields four genetically nonequivalent haploid cells (germ cells)
- **Cell biosynthesis:** Production of cellular macromolecules
  Transcription: Replication of single strand of DNA to complementary RNA
  Translation: Ribosome-mediated production of proteins from messenger RNA
- **Cell energy production:** Conversion of energy forms
  Glycolysis: Conversion of chemical energy of carbohydrates to ATP
  ATP synthesis: Process of prducing ATP from ADP from either:
    Chemical bond energy: Mitochondria (respiration)
    Light energy: Chloroplasts (photosynthesis)
- **Cell signaling:** Use of biomolecules initiated specific processes in adjacent or distant cells
  Endocrine: Hormonal signaling of distant cells
  Paracrine: Cell to adjacent cell signal
  Autocrine: Same cell stimulation signal
- **Cell death:** Control of multicellular development
  Apoptosis: Programmed cell death through a well-defined sequence of morphological changes
  Cell killing: Cell death caused by ancillary cells of the immune system
- **Cell transformation:** Genetic conversion of a normal cell into a cell having cancer-like properties (oncogenesis)

Cell replication by either mitosis or meiosis involves a number of stages, which constitute what is called the cell cycle (shown in Figure 2.8). A cell cycle is composed of two major phases: (i) the M phase, where the division of a parent cell into two daughter cells for a mitosis process takes place, and (ii) an interphase, which defines processes occurring within the cell before its division. The interphase itself is composed of three phases: the G1, S, and G2 phase. The main processes of these phases and their average time durations are also indicated in Figure 2.8. The actual time durations depend on the type of cell. The average time of 20–30 hours for the cell cycle is commonly observed in the cell cultures. In live organisms, however, most cells divide longer, and many highly specialized cells (e.g., neurons) typically do not divide.

The G1 phase, which is the first gap (G for gap) phase, represents the period during which proteins, lipids, and carbohydrates are synthesized to prepare a cell for DNA replication in the subsequent S phase. The S phase, where S stands for synthesis, represents the period during which the DNA is replicated (new copies of DNA are synthesized) in the nucleus, by enzyme (DNA polymerase)-catalyzed reaction. This process is the key step in cell replication where the genetic code is replicated. The G2 phase is the second gap phase

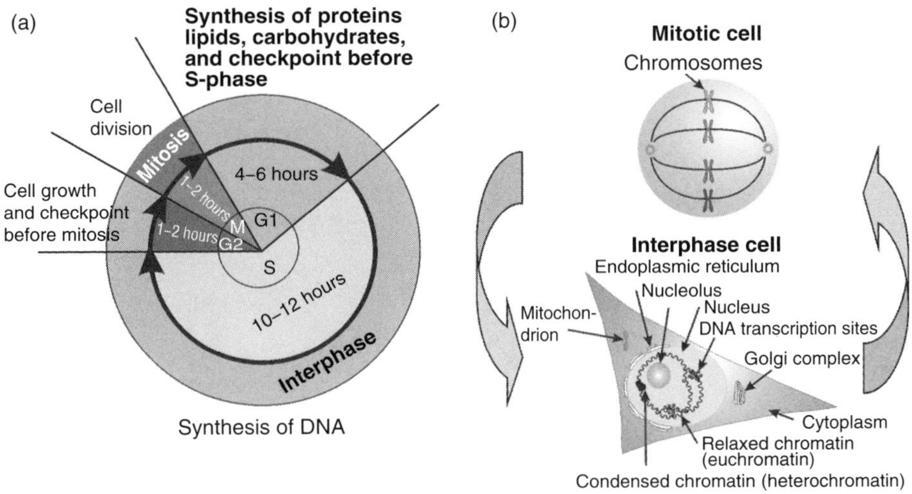

Figure 2.8. Cell cycle diagram (a) and shapes of mitotic and interphase cells (b).

that prepares conditions for cell division to occur in the M phase. In the M phase, the nuclear envelope disintegrates, genomic DNA condenses into dense chromosomes and splits apart, and then the cellular content divides between two daughter cells. The process of the cell division, and thus the physical separation of the new daughter cells, is called cytokinesis.

The process of duplication of DNA in the S phase of the cell cycle is described by a replication fork scheme, as illustrated in Figure 2.9. In this process a portion of the double helix is unwound by an enzyme called helicase. Then another enzyme, DNA polymerase, binds to one strand of the DNA and begins moving along it, reading in the 3′ to 5′ direction to assemble (synthesis goes 5′ to 3′) a "leading strand" of nucleotide using this strand as the template. However, this process requires a free 3′ OH group, which does not exist on the chain when the two separate. To begin synthesis, an RNA primer molecule synthesized on the site provides the free 3′ OH group to begin the DNA synthesis. A second type of DNA polymerase binds to the other template strand; however, it can only synthesize discontinuous segments of polynucleotides, which are called Okazaki fragments. These fragments are stitched together by another enzyme, DNA ligase, into what is called the "lagging strand." When the replication process is complete, each strand of the original DNA duplex remains intact since it served as the template for the synthesis of a complementary strand. For this reason, this mode of replication is termed "semiconservative" because one half of each new DNA is old, and the other half is new (Figure 2.9).

Cell Biosynthesis. Every process in a living cell involves proteins that are a major type of cellular macromolecules. Thus, an important function of the cell

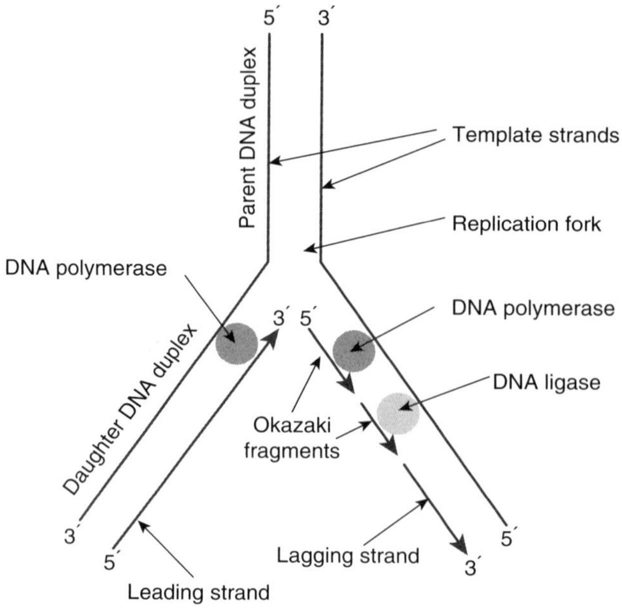

Figure 2.9. Replication fork scheme.

is the biosynthesis of proteins utilizing the constituent amino acids. The process of biosynthesis of proteins involves two steps: transcription and translation. In protein synthesis, three kinds of RNA cooperatively perform to lead to protein synthesis [Lodish et al., 2000; Stryer, 1995]. These are:

- Messenger RNA (abbreviated as mRNA), which carries the code (genetic information) copied on it in the form of a series of three-base code called "words" or codons. Each word is associated with a specific amino acid constituent of the protein.
- Transfer RNA (abbreviated as tRNA), which deciphers the code in mRNA and binds to a specific amino acid, carrying it to the growing end of the peptide chain.
- Ribosomal RNA (abbreviated as rRNA), which complexes with a set of proteins to form ribosomes. These complexes, moving along an mRNA molecule, catalyze the assembling of amino acids into protein chains in the process of translation.

Transcription is the process of copying the DNA template to produce mRNA, thereby transcribing the gene code from the DNA onto a strand of RNA. It takes place in the nucleus where the DNA unwinds in small segments, allowing access to an enzyme called RNA polymerase to the single-stranded DNA. The RNA polymerase moves along the whole length of a gene (a length of DNA), producing an unzipping action and making an accurate copy of the gene in the form of mRNA. The transcription process is highly regulated and controlled at several levels.

Translation is the process by which the template encoded in the mRNA strand leads to the synthesis of proteins in a specific sequence of amino acids, thereby translating the code. The mRNA passes out of the nucleus into the cytoplasm, where it attaches to ribosomes. The ribosomes "knit" together the amino acids in a sequence dictated by mRNA. As determined by the genetic code, specific tRNAs carrying the appropriate amino acid transfer the new amino acid to the growing polypeptide chain. Ribosomes may be free in the cytoplasm or associated with the endoplasmic reticulum (bound ribosomes). The completed protein chain from the bound ribosome is then threaded into the lumen of the endoplasmic reticulum, where it folds up into its 3D shape and may be further processed as it passes through the Golgi apparatus. Free ribosomes usually synthesize proteins for use within the cell, while bound ribosomes usually produce proteins that are intercalated into the plasma membrane, or are expelled (secreted) from the cell. The process of translation is a high-volume activity, because several ribosomes can be traveling along a single mRNA, producing the same protein.

Cell Signaling. Cell signaling refers to a two-step process. In the first step, a signaling molecule (e.g., a ligand) is bound by a receptor at a target cell. In the second step, the receptor is activated. Membrane-bound receptors respond to a large variety of extracellular signals, which range from light and odor to hormones, growth factors, and cytokines. The receptors are proteins, which can undergo a chemical change or a conformational change to switch from a dormant, inactive state to an active state. The receptors can transmit signals from the surface of the cell to the interior. If the signal transduction goes to the nucleus, up- or down-regulation of specific genes can occur. In such a case, the signaling ligand molecule binds to the extracellular domain of the membrane—that is, the membrane receptor—and subsequently triggers biochemical reactions that activate DNA transcription factors. For example, signaling introduced by a growth factor leads to a global response, such as growth, proliferation, and differentiation of cells. The different types of cell stimulation produced by cell signaling due to cellular regulators (i.e., cytokines and growth factors) and hormones are:

- *Autocrine.* Refers to the resulting cell stimulation when the receptor resides on the same cell where the ligand is expressed (produced).
- *Paracrine.* Describes cell stimulation when the growth factor diffuses from the cell to neighboring cells in the same organ.
- *Endocrine.* Describes cell stimulation when a factor is transported through the bloodstream from the place of synthesis of the signaling molecule to other tissues (distant cells) equipped with receptors that recognize the factor. This is the case with hormones.

Cell Death. Cell death is a necessary biological response for the control of multicellular development. Cell death occurs by two principal mechanisms: apoptosis or necrosis cell killing by injurious agents, which are listed in Table 2.4. These processes are further discussed here.

Apoptosis. This process is often called programmed cell death (PCD) because of genes that are activated induce cell death. It is also referred to as "suicide by a cell." A PCD is very orderly and necessary for normal cell turnover and to destroy cells that represent a threat to the organism, such as cells infected with a virus and cells with DNA damage. There are two mechanisms that cause apoptosis to occur. One is triggered by internal signals from within the cell, such as the byproducts of cellular respiration which may produce reactive oxygen species. The other is triggered by external signals, also called death activators, which bind to receptors at the cell surface. During apoptosis, cells undergo shrinkage, and their mitochondria break down, releasing cytochrome C, a protein that further mediates apoptotic development. Finally, the chromatin in the nucleus degrades and eventually breaks into small, membrane-wrapped fragments.

Necrosis. This process refers to cell death by injury. The term necrosis refers to a cell's response to overwhelming injury and is often called "accidental cell death." The injury may be caused by mechanical damage or by exposure to toxic chemicals. Death by injury may involve a set of changes such as swelling of cells and their organelles like mitochondria, with subsequent leakage of cellular contents, leading to inflammation of surrounding tissues.

Cell Transformation. This process involves the permanent, inheritable alteration of a cell resulting from the uptake and incorporation of foreign DNA. Normal cells can undergo this "transformation" as a result of exposure to viral or other cancer-causing agents such as mutagens. The result is the transformation of the normal cell into a cell having cancer-like properties (oncogenesis).

Cellular Processes of Current Interest. Some of the cellular processes that are receiving considerable interest today are:

- *Genomics.* Study of genetic code and its regulation.
- *Proteomics.* Study of protein structure and function as determined by both genetic code and post-translational modifications.
- *Metabolomics.* Study of the unique chemical signatures of specific cellular processes.
- *Cloning.* Insertion of foreign DNA into cells or duplication of an organism through genetic manipulation.

2.7. ORGANIZATION OF CELLS INTO TISSUES

A tissue is a multicellular bioassembly in which cells specialized to perform a particular task contact and interact specifically with each other. The functions of many types of cells within tissues are coordinated, which collectively allows an organism to perform a very diverse set of functions such as its ability to move, metabolize, reproduce, and conduct other essential functions. The various constituents forming tissue are shown in Table 2.5.

Figure 2.10 presents a schematic view of molecules and components that bind cells to cells and also bind cells to the extracellular matrix.

What follows is a brief description of the intercellular components.

Cell-Adhesion Molecules (CAM). These are cell-surface-bound proteins that mediate adhesion between cells of the same type (homophilic adhesion) as well as between cells of different types (heterophilic adhesion). Most CAMs are uniformly distributed within plasma membranes that contact other cells. The five principal classes of CAMs are: cadherins, immunoglobulins (Ig), selectin, mucins, and integrins. Figure 2.11 shows the major families of CAMs.

TABLE 2.5. List of the Constituents Forming Tissue

Animal tissue

↓

Cells	*Cell-adhesion molecules (CAM)*	*Extracelluar matrix (ECM)*
Often organized in a precise pattern.	Various integral membrane proteins that act as glues on the cell surfaces to bind cells to one another.	A complex network of proteins and complex carbohydrate polymers in the spaces between cells. The matrix helps bind the cells, acts as a reservoir for many hormones controlling cell growth and differentiation, and provides a lattice through which cells can move.

↓

Cell junctions
A cluster of cell adhesion molecules that stabilize cell–cell interaction and facilitate communications between neighboring cells.

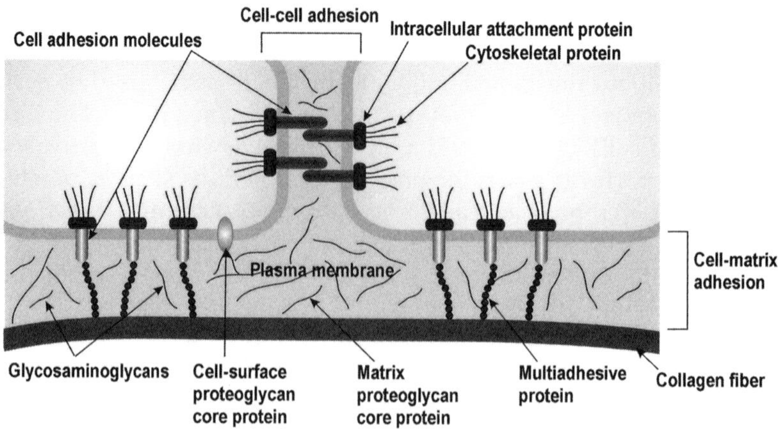

Figure 2.10. Cell-to-cell and cell-to-extracellular-matrix binding schematic view.

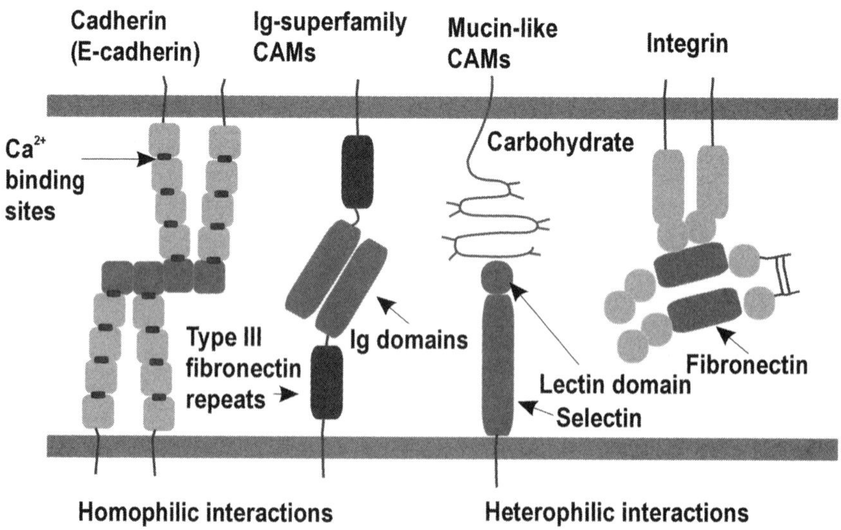

Figure 2.11. Major families of cell-adhesion molecules.

Extracellular Matrix (ECM). Cells in tissues are generally in contact with a complex network of secreted extracellular materials called the extracellular matrix. This matrix performs the function of holding cells and tissues together and provides an organized lattice within which cells can migrate and interact with one another. The extracellular matrix consists of a variety of polysaccharides and proteins that are secreted locally and assemble into an organized meshwork. Connective tissues, which largely consist of the extracellular matrix, form the architectural framework of an organism. An amazing diversity of tissue forms (skin, bone, spinal cord, etc.) is derived from a variation in the relative amount of the different types of matrix macromolecules and the manner in which they organize in the extracellular matrix. Connective tissues consist of cells (i.e., fibroblasts) sparsely distributed in the extracellular matrix, which is rich in fibrous polymers such as collagen. The cells are attached to the components of the extracellular matrix. In contrast, epithelial tissues consist of cells that are tightly bound together into sheets called epithelia. In this case, cells occupy most of the tissue's volume and the extracellular matrix content is relatively small.

The extracellular matrix consists of three major proteins: (i) highly viscous proteoglycans providing cushions for cells, (ii) insoluble collagen fibers, which provide strength and resilience, and (iii) multiadhesive matrix proteins, which are soluble and bind to receptors on the cell surface. Collagen is the single most abundant protein in all living species. Although there are at least 16 types of collagen, 80–90% belong to three types classified as Types I, II, or III, depending on the type of tissue in which they are found. These collagen molecules are packed together to form long, thin fibrils. Basal lamina, as tough matrix, forms a supporting layer underlying cell sheets and prevents cells from ripping apart.

Cell Junctions. Cell junctions occur at many points of cell–cell and cell–matrix contact in all tissues. There are four major classes of junctions:

- Tight junctions, which connect epithelial cells that line the intestine and prevent the passage of fluids, through the cell layers.
- Gap junctions, which are distributed along the lateral surfaces of adjacent cells and allow the cells to exchange small molecules for metabolic coupling among adjacent cells.
- Cell–cell junctions, which perform the primary function of holding cells into a tissue.
- Cell–matrix junctions, which also perform the primary function of holding cells into a tissue.

2.8. TYPES OF TISSUES AND THEIR FUNCTIONS

There are more than 200 distinguishable kinds of differentiated cells that organize to form a variety of tissues in the human body. This section describes some examples of the different types of tissues.

Epithelial Tissue. Epithelial tissues form the surface of the skin and line all the cavities, tubes, and free surfaces of the body. They are made of closely packed epithelial cells arranged in flat sheets. They function as the boundaries between cells and a cavity or space. They perform the function of protecting the underlying tissues, as in the case of skin. In the case of intestines, the columnar epithelium secretes digestive enzymes and absorbs products from the intestine. The specialized junctions between the cells enable these sheets to form barriers to the movement of water, solutes, and cells from one compartment of the body to another.

Muscle Tissue. The three kinds of muscle tissues are:

- Skeletal muscles, which are made of long fibers that contract to provide the locomotion force.
- Smooth muscle cells of the intestines, blood vessels, and other organs.
- The cardiac muscle of the heart.

Connective Tissue. The cells of connective tissues are embedded in the extracellular materials. Examples of supporting connective tissues are cartilage and bone. Examples of binding connective tissues are tendons and ligaments. Another type is fibrous connective tissues, which are distributed throughout the body and serve as the interstitial material of organs.

Nervous Tissue. These tissues are composed of nerve cells (neurons) and supporting and insulating cells (i.e., glial and Schwann cells). They specialize in the conduction of nerve impulses.

Tissues provide coordinated functions of the constituent cells. The functions of the tissues, however, are not just those provided by the constituent cells, but are also derived from intercellular communications and from the extracellular matrix (ECM) components. ECM acts as a reservoir for many hormones that control cell growth and differentiation. In addition, cells can move through ECM during the early stages of differentiation. ECM also communicates with the extracellular pathways, directing a cell to carry out specific functions. Through gap junction, two adjacent cells are metabolically coupled, because small molecules can pass from one cell to another.

2.9. VARIOUS ORGANS AND ORGAN SYSTEMS IN THE BODY

Just like several types of cells constitute a tissue, several tissues constitute an organ. Each organ performs one or more specialized function in the body, which are synergistically coordinated by its constituent tissues. For example, the heart contains muscle tissue that functions by pumping blood, fibrous tissue that constitute the heart valves, and special cells that maintain the rate

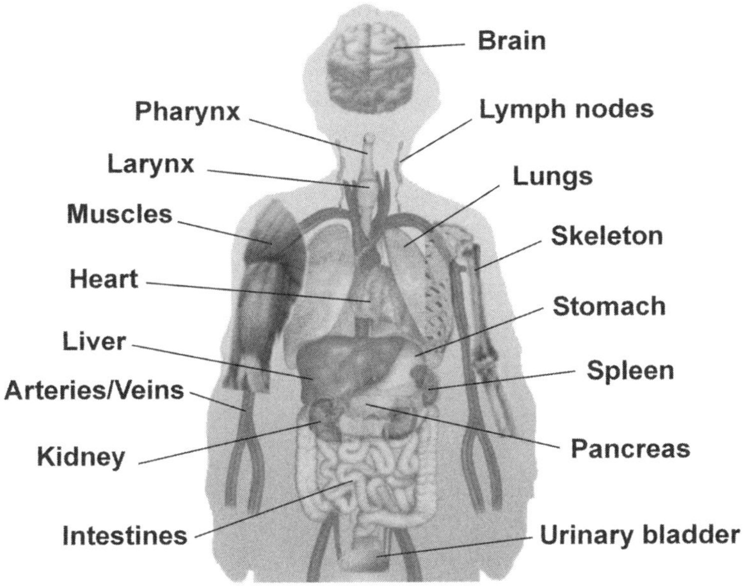

Figure 2.12. Schematic view of organs and tissues in the human body.

and rhythm of heartbeats. The various organs and tissues present in the human body are depicted in Figure 2.12. The following section briefly details the major organs and organ systems in the human body, along with the functions they perform.

Cardiovascular System. This system is a part of the circulatory system of the body, which regulates the flow of blood in the body. The main components of the human cardiovascular system are the heart, the veins, and the arteries. An average adult contains five to six quarts (roughly 4.7 to 5.7 liters) of blood, which consists of plasma, red blood cells, white blood cells, and platelets. The cardiovascular system is connected to all the major organs of the body, and it supplies essential nutrients, hormones, gases, and blood cells to these organs. Malfunctioning of this system causes various diseases such as atherosclerosis, stroke, abnormal blood pressure, aneurysm, thrombosis, ischemia, and so on.

This system is mainly comprised of the pulmonary circulation, which recycles deoxygenated blood from the heart (right side) through the lungs into oxygenated blood back to the heart (left side), and the systemic circulation, where the heart (left side) pumps oxygenated blood to the rest of the body. As the arteries of the systemic circulation are connected to all the organs of the body, they serve as major routes of nanoparticle-mediated drug delivery to potentially all diseased organs. However, the flow of nutrients, macromolecules, and cells through the systemic circulation is closely integrated with the reticuloendothelial system (RES), which is a major component of the immune system whose primary function is to recognize and destroy any malfunctioning or foreign agents circulating in the blood or tissues. Abnormal circulating agents are coated with specialized proteins, or opsonins, present in the blood. These "opsonized" agents are recognized by immune cells such as macrophages from the liver and spleen and are subsequently destroyed. In the context of nanomedicine, this RES sometimes acts as a barrier against the systemic delivery of nanoparticulate agents as they are recognized as foreign and are subsequently eliminated. Therefore, a significant research effort has been devoted toward the fabrication of RES evading, (i.e., "stealth") nanoparticles, which is discussed in Chapter 6.

Lymphatic System. The lymphatic system is the other part of the circulatory system, whose main function is to control drainage of lymph fluid formed from interstitial fluid, which lies within the interstices of all body tissues through lymphatic vessels to RES organs (i.e., lymph nodes). The lymphatic system is a major part of the immune system of the body and consists of lymphatic vessels, lymph nodes, which are oval-shaped tissues distributed throughout the body, and the spleen. For example, humans have about 500–600 nodes distributed throughout the body. These nodes serve as the reservoir for various immune cells, and they act as filters or traps for foreign particles and various pathogens. Mild to severe inflammation associated with lymph nodes, as a result of proliferation of immune cells present in this tissue, serves as an indicator for various disease conditions in the body, ranging from pathogenic infection to cancer. The lymph nodes are connected to one another, as well as with other major organs and with the systemic circulation, through the lymphatic vessels.

Bone Marrow. Bone marrow is the spongy tissue inside some of the large bones, such as hip and thigh bones. Human bone marrow contains large "primitive" undifferentiated cells, called marrow stem cells, which continuously produce new blood cells: white blood cells, red blood cells, or platelets.

Central Nervous System (CNS). The nervous system of vertebrate animals (including humans) is divided into the central nervous system (CNS) and peripheral nervous system. The central nervous system (CNS) is the largest part and includes the brain and spinal cord. The average adult human brain weighs ~1.3 kg (approximately 3 pounds). The brain contains about 100 billion nerve cells (neurons) that are interconnected into extensive network through over 10^{15} junctions (synapses). Besides neurons, the brain contains 10–50 trillion of supportive cells (neuroglia). The spinal cord is about 43 cm long in adult women and 45 cm long in adult men and weighs about 35–40 grams. The brain is the control center of the body, because it coordinates the functioning of other organ systems of the body. This organ controls and coordinates various functions such as memory, movement, sensing, speech, emotion, and so on. Malfunctioning of this system causes various diseases, such as stroke, dementia, Parkinson's disease, Alzheimer's disease, encephalitis, meningitis, brain tumor, and so on.

A characteristic anatomical feature of CNS is that it is separated from the blood circulatory system by the blood–brain barrier (BBB). The function of the BBB is to maintain a constant biochemical environment in the CNS by selectively allowing certain molecules and macromolecules to enter the brain, thus preventing the brain from the invasion of unwanted circulating agents. However, the BBB also acts as a major impediment towards nanoparticulate drug delivery in the brain, because most nanoparticles lack any specialized transport mechanism across the BBB. In order to circumvent this problem, various nanoparticles have been appropriately surface modified with BBB-traversing agents for brain delivery purposes, which is also discussed in Chapter 6.

Liver. The liver is located in the right upper quadrant of the abdominal cavity, to the right of the stomach and below the diaphragm. This organ is known as the "bioreactor" because many compounds (i.e., blood-clotting proteins) that are found in the circulation and other organs are synthesized by hepatocytes, the primary cell type of the liver. Additionally, this organ is also responsible for biochemical reactions that detoxify a number of endogenous and exogenous compounds. The liver, mainly composed of specific cells (hepatic cells, or hepatocytes), is known to participate in more than 500 distinct functions, including processing of digested food from the intestine, secretion of regulatory proteins to the bloodstream, metabolism of carbohydrates, control of the amount of fats, amino acids, and glucose present in the blood, elimination of elimination of exogenous particles and elimination of bacteria, decomposition products of red blood cells, producing and breaking down numerous hormones, and so on. The liver is also a primary component of the reticuloendothelial system (RES) of the body, which was discussed earlier. The Kupffer cells present in the liver are a class of macrophages, which play a critical role in the capture of foreign elements in the blood, which are then sequestered in the liver for degradation. Malfunctioning of this organ causes various diseases, such as hepatitis, cirrhosis, jaundice, liver cancer, and so on.

Lungs. Lungs are one of the largest organs in the body. Human lungs are a pair of spongy, air-filled organs located in two cavities (pleural) on either side of the heart. The trachea (windpipe) conducts inhaled air into the lungs through its tubular branches, called bronchi. The bronchi then divide into smaller and smaller branches (bronchioles), finally ending in clusters of microscopic air sacs called alveoli. The principal function of the lungs is respiration, which is the transport of oxygen from the atmosphere into the bloodstream, with concurrent release of carbon dioxide from the bloodstream into the atmosphere. This organ also regulates the pH of blood by facilitating alterations in the partial pressure of carbon dioxide. These organs also have nonrespiratory functions—for example, filtering out small blood clots formed in veins, secretion of immunoglobulin A in the bronchial secretion as a defense against respiratory infections, and converting several peptides into bioactive proteins. Malfunctioning of this organ causes various diseases, such as asthma, chronic obstructive pulmonary disease, lung cancer, and so on.

The lungs are directly connected to the environment via the inhalation route. Therefore this organ is protected from unwanted inhaled materials by components of the mucosal barrier, such as respiratory mucus and alveolar fluids (e.g., pulmonary surfactants). Detailed information about this biological barrier and its role in nanotechnology mediated drug delivery is discussed in Chapter 6.

Pancreas. The pancreas is a gland organ located in the upper part of the abdomen, behind the stomach. Pancreas is a dual-function gland contributing to the digestive and endocrine systems. The part of the pancreas with endocrine function is made up of approximately a million cell clusters called islets of Langerhans. Their function is to produce hormones (insulin and glucagon) that regulate carbohydrate, fat, and protein metabolism; they also produce somatostatin, which participates in the regulation of endocrine system, neurotransmission, and cell proliferation. The digestive role of the pancreas is carried out via secreting digestive enzymes (trypsinogen, chymotrypsinogen, elastase, carboxypeptidase, pancreatic lipase, and amylase) that break down complex food molecules into more simple substances that can be assimilated by the body. Malfunctioning of this organ causes various diseases, such as diabetes, pancreatitis, pancreatic cancer, and so on.

Kidneys. The kidneys are bean-shaped organs (11–14 cm in length, 6 cm wide, and 4 cm thick) located in the paravertebral gutter of the abdominal cavity. The major function of the kidney's is urinary, in which they filter blood and remove excess of minerals, metabolic byproducts, and toxins, along with the removal of excess water in the form of urine. About 190 liters of blood per day, on average, are filtered by kidneys. Other functions of this organ are endocrinal, which involves the regulation of acid–base balance, electrolyte concentrations, extracellular fluid volume, and the regulation of blood pressure by producing rennin, angiotensin, aldosterone, and other hormones. Kidneys also help in the control of blood pressure, and they play an important role in the metabolism of vitamin D, which is required for maintaining bones strong and healthy. In the context of nanomedicine, kidneys may play a critical role in the elimination of systemically administered nanoparticles from the body. However, the size of the circulating nanoparticles to be eliminated will be limited by the ultrasmall (less than 5 nm) pore size of the capillary filters present in the kidney. Malfunctioning of this organ causes various diseases, such as kidney failure, renal cancer, and so on.

2.10. TUMORS AND CANCERS

Tumors or neoplasms can be regarded as "abnormal" tissues, which can grow anywhere in the body. Tumors contain a multitude of cell types, each of them being defined by abnormal growth and proliferation. Understanding the molecular basis of cancer growth is of great significance to society. This understanding can perhaps lead to the prevention, early detection, and cure of cancer.

Tumor is a general term used to describe certain aberrations in normal cellular behavior. Tumors differ from their normal counterparts in growth control, morphology, cell-to-cell interactions, membrane properties, cytoskeletal structure, protein secretion, and gene expression. Neoplasms can be either benign or malignant. Malignancies consist of abnormal cells that multiply rapidly, many times in the absence of any growth-promoting factors, which are required for proliferation of normal cells. Furthermore, cancer cells may also be resistant to signals that program normal cell death (apoptosis). These cells also express elevated levels of certain cell-surface receptors and produce specific enzymes. The process by which normal cells can transition to tumor cells is known as transformation. In this process, genes that trigger cancer may become activated by diverse factors in a very complex process that is not yet clearly understood. These "oncogenes" interfere with normal regulatory processes, resulting in loss of replicative control in the cell.

Nonproliferating tumors that remain localized are called benign. In the case of a benign tumor, the tumor cells resemble and function like the normal cells. The benign tumors pose little threat to life and can be removed without affecting any normal functions of a tissue. Tumors in which the cells exhibit rapidly growing features and possess a high nucleus-to-cytoplasm ratio are called malignant tumors, neoplasms, or cancers. They often do not remain localized and they invade the surrounding tissues. Cancer cells spread through the body's circulatory system and establish secondary areas of growth. This behavior is called metastasis. Metastatic cells also break their contacts with other cells, thus adding to a degeneration of the tissue function.

The growth and metastasis of tumors is closely associated with the circulatory and lymphatic systems in the body. Tumor cells, owing to their high metabolic activity, require an elevated supply of nutrients from the circulating blood. In order to meet this demand, tumors have evolved the ability to generate around them new blood vessels, or neovasculature, in a process called angiogenesis. In addition, the vasculature associated with tumors has leaks, or "fenestraes," which allow higher amounts of circulating agents to leave the vasculature and "extravasate" into the tumor. This "leaky" vasculature, along with the impaired lymphatic drainage system of the tumors, give rise to the "enhanced permeability and retentivity" (EPR) effect of tumors.

HIGHLIGHTS OF THE CHAPTER

- Living systems are aggregates of water, amino acids, carbohydrates, fatty acids, ions, and other macromolecules called biopolymers (e.g., DNA, RNA, and proteins)
- All living creatures are made up of cells.
- Biological systems are classified as (1) viruses, which require host cells to replicate, (2) prokaryotic cells, which consist of a single closed compartment that lack a defined nucleus, and (3) eukaryotic cells, which have a complex structure including membrane-bound organelles.
- Cells undergo differentiation to form different types of cells and assemble to form tissues, which organize to form organs with different functions. Various organs integrate to form an organism.
- A plasma membrane consists of a lipid bilayer intercalated with proteins that form a semipermeable boundary around the cell. The membrane controls the transport of food, water, ions, and signals.
- The interior content of a eukaryotic cell, enclosed by the membrane, is subdivided into the cytoplasm and the cell nucleus. The cytoplasm includes the subcellular structures such as cytoskeleton and organelles, which are specialized subcellular structures [mitochondria, endoplasmic reticulum (ER), Golgi apparatus, lysosomes, and peroxisomes]. The aqueous component of the cytoplasm is called cytosol.
- The cytoskeleton consists of extensive meshwork of fibrillar proteins and is responsible for cell shape, the mechanical strength of a cell, muscular contraction and locomotion, as well as for anchoring, and intracellular transport of organelles.
- The nucleus harbors genomic DNA, distributed among the chromosomes, and is the site of DNA transcription and RNA processing (modifications). The nucleus communicates with the cytoplasm by means of nuclear pores.
- The mitochondria are responsible for the most of the energy production in the cell.
- The endoplasmic reticulum (ER) consists of flattened sheets, sacs, and tubes of membranes, of which there are of two types: smooth and rough. Ribosomes are present in the rough ER and are responsible for protein synthesis. The smooth ER is where the synthesis of lipids and carbohydrate moieties occurs.
- The Golgi apparatus is responsible for further modification of proteins and transport to other organelles and membranes of the cell.

- There are more than 200 different types of cells present in the human body. Some prominent ones are epithelial cells, which form the lining on the inner surfaces of the body; blood cells present in blood; nerve cells, which are responsible for communication within the body; neurons, which are cells in the brain; sensory cells, which detect sound and light; germ cells, which are responsible for cell reproduction; and stem cells, which are cells waiting to be assigned functions.
- The basic building blocks of cells are nucleic acids (DNA and RNA), proteins, polysaccharides (carbohydrates), and lipids. DNA consists of four bases: adenine, guanine, thymine, and cytosine. In RNA, the thymine is replaced by uracil. The sequence of these bases in the DNA represents the genetic makeup of the organism.
- Proteins are formed by the polymerization of essentially only 20 different types of amino acids.
- Some of important cellular processes are: (i) cell proliferation, which involves DNA duplication and division to produce new cells; (ii) gene expression that includes a process of transcription to produce RNA, followed by translation to synthesize new proteins; (iii) cell energy production in a usable form; (iv) cell signaling to initiate processes in adjacent and distant cells; (v) cell death, called apoptosis, which is programmed cell death and necrosis which is cell death due to injury; and (vi) cell transformation that produces a tumor.
- The constituents of a tissue are: (i) various cells; (ii) cell-adhesion molecules, which bind cells to one another; (iii) the extracellular matrix filling the space between cells; and (iv) the cell junctions.
- The extracellular matrix serves as a framework in which the cells are attached; it also provides a lattice through which cells can move, and it helps provide structural integrity to organs.
- Cell junctions stabilize cell–cell infrastructure and facilitate communication between neighboring cells.
- The circulatory system of the body, consisting of the cardiovascular system and the lymphatic system, coordinates the transport of blood and other physiological fluids through the body.
- The reticuloendothelial system (RES) and the lymphatic system recognize and eliminate abnormal cells and foreign or infectious agents present in the body.
- The central nervous system (CNS) in the brain coordinates much of the functioning of major organs of the body.
- The liver serves as a "bioreactor," because a number of important products are synthesized and stored in this organ. It also plays a critical role in the digestive system and immune systems of the body.
- The major function of lungs is respiratory, which is the transport of oxygen from the atmosphere into the bloodstream, with concurrent release of carbon dioxide from the bloodstream into the atmosphere.

- The pancreas coordinates two functions (a) regulation of metabolism via the production of hormones (insulin and glucagon) and regulation of the endocrine system by production of somatostatin and (b) regulation of digestion via secretion of various digestive enzymes.
- The kidneys are responsible for the filtration of blood and other physiological fluids and elimination of filtered metabolic and toxic compound products, via the urinary system.
- Tumors consist of cells with structural changes produced by transformation that show aberration from the normal behavior. Cancer is a special type of tumor that usually exhibits much more rapid multiplication (proliferation) of cells compared to normal cells and invasive growth.

EXERCISES

True or False

1. Five major types of biomolecules are DNA, RNA, proteins, lipids, and carbohydrates (saccharides).

2. DNA consists of four bases: adenine, guanine, thymine, and uracil.

3. The genomic DNA replication occurs during the S-phase of the cell cycle.

4. Organs are composed of several tissues.

5. Lymphatic system plays major role in oxygen exchange.

Multiple Choice

6. Major function of mitochondria is

(a) Synthesis of proteins
(b) Synthesis of ATP molecules
(c) Modification of proteins
(d) Metabolism of hydrogen peroxide

7. Which type of cell does not contain a nucleus?

(a) Erythrocyte
(b) Leukocyte
(c) Hepatocyte
(d) Schwann cell

8. A distinctive property of a cancer cell is

(a) Rapid proliferation and invasive growth
(b) Partial loss of functions

(c) Abnormal shape
(d) Evasion of apoptosis.

9. Select a subcellular structure absent in animal cells:
(a) Complex Golgi
(b) Peroxisomes
(c) Endoplasmic reticulum
(d) Cell wall

Answers: 1. T; 2. F; 3. T; 4. T; 5. F; 6. b; 7. a; 8. a; 9. d

REFERENCES

Alberts, B., Johnson, A., Lewis, J., Raff, M., Roberts, K., Watson, J. D., *Molecular Biology of the Cell*, 3rd edition, Garland, New York (2002).

Audesirk, T., Audesirk, G., Byers, B. E., *Biology: Life on Earth*, 9th edition, Prentice-Hall, Upper Saddle River, NJ (2011).

Horton, H. R., Moran, L. A., Scrimgeour, G., Perry, K. G., Rawn, J. D., *Principles of Biochemistry*, 4th edition, Prentice-Hall, Upper Saddle River, NJ (2006).

Lehninger, A. L., Nelson, D. L., Cox, M. M., *Lehninger Principles of Biochemistry*, 5th edition, W. H. Freeman, New York (2008).

Lodish, H., Berk, A., Kaiser C. A., Krieger, M., Scott, M. P., Bretscher, A., Ploegh, H., *Molecular Cell Biology*, 4th edition, W. H. Freeman, New York (2000).

Prasad, P. N., *Introduction to Biophotonics*, Wiley-Interscience, Hoboken, NJ (2003).

Solomon, S., *Introduction to General, Organic & Biological Chemistry*, McGraw-Hill, New York (1987).

Stryer, L., *Biochemistry*, 4th edition, W. H. Freeman, New York (1995).

Thibodeau, G. A., Patton, K. T., *Anatomy and physiology*, 5th edition, Mosby, St. Louis (2003).

Voet, D., Voet, J. G., Pratt, C. W., *Fundamentals of Biochemistry*, John Wiley & Sons, Hoboken, NJ (2002).

CHAPTER 3

NANOCARRIERS

Nanostructures play a central role in nanomedicine, because they are the nanocarriers of probes, therapeutic agents, or biological materials to be delivered to targeted sites such as a specific cell type, tissue, or organ. In addition, nanostructures can play a direct role, serving as imaging or sensing nanoprobes or producing therapeutic actions. A wide range of nanostructures is available. Some of these can be used for combined therapeutic and diagnostic actions, called theranostics. This chapter discusses the various nanostructures used as carriers in nanomedicine. Section 3.1 introduces the benefits of using a nanocarrier for theranostics and the challenges involved in design of a suitable nanocarrier. Section 3.2 presents the various nanostructures utilized for biomedical application, including an overview of the most important characteristics of these nanostructures, particularly nanoparticle formulations. Section 3.3 discusses virus-based nanocarriers. Section 3.4 covers polymeric nanocarriers. Section 3.5 discusses lipid-based nanocarriers. Section 3.6 presents dendrimers as nanocarriers. Section 3.7 describes various carbon-based nanocarriers. Section 3.8 discusses various inorganic nanoparticles for nanomedicine. Section 3.9 introduces PEBBLE, a multifunctional nanoplatform incorporating various imaging and sensing nanoprobes. Section 3.10 describes nanoclinics, another multifunctional nanoplatform for diagnostics and therapy. Section 3.11 presents nanoplexes, a nanostructure allowing for electrostatic complexation with genetic materials for gene delivery. Section 3.12 introduces

Introduction to Nanomedicine and Nanobioengineering, First Edition. Paras N. Prasad.

new-generation nanostructures that can be nontoxic, environmentally friendly, and batch processible at low cost.

3.1. NANOCARRIERS: DELIVERING PAYLOADS TO NEEDED SITES

Nanostructures ranging in size from a few nanometers to hundreds of nanometers have been used as nanocarriers of various payloads for diagnostics and therapy. These nanostructures can be biological objects (e.g., viruses), polymers, lipid-based, dendrimers, carbon-based structures, or rigid inorganic nanoparticles, such as ceramic, silica-based, metallic, and so on. Polymer-, ceramic-, and silica-based nanoparticles are more rigid, while micelles and liposomes are softer nanostructures. A nanocarrier approach offers the following advantages:

- Nanocarriers provide three different structural platforms for diagnosis and therapeutics: (a) An interior volume in which various probes and therapeutic agents can be encapsulated. (b) A surface to which targeting groups can be attached, which allows them to be carried to cells or biological sites expressing the appropriate receptors. In addition, the surface can bind to specific biological molecules for delivery (such as in gene therapy, discussed later). The surface can also be functionalized to introduce a hydrophilic (polar), hydrophobic (nonpolar), or amphiphilic character to enable dispersibility in a variety of fluid media. (c) Pores in a nanocarrier can be tailored to be of specific sizes to allow for the selective intake or release of biologically active molecules or activate therapeutic agents.
- Nanocarriers are generally not immunogenic, that is, they do not elicit any immune response when introduced into the body's circulatory system.
- The nanocarriers can be tailored with compositions (such as silica or organically modified silica) that are resistant to microbial attack. These compositions do not undergo enzymatic degradation and thus effectively protect the encapsulated probes or drugs.
- Nanocarriers can be appropriately surface functionalized to penetrate the cell membrane by endocytosis. This provides a convenient mechanism for intracellular diagnostics and therapy.
- Nanocarriers, such as silica-based nanobubbles (thin shells), are optically transparent. Therefore, light activation and optical probing can be readily accomplished.
- A nanocarrier matrix can protect the active agent payloads.
- A nanocarrier can be designed for controlled release of drugs, as well as to include diagnostic or contrast agents, which may be activated by an external stimulus.

However, to make impact on nanomedicine, some of the important characteristics of nanocarriers are as follows:

- They escape the capture by the reticuloendothelial system (RES) when introduced into the bloodstream. The purpose of the RES is to eliminate foreign objects in the bloodstream. This topic is discussed in Chapter 6, entitled "Crossing the Biological Barrier." For this purpose a common strategy is to introduce a polyethylene glycol (PEG) coating that allows prolonged circulation of nanoparticles in the bloodstream.
- They reach the intended biological target, such as cancer sites, to enhance site specific action. For this the nanocarriers should have the ability to be functionalized to couple to targeting groups and thus possess biorecognition ability. This aspect is covered in detail in Chapter 7, entitled "Biotargeting."
- They produce maximum cellular uptake for intracellular diagnostics and therapy. This requires tailoring the surface, size, and shape characteristics of the nanocarriers. This aspect is described below in Section 3.2.
- They exhibit no toxicity and preferentially exit the body either by biodegradation or by excretion. The nanotoxicity aspect is covered in Chapter 21.

At this point there is no single nanoformulation that serves as the most suitable nanocarrier, or so-called magic bullet, for all nanomedicine applications. Section 3.2 provides a general discussion of the various classes of nanoformulations which are then discussed in separate individual subsequent sections.

3.2. THE VARIOUS NANOFORMULATIONS FOR NANOMEDICINE

For nanomedicine applications, a diverse group of nanostructures with different chemical compositions, surface characteristics, and inherent functionalities for diagnostics and therapy provide a broad range of choices. These nanostructures can be biological, can be made of synthetic polymeric units, can have a dendrimer structure, can be a lipid-based system, or can be made of a more rigid inorganic structure platform to produce a combined therapeutic and diagnostic platform, now more commonly referred to as "theranostic" nanostructures. The term theranostic is used throughout this book to refer to combined features of diagnostics and therapeutics.

A representative list of nanoparticles being used at our Institute for Lasers, Photonics and Biophotonics (ILPB) (other than viruses, lipid-based nanoparticles, and dendrimers) is provided in Figure 3.1. The chemical methods for their preparations are discussed in Chapter 4. The specific inherent

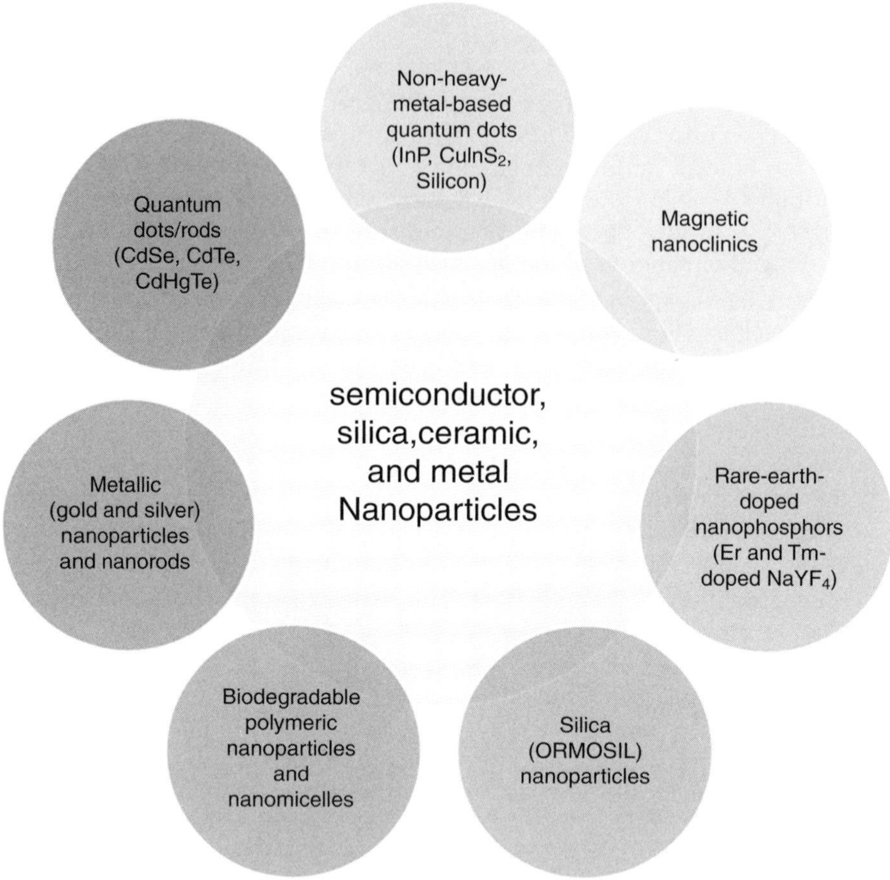

Figure 3.1. Nanomaterials synthesized at ILPB.

functionalities expressed by these nanoparticles are discussed in Chapter 5. Another type of nanoparticle, combining two or more functionalities or enhancing a specific functionality, involves a core–shell or a core–shell–shell (multiple shells) type of structure built hierarchically. They also provide an opportunity to incorporate different therapeutic or diagnostic payloads in core or different shells, thus making them suitable carriers for multiple therapeutic or diagnostic agents for multimodal theranostics.

These nanoparticle formulations are discussed in subsequent sections together with other nanostructures such as dendrimers and carbon-based nanostructures. A logical question is why one needs to have so many choices for nanoparticles. For nanomedicine applications, there are so many factors to consider in determining the suitability of a nanoparticle platform for a specific application. For example, for *in vitro* laboratory-based diagnosis, the toxicity of the nanoparticle may not be a concern, but their appropriate functionalities

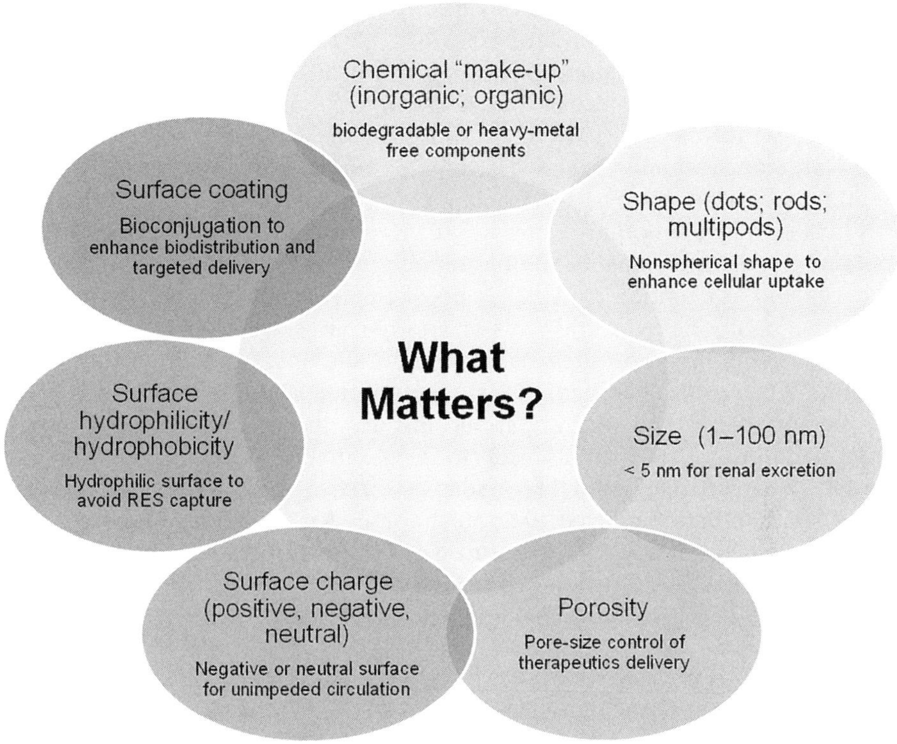

Figure 3.2. Tailoring of the nanoparticle platform to produce what matters.

for high sensitivity and high-throughput diagnostics may be of prime importance. However, for *in vivo* diagnosis and therapy, toxicity is of prime concern. Similarly, in order for the nanoparticles to cross a specific biological barrier, discussed in Chapter 6, it must have certain size and surface characteristics. Figure 3.2 lists what matters in the tailoring of a nanoparticle platform for a specific application.

It is clear that chemical composition, surface group, surface charge, size, and shape of a nanocarrier all play important roles. Thus, a precise control of production of nanocarriers to achieve monodispersity (all particles same) and reproducibility is vital for their applicability to nanomedicine.

3.3. VIRUSES AS NANOCARRIERS

As covered in Chapter 2, viruses are nanosize (≤100 nm) biological objects that can also serve as nanocarriers. However, for the purpose of using them as carriers, they have to be inactivated so that they are not harmful. Various forms of virus-derived nanoparticles have been used for nanomedicine applications. An example is a class of virus-like particles (VLPs) that retain the self-assembled viral envelope protein, but are devoid of the accompanying genetic materials of the virus [Manchester and Singh, 2006]. These VLPs are thus inactive forms of viruses. They still offer the advantages of providing a nanoplatform containing reactive groups to which many peptide groups can be attached for introducing targeting, imaging, and drug delivery payloads.

Virosomes are another example of virus-derived nanoparticles about 150 nm in size, which have been investigated for targeted drug delivery [Bhattacharya and Mazumder, 2011]. They consist of a virus envelope, but to which additional components of the virus or another virus are attached or inserted. However, generally virosomes include membrane lipids and viral spike glycoproteins, but are still devoid of the genetic materials of the virus. The virosomes are biodegradable and biocompatible and are also claimed to be nontoxic [Huckriede et al., 2005]. In addition to drug delivery, their potential for vaccines have been investigated [Huckriede et al., 2005]. This application is exemplified in Chapter 15, entitled "Nanotechnology for Infectious Diseases."

Virosomes containing glycoproteins on their envelope, which have receptor-binding and membrane-fusion properties, offer the natural ability of cellular delivery of the encapsulated payloads (e.g., drugs).

3.4. POLYMERIC NANOCARRIERS

Polymeric nanocarriers have been extensively used in various nanoarchitectures [Shi et al., 2009] as shown in Figure 3.3. The richness in design and synthesis and in grafting of polymeric materials allows for a great deal of flexibility to tailor their chemical composition, size, morphology, and surface functional-

(a) (b) (c)

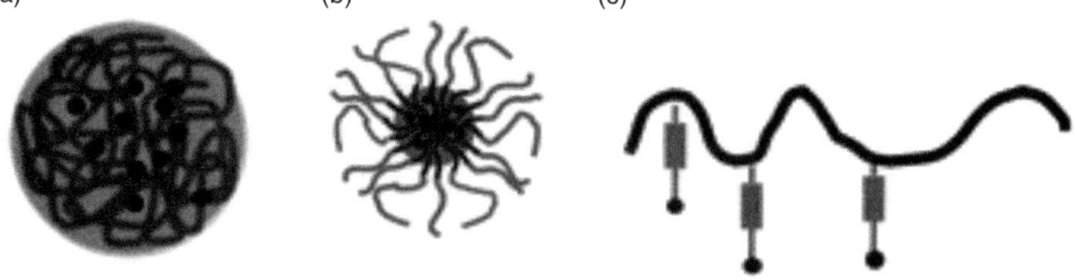

Figure 3.3. Nanocarriers incorporating drugs: (a) Solid nanosphere from polymers or other materials incorporating drugs inside. (b) Core-shell micellar polymeric nanoparticles self-assembled from well-defined synthetic amphiphilic polymers where drugs are physically or chemically incorporated in the inner core. (c) Water-soluble polymer–drug conjugate with biologically defined linkers between the drugs and polymers. (Reproduced with permission from Shi et al. [2009].)

ity for delivery of sensing, imaging, and therapeutic agents. Thus, a wide array of polymeric nanocarriers has been used. A solid polymeric nanoparticle can have payloads either physically encapsulated in the interior or chemically conjugated on the surface. The release of an encapsulated payload can be controlled either by polymer degradation by intracellular enzymes or activation by a stimulus. An example of stimulus activation is a hydrogel or thermogel, where activation may be provided by swelling in response to the changing pH of the cellular environment or by external thermal activation. Polyacrylate or polyacrylic acid nanoparticles are one example that has been widely used.

Biodegradable polymeric nanoparticles have attracted much attention recently for drug and gene delivery systems. Nanoparticles made from poly(amino esters), polyanhydrides, poly(lactides), chitosan, and other biodegradable polymers can be degraded by hydrolysis *in vitro* and *in vivo* under particular conditions and thus produce minimum long-term toxicity risks. For example, microspheres of PLGA (poly(lactic-co-glycolic acid)), an FDA-approved biodegradable polyester, have been successfully used to encapsulate therapeutic genes by double emulsion solvent evaporation method [Cun et al., 2010]. The PLGA nanoformulations have been explored for a large number of applications and are widely covered in this book. Langer's group described the incorporation of PBAE (poly(β-amino ester)) in PLGA particles, which can effectively enhance the loading efficiency of a certain payload such as plasmid DNA for gene delivery. A major drawback of the PLGA nanoparticles, however, is that complicated procedures were required for PLGA nanoparticle synthesis due to the hydrophobicity of PLGA. Hydrophilic coatings, such as polyvinyl alcohol (PVA) and polyethylene glycol (PEG), thus become crucial. The reaction efficiencies for these side-chain modifications on PLGA backbone, however, were not very high.

Another promising biodegradable nanocarrier is chitosan. Chitosan is a polysaccharide that is primarily extracted from crab shell. It is a highly positively charged material, depending on the degree of deacetylation of the polymer, and can bind to genetic material such as DNA and siRNA by electrostatic interactions. This is useful in gene therapy, a topic covered in Chapter 14. Our group has demonstrated an efficient delivery of ssRNA and shRNA *in vivo* for infectious disease therapy using chitosan nanoparticles as the carrier, which is discussed in Chapter 15.

Another important type of polymeric nanostructure involves a polymeric core-shell structure that utilizes a diblockamphiphilic copolymer containing well-defined length segments of a hydrophobic block and a hydrophilic block [Shi et al., 2009]. Such block copolymers can self-assemble into a core–shell structure with a hydrophobic inner core that can incorporate hydrophobic payloads (many drugs are hydrophobic). Another variation of a polymeric core–shell structure allows for a polymer core decorated with polymeric brushes which can act as biosensors or contain imaging or diagnostic agents [Tokarev et al., 2009].

3.5. LIPID-BASED NANOCARRIERS

Lipid-based nanocarriers utilize lipid (surfactant)-type molecules consisting of a polar head group and a nonpolar long alkyl chain tail [Shi et al., 2009; Mulder et al., 2009]. Two major formulations are used for micellar nanoparticles and bilayer lipid structures containing liposomes, as shown in Figure 3.4. In the micellar formulation, these nanoshells of lipids self-assemble in an aqueous environment to form a micellar nanocavity with a nonpolar inner core (cavity) and polar head outer surface. Various hydrophobic drugs and diagnostic or imaging agents can thus be encapsulated in the nonpolar cavity. To overcome the poor structural integrity of the micellar nanoparticles, polymeric phospholipid micellar structures are now being used. These provide significantly improved structural stability. Many hydrophobic drugs are encapsulated in micellar structures to improve their biodistribution. Furthermore, the use of PEG groups on the surface of a micellar carrier (by chemical conjugation) leads to prolonged circulation and ultimately increased targeting ability.

Liposomes [Mulder et al., 2009] are nanoparticles consisting of one or several lipid bilayers with an aqueous phase inside, as well as in between the lipid bilayers. They generally utilize a phospholipid. The aqueous interior has been used to incorporate a water-soluble drug in order to increase its bioavailability. Various hydrophilic payloads can be separately included in the hydrophilic interior or in between the bilayers. Hydrophobic payloads can be included within the bilayers that provide the hydrophobic regions. Liposome nanoparticles have been extensively utilized for delivery of MRI contrast agents and fluorescent probes, as well as for drug delivery. Pegylated liposomes based upon pegylated (PEG group containing) phospholipids have been utilized, again to improve circulation.

Another type of lipid nanostructure receiving considerable attention is solid lipid nanoparticles (SLN), which are made of lipids that are solid at room

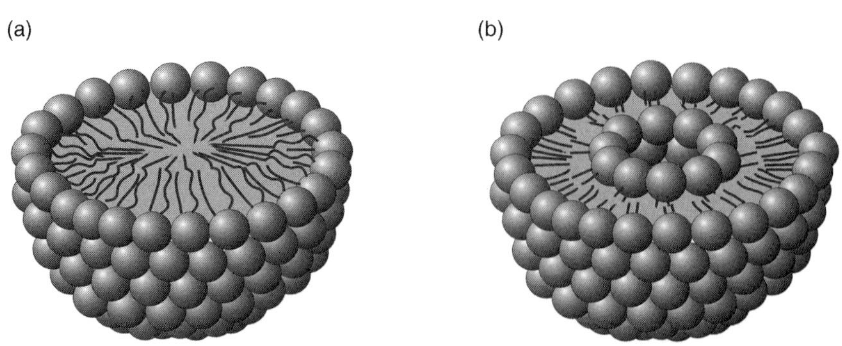

Figure 3.4. Lipid-based nanocarriers: (a) micelle, (b) liposome.

temperature as well as at body temperature [Rai et al., 2008; Basu et al., 2010]. They are produced in a wide range of sizes from 10 to 1000 nm. SLN thus possess a solid lipid core that is stabilized by surfactants. They hold promise for both oral and parenteral drug delivery. Compared to polymeric nanoparticles, they have advantages of being highly biocompatible and biodegradable with considerably reduced side effects compared to other lipid/micellar nanostructures, such as liposomes discussed earlier; also, they are more rigid, and thus they are physically more stable and offer better protection to an encapsulated drug against chemical degradation. Also, there is no or little water access in the inner core [Rai et al., 2008]. This book provides a number of examples of application of SLN (e.g., in Chapters 15 and 20).

3.6. DENDRIMERS

The word *dendrimer* is a composite of two Greek words: "dendron," meaning tree or branch, and "meros," meaning part. It represents a vast class of supramolecular architectures that contain highly ordered and branched nanostructures. Dendritic nanostructures hold promise for many applications of nanomedicine ranging from sensing to drug delivery.

Dendrimers are regularly branched globular macromolecules made by a controlled step-by-step iterative synthetic approach (Grayson and Fréchet, 2001). Dendrimers thus differ from hyperbranched polymers by their structure perfection, containing an exact number of concentric layers of branching points, referred to as *generations*. A dendrimer can be described as consisting of three principal regions (Grayson and Fréchet, 2001; Dykes, 2001):

- A core or focal moiety that forms the site for branching.
- Layers of branched repeat units that emanate from this core and define the generation of a dendrimer.
- The end groups on the outer layer of repeat units, defining the periphery, often serving as a multivalent surface for the dendrimer.

Dendrimers are multibranched structures produced by a highly controlled multistep synthesis, using a core structure on which is grafted a chemically active branch unit to sequentially branch out in a divergent way (like a tree branching out) [Lee et al., 2005]. Thus, the branching out chemistry (described in Chapter 4) can be used to produce successively built layers (called generations, like generation 1 describing the first layer). Dendrimers pioneered by Tomalia [Caminati et al., 1990; Tomalia et al., 2007] and Fréchet [Hawker et al., 1990; Lee et al., 2005] have been extensively used as nanocarriers for diagnostic and therapeutic agents [Rolland et al., 2009]. The advantage of a dendritic carrier is its multivalency—that is, multiple sites it provides for the conjugation of various payloads as shown in Figure 3.5.

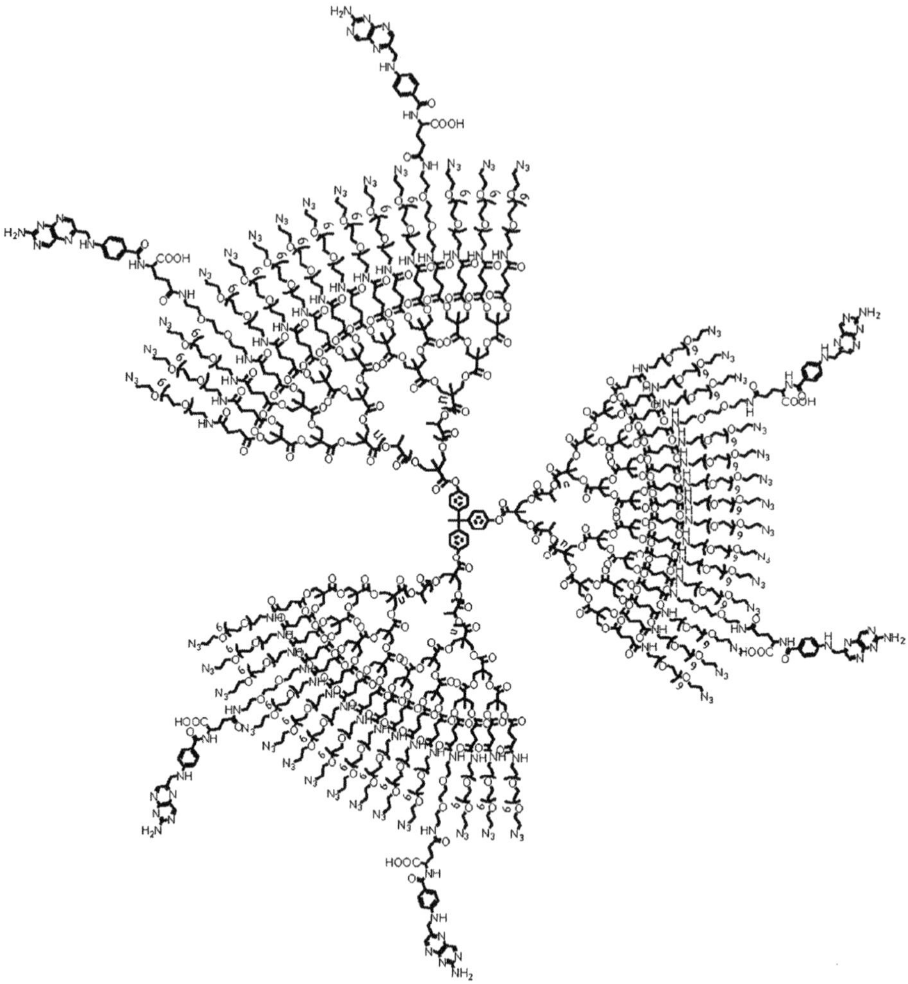

Figure 3.5. Anatomy of a folate-functionalized dendrimer for doxorubicin conjugation. (Reproduced with permission from Cao et al. [2011].)

Some noteworthy examples on use of dendrimer structures for nanomedicine are contributions from the research groups of Fréchet [Lee et al., 2005; Gillies and Fréchet, 2005; Lee et al., 2006], Tomalia [Tomalia et al., 2007; Menjoge et al., 2010], and Baker [Majoros et al., 2005; Majoros et al., 2008; Patri et al., 2002].

A dendrimer can be employed to encapsulate a payload in the core or the payload can be conjugated to a branch point. Furthermore, the loading can be easily tuned by controlling the generation number. The release of a payload can be achieved by using a degradable chemical linker attaching the payload to a branch. A highly controlled multistep synthesis allows for highly monodispersed dendrimers, where the end groups can be functionalized for conjugation to PEG or a targeting group. These features of dendrimers for nanomedicine are further elaborated upon in Chapter 4.

3.7. CARBON NANOSTRUCTURES

Carbon nanostructures form a major group of nanocarriers utilized for nanomedicine applications. These nanostructures are an exciting group of nanomaterials that have been investigated for a wide variety of applications, and thus extensive literature exists on them. Here they will only be briefly discussed; however, their applications are covered in many chapters in this book.

Fullerene C_{60}. This is a 3D nanostructure, composed of 60 carbon atoms involved in the formation of double and single bonds. Its structure is shown in Figure 3.6a. The overall shape is like a soccer ball or a geodesic dome (popularized by Buckminster Fuller). The synonymous names buckminsterfullerene, fullerene, and "bucky ball" arise from this similarity. The surface of C_{60} can be chemically functionalized to make it water-dispersible and to introduce various linking groups for introducing targeting and PEG groups as well as for attaching drug sensing groups and imaging probes [Montellano et al., 2011]. This multifunctionalization ability can be utilized to prepare multimodal C_{60} nanocarriers that can cross the cell membranes and efficiently deliver active molecules to the target biological site.

Graphene. Graphene is a form of carbon that has a two-dimensional planar sheet structure, one atom thick (monolayer), consisting of carbon atoms bonded in a cyclic form with alternate single and double bonds [Geim, 2009]. This type of arrangement of alternate single and double bonds in a cyclic form is also known as aromaticity. Figure 3.6 also shows the graphene structure. It

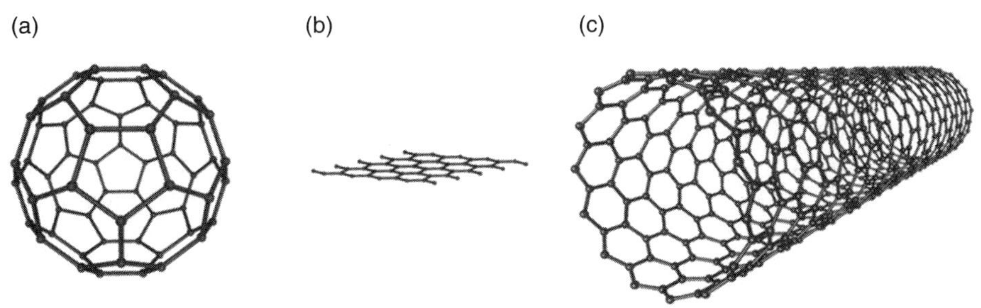

Figure 3.6. Structure of (a) fullerene C_{60}, (b) graphene, and (c) single wall carbon nanotube. (http://en.wikipedia.org/ created by Michael Ströck under the terms of the GNU Free Documentation License.)

can be described as the planar sheet building block of graphite that consists of a stack of planar graphenes. In fact, graphene can be prepared by peeling of layers of graphite to isolate a single layer.

Graphene can also be functionalized to create carbon–oxygen bonds and thus form graphene oxide, paving the way to introduce other reactive chemical units, such as amines which can form linking sites for attachment of sensing groups, PEGylating units, and drugs [Liu et al., 2008]. Graphene-based biosensors have been developed [Feng and Liu, 2011]. Graphene has also been explored as a nanocarrier for drug or gene delivery [Zhang et al., 2010; Feng et al., 2011]. An example of graphene for therapy is *in vivo* cancer treatment [Yang et al., 2010].

Carbon Nanotubes. Carbon nanotubes are cylindrical nanostructures of carbon and can be envisioned as rolled sheets of graphene. Figure 3.6 also depicts carbon nanotubes. These tubes have a hollow cylindrical structure, thus providing opportunities to include substances inside the tube or attach them to its surface. The carbon nanotubes can be single-walled (single-walled nanotubes, abbreviated as SWNT) where the cylinder can have a diameter of ~1 nm and a length from a few nanometers to many millimeters. Another form is a multi-walled carbon nanotube (MWNT), which consists of multiple rolled layers of graphene sheets. These nanotubes exhibit extraordinary thermal, mechanical, and electrical, properties, making them very attractive for biosensing and tissue engineering applications, which are discussed in this book in the respective chapters.

Carbon nanotubes have also been extensively explored for drug delivery [Liu et al., 2011]. Chemotherapeutic agents, such as paclitaxel and cisplatin, have been chemically conjugated to carbon nanotubes for both *in vitro* and *in vivo* drug delivery studies [Bhirde et al., 2010].

Nanodiamonds. Nanodiamonds are diamond nanoparticles with a small diameter (typically around 5 nm), which are most often formed by detonation. Because of their excellent mechanical optical properties, high surface areas, and nontoxicity, they have attracted considerable attention for biomedical applications [Mochalin et al., 2012; Purtov et al., 2010]. Nanodiamond properties can be modified in a number of ways, such as surface doping and interior doping, as well as chemical functionalization of the surface to introduce various chemical linkers [Mochalin et al., 2012]. The applications have ranged from biomedical imaging [Mochalin and Gogotsi, 2009], to magnetic sensing [Maze et al., 2008], to delivery of drugs and genetic materials [Purtov et al., 2010].

Carbon Dots. These are amorphous carbon nanoparticles of size less than 5 nm produced originally by laser ablation [Sun et al., 2006], which exhibit emission in the visible spectral range upon surface functionalization with polymer chains. Thus they are a new class of photoluminescent nanoparticles which exhibit promise for optical bioimaging. Various facile synthesis methods have been developed in recent years, which makes them potentially more accessible than other bioimaging agents [Baker and Baker, 2010]. Our group, in collaboration with Li, Liu, and co-workers, showed that these photoluminescent carbon dots can be used as biocompatible nanoprobes for targeting cancer cells *in vitro* [Li et al., 2010].

3.8. INORGANIC NANOPARTICLES

A variety of inorganic nanoparticles have been studied for biomedical applications. Figure 3.2 gives some examples of the inorganic nanoparticles being pursued at our Institute. These nanoparticles offer the advantage of being extremely robust and thus can serve as a very stable nanocarrier for nanomedicine. They are highly resistant to enzymatic degradation and can be prepared in ultrasmall sizes that can be more suitable for renal excretion. In addition, the electronic, optical, and magnetic properties of these inorganic materials can be tailored by judicious control of their crystal phase, size, shape, and surface characteristics. Multifunctionality can be readily introduced, making them highly suitable for combined therapy and diagnostics (theranostics). In some cases, these nanoparticles can themselves be used for theranostics, as opposed to just being a delivery vehicle.

With some of the inorganic nanoparticles, particularly those containing heavy metal atoms (e.g., Cd), there is toxicity concern that requires a biocompatible coating on the nanoparticle surface. Also, there is an emphasis now, as in our lab, to replace these heavy-metal-containing nanoparticles by non-heavy-metal-containing ones. This section provides a brief description of some of the important types of inorganic nanoparticles.

Semiconductor Nanoparticles. These nanoparticles have received considerable attention because of their size-dependent optical properties, which are derived from the quantum confinement effect on electrons as discussed in Chapter 5. For this reason these nanoparticles are called quantum dots when spherical in shape and are called quantum rods when cylindrical in shape [Prasad, 2004]. The most widely studied quantum dots are Cd-based semiconductors such as CdSe and CdTe. Their preparation and surface functionalization are described in Chapter 4, entitled “Nanochemistry of Nanocarriers.” These quantum structures have been extensively used for biomedical imaging as discussed in Chapter 8, as well as for delivery of payloads for therapy, which is also discussed in several chapters in this book.

Because of the toxicity concern of Cd, other types of quantum dots such as $CuInS_2$ and Si nanocrystals are now being explored at our Institute, as well as elsewhere, particularly for *in vivo* applications.

Metallic Nanostructures. As described in Chapter 1, metallic nanostructures exhibit optical properties different from the corresponding bulk metals.

This behavior arises from the presence of new optical absorption involving a plasmonic resonance [Prasad, 2004]. This plasmonic property is discussed in detail in Chapter 5. For this reason, the metallic nanostructures are often called plasmonic nanostructures. These can be in the shape of spherical nanoparticles, nanorods, and multipods (multibranched rods). The most extensively used metallic nanostructures for nanomedicine application are gold nanoparticles and gold nanorods. Because of strong optical absorption and subsequent heating, these nanoparticles are used as photothermal nanoheaters for thermal therapy as discussed in Chapter 13, entitled "Nanotechnology for Cancer." The strong confinement of electromagnetic fields around these nanostructures has been utilized for biosensing as presented in Chapter 9, entitled Biosensing." Silver nanoparticles have also been extensively used for their antimicrobial feature, described in Chapter 15, entitled "Nanotechnology for Infectious Diseases," and in Chapter 20, entitled "Nanodentistry."

Due to their inertness and biocompatibility, we have used appropriate surface-coated gold nanorods for delivery of genetic materials, as described in Chapter 14, entitled "Gene Therapy."

Magnetic Nanoparticles. Magnetic nanoparticles such as iron oxides (e.g., Fe_3O_4) ranging in size from 5 to 50 nm have been used for a wide variety of biomedical applications, from bioimaging, to biosensing, to magnetically assisted drug/gene delivery as well as for magnetically directed tissue engineering. These applications are discussed in respective chapters. The magnetic properties of magnetic nanoparticles are discussed in Chapter 5.

Silica-Based Nanoparticles. Silica-based nanomaterials are attractive in nanomedicine due to their high biocompatibility, controllable pore size, tunable diameter, and ability to encapsulate water-insoluble compounds such as drugs and organic dyes. The silica matrix is composed of basic cross-linked silicate structure as shown in Figure 3.7a. The silicate can be organically modified with different functional groups (R = carboxyl, thiol, or amino, Figure 3.7b) to facilitate the colloidal stability, degradability, and bioconjugation of silica nanoparticles.

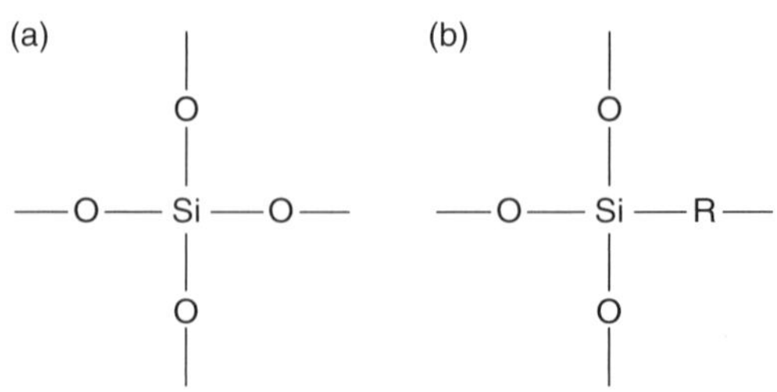

Figure 3.7. Chemical structure of (a) silica and (b) ORMOSIL.

Organically modified silica (ORMOSIL) nanoparticles have recently shown enormous promise as diagnostic and therapeutic agents. These nanoparticles can be synthesized by condensation–polymerization of their organosilane precursor molecules, where one or more of the alkoxy "arms" of the tetraalkoxysilane precursor has been replaced by a functional group. When this synthetic pathway is carried out in a microemulsion, highly monodispersed and ultrafine (diameter less than 50 nm) nanoparticles are precipitated. These can be purified as an aqueous suspension through simple steps such as centrifugation or dialysis (their chemistry will be discussed in Chapter 4). Since these particles are mesoporous, they can serve as sustained release vehicles for encapsulated biomolecules and can be further modified to incorporate imaging/diagnostic probes, genetic materials, and biorecognition molecules. ORMOSIL can thus serve as an ideal nanoplatform for the assembly of 'multimodal' nanoparticles for targeted diagnostics and therapeutics.

Our group has extensively demonstrated the utility of ORMOSIL nanoparticles in various applications in medicine, which includes the potential to serve as (a) targeted probes for optical bioimaging of cancer; (b) carriers of photosensitizing anticancer drugs for photodynamic therapy (PDT) in cancer; and (c) carriers of genetic materials for nonviral gene delivery [Roy et al., 2005; Cinteza et al., 2006; Kumar et al., 2010].

Fluoride/Oxide/Phosphate Nanoparticles. A variety of fluoride, oxide, and phosphate nanoparticles have been utilized in nanomedicine. The fluoride lattice, doped with rare-earth elements provides a matrix for enhanced upconversion of light as describe in Chapter 5. We have used these nanoparticles for up-conversion bioimaging (will be discussed in Chapter 8).Their surface can also be modified to introduce multimodal imaging and drug delivery.

Oxide nanoparticles such as TiO_2 and CeO_2 have been used for their antimicrobial and free radical scavenging properties, respectively, as presented in Chapters 15, 16, and 20.

Phosphate nanostructures have been used for gene delivery, as discussed in Chapter 14, and for tissue engineering, as presented in Chapter 18.

3.9. PEBBLE

A probe encapsulated by biologically localized embedding (PEBBLE), introduced by Kopelman and co-workers [Clark et al., 1999; Monsoon et al., 2003], was a major advancement in the formulation of "multifunctional" nanocarriers for intracellular delivery of multiple sensing and therapeutic agents. PEBBLEs are nanoscale devices consisting of sensor molecules entrapped in a chemically inert matrix. It enables optical measurement of changes in intracellular calcium levels, pH, and other biologically significant chemicals. Figure 3.8 shows a schematic diagram of a multifunctional PEBBLE nanosensor. The matrix materials used for the production of PEBBLEs are also shown in the figure.

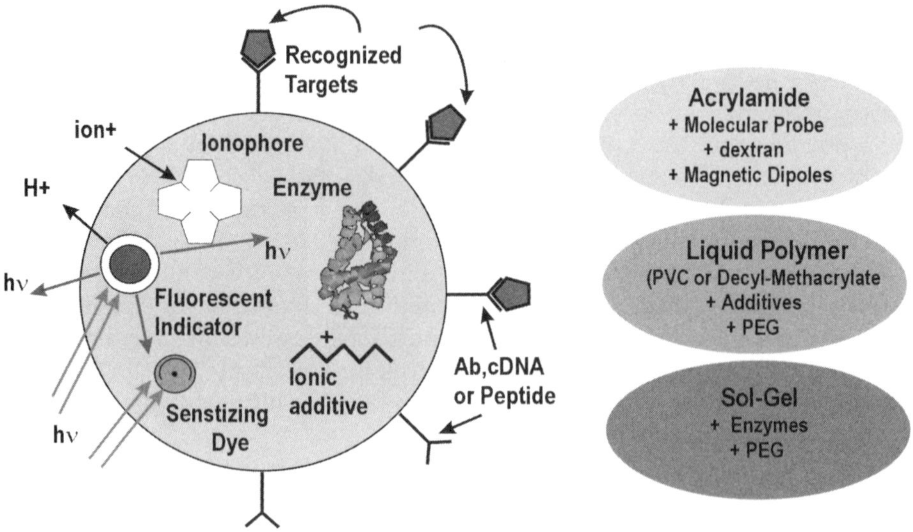

Figure 3.8. Schematic of a PEBBLE nanosensor with various functionalities shown. Current matrix materials are shown on the right. (From http://www.umich.edu/~koplab/research2/analytical/EnterPEBBLEs.html.)

Examples of the matrix media used for PEBBLE technology are polyacrylamide hydrogel, sol–gel silica, and cross-linked decyl methacrylate. These matrices have been used by Kopelman's group to fabricate sensors for H^+, Ca^{2+}, Na^+, Mg^{2+}, Zn^{2+}, Cl^-, NO_2^-, O_2, NO, and glucose. The PEBBLE size ranges from 30 to 600 nm. The matrix porosity allows for the entrapment and sensing of the analyte.

Kopelman and co-workers have demonstrated the following benefits of the PEBBLE technology:

- It protects the cells from any toxicity associated with the sensing dye.
- It provides an opportunity to combine multiple sensing components (dyes, ionophores, etc.) and create complex sensing schemes.
- It insulates the indicator dyes from cellular interferences such as protein binding.

3.10. NANOCLINICS

Our Institute for Lasers, Photonics, and Biophotonics developed nanocarriers called nanoclinics. These have a complex surface functionalized silica nanoshell containing various probes for diagnostics and drugs for targeted delivery [Levy et al., 2002]. These "nanoclinics" (Figure 3.9), also introduced in Chapter

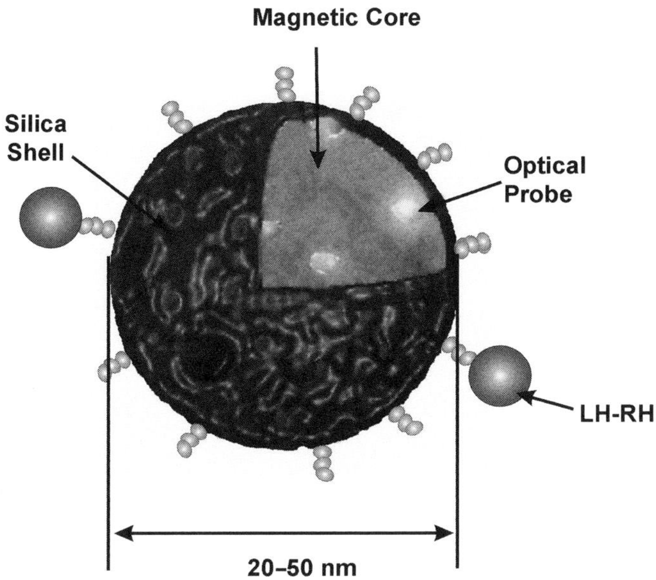

Figure 3.9. Illustrated representation of a nanoclinic.

1, provide a new dimension to targeted diagnostics and therapy. They are produced using (a) multistep nanochemistry in a reverse micelle nanoreactor (see Chapter 4 for this method) and (b) surface functionalized with known biotargeting agents.

Nanoclinics are ~30-nm silica shells that can encapsulate various optical, magnetic, or electrical probes, as well as externally activatable therapeutic agents (see Figure 3.9). These nanoclinics are small enough to enter the cell, in order for them to function from within the cell. Through the development of nanoclinics (functionalized nanometer-sized particles that can serve as carriers), new therapeutic approaches to disease can be accomplished from within the cell. At our Institute, nanoclinic prototypes were produced through the integration of a ferrofluid, nanotechnology, and a peptide hormone analogue targeting agent [Levy et al., 2002]. This multilayered nanosized structure, consisted of an iron oxide core, a two-photon optical probe for bioimaging (discussed in Chapter 5), and a silica shell with a targeting peptide hormone analogue (luteinizing hormone-release hormone analogue: LH-RH) covalently coupled to the surface of the shell. High-resolution transmission electron microscopy (TEM) shows that the structure of the nanoparticles is composed of a crystalline core corresponding to Fe_2O_3 and an amorphous silica layer (bubble).

The selective interaction and internalization of these nanoclinics within cells was visualized using two-photon laser scanning microscopy, allowing for real-time observation of the uptake of the nanoclinics [Bergey et al., 2002].

The selective uptake of targeted nanoclinics was readily demonstrated by their uptake by (KB) oral epithelial carcinoma cells (LH-RH receptor positive), whereas a similar accumulation was not observed in LH-RH-negative nanoclinic studies or LH-RH positive nanoclinics incubated with receptor negative cells (UCI-107).

The multifunctional nanoclinics, containing the magnetic Fe_2O_3 nanoparticles, also demonstrated the novel ability to selectively lyse targeted cells when exposed to a DC magnetic field as discussed in Chapter 13. More recent formulations of nanoclinics by our group have used an organically modified silica (ORMOSIL), nanostructure prepared by a sol–gel chemistry as described in Chapter 4. The ORMOSILs have one of the –Si–O–Si– bonds replaced by an organic group directly coupled to Si.

3.11. NANOPLEXES

The nanoclinic concept has been further evolved into a nanoplex platform as shown in Figure 3.10. Here, a nanoclinic is prepared with a resulting net positive charge on the surface which can then be utilized to electrostatically bind

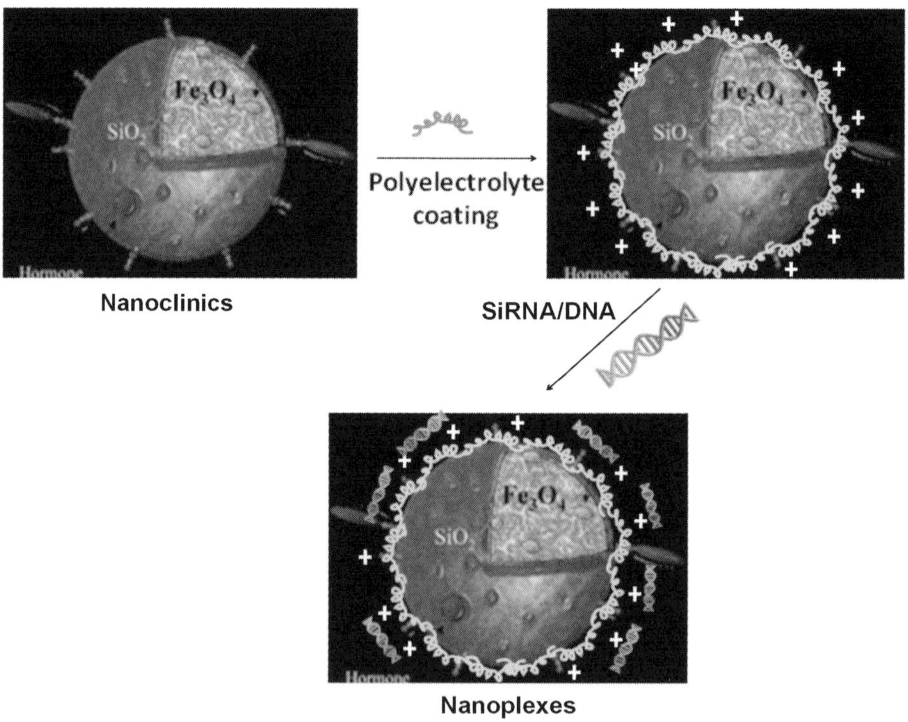

Figure 3.10. Translation of nanoclinics to nanoplexes.

a negatively charged therapeutic or diagnostic element for targeted delivery. This approach has been widely used by our group for nonviral gene delivery (discussed in Chapter 14). ORMOSIL-based nanoparticles can be synthesized with a variety of surface charges and are stable (nonaggregating) in aqueous environments. The genetic payload (negatively charged DNA) is bound to the cationic groups present on the surface of the nanoparticles. The nonaggregating potential, coupled with the designed ability to bind and protect DNA, provides a second platform for gene transfer as presented in Chapter 14, entitled "Gene Therapy."

3.12. NEW-GENERATION NANOCARRIERS

A great deal of emphasis is now being placed on the development of new nanomaterials that are nontoxic, environmentally friendly, and are made from natural resources and earth abundant raw materials that can be cost effectively produced in bulk. Some examples of these new generation materials are presented in Figure 3.11.

Silica and silica-based nanoparticles, as well as copper- (e.g., copper sulfide) and zinc-based (e.g., zinc oxide) nanostructures, are examples of nanostructures made from earth-abundant materials, where the cost of the raw materials can be very low and environmentally friendly.

Agro-materials are also natural and can provide an opportunity to make nanoparticles utilizing resources that are indigenous to a region. An examples is guar gum, a nonionic natural polysaccharide derived from the seeds of *Cyamopsis tetragonolobus*, family *Leguminosae*. Its nanoparticle formulation has been used for controlled release of tamoxifen citrate, the most widely used drug for the treatment of estrogen receptor-positive breast cancer [Sarmah et al., 2011]. Another example is a crosslinked nanoparticle formulation of gum arabic, an emulsifier, used to contain fluorescent probes [Prata et al., 2008]; also nano-biocarriers using gum arabic have been reported [Gomes et al., 2010]. Rosin, an antifouling agent, is a third example that has been used to make nanofibers for use in membrane technologies [Baek et al., 2011]. These agro-materials are indigenous to the African continent.

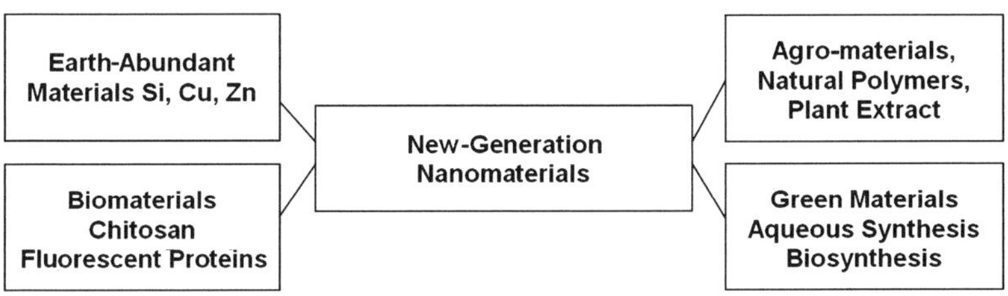

Figure 3.11. New-generation nanocarriers for nanomedicine.

Chitosan nanoparticles are an example of nanomaterials made from a natural biomaterial. Chitosan has already been discussed in Section 3.3.

Green materials refer to those materials that are produced by green chemistry, which is environmentally friendly. Many nanomaterials are produced using organic-solvent-based chemistry that often involves toxic organic substances. Considerable emphasis is now being placed on using aqueous synthesis. Our group has developed an aqueous-phase synthesis approach to prepare highly luminescent core/shell quantum dots for *in vitro* and *in vivo* bioimaging [Law et al., 2009]. The use of green chemistry and low reaction temperatures provide an alternative route, as opposed to the commonly used organometallic synthesis method. Another approach is to use bacterial biosynthesis of functional materials. An example is the bacterial synthesis of polyhydroxyalkanoic acid (PHA) [Prasad, 2003].

HIGHLIGHTS OF THE CHAPTER

- Nanocarriers provide three different structural platforms: (1) interior volume, (2) surface, and (3) pores, for diagnosis and therapeutics applications.
- Nanocarriers are nonimmunogenic and can be tailored with compositions that are resistant to microbial attack. Nanocarriers can be surface-functionalized to penetrate the cell membrane by endocytosis.
- The use of PEG groups on the surface of nanocarriers leads to prolonged *in vivo* circulation and increased targeting ability.
- Tailoring of nanostructure characteistics is vital for their suitability as nanocarriers.
- Chemical composition, surface groups, surface charge, and size and shape of a nanocarrier all play important roles. Thus a precise control of production of nanocarriers to achieve monodispersity (all particles same) and reproducibility is vital for their applicability to nanomedicine
- Various forms of virus-derived nanoparticles have been used as nanocarriers. An example is a class of virus-like particles (VLPs) that retain the self-assembled viral envelope protein, but are devoid of the accompanying genetic materials of the virus.
- Virosomes are another example of virus-derived nanoparticles that have been investigated for targeted drug delivery. They consist of a virus envelope, but to which additional components of the virus or another virus are attached or inserted.
- Polymeric nanocarriers allow for a great deal of flexibility to tailor their chemical composition, size, morphology, and surface functionality for delivery of sensing, imaging, and therapeutic agents.
- Biodegradable PLGA nanoparticles are approved by FDA for clinical use and have been extensively used for carrying contrast agents and for controlled-release of therapeutic agents.

- Lipid-based nanocarriers utilize lipid (surfactant)-type molecules consisting of a polar head group and a nonpolar long alkyl chain tail.
- Lipid-based nanocarriers are micellar nanoparticles, bilayer lipid structures, liposomes, and solid lipid nanoparticles.
- Liposomes are nanoparticles consisting of one or several lipid bilayers with an aqueous phase inside, as well as in between the lipid bilayers.
- Solid lipid nanoparticles (SLNs) are made of lipids that are solid at room temperature as well as at body temperature.
- Dendrimers are multibranched structures produced by a highly controlled multistep synthesis.
- A dendrimer can be employed to encapsulate a payload in the core, and it can be conjugated to a branch point; it also offers attachment to the periphery.
- Carbon-60, graphene, carbon nanotube, nanodiamond, and carbon dots are the major carbon nanostructures utilized for nanomedicine applications.
- The shape of carbon-60 is like a soccer ball and it is often referred to as fullerene, buckminsterfullerene, or buckyballs.
- Graphene has a two-dimensional planar sheet structure, one atom thick (monolayer), consisting of carbon atoms bonded in a cyclic form with alternate single and double bonds.
- Carbon nanotubes are cylindrical nanostructures of carbon and can be envisioned as rolled sheets of graphene. These nanotubes exhibit extraordinary thermal, mechanical, and electrical properties, making them very attractive for biosensing and tissue engineering applications.
- Nanodiamonds are diamond nanoparticles with a small diameter around 5 nm, which are formed by detonation. Because of their excellent mechanical and optical properties, high surface areas, and nontoxicity, they have attracted considerable attention for biomedical applications.
- Carbon dots are carbon nanoparticles produced originally by laser ablation which exhibit emission in the visible spectral range upon surface functionalization with polymer chains.
- The electronic, optical, and magnetic properties of inorganic materials can be tailored by judicious control of their crystal phase, size, shape, and surface characteristics.
- Semiconductor nanoparticles, quantum dots, and quantum rods have received considerable attention because of their size-dependent optical properties that are derived from the quantum confinement effect on electrons.
- Metallic nanostructures (e.g., gold nanoparticles, gold nanorods) exhibit optical properties different from their bulk form that is derived from the presence of plasmon resonance.
- Silica-based nanoparticle can be synthesized by condensation–polymerization of their organosilane precursor molecules. Organically

modified silica nanoparticles (ORMOSIL) utilize a precursor in which one of the alkoxy "arms" of the tetraalkoxysilane precursor has been replaced by a hydrocarbon group. Since these particles are mesoporous, they can serve as sustained release vehicles for encapsulated biomolecules.

- A variety of fluoride, oxide, and phosphate nanoparticles have also been used as nanocarriers. Fluoride lattice doped with rare-earth elements provides a matrix for enhanced upconversion of light.
- A probe encapsulated by biologically localized embedding (PEBBLE) is a nanoscale device consisting of sensor molecules entrapped in a chemically inert matrix.
- PEBBLEs enable optical measurement of changes in intracellular calcium levels, pH, and other biologically significant chemicals.
- "Nanoclinics" have a complex surface functionalized silica nanoshell containing various probes for diagnostics and drugs for targeted delivery.
- Nanoclinics are ~30-nm silica shells that can encapsulate various optical, magnetic, or electrical probes, as well as externally activatable therapeutic agents.
- The multifunctional nanoclinics, containing the magnetic Fe_2O_3 nanoparticles, also demonstrated the novel ability to selectively lyse targeted cells when exposed to a DC magnetic field.
- The nanoclinic concept has been further evolved into a nanoplex platform.
- Organically modified silica (ORMOSIL)-based nanoparticles can be synthesized with a variety of surface charges and are stable (nonaggregating) in aqueous environments.
- New-generation nanomaterials, which are nontoxic and environmentally friendly, are envisioned to be made from natural resources and earth-abundant raw materials using green chemistry.

EXERCISES

True or False

1. PLGA polymer is highly water-soluble.
2. Dendrimer is one of the examples of multibranch-structured polymer.
3. Coating with polyethylene glycol (PEG) layer can facilitate the circulation half-life *in vivo*.
4. PEPPLE combines imaging, sensing, and therapeutic functionalities into one system.
5. Liposome is a rigid nanoparticle.

Multiple Choice

6. Which of the following nanoparticle is *not* biodegradable?
 (a) Chitosan
 (b) PLGA
 (c) Silica
 (d) Liposome

7. Which of the following material can be used to prolong the circulation time *in vivo*?
 (a) PVA
 (b) PEG
 (c) siRNA
 (d) Quantum dot

8. What is "theranostic" referring to?
 (a) Combined biodegradable and biocompatible platform
 (b) Combined therapeutic and magnetic platform
 (c) Combined transparent and light activable platform
 (d) Combined therapeutic and diagnostic platform

Answers: 1. F; 2. T; 3. T; 4. T; 5. F; 6. c; 7. b; 8. d

REFERENCES

Baek, W., Nirmala, R., Barakat, N. A. M., El-Newehy, M. H., Al-Deyab, S. S., Kim, H. Y., Electrospun cross linked rosin fibers, *Appl. Surf. Sci.* **258**, 1385–1389 (2011).

Baker, S. N., Baker, G. A., Luminescent carbon nanodots: Emergent nanolights, *Angew. Chem. (Int. Ed.)* **49**, 6726–6744 (2010).

Basu, B., Garala, K., Bhalodia, R., Joshi, B., Mehta, K., Solid lipid nanoparticles: A promising tool for drug delivery system, *J. Pharm. Res.* **3**, 84–92 (2010).

Bergey, E. J., Levy, L., Wang, X., Krebs, L. J., Lal, M., Kim, K.-S., Pakatchi, S., Liebow, C., Prasad, P. N., DC magnetic field induced magnetocytolysis of cancer cells targeted by LH-RH magnetic nanoparticles *in vitro*, *Biomed. Microdevices* **4**, 293–299 (2002).

Bhattacharya, S., Mazumder, B., Virosomes: A novel strategy for drug delivery and targeting, *BioPharm. Int. Suppl.* **24**, S9–S14 (2011).

Bhirde A. A., Patel, S., Sousa, A. A., Patel, V., Molinolo, A. A., Ji, Y., Leapman, R. D., Gutkind, J. S., Rusling, J. F., Distribution and clearance of PEG-single-walled carbon nanotube cancer drug delivery vehicles in mice. *Nanomedicine (Lond).* **5**, 1535–1546 (2010).

Caminati, G., Turro, N. J., Tomalia, D. A., Photophysical investigation of starburst dendrimers and their interactions with anionic and cationic surfactants, *J. Am. Chem. Soc.* **112**, 8515–8522 (1990).

Cao, W., Zhou, J., Mann, A., Wang, Y., Lei Zhu, L., Folate-functionalized unimolecular micelles based on a degradable amphiphilic dendrimer-like star polymer for cancer cell-targeted drug delivery, *Biomacromolecules* **12**, 2697–2707 (2011).

Cinteza, L. O., Ohulchanskyy, T. Y., Sahoo, Y., Bergey, E. J., Pandey, R. K., Prasad, P. N., Diacyllipid micelle-Based nanocarrier for magnetically guided delivery of drugs in photodynamic therapy, *Mol. Pharm.* **3**, 415–423, (2006).

Clark, H. A., Hoyer, M., Philbert, M. A., Kopelman, R., Optical nanosensors for chemical analysis inside single living cells. 1. Fabrication, characterization, and methods for intracellular delivery of PEBBLE sensors, *Anal. Chem.* **71**, 4831–4836 (1999).

Cun, D. M., Foged, C., Yang, M. S., Frokjaer, S., Nielsen, H. M., Preparation and characterization of poly(DL-lactide-co-glycolide) nanoparticles for siRNA delivery, *Int. J. Pharm.* **390**, 70–75 (2010).

Dykes, G. M., Dendrimers: A review of their appeal and applications, *J. Chem. Technol. Biotechnol.* **76**, 903–918 (2001).

Feng, L., Liu, Z., Graphene in biomedicine: Opportunities and challenges, *Nanomedicine (Lond.)* **6**, 317–324 (2011).

Feng, L. Z., Zhang, S., Liu, Z., Graphene based gene transfection, *Nanoscale* **3**, 1252–1257 (2011).

Geim, A. K., Graphene: Status and prospects, *Science* **324**, 1530–1534 (2009).

Gillies, E. R., Fréchet, J. M. J., Dendrimers and dendritic polymers in drug delivery, *Drug Discovery Today* **10**, 35–43 (2005).

Gomes, J. F., Rocha, S., do Carmo Pereira, M., Peres, I., Moreno, S., Toca-Herrera, J., Coelho, M. A., Lipid/particle assemblies based on maltodextrin–gum arabic core as bio-carriers, *Colloids Surf. B: Biointerfaces* **76**, 449–455 (2010).

Grayson, S. M., Fréchet, J. M. J., Convergent dendron and dendrimers: From synthesis to applications, *Chem. Rev.* **101**, 3819–3867 (2001).

Hawker, C. J., Fréchet, J. M. J., Preparation of polymers with controlled molecular architecture. A new convergent approach to dendritic macromolecules, *J. Am. Chem. Soc.* **112**, 7638–7647 (1990).

Huckriede, A., Bungener, L., Stegmann, T., Daemen, T., Medema, J., Palache, A. M., Wilschut, J., The virosome concept for influenza vaccines, *Vaccine* **23**, 526–538 (2005).

Kumar, R., Roy, I., Ohulchanskky, T. Y., Vathy, L. A., Bergey, E. J., Sajjad, M., Prasad, P. N., *In vivo* biodistribution and clearance studies using multimodal organically modified silica nanoparticles, *ACS Nano* **4**, 699–708, (2010).

Law, W.-C., Yong, K.-T., Roy, I., Ding, H., Hu, R., Zhao, W., Prasad, P. N., Aqueous-phase synthesis of highly luminescent CdTe/ZnTe core/Shell quantum dots optimized for targeted bioimaging, *Small* **5**, 1302–1310 (2009.)

Lee, C. C., Gillies, E. R., Fox, M. E., Guillaudeu, S. J., Fréchet, J. M. J., Dy, E. E., Szoka, F. C., A single dose of doxorubicin-functionalized bow-tie dendrimer cures mice bearing C-26 colon carcinomas, *Proc. Natl. Acad. Sci. USA* **103**, 16649–16654 (2006).

Lee, C. C., MacKay J. A., Fréchet, J. M. J., Szoka F. C., Designing dendrimers for biological applications, *Nat. Biotechnol.* **23**, 1517–1526 (2005).

Levy, L., Sahoo, Y., Kim, K.-S., Bergey, E. J., Prasad, P. N., Nanochemistry: synthesis and characterization of multifunctional nanoclinics for biological applications, *Chem. Mater.* **14**, 3715–3721 (2002).

Li, Q., Ohulchanskyy, T. Y., Liu, R., Koynov, K., Wu, D., Best, A., Kumar, R., Bonoiu, A., Prasad, P. N., Photoluminescent carbon dots as biocompatible nanoprobes for targeting cancer cells *in vitro*, *J. Phys. Chem. C* **114**, 12062–12068 (2010).

Liu, Z., Robinson, J. T., Sun, X. M., Dai, H. J., PEGylated nanographene oxide for delivery of water-insoluble cancer drugs, *J. Am. Chem. Soc.* **130**, 10876–10877 (2008).

Liu, Z., Robinson, J. T., Tabakman, S. M., Yang, K., Dai, H. J., Carbon materials for drug delivery and cancer therapy, *Materials Today* **14**, 316–323 (2011).

Lodish, H., Berk, A., Kaiser C. A., Krieger, M., Scott, M. P., Bretscher, A., Ploegh, H., *Molecular Cell Biology*, 4th edition, W. H. Freeman, New York (2000).

Majoros, I. J., Thomas, T. P., Mehta, C. B., Baker, Jr., J. R., Poly(amidoamine) dendrimer-based multifunctional engineered nanodevice for cancer therapy, *J. Med. Chem.* **48**, 5892–5899 (2005).

Majoros, I. J., Williams, C. R., Baker, J. R. Jr., Current dendrimer applications in cancer diagnosis and therapy, *Curr. Topics Med. Chem.* **8**, 1165–1179 (2008).

Manchester, M., Singh, P., Virus-based nanoparticles (VNPs): Platform technologies for diagnostic imaging, *Adv. Drug Delivery Rev.* **58**, 1505–1522 (2006).

Maze, J. R., Stanwix, P. L., Hodges, J. S., Hong, S., Taylor, J. M., Cappellaro, P., Jiang, L., GurudevDutt, M. V., Togan, E., Zibrov, A. S., Yacoby, A., Walsworth, R. L., Lukin, M. D., Nanoscale magnetic sensing with an individual electronic spin in diamond, *Nature* **455**, 644–647 (2008).

Menjoge, A. R., Kannan, R. M., Tomalia, D. A., Dendrimer-based drug and imaging conjugates: Design considerations for nanomedical applications, *Drug Discovery Today* **15**, 171–185 (2010).

Mochalin, V. N., Gogotsi, Y., Wet chemistry route to hydrophobic blue fluorescent nanodiamond, *J. Am. Chem. Soc.* **131**, 4594–4595 (2009).

Mochalin, V. N., Shenderova, O., Ho, D., Gogotsi, Y., The properties and applications of nanodiamonds, *Nat. Nanotechnol.* **7**, 11–23 (2012).

Monsoon, E., Brasuel, M., Philbert, M. A., Kopelman, R., *Biomedical Photonics Handbook*, T. Vo-Dinh, ed., CRC Press, Boca Raton, FL, pp. 59-1–59-14 (2003).

Montellano, A., Da Ros, T., Bianco, A., Prato, M., Fullerene C60 as a multifunctional system for drug and gene delivery, *Nanoscale* **3**, 4035–4041 (2011).

Mulder, W. J. M., Strijkers, G. J., van Tilborg, G. A. F., Cormode, D. P., Fayad, A., Nicolay, K., Nanoparticulate assemblies of amphiphiles and diagnostically active materials for multimodality Imaging, *Acc. Chem. Res.* **42**, 904–914 (2009).

Patri, A. K., Majoros, I. J., Baker, J. R., Dendritic polymer macromolecular carriers for drug delivery, *Curr. Opin. Chem. Biol.* **6**, 466–471 (2002).

Prasad, P. N., *Introduction to Biophotonics*, Wiley Interscience, Hoboken, NJ (2003).

Prasad, P. N., *Nanophotonics*, Wiley-Interscience, Hoboken, NJ (2004).

Prata, A. S., Zanin, M. H. A., Ré, M. I., Grosso, C. R. F., Release properties of chemical and enzymatic crosslinked gelatin-gum arabic microparticles containing a fluorescent probe plus vetiver essential oil, *Colloids and Surfaces B: Biointerfaces* **67**, 171–178 (2008).

Purtov, K. V., Petunin, A. I., Burov, A. E., Puzyr, A. P., Bondar, V. S., Nanodiamonds as carriers for address delivery of biologically active Substances, *Nanoscale Res. Lett.* **5**, 631–636 (2010).

Rai, S., Paliwal, R., Gupta, P. N., Khatri, K., Goyal, A. K., Vaidya, B., Vyas, S. P., Solid lipid nanoparticles (SLNs) as a rising tool in drug delivery science: One step up in nanotechnology, *Curr. Nanosci.* **4**, 30–44 (2008).

Rolland, O., Turrin, C.-O., Caminade, A.-M., Majoral, J.-P., Dendrimers and nanomedicine: Multivalency in action, *New J. Chem.* **33**, 1809–1824 (2009).

Roy, I., Ohulchanskyy, T. Y., Bharali, D. J., Pudavar, H. E., Mistretta, R. A., Kaur, N., Prasad, P. N., Optical tracking of organically modified silica nanoparticles as DNA carriers: A nonviral, nanomedicine approach for gene delivery, *Proc. Natl. Acad. Sci. USA* **102**, 279–284 (2005).

Sarmah, J. K., Mahanta, R., Bhattacharjee S. K., Mahana, R., Biswas, A., Controlled release of tamoxifen citrate encapsulated in cross-linked guar gum nanoparticles, *Int. J. Biol. Macromol.* **49**, 390–396 (2011).

Shi, M., Lu, J., Shoichet, M. S., Organic nanoscale drug carriers coupled with ligands for targeted drug delivery in cancer, *J. Mater. Chem.* **19**, 5485–5498 (2009).

Sun, Y. P., Zhou, B., Lin, Y., Wang, W., Fernando, K. A. S., Pathak, P., Meziani, M. J., Harruff, B. A., Wang, X., Wang, H. F., Luo, P. J. G., Yang, H., Kose, M. E., Chen, B. L., Veca, L. M., Xie, S. Y., Quantum-sized carbon dots for bright and colorful photoluminescence, *J. Am. Chem. Soc.* **128**, 7756 (2006).

Tokarev, I., Minko, S., Multiresponsive hierarchically structured membranes: New challenging biomimetic materials for biosensors, controlled release, biochemical gates and nanoreactors, *Adv. Mater.* **21**, 241–247 (2009).

Tomalia, D. A., Reyna, L. A., Svenson, S., Dendrimers as multi-purpose nanodevices for oncology drug delivery and diagnostic imaging, *Biochem. Soc. Trans.* **35**, 61–67 (2007).

Yang, K., Zhang, S., Zhang, G. X., Sun, X. M., Lee, S. T., Liu, Z., Graphene in mice: Ultrahigh in vivo tumor uptake and efficient photothermal therapy. *Nano Lett.* **10**, 3318–3323 (2010).

Zhang, L. M., Xia, J. G., Zhao, Q. H., Liu, L. W., Zhang, Z. J., Functional graphene oxide as a nanocarrier for controlled loading and targeted delivery of mixed anticancer drugs. *Small* **6**, 537–544 (2010).

CHAPTER 4

NANOCHEMISTRY OF NANOCARRIERS

Nanochemistry has emerged as a powerful method for creating various nanostructures, which can serve as nanocarriers in nanomedicine applications. Nanochemistry also provides the means to produce appropriate surface modifications of nanostructures to make them dispersable in biological fluids, improve their biocompatibility, and provide linking groups for conjugation of the nanostructures with targeting groups, chemical groups such as PEG to prolong circulation, and other nanoprobes and therapeutic payloads. Furthermore, nanochemistry enables the precise control of conditions required to produce monodispersed nanostructures (all nanoparticles within a very narrow size distribution range) that can exhibit uniform physical, chemical, and biological activities.

This chapter describes nanochemistry and its application to nanomedicine. Section 4.1 describes nanochemistry and its impact on nanomedicine. It also introduces the two general classes of approaches to creating nanostructures: (i) top-down and (ii) bottom-up methods. Section 4.2 provides selected examples of the top-down approaches. Specifically, the methods of Mechanical Milling, Dip-Pen Lithography, and Particle Replication in Non-Wetting Templates (PRINT) are discussed in the subsections of Section 4.2. Selected bottom-up approaches are covered in Section 4.3, including Dendrimer Synthesis, Microemulsion Chemistry, Hot-Colloidal Synthesis, and Reprecipitation. Section 4.4 shows an example of nanochemistry that employs a combination of both the bottom-up and top-down approaches.

Introduction to Nanomedicine and Nanobioengineering, First Edition. Paras N. Prasad.

The surface modification chemistry applied to nanoparticles is described in Section 4.5. The specific examples provided are ligand exchange, silica coating, polymer coating, and phospholipid encapsulation. The bioconjugation strategies used for coupling various targeting groups, diagnostic probes, and therapeutic units to nanoparticles are described in Section 4.6.

4.1. NANOCHEMISTRY AND NANOMEDICINE

Nanochemistry is an active new field that deals with the confinement of chemical reactions to the nanometer length scale to produce chemical products that have nanometer dimensions (generally in the range of 1–100 nm) [Murray et al., 2000]. The challenge is to be able to use chemical approaches that reproducibly provide precise control of composition, size, and shape of the nano-objects formed. These nanomaterials exhibit new electronic, optical, and other physical properties that depend on their composition, size, and shape. Nanoscale chemistry also provides an opportunity to design and fabricate hierarchically built multilayer nanostructures to incorporate multifunctionality at the nanoscale.

Some important features derived from nanochemistry are as follows:

- Preparation of a diverse group of nanoparticles from metals, semiconductors, glasses, and polymers.
- Preparation of hierarchically built multilayer, core-shell-type nanoparticles.
- Modification, introduction of chemical linking groups and self-assembly of structures on a nanopatterned surface.
- Assembly of nanoparticles into periodic or aperiodic functional structures.
- *In situ* preparation and attachment of nanoscale probes and sensors.

A number of approaches provide nanoscale control of chemistry. Some of these methods are listed in Figure 4.1 and discussed in this chapter.

4.2. TOP-DOWN APPROACHES

A top-down approach, in preparation of nanomaterials, involves a process that can be scaled down to produce nanosize domains. There are several examples of top-down approach, each with its own merit, thus making it suitable for some specific applications. In this section, only selected examples of top-down approaches, applied specifically in Nanomedicine, will be presented. In particular, the photolithographic processes that are used to produce state-of-the-art computer chips with nanoscale components, as well as related processes for nanoscale patterning on surfaces, are outside the scope of this section and

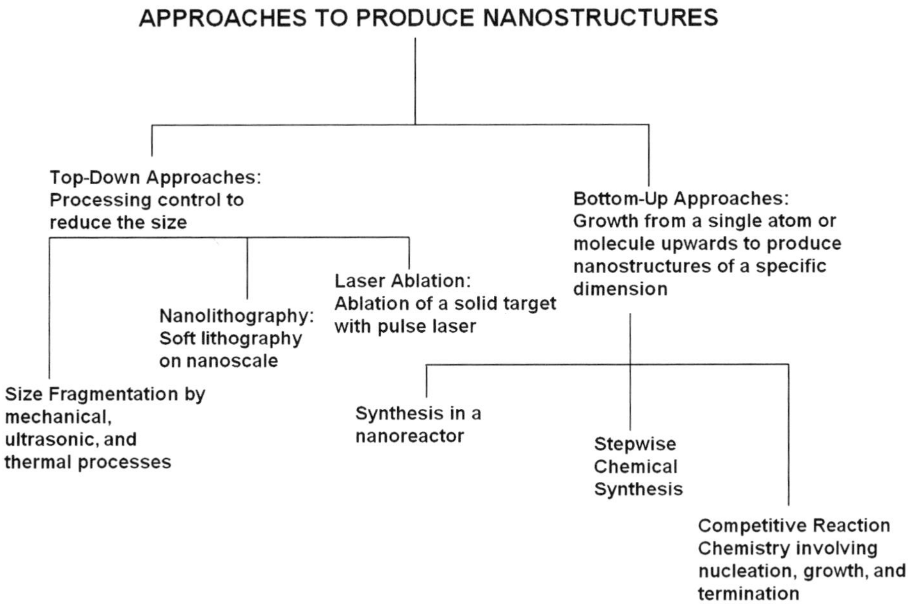

Figure 4.1. The two major classes of approaches used to produce nanostructures.

are not considered. Instead, we focus on methods of producing free-standing nanostructures, which are relevant for nanomedicine.

4.2.1. Mechanical Milling

Mechanical milling is utilized to reduce the size of larger particles. It has been applied to many drugs that exhibit poor solubility or are poorly absorbed. Reducing their size to the nanoscale (≤100 nm) improves their dissolution rate and oral absorption. A wet-milling process is usually more effective than dry milling, because it minimizes any thermal degradation of the drug that is suspended in an aqueous medium.

A representative schematic of the wet-milling technique is shown in Figure 4.2 [Tanaka et al., 2009]. The milling chamber is filled with fine beads of a hard material such as zirconia of ~50-μm diameter and then mixed with a slurry containing the drug and a dispersing agent (e.g., a common surfactant such as Tween 80). The large drug particles are then milled by impact with the zirconia beads. A centrifugal separator at the top of the chamber returns large particles and the zirconia beads to the mill while allowing fine particles to exit.

4.2.2. Dip-Pen Nanolithography

Dip-pen nanolithography (DPN) is another example of a top-down approach, developed by Mirkin and co-workers [Piner et al., 1999; Hong et al., 1999],

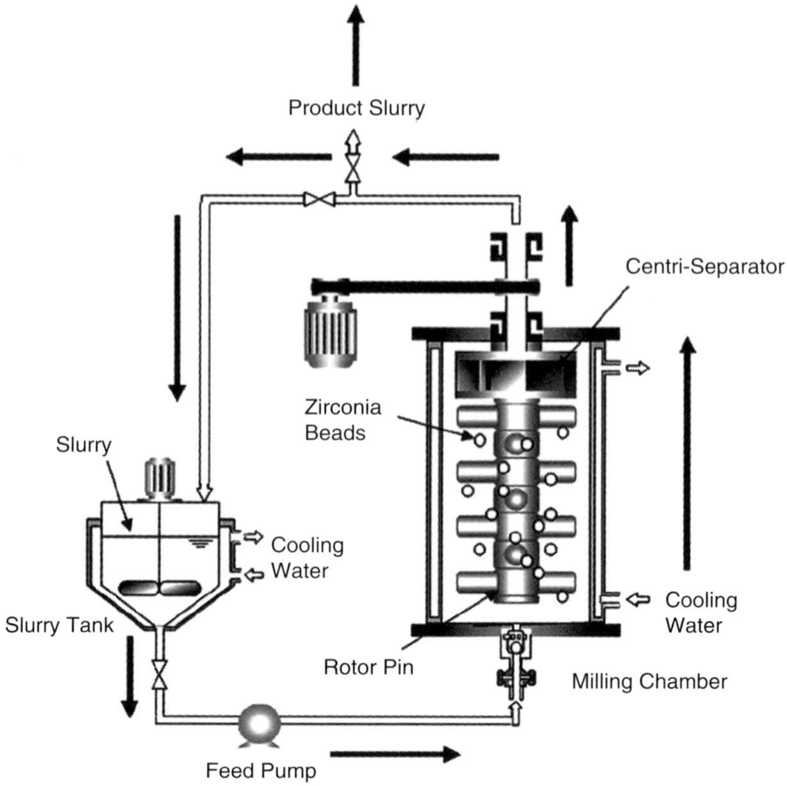

Figure 4.2. Schematics of Wet-Milling Process by ULTRA APEX MILL. [Reproduced with permission from Tanaka et al. (2009).]

which allows nanopatterning of molecules to produce structures that are one molecular unit thick. It utilizes the tip of an atomic force microscope (AFM) as a pen, with molecules to be laid down on a substrate acting as an ink, to print an array of soft and hard materials. Condensation between the tip and the substrate forms a water-filled capillary that moves with the tip and helps the transport and subsequent chemisorption of materials with low water solubility to be deposited on the substrate. This process is schematically described in Figure 4.3. Mirkin and co-workers originally demonstrated the deposition of alkanethiol (octadecanethiol) in a pattern on a gold (111) substrate. By appropriately selecting the molecule to be transported, the nature of the surface can be controlled. For example, using 16-mercaptohexadecanoic acid yields a hydrophilic surface, while octadecanethiol produces a hydrophobic surface.

In recent years, the application of this technique has been expanded to pattern many insulating, semiconducting, and metallic substrates using a variety of inks that include small organic molecules, polymers, DNA, proteins, peptides, colloidal nanoparticles, metal ions, and sols [Salaita et al., 2007].

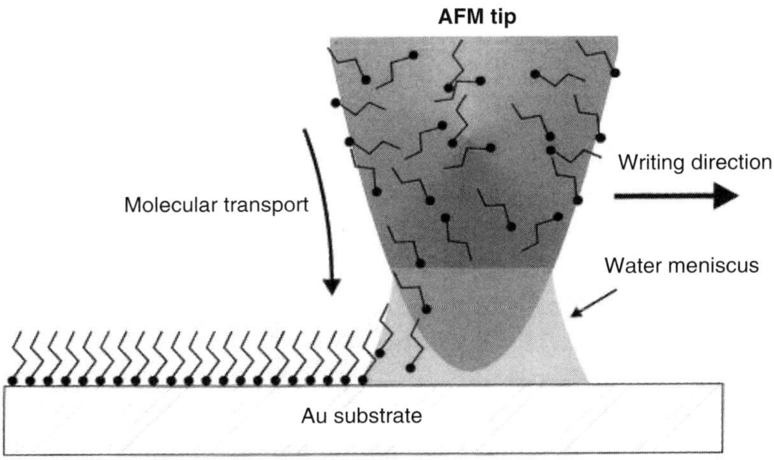

Figure 4.3. Schematic representation of dip-pen nanolithography. (Reproduced with permission from Piner et al. [1999].)

4.2.3. PRINT Process

A top-down lithographic process, developed by DeSimone for producing polymeric nanoparticles of precisely controlled shape and size, is known as Particle Replication in Nonwetting Templates, PRINT [Canelas et al., 2009]. This polymer process is schematically represented in Figure 4.4. A roller sheet spreads the particle precursor solution into the cavities of a fluoropolymer mold. The excess particle precursor solution is removed by a high-surface-energy polymer sheet. This particle precursor then is converted into solid nanoparticles by using the appropriate chemistry. The solid polymeric nanoparticles are removed from the cavities of the mold by bringing the mold into contact with an adhesive layer. This adhesive layer may also introduce chemical modification of the nanoparticles and eventually be dissolved away. As the size and shape of the cavity can be precisely controlled using lithographic techniques, this method has the merit to yield highly monodispersed polymeric nanoparticles in a broad range of sizes (20 nm to microns) and shapes. The method can also be utilized to incorporate various proteins, DNA, and drugs (such as chemotherapeutic agent, doxorubicin) into a biocompatible polymeric nanoparticle such as those of poly(ethylene glycol) (PEG).

4.2.4. Laser Ablation

This method utilizes ultrashort and high-intensity focused laser beams to ablate materials from a rotating solid target that may be immersed in a liquid to provide both cooling of the target and dispersion of the nanoparticles produced [Kabashin et al., 2010]. It is schematically represented in Figure 4.5. The intensity at the focused spot and the speed of rotation of the solid target can

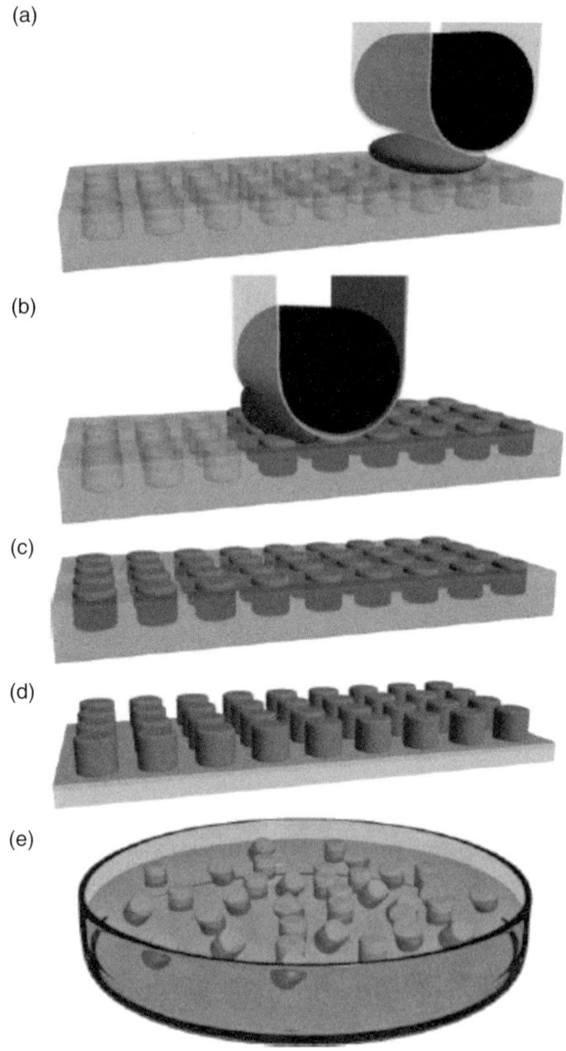

Figure 4.4. Schematic representation of the Particle Replication in Nonwetting Templates (PRINT) process. (a) Empty mold high-surface-energy polymer sheet (clear) roller (black) is brought into contact with the particle precursor solution and the mold. (b) Roller evenly distributes particle precursor solution into the cavities of mold. Excess particle precursor solution is wicked away by the high-surface-energy polymer sheet. (c) Particles are cured in the mold. (d) Particles are removed from the mold. (e) Particles are collected or harvested using a number of different film-based techniques and are ultimately dispersed in solution. [Reproduced with permission from Canelas et al. (2009).]

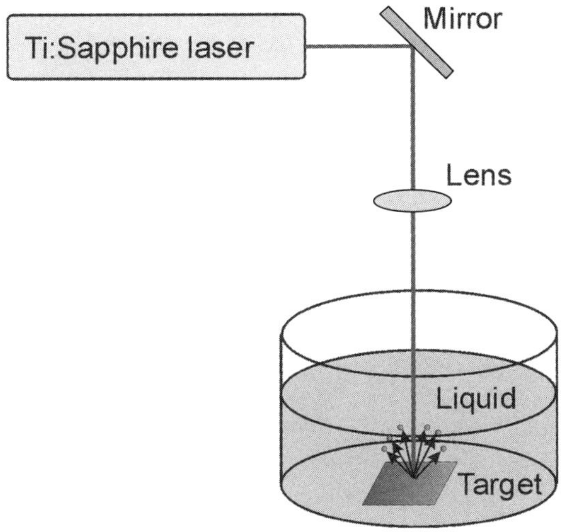

Figure 4.5. Schematics of laser–target interaction and material nanostructuring.

be adjusted to produce nanoparticles in a given size range. The method has been extensively utilized to produce nanoparticles of hard materials such as ceramics, inorganic semiconductors, and metals. The advantage of this method is that the nanoparticles produced have clean surfaces, in contrast to surfactant-coated surfaces usually obtained in a wet chemical process. Particles produced can be quite small (<5 nm). A limitation of this process is that the produced particles may have a broad size distribution. Gold nanoparticles produced by this method have been shown to be dispersible in water without the need of added surfactant [Kabashin et al., 2010].

4.3. BOTTOM-UP APPROACHES

This section describes a number of chemical approaches using wet chemistry to produce organic nanostructures and nanoparticles of various compositions, sizes, and shapes. This section provides selected examples of these approaches. As discussed in Chapter 3, organic nanostructures utilized in nanomedicine take on a variety of structures, compositions, and sizes. Although linear block polymeric structures with controlled segments will not be described, the dendrimers (dendritic structures) will be discussed because they have been extensively investigated for drug delivery (discussed in Chapter 3).

4.3.1. Dendrimers

The chemical synthetic strategies used to produce dendrimers can be classified into two categories that are described below.

Divergent Approach. This approach, developed by Tomalia et al. (1990) and Newkome et al. (1996), involves an outward growth from a core by repetition of a series of coupling and activation steps, each adding one layer (or generation) to the structure. The core is a monomer unit with a multiple coupling (branching) site that reacts with the complementary reactive group (unprotected) of other monomers. This is the coupling step. The other reactive groups on the monomers are protected, until needed, to produce a controlled growth. This is shown schematically in Figure 4.6. The protected groups can be chemically activated by conversion to a reactive chemical group, for coupling to a second set of molecules to produce the next-generation dendrimer. This step is the activation step. A dendrimer with a well-defined generation can thus be produced. With an appropriate choice of coupling and activation steps, this divergent approach has the advantage that it can lead to large-scale preparation of dendrimers, as the quantity of the dendrimer sample essentially doubles with each added generation. A major issue, however, is the likelihood of incomplete functionalization or additional side reactions, which grows exponentially with each generation. Thus, structural defects are likely.

Convergent Approach. This approach, introduced by Fréchet and co-workers [Hawker and Fréchet, 1990; Grayson and Fréchet, 2001], starts with inward growth by coupling end groups to each branch of the monomer. First, wedge-shaped dendritic fragments called *dendrons* are produced, which are then coupled to produce the globular dendritic molecule. The dendron unit and the resulting dendrimer are illustrated in Figure 4.7 [Andronov and Fréchet, 2000]. In this diagram, the core is highlighted by a black circle, the rings of branching units are represented by gray circles, and the end groups are represented by rectangles. The advantage offered by the convergent approach is that it offers greater structural control, because each growth step requires a relatively low number of coupling reactions. Another advantage is the ability to precisely place functional groups throughout the structure. A disadvantage, however, is that it is less readily scalable when compared to the divergent approach.

A dendrimer nanostructure offers a number of potential advantages for nanomedicine. A major feature offered by a dendritic architecture is the ease of incorporating multifunctionality; the core, the branching unit, and the peripheral groups provide sites to incorporate separate or synergistic functions for imaging, sensing, and controlled drug release. In addition, molecules can be physically trapped as inclusions inside the dendritic structure by chemically sealing the periphery.

Another advantage is that dendritic structures provide isolation of the core from the surrounding environment, thereby reducing environmentally induced nonradiative quenching of an emitter at the core. This feature is important for optical bioimaging, for which a dye molecule needs to be insulated from the aqueous environment, which can quench emission.

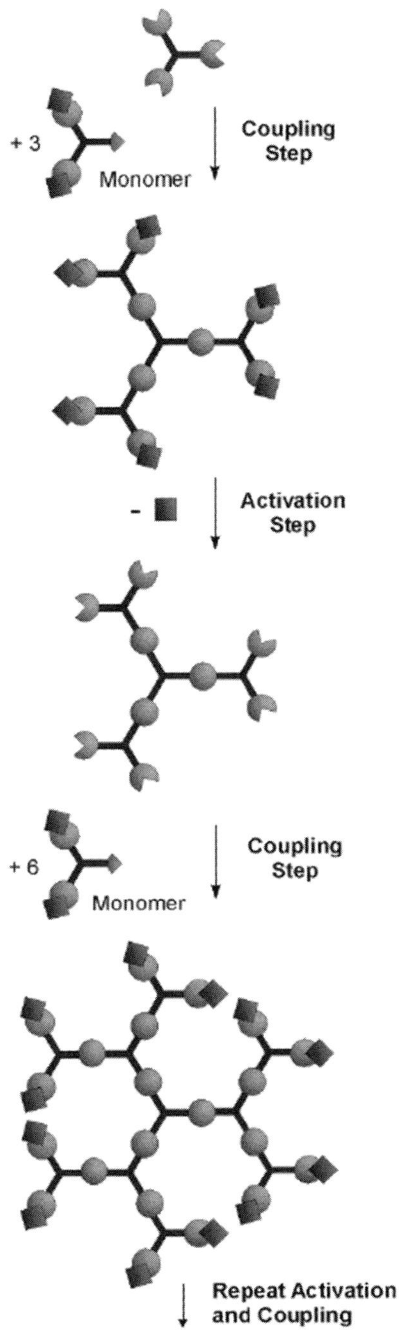

Figure 4.6. Cartoon representation of the divergent approach. [Reproduced with permission from Grayson and Fréchet (2001).]

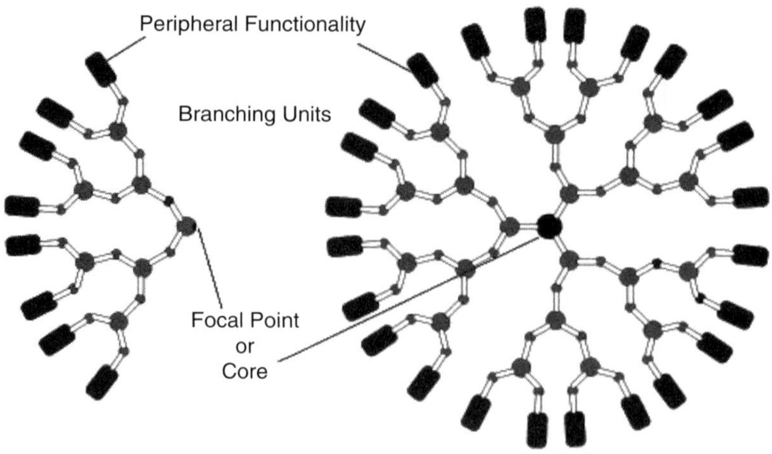

Figure 4.7. Schematic diagram of a dendron and a dendrimer, highlighting the focal point or core (black), surrounded by rings of branching units (gray circles) and end groups (rectangles). [Reproduced with permission from Andronov and Fréchet (2000).]

Placing many fluorescent sensing dyes on the periphery in a dendritic structure can also serve to amplify the sensitivity of a sensor. Balzani et al. (2000) demonstrated this by preparing a poly(propylene amine) dendrimer containing 32 dansyl units at its periphery. The dansyl molecules serve as fluorescent sensors for metal ions, such as Co^{2+}, by quenching of fluorescence. They showed that the sensor was able to respond to very low concentrations of Co^{2+}.

4.3.2. Microemulsion Chemistry

A microemulsion may be defined as a thermodynamically stable, optically isotropic dispersion of two immiscible liquids (e.g., water and oil), with at least one type of surfactant. Surfactants are amphiphilic molecules having a polar (hydrophilic) head group and a long nonpolar (hydrophobic or lipophilic) tail. They have been already described in Chapter 3. Such molecules residing at the oil–water interface considerably reduce the interfacial tension. These systems form dispersions of nanosize droplets of oil-in-water (o/w) if water is the majority bulk phase. The microemulsion droplets act to traps nonpolar reactants within them and act as micelle nanoreactors. If water is dispersed in a large volume of oil, the surfactants assemble around them to form a reverse (water in oil, w/o) microemulsion, in which the polar groups are directed into the aqueous nanodroplets, to form an aqueous (or polar) reverse micelle nanoreactor where polar molecules can react. The surfactants present in the microemulsion not only serve to stabilize the nanoreactors for chemical syntheses, but also act as steric stabilizers to inhibit the aggregation of reacting species during the reaction period. In our case, the continuous oil phase often consists of isooctane with sodium bis(2-ethylhexyl) sulfosuccinate (AOT) serving as the surfactant. In addition to water, aqueous solutions containing a variety of functionalities, including organic dyes, nanoparticles, and hydrophobic drug, can be solubilized within the reverse micelles. The size of the micelle, and subsequently the volume of the aqueous pool contained within the micelle, is governed by the water-to-surfactant ratio, also termed W_0, where $W_0 = [H_2O]/[\text{surfactant}]$. Continuous exchange of the micellar contents through dynamic collisions enables the reaction to proceed. Since the reaction is confined within the nanocavity of the micelle, however, growth of the nanoparticle beyond the dimensions of the cavity is inhibited. In the final stage of the synthesis, the unreacted precursors are removed by dialysis.

An example is the synthesis of silica and ORMOSIL nanoparticles by sol–gel processing in a reverse micelle nanoreactor. Sol–gel processing basically refers to the hydrolysis and condensation of alkoxide-based precursors such as $Si(OEt)_4$ (tetraethyl orthosilicate, or TEOS) to prepare silica. The base-catalyzed alkoxide sol–gel route has been exploited by a number of authors to synthesize nanosize silica particles. In most cases, oil-soluble metal alkoxide is added to microemulsion containing solubilized ammoniacal water. In order for reaction to occur, the alkoxide must diffuse through the continuous oil phase to the reverse micelles. It must then cross the surfactant film before reaching the water pool where the hydrolysis and condensation reaction takes place. The reaction takes place only at the interface where the TEOS molecule changes from nonpolar to polar upon stepwise hydrolysis. Under such conditions, TEOS molecules remain adsorbed at the interface and the hydrolysis continues without interruption to form tetrahydroxysilane [$Si(OH)_4$]. $Si(OH)_4$ molecules then condense with other silicon hydroxide $Si(OH)_4$ molecules, and then this polycondensation reaction continues to form hydrated silica particles. The entire process can be represented schematically in Figure 4.8.

Our group has developed a reverse micellar system to synthesize ORMOSIL nanoparticles using triethoxyvinylsilane (VTES) and (3-aminopropyltriethoxysilane) APTES as the silicate precursors [Roy et al., 2005]. A schematic of our method along with the involved chemistry is presented in Figure 4.9.

4.3.3. Hot-Colloidal Synthesis

Semiconductor nanocrystals or quantum dots (e.g., CdSe, CdTe, InP, PbS.) are most often prepared through hot colloidal synthesis, which employs three components: (i) a high-boiling-point solvent, (ii) ligands (also called surfactants or capping groups), and (iii) organometallic precursors. A high-boiling-point solvent is required to achieve temperatures high enough for crystallization of the inorganic nanoparticles. The ligands or surfactants adsorb to the surface of the nanoparticles, limiting their growth. The organometallic precursors provide the inorganic elements (e.g., Cd and Se) from which the particles form.

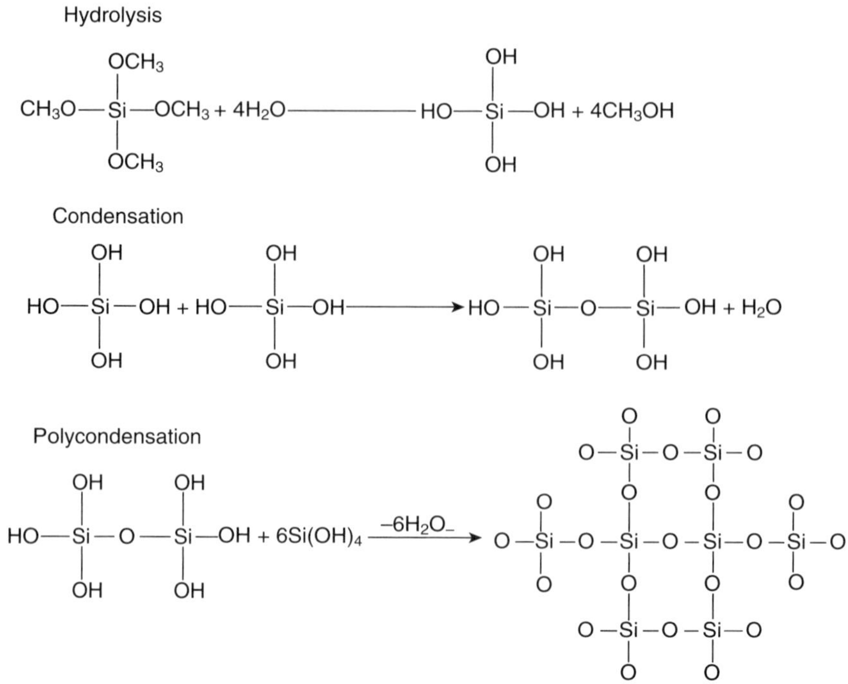

Figure 4.8. The process of formation of silica nanoprticles involving hydrolysis, condensation, and polycondensationprocess.

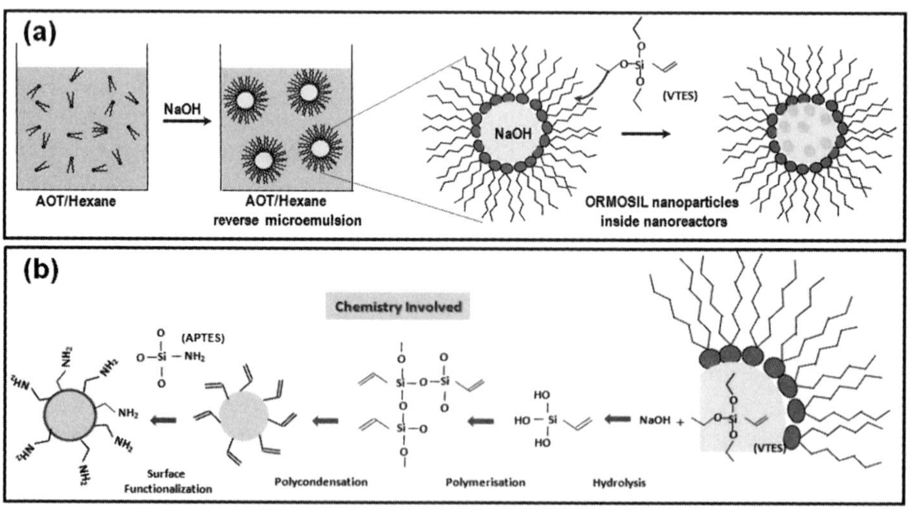

Figure 4.9. (a) Schematic presentation of synthesis of ORMOSIL nanoparticles using AOT/hexane reverse microemulsion system and (b) the detailed chemistry involved.

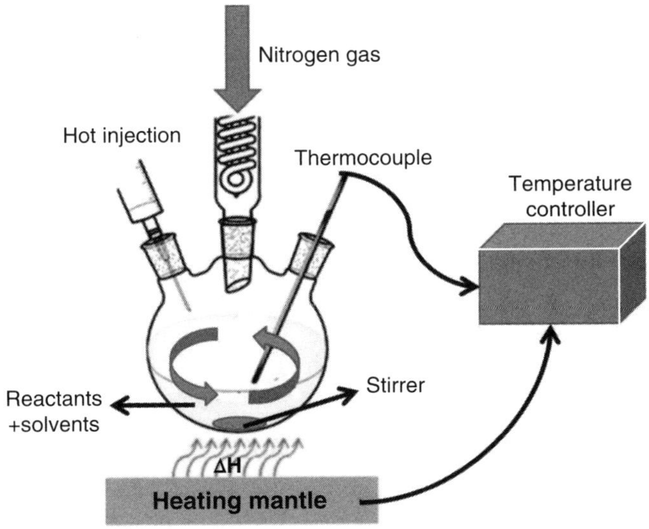

Figure 4.10. Schematic of typical laboratory-scale apparatus for preparing QDs. (Courtesy of Y. Li and M. T. Swihart.)

Typically, a solution containing surfactants is heated to 250 to 350°C. Organometallic precursors are then quickly injected into the hot solution [Murray et al., 1993; Peng et al., 1997]. Figure 4.10 illustrates a typical laboratory-scale apparatus for preparing QDs.

Injection of precursors into the hot solvent leads to a rapid burst of particle nucleation, followed by a period of particle growth without further nucleation. This separation of nucleation and growth is essential for producing uniform nanoparticles. The size of nanocrystals, and thereby their emission color, can be controlled by the temperature, identity of solvent and surfactants, and aging time. Many variations on the above approach are possible. For example, one or both precursors may be added prior to heating. Organometallic precursors can be formed *in situ* by reaction of an inorganic oxide (e.g., CdO), chloride, or metal with ligands in the growth solution. Sometimes the high-boiling-point solvent also serves as the surfactant, in which case it is usually called a coordinating solvent.

Procedures for synthesizing QDs have been extensively studied over the past two decades, and many variations and extensions of this general approach have been demonstrated. Anisotropic growth of QD (i.e., quantum rod) can be accomplished by introducing surfactants, such as phosphonic acids, that preferentially bind to a particular crystal facet of the growing nanoparticles. The nanocrystal then grows preferentially along a particular direction in the crystal lattice, producing an anisotropic (rod or wire) structure [Peng et al., 2000]. More complex structures, such as tetrapods, can be obtained by abruptly changing reaction conditions from those favoring one growth mode to those favoring another. Sequential addition of multiple precursor combinations can be used to produce core-shell QDs and even multilayered or graded-composition QDs. A shell of higher-bandgap semiconductor on a core of lower-bandgap material increases quantum confinement and often dramatically improves photoluminescence quantum yield.

Although it was originally developed for QD synthesis, the general hot-colloidal approach can also be used to produce metallic (Au, Ag), magnetic (Fe_3O_4, FePt), and other nanocrystalline materials. The nanocrystals prepared by hot colloidal synthesis are dispersible in organic solvents such as toluene or chloroform because of the presence of nonpolar (hydrophobic) ligands or surfactants bound to their surface. Further surface modification is required to disperse them in water or biological media.

4.3.4. Seed-Mediated Synthesis of Anisotropic Metallic Nanostructures

Anisotropic nanostructures of noble metals, in which the crystal structure is isotropic, can be produced using ionic surfactants under appropriate conditions. The prototypical example of this is preparation of gold nanorod (GNR) using cetyltrimethylammonium bromide (CTAB), which can self-assemble into cylindrical nanostructures at appropriate concentration in water. A seed-mediated synthesis was proposed and demonstrated by Murphy's group and El-Sayed's group to produce a high yield of monodispersed GNR in CTAB solution [Jana et al., 2001; Nikoobakht et al., 2003; Gole et al., 2004]. In practice, gold salt is dissolved in a concentrated CTAB solution containing some additives (e.g., silver nitrate) that can be used to tune the aspect ratio (length vs. width) of GNR. A mild reducing agent (e.g., ascorbic acid) is used, which will not fully reduce the gold salt to metallic gold in the absence of seed particles, but which can reduce the gold salt in the presence of existing gold particles or surfaces. When seed particles (2 to 3-nm-diameter gold nanoparticles) are added to the solution, they grow anisotropically by preferential deposition of gold onto nanocrystal surfaces that are not passivated by CTAB. The growth mechanism, as proposed by Murphy et al., is illustrated in Figure 4.11.

4.3.5. Reprecipitation Method

Many of the commonly used pharmaceutical agents including various drugs are hydrophobic in nature. Delivery of these drugs requires special formulations to disperse them in aqueous solution, using surfactants or other nanoparticle-based delivery vehicles. Upon systemic administration, such drug-doped carriers are preferentially taken up by tumor tissues by virtue of the "enhanced permeability and retention effect." This is the property of such tissues to engulf and retain circulating macromolecules and particles or in some cases the byproducts produced during preparation of these surfactant-based drug dispersions. These tend to increase the systemic toxicity of the drug formulation. There is therefore increasing interest in the development of novel

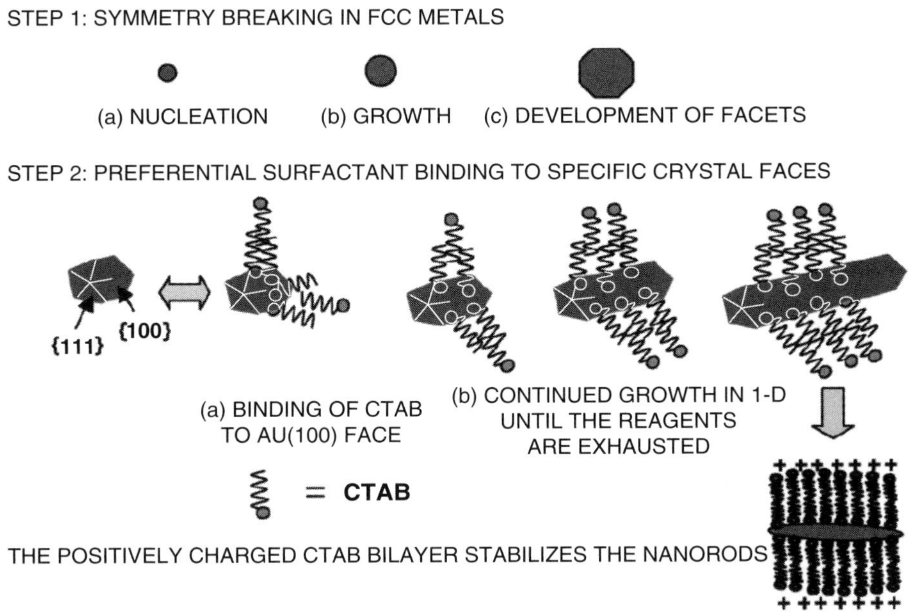

Figure 4.11. Seed-mediated synthesis of gold nanorod using CTAB. (Reproduced with permission from Murphy et al. [2005].)

drug formulation and delivery methods, without the addition of any external agents such as surfactants or other carrier vehicles. One highly suitable method for dispersion of hydrophobic compound in water is the "reprecipitation method." To demonstrate this concept, we prepared a nanocrystalline formulation of a hydrophobic drug for photodynamic therapy and compared its efficacy with the conventional surfactant-based formulation.

This method for the delivery of water-insoluble photosensitizer (PS) drugs does not involve incorporation of any additional stabilizing agents such as surfactants or other delivery vehicles. The reprecipitation method involves injection of the drug, dissolved in a semipolar solvent like DMSO, into water. The drug molecules precipitate from the mixed solvent and self-assemble as pure nanosized crystals that remain stably dispersed in water. We synthesized nanocrystals of 2-devinyl-2-(1-hexyloxyethyl)pyropheophorbide (HPPH) by the reprecipitation method. HPPH is an effective PS which is in phase I/II clinical trials at Roswell Park Cancer Institute, Buffalo. The resulting nanocrystals are monodispersed with a diameter of 100 nm. Though the fluorescence and photodynamic activity of the drug nanocrystals were substantially quenched in aqueous media, both recovered under *in vitro* and *in vivo* conditions. The recovery of fluorescence and photodynamic activity inside cells was verified by confocal microscopy and cellular phototoxicity assays. The recovery of drug activity and fluorescence is attributed to the interaction of the nanocrystals with blood serum or other intracellular components (e.g., serum albumin), resulting in conversion of the drug nanocrystals into the molecular form. This was confirmed by observing a similar recovery in the presence of fetal bovine serum (FBS) or bovine serum albumin (BSA). Efficacy of the nanocrystal formulation *in vitro* as well as *in vivo* was found to be comparable with that of the same drug formulated in the conventional delivery vehicle, Tween-80.

4.4. COMBINATION OF BOTTOM-UP AND TOP-DOWN APPROACHES

An example of a combination of the two approaches is provided by the synthesis of silicon nanocrystals, which are a new generation of biocompatible quantum dots for various nanomedicine applications [Swihart, 2007]. First, using a bottom-up approach, silicon powders are synthesized by laser-induced heating of silane to a temperature where it dissociates. This is done in a reactor like that shown schematically in Figure 4.12A. A continuous-wave CO_2 laser beam is focused above the central reactant inlet, through which a flow of silane and hydrogen enters the reactor. Silane absorbs the laser energy at a wavelength of 10.6 μm and is thereby heated to a temperature where it decomposes and silicon particles nucleate. Helium and hydrogen flows confine the reactant and nanoparticles to the center of the reactor. Hydrogen also serves to increase the temperature at which particle nucleation occurs, and to decrease the particle growth rate, since it is a byproduct of silane dissociation and particle formation. The resulting particles are collected by filtration.

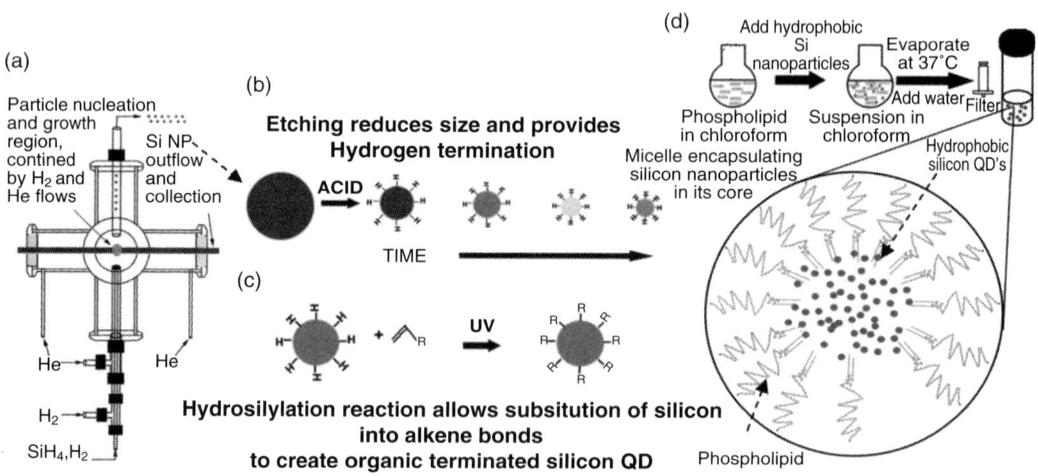

Figure 4.12. Schematic illustrations of (A) laser pyrolysis, (B) etching, (C) hydrosilylation, and (D) micelle encapsulation. (Reproduced with permission from Erogbogbo and Swihart [2010].)

In the second step, which involves a top-down approach, the particles collected from the reactor are etched with a mixture of HF and HNO_3 to reduce their size, as depicted in Figure 4.12B. The silicon nanoparticles are dispersed via sonication in methanol, and the acid mixture is added to the resulting dispersion. The size of the particles decreases; as a result, the particles exhibit bright, visible photoluminescence. The color of the photoluminescence changes from red to yellow to green as the etching proceeds. The etching rate decreases with an increase in the amount of methanol used to disperse the particles initially. When a desired emission color is reached, the etching is slowed by adding more methanol. The particles are collected on a poly(vinylidene fluoride) (PVDF) membrane filter and washed with a large amount of the methanol–water mixture to remove any adsorbed acid mixture. The particles are finally rinsed with pure methanol.

The freshly etched particles are then transferred into a glove box to prevent oxidation, which provides an environment for hydrosilylation. A Rayonet photochemical reactor, equipped with UV tubes, is used to initiate the hydrosilylation reaction. This reaction is depicted in Figure 4.12C. The reaction time required varies substantially, depending on the compound being attached to the particles and the particle size. Produced hydrophobic particles are thus dispersed in chloroform or toluene. To make them water dispersible, they are encapsulated in phospholipid micelles, as discussed further in the next section. The hydrophobic Si QDs encapsulated in the core of the phospholipid micelles are depicted in Figure 4.12D.

4.5. NANOPARTICLE SURFACE MODIFICATION

Nanoparticles prepared using any of the above methods may require surface modification, by either a physical or a chemical method, to make them more suitable for a specific nanomedicine application. Many of the methods produce nanoparticles in an organic solvent, and thus the nanoparticles are surrounded by ligands which produce a hydrophobic surface. To make these nanoparticles dispersible in a biological medium or a buffer solution, they require surface modification by ligand exchange or encapsulation to make them hydrophilic. To make them more biocompatible and/or reduce their toxicity, the nanoparticles may require additional coating with a biocompatible layer [Yong et al., 2009a]. As discussed in the previous chapter, the surface charge on these nanoparticles also plays a key role in their cellular uptake. Appropriate surface modification can lead to a judicious control of the charge. It is also necessary to prolong the circulation of the nanoparticles in an intravenous delivery to enhance the pharmacokinetics and produce preferential localization at the desired biological site (e.g., cancer cells). Finally, linking of appropriate reactive groups on the surface of the particles may be required for chemical conjugation of sensing probes, imaging groups, or a therapeutic payload. This section provides a brief description of the strategies used for each.

Ligand Exchange. Nanoparticles synthesized in organic media are passivated with a hydrophobic layer (oil like) in order to maintain their colloidal stability in organic solvents such as chloroform, toluene, or hexane. Ligand exchange involves the replacement of this hydrophobic layer with a hydrophilic (water-like) moiety. The ligand usually contains a short carbon chain with one surface-anchoring end and one hydrophilic end. During the ligand exchange process, a hydrophobic suspension will be mixed with an excess of the reactive ligand which binds to the nanoparticle surface. Under vigorous stirring, the hydrophobic ligands will be detached from the surface and displaced by the hydrophilic ones that impart water dispersibility to the nanoparticles. At the same time, however, incomplete ligand exchange and poor surface passivation may lead to formation of surface defects (e.g., surface electron trap state) that have the potential to compromise the optical and electronic properties of the nanoparticles.

Silica Coating. A widely used method for the production of a silica or ORMOSIL coating is reserve microemulsion synthesis as described earlier. The same strategy can be applied to encapsulate hydrophobic nanoparticles. Our group demonstrated the co-encapsulation of hydrophobic QDs (CdSe/CdS/ZnS) and magnetic nanoparticles (Fe_3O_4) within an ORMOSIL shell [Law et al., 2008]. The silica was organically modified with different functional groups such as amine (NH_2) groups produced using 3-aminopropyltriethoxysilane (APTES) (see Figure 4.13). This method offers precise control over the particle size and the freedom to further functionalize the surface.

Polymer Coating. In polymer coating, a variety of amphiphilic polymers, which contain both hydrophobic and hydrophilic groups, including poly(acrylic acid) (PAA), poly(vinylpyrrolidone) (PVP), *Pluronics*®, and β-cyclodextrin are commonly used (see Figure 4.14). The hydrophobic part of the polymer interacts with the ligand on the nanoparticle surface, serving as an interface between the hydrophobic core and the hydrophilic solution. In some cases, the encapsulation is followed by cross-linking on the surface to form a protective shell

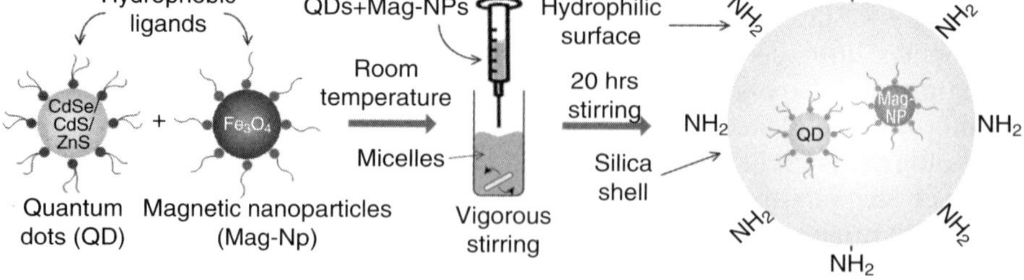

Figure 4.13. Synthesis of co-encapsulated ORMOSIL nanoparticles. (Reproduced with permission from Law et al. [2008].)

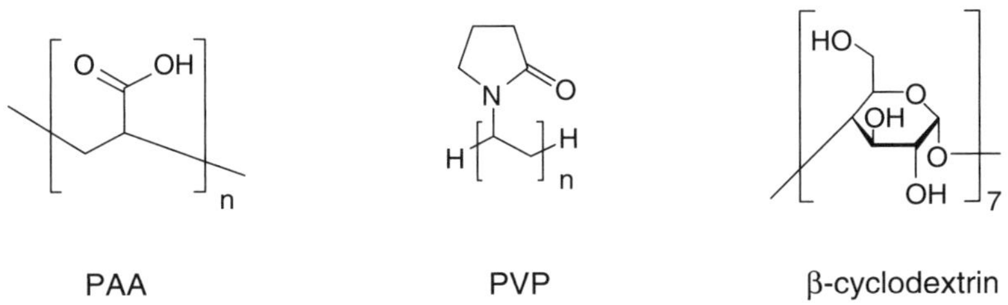

Figure 4.14. Examples of amphiphilic polymers (unit cell).

around the nanoparticle. The use of lysine-cross-linked near-infrared quantum dots for long-term *in vivo* distribution, clearance, and toxicity studies has been successfully demonstrated [Yong et al., 2009a]. An intravenous injection of ~10.5 mg/mL QDs into mice had no apparent adverse effect to their health as observed for 3 months. *Pluronic®* tri-block copolymer-modified QD was also used to demonstrate antibody-mediated noninvasive tumor detection in a small animal model [Ding et al., 2011].

Phospholipid Encapsulation. PEGylated phospholipids are a kind of amphiphilic polymer. This block copolymer micelle has been emphasized by the biomedical research community over the past decade because of its excellent biocompatibility and strong resistance against nonspecific interaction with biomolecular degradations. It is composed of a hydrophobic part that can interact with the surface ligand, an adjustable poly(ethylene glycol) (PEG) chain, which is hydrophilic and also serves as a repellent in the reticuloendothelial system (RES) *in vivo,* and an outer functional group (amine, carboxyl, maleimide and biotin) for a variety of functionalization and bioconjugation. Dubertret et al., 2002 demonstrated the encapsulation of single nanocrystals in phospholipid micelles. The micelle impeded the oxidation of the hydrophobic core from the outside aqueous media so that the nanoparticle can remain intact in the biological buffer.

4.6. FUNCTIONALIZATION AND BIOCONJUGATION

Cationic/Anionic Polymer. Due to the charged nature of many biomolecules, bioconjugation can often be achieved through the use of electrostatic interactions. A layer-by-layer method is usually utilized to deposit a charged polymer, or polyelectrolyte, on the nanoparticle surface. The use of gold nanorod (GNR) coated with multilayers of a polyelectrolyte as a biocompatible contrast agent for dark-field microscopy has been reported by our group [Ding et al., 2007]. The as-synthesized GNRs were capped with a bilayer of cetyltrimethylammonium bromide (CTAB), which is positively charged. Because CTAB is highly toxic, the positively charged surface of the GNRs was sequentially covered with a negatively charged layer of poly (3,4-ethylenedioxythiophene)-poly(styrenesulfonate) (PEDT/PSS) and finally a positively charged layer of poly(diallyldimethylammonium chloride) (PDDAC). The PDDAC-capped GNR was used for transferrin conjugation by electrostatic force. This enabled targeting of cancer cells upon which transferrin receptors are overexpressed [Yong et al., 2009b]. Cationically modified nanoparticles, such as poly-L-lysine (PLL), polyethylenimine (PEI), β-cyclodextrin, and chitosan, have attracted remarkable attention due to their high gene delivery efficiency. Our group introduced a novel approach with the use of gold nanorods coated to carry a positive charge to form a DARPP-32 siRNA nanoplex (by the electrostatic force) that targets the dopaminergic signaling pathway in the brain [Bonoiu et al., 2009]. The suppression of this signaling pathway can facilitate drug addiction therapy as discussed in Chapter 14.

Tetraazacyclododecanetetraacetic Acid (DOTA). DOTA, a macrocyclic agent for metal ion chelation, has become an increasingly important component in diagnostic and therapeutic applications. DOTA has become a powerful tool for imaging and radiotherapy, because a large number of metal ions or radioactive isotopes (Gd^{3+}, Eu^{3+}, ^{64}Cu, ^{68}Ga, ^{90}Y) can be chelated with high thermodynamic stability. The amine-reactive DOTA complex (NHS–DOTA) is commercially available, which enables easy functionalization of nanoparticle with DOTA. Our group developed a multimodal imaging probe by conjugating a DOTA–gadolinium derivative to a quantum dot surface [Liu et al., 2011] to create a nanoprobe that exhibited optical and magnetic properties. Another DOTA-based molecular probe was developed by Gambhir and Chen [Cheng et al., 2007]. The ^{64}Cu-labeled DOTA bioconjugate was synthesized and injected into a melanoma tumor mice model. The high tumor uptake was evaluated through the use of a MicroPET imaging system (see Figure 4.15).

Environment-Sensitive Linker. A great deal of research interest has been focused on designing linkers that incorporate different functional groups with

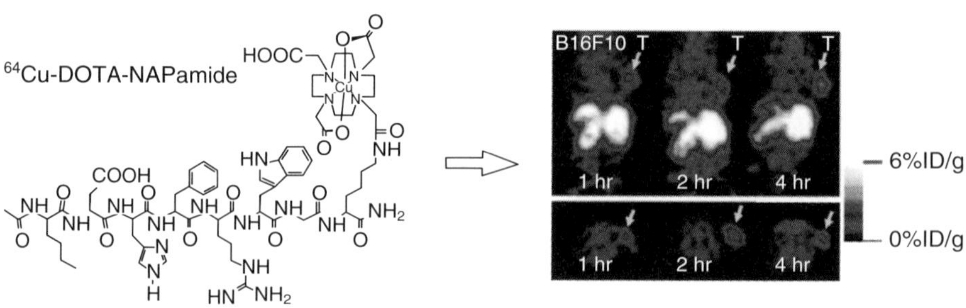

Figure 4.15. Schematic structure of ^{64}Cu-labeled DOTA bioconjugate and the MicroPET image. Arrows indicate the location of tumors (T). (Reproduced with permission from Cheng et al. [2007].)

the ability to be cleaved under specific environmental conditions (e.g., pH, reduction, photoillumination). Nanoparticles modified with environment-sensitive linkers, or sensitive polymer coatings, can be readily applied for triggered drug/gene release at a particular site. The nanoparticles are usually taken up by the cell through endocytosis and enclosed in an endosomal compartment in which the pH value drops progressively from 7.4 to 5.5. In this environment, a pH-sensitive linker can degrade and trigger the release of the enclosed biomaterial. Reduction-sensitive linkers contain a disulfide linkage in the main chain. This disulfide linkage remains very stable in the circulation and extracellular environment. When the linker is exposed to a high concentration of thiol-reductive substance (e.g., glutathione), it is prone to rapid degradation within minutes. This makes the reduction-sensitive linker-modified nanoparticles a promising tool for drug and gene delivery.

HIGHLIGHTS OF THE CHAPTER

- Nanochemistry involves the preparation of a diverse group of nanoparticles of metals, semiconductors, glasses, and polymers.
- A top-down approach is a process that can be scaled down to produce nanosize domains.
- Mechanical milling is utilized to reduce the size of larger particles. It has been applied to many drugs that exhibit poor solubility or are poorly absorbed.
- Dip-pen nanolithography (DPN) allows nanopatterning of molecules to produce structures that are one molecular unit thick. It utilizes the tip of an atomic force microscope (AFM) as a pen.
- The Particle Replication in Nonwetting Templates (PRINT) process produces polymeric nanoparticles of precisely controlled shape and size. The particle precursor is spread into the cavities of a fluoropolymer mold and then converted into solid nanoparticles by using the appropriate chemistry.
- Laser ablation utilizes ultrashort and high-intensity focused laser beams to ablate materials from a rotating solid target that may be immersed in a liquid to provide both cooling of the target and dispersing the nanoparticles produced.
- A bottom-up approach is a process to assemble smaller components into more complex particles.
- Dendrimers are regularly branched globular macromolecules made by a controlled step-by-step iterative synthetic approach.
- Synthesizing dendrimers by the Divergent Approach involves an outerward growth from a core by repetition of a series of coupling and activation steps, each producing one layer buildup at a time.

- Synthesizing dendrimers by the Convergent Approach starts with an inward growth by coupling end groups to each branch of the monomer.
- A widely used method to produce nanoparticles is a chemical reaction in the confined nanoscale cavities created in microemulsions.
- Microemulsion systems form dispersions of nanosize droplets of water-in-oil (w/o) or oil-in-water (o/w), depending on which one is the larger (bulk phase). In the case of w/o system, the aqueous phase is dispersed as nanosized water droplets encapsulated by a monolayer film of surfactant molecules to provide reverse micelle nanoreactors.
- Sol–gel processing of organically modified trialkoxy silanes in a reverse micelle nanoreactor provides a convenient way of making ORMOSIL nanoparticles.
- Hot-Colloidal Synthesis has extensively been used for the synthesis of quantum dots (QDs) and quantum rods (QRs). The reactants are injected into a hot solution to initiate the reaction.
- The Reprecipitation Method is usually used to make hydrophobic drug dispersive in aqueous solution through the injection of a drug dissolved in a miscible solvent like DMSO in water; the drug molecules self-assemble as pure nanosized crystals and remain stable in water.
- The synthesis of silicon nanocrystals combines bottom-up and top-down approaches together. Using a bottom-up approach, silicon powders are synthesized by laser-induced heating of silane to temperatures where it dissociates. The particles collected from the reactor are etched (top-down approach) with an appropriate mixture of HF and HNO_3 to reduce the size of the particles.
- To make hydrophobic nanoparticles dispersable in a biological medium or a buffer solution, a surface modification of the nanoparticles is necessary.
- Ligand exchange involves the replacement of this hydrophobic layer with a hydrophilic (water-like) moiety. The ligand usually contains a short carbon chain with one surface-anchoring end and one hydrophilic end.
- A widely used method for the production of a silica coating of a nanoparticle utilizes reverse micelle sol–gel processing.
- Polymer coating utilizes a variety of amphiphilic polymers that contain both the hydrophobic and hydrophilic group. The hydrophobic part of the polymer interacts with the ligand on the nanoparticle surface, serving as an interface between the hydrophobic core and the hydrophilic solution.
- Phospholipid micelles have been emphasized in the biomedical research community over the past decade because of their excellent biocompatiability and strong resistance against nonspecific interaction with biomolecules.

- Bioconjugation to a charged biomolecule or a polyelectrolyte coating on a nanoparticle can be achieved through the use of electrostatic force.
- DOTA, a macrocyclic agent for metal ion chelation, has become an increasingly important component in diagnostic and therapeutic applications. Metal ions or radioactive isotopes (Gd^{3+}, Eu^{3+} ^{64}Cu, ^{68}Ga, ^{90}Y) can be chelated with high thermodynamic stability.
- Research interest has been focused on designing linkers that incorporate different functional groups, with the ability to be cleaved under a specific environmental condition. The nanoparticle modified with the environmentally sensitive linker-sensitive polymer coatings can be readily applied for triggered drug/gene release at a particular site.

EXERCISES

True or False

1. In a seed-mediated synthesis of metallic nanostructures, the shape of gold nanoparticles can be tuned by adding different concentrations of silver nitrate.

2. The Laser Ablation technique utilizes a low-power laser to produce nanoparticles.

3. The shape and size of polymers can be precisely controlled and synthesized by Particle Replication in Nonwetting Templates (PRINT) technology.

4. In the microemulsion synthetic process, the formation of the nanosize droplets of water-in-oil or oil-in-water is mainly depending on the reaction temperature.

5. High temperature (over 250°C) is compulsory for synthesizing cadmium-based quantum dots.

6. Bioconjugations of nanoparticles cannot be accomplished by electrostatic force.

7. siRNA is positively charged in nature.

Multiple Choice

8. Which of the following is a bottom-up approach for nanoparticle synthesis:

(a) PRINT

(b) Laser ablation

(c) Hot-colloidal synthesis

(d) Dip-pen nanolithography

9. Which of the following chemical could be used as a surfactant to form a microemulsion nanocavity?
   (a) Poly(acrylic acid)
   (b) L-Cysteine
   (c) Ethanol
   (d) Sodium hydroxide

10. Which of the following is not an advantage of using phospholipid encapsulation?
   (a) It facilitates the water dispersity of the nanoparticle.
   (b) It increases the circulation half-life of the nanoparticle *in vivo*.
   (c) It increases the oxidation rate of the nanoparticle in the core.
   (d) It provides the functionalities of the nanoparticle.

Answers: 1. F; 2. F; 3. T; 4. F; 5. F; 6. F; 7. F; 8. c; 9. a; 10. c

REFERENCES

Andronov, A., Fréchet, J. M. J., Light-harvesting dendrimers, *Chem. Commun.* 1701–1710 (2000).

Balzani V., Ceroni, P., Gestermann, S., Kauffmann, G., Gorka, M., and Vögtle, F., *Chem. Commun.*, 853–854 (2000).

Bonoiu, A. C., Mahajan, S. D., Ding, H., Roy, I., Yong, K. T., Kumar, R., Hu, R., Bergey, E. J., Schwartz, S. A., Prasad, P. N., Nanotechnology approach for drug addiction therapy: Gene silencing using delivery of gold nanorod-siRNA nanoplex in dopaminergic neurons, *Proc Natl. Acad. Sci. USA* **106**, 5546–5550 (2009).

Canelas, D. A., Herlihy, K. P., DeSimone, J. M., Top-down particle fabrication: Control of size and shape for diagnostic imaging and drug delivery, *Nanomed. Nanobiotechnol.* **1**, 391–404 (2009).

Cheng, Z., Xiong, Z. M., Subbarayan, M., Chen, X. Y., Gambhir, S. S., Cu-64-labeled alpha-melanocyte-stimulating hormone analog for MicroPET imaging of melanocortin 1 receptor expression, *Bioconj. Chem.* **18**, 765–772 (2007).

Ding, H., Yong, K.-T., Roy, I., Pudavar, H. E., Law, W. C., Bergey, E. J., Prasad, P. N., Gold nanorods coated with multilayer polyelectrolyte as contrast agents for multimodal imaging, *J. Phys. Chem. C* **111**, 12552–12557 (2007).

Ding, H., Yong, K.-T., Law, W.-C., Roy, I., Hu, R., Wu, F., Zhao, W., Huang, K., Erogbogbo, F., Bergey, E. J., Prasad, P. N., Non-invasive tumor detection in small animals using novel functional pluronic nanomicelles conjugated with anti-mesothelin antibody, *Nanoscale* **3**, 1813–1822 (2011).

Dubertret, B., Skourides, P., Norris, D. J., Noireaux, V., Brivanlou, A. H., Libchaber, A., *In vivo* imaging of quantum dots encapsulated in phospholipid micelles, *Science* **298**, 1759–1762 (2002).

Erogbogbo, F., Swihart, M. T., Imaging pancreatic cancer with folic acid terminated luminescent silicon nanocrystals, *AIP Conf. Proc. (Bonsai Project Symposium: Breakthroughs in Nanaoparticles for Bio-Imaging)* **1275**, 35–39 (2010).

Gole, A., Murphy, C. J., Seed-mediated synthesis of gold nanorods: Role of the size and nature of the seed, *Chem. Mater.* **16**, 3633–3640 (2004).

Grayson, S. M., Fréchet, J. M. J., Convergent dendron and dendrimers: From synthesis to applications, *Chem. Rev.* **101**, 3819–3867 (2001).

Hawker, C. J., Fréchet, J. M. J., Preparation of polymers with controlled molecular architecture: A new convergent approach to dendritic macromolecules, *J. Am. Chem. Soc.* **112**, 7638–7647 (1990).

Hong, S., Zhu, J., Mirkin, C. A., Multiple ink nanolithography: Toward a multiple pen nano-plotter, *Science* **286**, 523–525 (1999).

Jana, N. R., Gearheart, L., Murphy, C. J., Seed-mediated growth approach for shape-controlled synthesis of spheroidal and rod-like gold nanoparticles using a surfactant template, *Adv. Mater.* **13**, 1389 (2001).

Kabashin, A. V., Delaporte, P., Pereira, A., Grojo, D., Torres, R., Sarnet, T., Sentis, M., Nanofabrication with pulsed lasers, *Nanoscale Res. Lett.* **5**, 454–463 (2010).

Law, W.-C., Yong, K.-T., Roy, I., Xu, G., Ding, H., Bergey, E. J., Zeng, H., Prasad, P. N., Optically and magnetically doped organically modified silica nanoparticles as efficient magnetically guided biomarkers for two-photon imaging of live cancer cells, *J. Phys. Chem. C* **112**, 7972–7977 (2008).

Liu, L., Law, W.-C., Yong, K.-T., Roy, I., Ding, H., Erogbogbo, F., Zhang, X., Prasad, P. N., Multimodal imaging probes based on Gd-DOTA conjugated quantum dot nanomicelles, *Analyst* **136**, 1881–1886 (2011).

Murphy, C. J., Sau, T. K., Gole, A. M., Orendorff, C. J., Gao, J. X., Gou, L. F., Hunyadi, S. E., Li, T., Anisotropic metal nanoparticles: Synthesis, assembly, and optical applications, *J. Phys. Chem. B* **109**, 13857–13870 (2005).

Murray, C. B., Norris, D. J., Bawendi, M. G., Synthesis and characterization of nearly monodisperse CdE (E = S, Se, Te) semiconductor nanocrystallites, *J. Am. Chem. Soc.* **115**, 8706, (1993).

Murray, C. B., Kagan, C. R., Bawendi, M. G., Synthesis and characterization of monodisperse nanocrystals and close-packed nanocrystal assemblies, *Annu. Rev. Mater. Sci.* **30**, 545–610 (2000).

Newkome, G. R., Moorefield, C. N., Vögtle, F., *Dendritic Molecules: Concepts, Syntheses, Perspectives*. Wiley-VCH (1996).

Nikoobakht, B., El-Sayed, M. A., Preparation and growth mechanism of gold nanorods (NRs) using seed-mediated growth method, *Chem. Mater.* **15**, 1957–1962 (2003).

Peng, X., Schlamp, M. C., Kadavanich, A. V., Alivisatos, A. P., Epitaxial growth of highly luminescent CdSe/CdS core/shell nanocrystals with photostability and electronic accessibility, *J. Am. Chem. Soc.* **119**, 7019 (1997).

Peng, X., Manna, L., Yang, W. D., Wickham, J., Scher, E., Kadavanich, A., Alivisatos, A. P., Shape control of CdSe nanocrystals, *Nature* **404**, 59–61 (2000).

Piner, R. D., Zhu, J., Xu, F., Hong, S., Mirkin, C. A., Dip-pen nanolithography, *Science* **283**, 661–663 (1999).

Roy, I., Ohulchanskyy, T. Y., Bharali, D. J., Pudavar, H. E., Mistretta, R. A., Kaur, N., Prasad, P. N., Optical tracking of organically modified silica nanoparticles as DNA carriers: A nonviral, nanomedicine approach for gene delivery, *Proc. Natl. Acad. Sci. USA* **102**, 279–284 (2005).

Salaita, K., Wang, Y., Mirkin, C. A., Applications of dip-pen nanolithography, *Nat. Nanotechnol.* **2**, 145–155 (2007).

Swihart, M. T., Silicon nanoparticles for biophotonics, Chapter 4 in *Nanotechnology in Biology and Medicine: Methods, Devices, and Applications*, edited by T. V. Dinh, ed., CRC Press, Boca Raton, FL (2007).

Tanaka, Y., Inkyo, M., Yumoto, R., Nagai, J., Takano, M., Nagata, S., Nanoparticulation of poorly water soluble drugs using a wet-mill process and physicochemical properties of the nanopowders, *Chem. Pharm. Bull.* **57**(10), 1050–1057 (2009).

Tomalia, D.A., Naylor, A.M., Goddard, W.A., III, Starburst dendrimers: Molecular-level control of size, shape, surface chemistry, topology, and flexibility from atoms to macroscopic matter, *Angew. Chem. Int. Ed. Engl.* **29**, 138–175 (1990).

Yong, K. T., Roy, I., Ding, H., Bergey, E. J., Prasad, P. N., Biocompatible Near-Infrared Quantum dots as ultrasensitive probes for long-term *in vivo* imaging applications, *Small* **5**, 1997–2004 (2009a).

Yong, K. T., Roy, I., Swihart M. T., Prasad, P. N., Multifunctional nanoparticles as biocompatible targeted probes for human cancer diagnosis and therapy, *J. Mater. Chem.* **19**, 4655–4672 (2009b).

CHAPTER 5

MULTIFUNCTIONALITIES FOR DIAGNOSTICS AND THERAPY

This chapter provides an account of various functionalities that can be incorporated in nanoparticles for nanomedicine applications. Where necessary, the physical principles are explained at a level useful for multidisciplinary readership. Section 5.1 provides a listing of some of the important functionalities. Section 5.2 discusses optical functionalities introduced by various linear and nonlinear optical processes that can be utilized for optical bioimaging and biosensing as well as for light-activated therapy. Section 5.3 describes optical nanoprobes that exhibit optical properties unique to their nanostructures. The examples provided are (a) organic dyes containing nanoparticles, (b) quantum dots and rods with optical properties dependent on their nanoscale size, and (c) metallic nanostructures with novel plasmonic absorption bands not present in their bulk form. Section 5.4 describes the magnetic functionalities used for multiple applications such as (a) enhanced contrast magnetic resonance imaging (MRI), (b) magnetic sensing, (c) magnetically directed tissue engineering, and (d) magnetically directed and induced therapy. Section 5.5 presents thermal functionalities which can be imparted to a nanoparticle formulation for thermal imaging or to produce nanoheaters for thermal therapy. Two approaches to induce thermal therapy discussed in this section are heating produced (i) by coupling magnetic nanoparticles to an alternating magnetic field or RF field and (ii) by nonradiative relaxation of excitation produced by light absorption (photothermal effect).

Introduction to Nanomedicine and Nanobioengineering, First Edition. Paras N. Prasad.

Section 5.6 introduces radioactive functionalities for both bioimaging and radiotherapy. The examples of radio bioimaging described here are: position emission tomography (PET) and single photon emission computed tomography (SPECT). In addition, radioactive functionality-based therapy discussed is the neutron capture therapy (NCT). The functionality described in Section 5.7 is biological, which can be introduced to make a nanoplatform for diagnostics and /or therapy (i) biocompatible, (ii) exhibit enhanced circulation, (iii) target a specific biological site and (iv) exhibit efficient clearance from the body. Finally, Section 5.8 illustrates how a combination of two or more of these functionalities can be introduced in the same nanoparticle formulation to derive multifunctionality.

5.1. THE VARIOUS FUNCTIONALITIES

Nanoparticles for nanomedicine applications can be prepared with various useful functionalities, which are listed in Table 5.1. A nanostructure platform enables the enhancement of these functions either by manipulation of interactions manifesting them or by simply increasing the local concentration of the species exhibiting these functions. In addition, one can introduce multiple functionalities in the same nanoparticle or nanostructure. In this case, appropriate tailoring of the nanoplatform not only minimizes interference between two or more functionalities, but can also bring synergism between them. A combination of two functionalities may also impart a new property to the nanoplatform. These functionalities can be utilized for biomedical imaging, sensing, various diagnostics, or therapeutics. Of course, an important new direction for nanomedicine is the use of nanoplatforms for "theranostics," whereby both diagnostic and therapeutic functions are combined. In the sections below, these functionalities are described and examples of multifunctionalities are discussed.

TABLE 5.1. Various Nanomedicine Functionalities

Optical	Magnetic	Thermal	Radioactive	Biological
Imaging	MRI	Thermal mapping	Radio imaging	Targeting
Sensing	Sensing	Thermal activation	PET	Enzymatic release
Light-activated therapy	Magnetic separation	Photothermal activation	Neutron	
	Magnetic therapy		Capture therapy	

5.2. OPTICAL FUNCTIONALITIES

The optical functionalities utilize either light scattering or optical transitions between energy states of a nanostructure. A review of key concepts in this area—that is, energy levels, optical transitions, and scattering—can be found in *Introduction to Biophotonics* [Prasad, 2003]. Here we briefly describe the various optical excitation, energy transfer, photoemission, and relaxation processes.

Optical Excitation and Subsequent Decay. Briefly, the process of light absorption involves the absorption of photons by a material, which creates an excitation from a lower to a higher energy level. The energy levels relevant to processes most commonly used in nanomedicine are primarily electronic; but for detailed molecular imaging of cells and tissues or molecular interactions in them, the vibrational states may be employed (e.g., in CARS imaging). Most optical processes are generated by linear absorption in which the energy of one photon absorbed is sufficient to create an optical excitation leading to population of an excited state. However, nonlinear or multiphoton excitation in which the absorption of more than one photon of lower energy (or longer wavelength) together provides sufficient energy to produce excitation to a desired higher energy state. These multiphoton absorption processes, together with other types of multiphoton processes, can provide useful optical functions as discussed at a later stage. The material (a molecule or a nanoprobe) absorbing the photons and thus undergoing excitation to a higher energy level possesses excess energy, which it expends in order to return to its normal lower energy state (called ground state) by a number of pathways. First, it may emit a photon of shifted energy (different wavelength); the general term for this process is radiative transition or emission or luminescence. In the case of an electronic level transition (where the electrons jump from one energy level to another), if there is no change in the spin of electrons (such as emission from an excited singlet, zero spin, state to the lower singlet state), this emission is spin-allowed and fast, with a lifetime decay in nanoseconds; the emission is called fluorescence. In the case where the emission is from a triplet level e.g., T_1 with spin 1 to the ground singlet level S_0 (spin equal to zero), this emission is spin-forbidden, thus a weak process with a long lifetime, which could be in milliseconds to even seconds; this emission is called phosphorescence, which is typically at a markedly longer wavelength than the corresponding fluorescence. In both of these processes, there is an energy shift to a lower value of energy, from that deposited during the absorption of photons; the excess energy initially deposited by photon absorption is dissipated by nonradiative relaxation pathways such as vibrational relaxation, internal conversion, and intersystem crossing. These processes are often represented by the Jablonski diagram shown in Figure 5.1.

Linear absorption spectra are useful for the identification of chemical functional units and the quantitative analysis of various molecular species, which

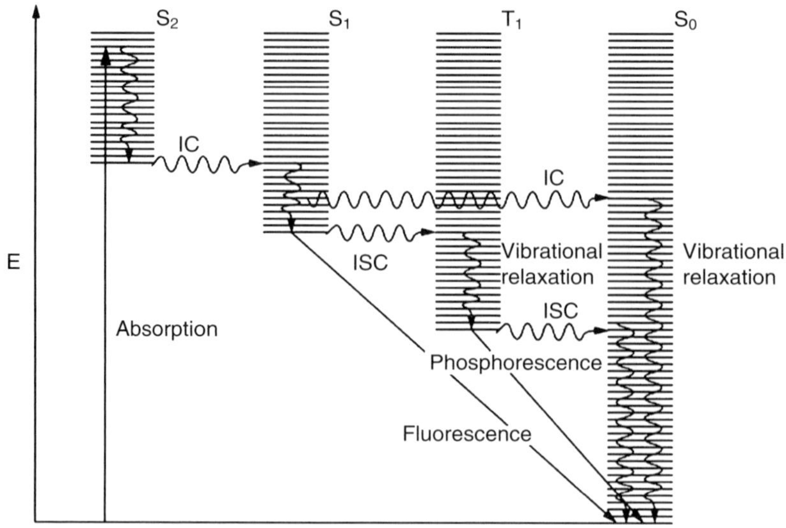

Figure 5.1. The Jablonski diagram describing the possible fates of excitation. (IC is internal conversion and ISC is intersystem crossing.)

are intrinsic components of cellular or tissue structures that are produced during a physiological process/disease manifestation or generated by drug action. Fluorescence-detection-based microscopy, discussed in Chapter 8, is the most commonly used optical imaging technique. Nonradiative processes are also of significant value. For example, the heat generated by appropriate nonradiative pathways can be utilized for photothermal therapy as discussed in a later section in this chapter. The spin-forbidden excited triplet states having a longer lifetime provide more suitable pathways for photochemistry, which can also be utilized for diagnostics or therapy.

Forster (or Fluorescence) Resonance Energy Transfer (FRET). Another important type of nonradiative process is energy transfer in which an initially absorbing molecule can transfer its excess energy (excitation energy) to another type of nearby molecule. The most common type of energy transfer is the Forster (or fluorescence) resonance energy transfer (FRET), during which there is no net change of spin. This process has been utilized for sensing and imaging as well as for monitoring protein–protein or DNA–protein interactions. FRET occurs when energy passes nonradiatively from one excited molecule (the donor) to a second molecule (the acceptor) under three conditions: (i) The emission spectrum of the donor overlaps the excitation spectra of the acceptor. (ii) The efficiency of this transfer is only significant if the two molecules are very close to each other (1–10 nm); this energy transfer (E) varies inversely with the sixth power of the distance (r) between the donor and the acceptor. (iii) The donor and the acceptor should be in a particular orientation with respect to each other, which is characterized by an orientation factor [dos Remedios and Moens, 1995].

In vitro FRET spectroscopy has been used extensively in biology to study molecular structure, conformational changes, and molecular associations (Lakowicz, 2006). Over the last decade, FRET methods have been developed for microscopic analysis of molecular associations and conformations within living organisms. This growth in popularity has been fueled by systems biology and proteomics interests in understanding protein interactions in living cells.

A family of proteins called fluorescent proteins has been widely utilized for FRET studies. The first in this family to be discovered was green fluorescent protein (GFP), isolated from photogenic cells of the jellyfish, *Aequoria Victoria* [Chalfie and Kain, 1998]. GFP consists of 238 amino acids, has a cylindrical fold with dimensions of ~4 nm long and ~2 nm in diameter, and, thus, can be considered as nanoemitters. The absorbing, and subsequently fluorescing, fluorophore unit is located within this very compact and folded structure that protects it from the bulk solvent to impart GFP a number of unique characteristics [Chalfie and Kain, 1998]. Here are some of these characteristics:

- GFP is a very robust protein, resistant to denaturation.
- The intrinsic fluorescence of the protein is due to a chromophore, formed in the protein itself by the cyclization and oxidation of residues 65–67, Ser–Tyr–Gly [Periasamy, 2001].
- The encapsulation of the chromophore within the protein reduces non-radiative decays, induced by interaction with external environment, and leads to high quantum yield of fluorescence.
- GFP can be coupled (fused) to another protein.
- GFP is a noninvasive fluorescent marker for living cells, thus lending itself to a wide range of applications where GFP may function as a cell lineage tracer, a reporter of gene expression, or a measure of protein–protein interactions.

GFP is generally introduced into a living cultured cell or into specific cells of a living organism as a GFP gene. This method involves recombinant DNA technology whereby the gene corresponding to GFP is inserted into a cell to synthesize GFP. In other words, the introduced GFP genes express GFP that emit green when irradiated and can be used to image the transfected cells and tissues containing them.

In yet another method, the gene for GFP can be fused to the gene for another protein of interest. This recombinant DNA encodes one long chimeric protein that contains both proteins in their entirety. The cells transfected with this recombinant DNA synthesize this chimeric protein. The green fluorescence from the chimeric protein can thus be used to image subcellular localization of the protein of interest.

A number of mutants of the GFP gene have been produced which exhibit enhanced emission and emit at different wavelengths. These variants of GFP include Blue FP(BFP), CyanFP(CFP), Yellow FP(YFP), and red fluorescent protein (RFP) as well as the fluorescence-enhanced variants of the above-mentioned variants (e.g., eGFP, eBFP, etc.).

The development of the spectral variants of green fluorescent protein (GFP) has greatly accelerated usage of the FRET microscopy in living cells by allowing tagging of proteins by genetic manipulation (see the literature in this URL: http://www.kcci.virginia.edu/Literature/index.php). Together, the tools of GFP combined with quantitative FRET microscopy, discussed in Chapter 8, promise to provide unprecedented new insights into understanding the dynamics and localization of molecular interactions inside living cells.

Suitable fluorophore donor–acceptor FRET pairs are the key to the success of a FRET application. Many visible fluorescent proteins (FPs) as well as nanoemitters, discussed below in Section 5.3, have been used [Periasamy and Day, 2005; Periasamy and Clegg, 2010; Shaner et al., 2008] .

An important application of the FRET process is in photodynamic therapy, discussed in Chapter 13, where an important mechanism involves FRET from the excited T_1 triplet state of a photosensitizer to the ground triplet state of an oxygen molecule as a result of which the photosensitizer returns to its ground singlet S_0 state while the oxygen molecule is excited to the S_1 singlet state, which is highly cytotoxic to cells.

Multiphoton Processes. Multiphoton processes also play an important role in nanomedicine. These processes involve interactions of several photons to generate an excitation. However, there are also multiphoton processes that do not lead to any excitations; hence no energy is deposited in the material. These nonresonant processes occur by nonlinear optical interactions in molecules. A detailed description of these processes is beyond the scope of this book, and the readers are referred to two monographs, *Introduction to Biophotonics* [Prasad, 2003] and *Introduction to Nonlinear Optical Effects in Molecules and Polymers* [Prasad and Williams, 1999]. Figure 5.2 provides a schematic representation of some of these processes.

The two-photon excitation presented in Figure 5.2 involves simultaneous absorption of two photons. The excitation initially generated can subsequently relax to produce fluorescence from the lowest excited singlet state. This excitation process is quadratically dependent on the intensity of light being absorbed, thus very localized near the focus of the light beam. This feature offers advantages as one can do point-by-point excitation by scanning the focal point of the beam in three-dimensions. For this reason, two-photon microscopy that utilizes the feature of two-photon excitation has gained wide popularity for high-resolution, three-dimensional imaging. Coherent anti-Stokes Raman scattering is another multiphoton-induced nonlinear optical process in which two incident photons of different frequencies, ν_I and ν_S, generate a photon of a new frequency $2\nu_I - \nu_S$, where the difference $\nu_I - \nu_S$ matches the frequency

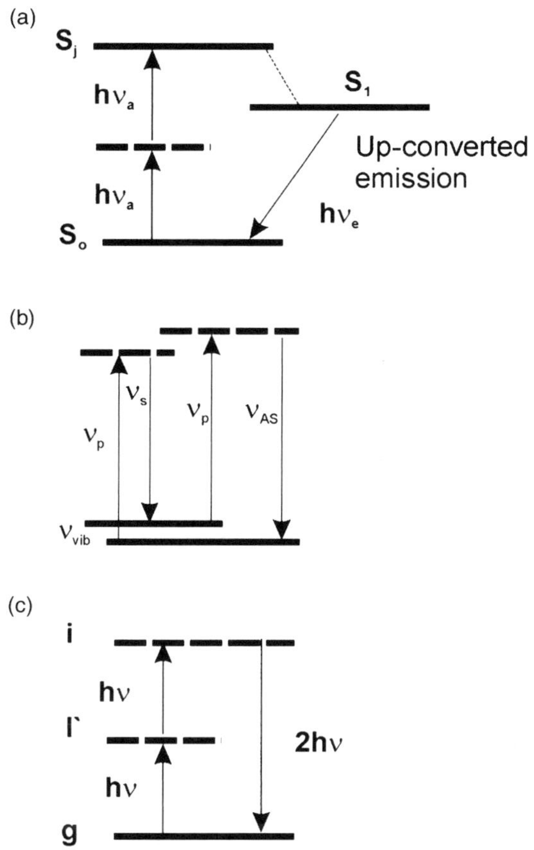

Figure 5.2. (a) Two-photon absorption and subsequent up-converted emission. (b) Coherent anti-Stokes Raman scattering (CARS). (c) Second harmonic generation processes diagrams.

of a Raman-active vibration of a constituent chemical group of the medium (a biological specimen). This process again is dependent on the higher power of the intensities of incident photons and has been utilized for high-resolution three-dimensional vibrational imaging.

Second-harmonic generation (SHG) is an example of a nonresonant nonlinear optical process in which two photons of a given frequency are combined to generate a new photon of twice the frequency (frequency doubling) without depositing any energy into the medium. This process is different from the two-photon excited fluorescence in which the two photons are actually absorbed. Also, in the case of SHG, the generated photon is always twice the frequency, whereas in the case of two-photon excited fluorescence, the emitted photon has the frequency determined by the energy gap between the excited emitting level and the ground level of the emitter.

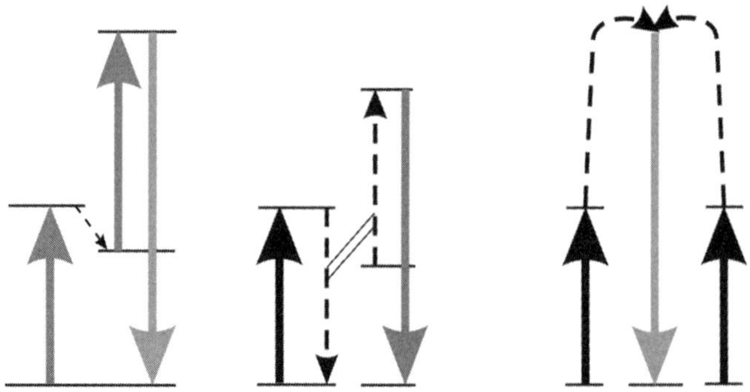

Figure 5.3. Upconversion processes in rare-earth doped nanocrystals (ESA is excited state absorption, ETU is energy transfer upconversion).

Another important type of multiphoton process, which is different from considered above nonlinear processes, involves a sequence of linear absorptions through multiple steps. An important class of emitters exhibiting these types of processes are rare-earth phosphors containing rare-earth ions such as erbium (Er^{3+}), ytterbium (Yb^{3+}), and thulium (Tm^{3+}). Examples of these processes are shown in Figure 5.3. Thus, like the two-photon excitation described earlier, these processes produce up-converted emission; that is, the emitted photon is of higher frequency (shorter wavelength) compared to the absorbed photon. However, the advantage offered by this sequential multiphoton absorption is that it does not require a high intensity, whereas a pulse laser beam is necessary for simultaneous two-photon absorption. Thus, one can use a low-cost continuous-wave laser for sequential multiphoton excited upconversion.

An important factor for optical functionalities to be useful in nanomedicine is light penetration in biological systems. As we know, the human body is not transparent, and thus visible light does not pass through the body. The light is absorbed and scattered as it passes through cells and tissues. Figure 5.4 gives a plot of light absorption and scattering in biological materials. As can be seen, maximum transmission or deepest penetration into biological materials is by light in the wavelength range of 800–1000 nm, which is the near-IR spectral range. Thus, for light-based imaging/sensing/diagnostic as well as therapy, the most effective wavelength range allowing for deeper tissue penetration is 800–1000 nm.

5.3. OPTICAL NANOPROBES

This section describes how nanoparticles can be designed to provide more effective optical functions for nanomedicine. This is the area of nanophotonics,

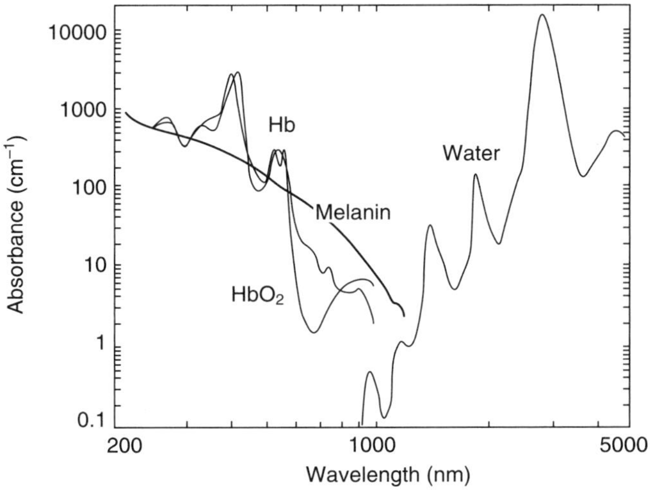

Figure 5.4. Light absorption and scattering in biological materials. (Reproduced with permission from Lim and Soter [1993].)

TABLE 5.2. Various Optical Nanoprobes

Organic dye-containing nanoparticles	Quantum dots and rods	Upconverting nanophosphors	Metallic nanoparticles

which deals with nanoscale manipulation of light excitation and control of excitation energy flow (excitation dynamics). In this section, a brief description is provided. For details, the reader is referred to the monograph *Nanophotonics* by this author [Prasad, 2004].

The major types of nanoparticles containing various optical functionalities are listed in Table 5.2. A major class is optically clear polymeric, silica, or ORMOSIL nanoparticles as well as dendritic structures containing organic dye molecules, which can strongly absorb light of a specific wavelength and fluoresce strongly (high fluorescence quantum yield) if they are to be used for fluorescence-based imaging or sensing. These dyes can either be physically entrapped within the nanoplatform volume or chemically attached to the nanoplatform. Typical examples of an ORMOSIL nanoparticle and a dendritic structure containing chemically conjugated dyes are shown in Figure 5.5. The advantage provided by the nanoplatform is that the local concentration of the emitting dye molecules is enhanced with the increased loading density of

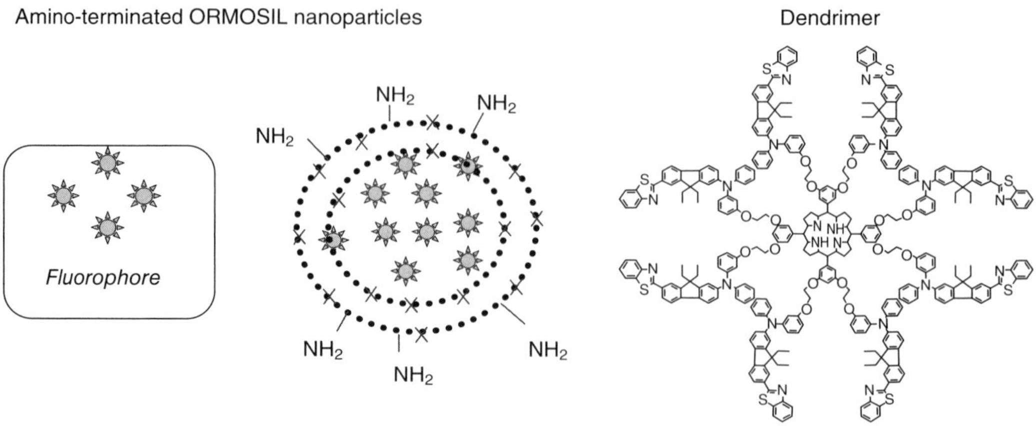

Figure 5.5. ORMOSIL nanoparticle and a dendritic structure containing chemically conjugated dyes.

the dye in the nanoplatform. In addition, an appropriate design of the surface of the nanoparticle (or the periphery of the dendritic structure) can isolate the dye from the surrounding aqueous environment of a biological medium, which acts to reduce the fluorescence efficiency of the dye. Furthermore, the encapsulating environment of a nanoplatform may protect the dye molecule from degradation.

A major problem encountered with organic dyes at higher level of loading is concentration quenching of fluorescence caused by new nonradiative pathways introduced by aggregation of dye molecules (e.g., forming dimers, trimers, etc., types of clusters of dye molecules). This is especially problematic for two-photon microscopy, discussed in Chapter 8, because the relatively weaker two-photon excitation process, discussed earlier, often requires the use of very high concentration of dye molecules. The nanophotonics concept of controlling the excitation dynamics on nanoscale (i.e., how the energy of the excited state can be appropriately channeled to enhance the desired pathway such as fluorescence) can be very valuable. We have demonstrated this concept by the design of two-photon molecules shown in Figure 5.6, which actually exhibit the opposite behavior; that is, their fluorescence efficiency is increased significantly by aggregation as shown. The reason is that the center part of the molecule contains a floppy polynuclear (fused aromatic rings; see the three-fused-ring anthracene unit in Figure 5.6) aromatic structure, which in the isolated molecular form can exhibit large torsional oscillation leading to nonradiative pathway to dissipate the excitation energy. When these molecules aggregate at high concentrations, they form stacks whereby their torsional motion is significantly hindered, thereby reducing the nonradiative process and thus enhancing the radiative channel of energy flow to produce significantly brighter fluorescence. We have used these dye aggregates incorporated

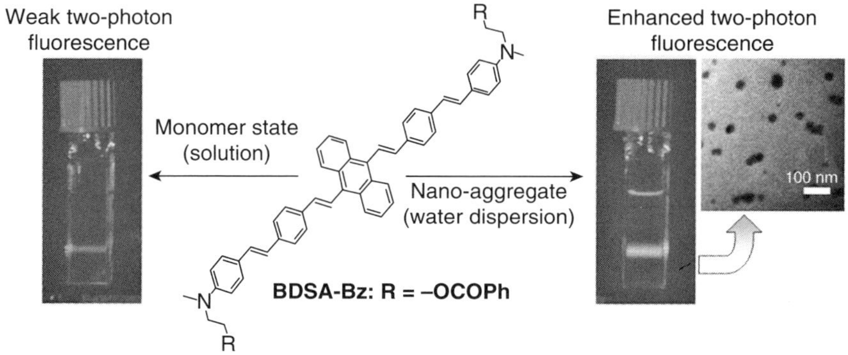

Figure 5.6. Aggregation enhanced two-photon dye emitter.

in ORMOSIL structures for two-photon imaging as well as for two-photon photodynamic therapy of cancer.

An important class of nanoemitters, which show a true manifestation of nanosize confinement, are nanocrystals of inorganic semiconductors. These nanocrystals, with sizes typically ranging from 1 to 10 nm, exhibit manifestations of properties of electrons and holes confined in a three-dimensional box and are called quantum dots. Unlike a bulk semiconducting material, which contains an electron-filled lower-energy valence band and an empty conduction band, whereto an electron can be photoexcited, the quantum dots contain discrete (quantized) energy levels. This behavior is shown in Figure 5.7.

In a bulk semiconductor, the gap between the top of the valence band and the bottom of the conduction band, called the bandgap, is fixed. In contrast, the gap between the highest quantized level of the valence band and the bottom quantized level of the conduction band is dependent on the size of the quantum dot. Thus, the optical transition (absorption and the corresponding emission) becomes size-dependent, with the smaller-sized quantum dot emitting at a shorter wavelength (toward the blue) while the larger size emits at a longer wavelength (toward the red). There is another manifestation of the nanoscale confinement: Silicon in the bulk form is an indirect bandgap semiconductor, in which case (as shown in Figure 5.7) the bottom of the conduction band is not exactly aligned with the top of the valence band. In this case, radiative transition such as light emission by return of the excited electron from the bottom of the conduction band to the top of the valence band is not strictly allowed. This makes light emission in bulk silicon very weak. In contrast, silicon nanocrystal quantum dots that have discrete localized states show good light emission, the color (wavelength) of which is dependent on the size of the silicon quantum dots. This size-dependent feature of silicon quantum dots is shown in Figure 5.8.

Quantum dot nanoemitters have been extensively used in recent years for fluorescence (luminescence) microscopy as well as nanoplatforms for delivery

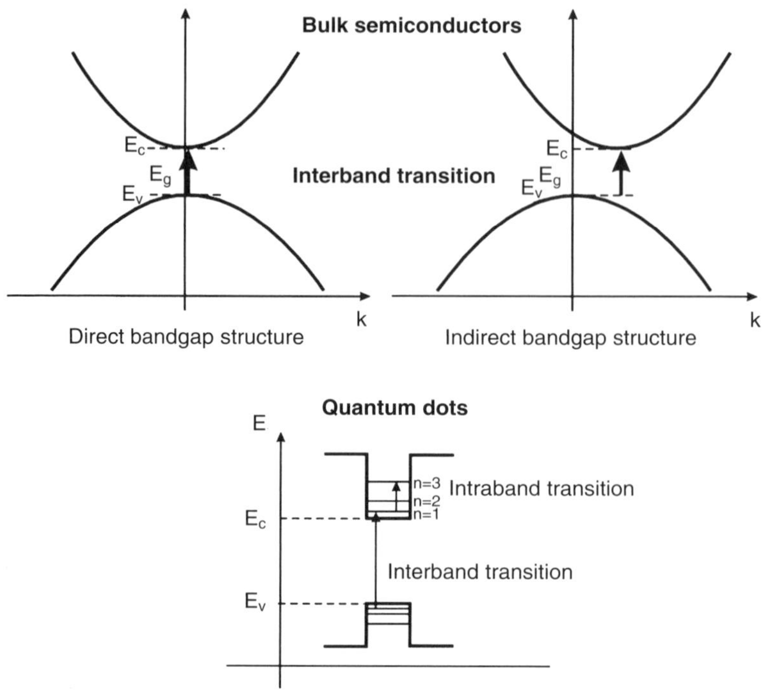

Figure 5.7. Energy diagrams for bulk semiconductor and quantum dot nanoparticles.

Figure 5.8. Light emission of silicon QDs with increasing size.

of therapeutics with its luminescence being utilized for optical tracking. Quantum dots offer a number of advantages for bioimaging. With appropriate selection of the semiconducting materials and their nanosizes, a wide spectral range can be covered for bioimaging. These quantum dots can be excited at the same wavelength, even though their emissions are at widely different wavelengths. For bioimaging, some major advantages offered by quantum dots compared to organic dyes are [Prasad, 2003]:

- Quantum dot emissions are considerably narrower, typically 20–30 nm, compared to organic fluorophores which exhibit broad emissions (>100 nm). Thus, the complication in simultaneous quantitative multichannel detection, posed by cross-talks between different detection channels derived from spectra overlap, is significantly reduced. Therefore, imaging multiplexing can readily be accomplished.
- The quantum dots do not readily photobleach. In contrast, chemical degradation by photobleaching is a major concern in the case of organic dyes, which severely limits repetitive imaging or multiple scans in imaging.

Quantum rods (QR) are rod-shaped nanocrystals of semiconductors, which have the advantage of providing a larger size surface area for attaching multiple units. In addition, they show polarized emission, which can be useful for polarization-based microscopy.

Another kind of nanoemitter useful for light activation and optical imaging are rare-earth containing nanophosphors described above. These nanophosphors exhibit sequential multiphoton excited up-converted emission. A highly suitable nanophosphor is nanocrystals (10–20 nm in size) of $NaYF_4$ co-doped with Yb^{3+} and Tm^{3+} ions that exhibit upconverted emission at 802 nm when excited at 975 nm. Thus, both the excitation and the emission wavelengths are in the optical window of maximum transparency in biological media. As discussed in a later chapter dealing with bioimaging, we have successfully used these nanophosphors for deep tissue *in vivo* imaging.

Another important class of optical nanoprobes used for sensing, imaging, and therapy are metallic nanostructures, also called plasmonic, or nanoplasmonic structures. They also exhibit strong manifestation of nanoscale confinement. The two types of plasmonic structures useful for nanomedicine are shown in Figure 5.9. In each case, new optically generated excitation takes

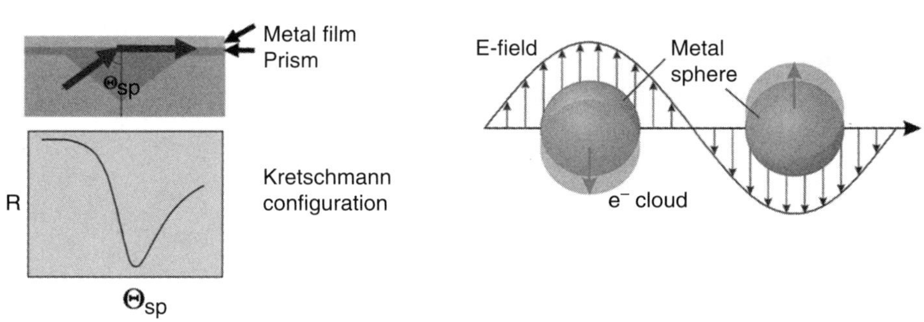

Figure 5.9. Nanoplasmonic structures and excitations.

place at the interface between metal and the surrounding. These excitations, called surface plasmon resonance (SPR), consist of oscillations of the conduction electrons in the metal at the frequency of electromagnetic (optical) wave generating them. These SPR excitations do not exist in bulk metal. The first plasmonic structure in Figure 5.9 is that of a metal film, about 50 nm thick, deposited on a dielectric. In this case the SPR excitation wave is delocalized (spread in two dimensions) over the entire film surface. Hence, their propagation is characterized by a wave vector, **k**. The relation between the frequency (w) of SPR oscillation and the wave vector (**k**), called dispersion, is different from that of a free-space propagating light wave. In order to compensate for this mismatch, one has to use an optical geometry such as provided by the prism coupling, also known as the Kretschmann configuration. In this configuration the light incident at a specific angle, θ_{sp}, along the film lateral dimension provides the matching of the wave vector **k** and thus couples to produce the SPR excitation. Hence, at this angle no light is reflected from the metal surface. The coupling angle θ_{sp} is very sensitive to any species adherent to the metal film, and this feature is utilized in SPR biosensors discussed in a later chapter.

The other class of plasmonic structures consists of metallic nanoparticles, nanoshells, and nanorods of dimensions <100 nm. Figure 5.9 shows a metallic nanosphere. In this case, light of a specific frequency, dependent on the nature of the metal, produces plasmon excitation. This surface plasmon resonance is localized around the metal nanoparticle and for this reason is also called localized surface plasmon resonance (LSPR). For gold nanoparticles (of size ≤50 nm), this resonance is around ~520 nm, where gold can absorb and scatter light strongly, making a colloidal dispersion of gold bulk red. For smaller particles the absorption is strong, which has been used for photothermal therapy. For large particles (≥100 nm) the plasmonic band enhanced scattering dominates; this feature has been utilized for dark-field imaging where the scattering at the plasmonic resonance wavelength provides the contrast. In the case of a metallic nanorod, two types of localized surface plasmon resonances exist. One along the transverse cross section, also called transverse plasmon resonance, is nearly the same frequency as the case of spherical nanoparticles. The other oscillation, called the longitudinal SPR, is along the length of the rod. The longitudinal SPR frequency is dependent on the aspect ratio (length to width) of the nanorod, and they can be changed from the visible to IR by increasing the aspect ratio. This feature has been utilized for multiplexed dark-field imaging.

In the case of a nanoshell consisting of a metallic coating along a dielectric (such as a polystyrene or a glass microbead), the plasmon resonance in the nanoshell can also be varied by varying the thickness of the shell.

5.4. MAGNETIC FUNCTIONALITY

As listed in Table 5.1, magnetic functionality plays a vital role in nanomedicine by providing multiple functions ranging from imaging, to sensing, to bioseparation, to drug delivery as well as being useful for magnetic therapy. Magnetic functionality involves magnetic interactions induced by paramagnetic ions, magnetic nanoparticles or nanomagnets. The basic underlying structure of such magnetic elements involves unpaired electron spin that imparts to them their magnetic characteristics.

Magnetic functionality plays a major role in modern diagnostics, especially magnetic resonance imaging (MRI), which allows one to observe morphological details of specific tissue region. MRI is routinely used for a variety of disease or injury profiling. MRI utilizes the nuclear magnetic resonance of spin states of hydrogen nucleus (protons) present in water. A contrast agent that is a paramagnetic material (ion or a magnetic nanoparticle) is introduced, which can localize in the tissue region of interest where it changes either the spin–lattice relaxation (T_1 – contrast) or the spin-spin relaxation (T_2 – contrast) times of protons. This change provides the contrast to differentiate the tissue of interest from others. Specifically, there is a magnetic dipole–dipole interaction between the magnetic moment of paramagnetic atoms and the much smaller magnetic moments of the protons in neighboring water molecules. Random fluctuations in these dipolar interactions affect both the longitudinal T_1 and the transverse T_2 relaxation times of the water protons. Contrast agents function by changing the T_1 and T_2 relaxation times of nuclei in the target tissue. Thus they are described as either T_1 agents or T_2 agents, depending on whether the relative change in relaxation times is greater for T_1 or T_2. The contrast provided by the contrast agent is measured by the quantity relaxivity (r_1 or r_2), which is the change in $1/T_1$ or $1/T_2$ of protons per concentration of the contrast agent [Geraldes and Laurent, 2009].

Typical paramagnetic ions used as contrast agents are manganese (Mn^{2+}), iron (Fe^{3+}), and gadolinium (Gd^{3+}). A nanoparticle formulation containing paramagnetic ions provides the following advantages:

- Targeted delivery to the biological site to maximize contrast.
- Enhanced local concentration of the paramagnetic ions to increase the relaxivity and thus enhance sensitivity.
- Reduced toxicity of the paramagnetic ion.

An example of a nanoparticle formulation containing Gd is gadolinium oxide nanoparticles; another is where Gd^{3+} ions are doped in a $NaYF_4$ nanocrystal. In both of these nanocrystals, only the gadolinium ions near the surface of the nanoparticles can affect the relaxation times of proton in water surrounding them. Another type of formulation prepared in our laboratory utilizes Gd^{3+} ions coupled to a polymeric micelle nanoparticle surface through a chelating organic ligand group; this is shown in Figure 5.10 and has successfully been utilized for combined MRI and optical imaging.

Another type of contrast enhancer that has also been used for a variety of applications in nanomedicine is a magnetic nanoparticle made from ferromagnetic materials [Furlani, 2010]. The most commonly used material for

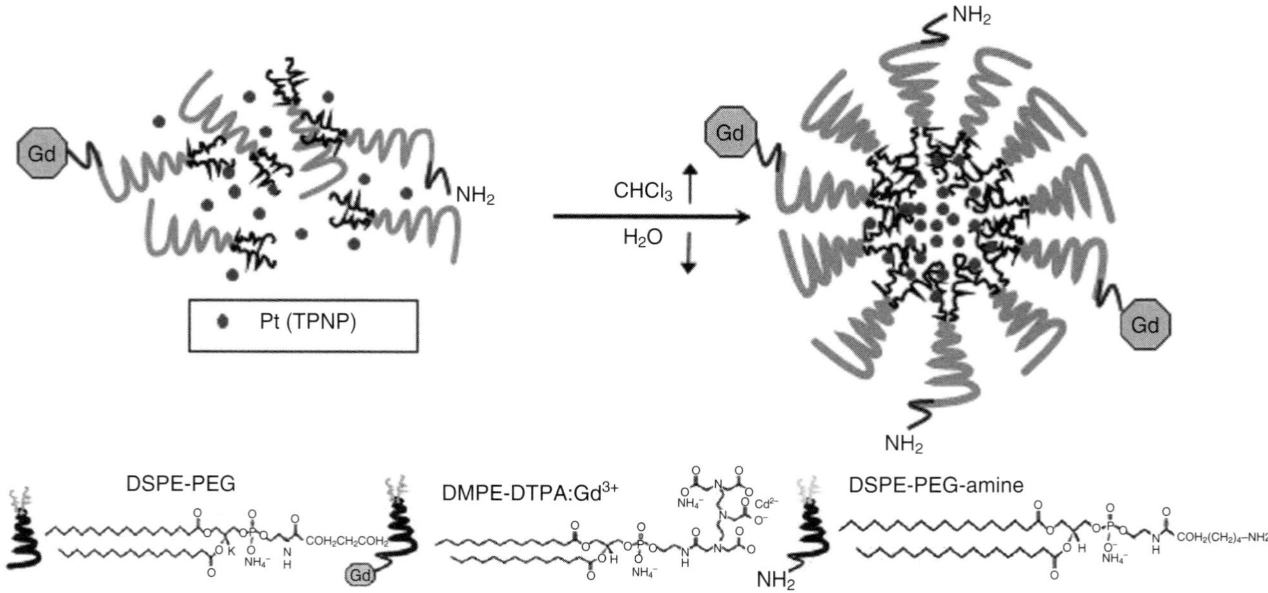

Figure 5.10. Synthesis of phospholipid nanomicelles encapsulating phosphorescent molecule of Pt(TPNP) and surface functionalized with gadolinium. (Reproduced with permission from Kumar et al. [2010b].)

bioapplications is iron oxide (Fe_3O_4), which is ferromagnetic in the bulk form and thus possesses a large magnetization. However, when reduced to nanosize (such as 30 nm), such particles do not exhibit ferromagnetization but instead are superparamagnetic, meaning that they still show much stronger magnetic behavior than that of a typical paramagnetic material. This form of iron oxide nanoparticles (Fe_3O_4), also called magnetite, has been utilized for bioseparation, biosensing, and magnetic-field-assisted drug and gene delivery, as well as for magnetic therapy of cancer [Pankhurst et al., 2003, 2009; Berry, 2009; Berry and Curtis, 2003]. These particles have many advantages for such applications. They can be synthesized in sizes that range from a few nanometers to microns with a narrow size distribution, which makes them ideal for probing and manipulating biomaterials such as proteins (5–50 nm), viruses (20–450 nm), genes (2 nm wide and 10–100 nm long), or whole cells (10–100 μm) [Pankhurst et al., 2003]. In addition, they are nontoxic and well-tolerated by living organisms when properly synthesized and functionalized. They can be custom-tailored with appropriate surface treatments that enhance biocompatibility and enable coating with affinity biomolecules for highly specific binding to a target biomaterial [Pankhurst et al., 2009; Berry, 2009; Arrueboa et al., 2007; Majewski and Thierry, 2007; Safarik and Safarikova, 2002].

The ability to selectively bind magnetic particles to target biomaterials combined with their response to a magnetic field is a key motivation for their use for many application. The particles can be manipulated using an external magnetic field; and once bound to a biomaterial, they enable noninvasive control of it for steering, sorting, and so on. Furthermore, the unique superparamagnetic behavior of magnetic nanoparticles provides flexibility in their use for biotransport. Specifically, superparamagnetic particles are easily magnetized by an applied magnetic field, but revert back to an unmagnetized state instantaneously once the field is removed. Thus, they experience a magnetic force in a magnetic field when their induced dipole moment couples to the local field gradient, but the force reduces to zero once the field is removed, which facilitates biosorting and separation processes [Furlani, 2010]. In addition, the low intrinsic magnetic susceptibility of most fluids and biomaterials provides substantial contrast between labeled and unlabeled material, which enables efficient magnet coupling to immersed labeled biomaterial, which, in turn, enables a high degree of selectivity and control.

A colloidal suspension of nanoparticles is utilized for most applications. For bioseparation, the magnetic nanoparticles contain capture groups that can detect and bind to (magnetically label) a targeted biological species. Then a magnet can be used to attract and thus collect the magnetically labeled biomaterial, thereby enabling magnetic separation or sorting for various biomedical applications.

Drug and gene delivery utilize magnetic nanoparticles as nanocarriers where field-directed targeting to a biologic site is achieved by using a magnetic field. Figure 5.11 shows an example of magnetic field enhanced cellular uptake of the Fe_3O_4 magnetic nanoparticles.

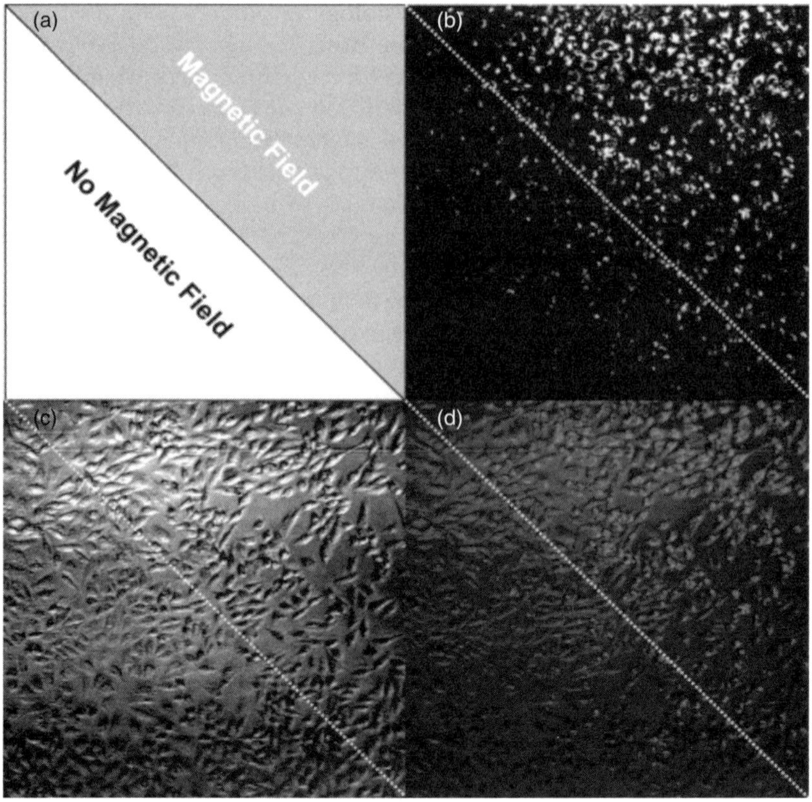

Figure 5.11. Magnetically guided two-photon excited fluorescence microscopic images of cells stained with PEG-PE micelles coentrapping two-photon fluorescent dye (H2) and magnetic nanoparticles. A magnet was applied on the top right triangular area of the sample. (a) Schematic illustration of the magnetic field applied area. (b) Fluorescence from H2-labeled nanoparticles. (c) Transmission. (d) Combined images for fluorescence and transmission. (Reproduced with permission from Zheng et al. [2007].)

Finally, magnetic nanoparticles can be directly used for cancer therapy. A widely investigated mechanism of cancer therapy using magnetic nanoparticles involves heating produced by coupling these particles to an AC magnetic field. Another mechanism, demonstrated by our group, utilizes alignment of these magnetic nanoparticles in a DC magnetic field such as the ones used in MRI machines; and when in large concentrations, this aligned nanoparticle chain mechanically disrupts the cellular structure.

5.5. THERMAL FUNCTIONALITY

This functionality can be utilized both for diagnostics and therapy. For diagnostics, thermal mapping of suspected diseased sites (such as tumor) can provide useful information on the physiological activities. We have recently described how optical techniques can be utilized to obtain thermal mapping [Kachynski et al., 2005]. This method has an advantage over regular thermal mapping using an infrared camera, because it provides thermal images with optical resolution. The principle utilizes anti-Stokes emission on the shorter wavelength side of the excitation wavelength. This anti-Stokes emission originates from the thermally excited vibrational levels. Nanoprobes for this purpose can include organic dyes, in which case these vibrational levels are molecular vibrations of the organic molecule. Alternatively, the nanoparticles can be rare-earth ions containing nanoemitters, in which case the thermally generated vibrational excitation involve lattice vibrations. By judicious selection of the energy gap between the emitting level and the excitation energy, a resolution of temperature difference of 0.1°C can be achieved.

Thermal functionality has also been utilized for thermal therapy [Berry, 2009; El-Sayed, 2010]. This can be direct thermal therapy such as the thermal destruction destroying of tumor thermally. Alternatively, it can be used to sensitize the cells for another therapeutic mode. For both these purposes a number of methods are available. Nanoparticles can be designed to couple to an alternating magnetic field, rf field, or light to absorb energy and effect local heating. Light absorption has the advantage that through photothermal effect produced by nonradiative processes discussed in Section 5.2, one can highly localize heating to the tumor region, thus minimizing any collateral damage to the neighboring healthy tissues. A major issue to deal with is light penetration to reach the tumor (or diseased site). For this reason, a nanoparticle absorbing in the near-IR region (800–1000 nm) of maximum optical transparency in biological systems is more suitable. West and Halas [2000] have used core-shell metallic structures (nanoshells) appropriately designed to have plasmon resonance (discussed in Section 5.3) in this region for cancer therapy. El-Sayed and co-workers [Huang et al., 2006] have used gold nanorods with longitudinal plasmon resonance (also discussed in Section 4.3) in this optical region for photothermal cancer therapy.

5.6. RADIOACTIVE FUNCTIONALITY

Nanoparticles can be designed to contain radioactive functionality whereby they either contain a radioactive nucleus (radionucleotide) or a nucleus that can capture higher-energy radiation or an elementary particle (such as a neutron) to become radioactive. These nanoparticles again provide the advantage of increased local concentration by using targeting and high-density loading. A major function performed by these radioactive nanoparticles is in medical imaging, although new modalities of cancer therapy using neutron capture are promising direction.

The two examples of radio imaging provided here are positron emission tomography (PET) and single-photon emission computed tomography (SPECT). PET is a medical imaging technique in which a short-lived positron-emitting radioactive tracer isotope is localized at the tissue site of imaging. The emitted positron annihilates with an electron in the surrounding medium to produce a pair of γ rays that propagate in the opposite directions and are imaged by a γ-ray sensitive camera [Bailey et al., 2005]. A computed tomographic method is used to construct a three-dimensional image. A number of radionucleotide isotopes have been utilized which are incorporated in a biologically active molecule. An example is the ^{18}F isotope with half-life of 110 minutes; it is incorporated in fluorodeoxyglucose, which has been extensively used for mapping of metabolic functional imaging [Anand et al., 2009].

In PET imaging, two coincident 511-keV gamma photons are produced upon annihilation of the emitted positron by a nearby electron [Bailey et al., 2005]. Because these photons are always coincident, three-dimensional tomographic information can be generated about isotope concentration, and location by simultaneous detection of the gamma photons with a cylindrical detector. Here the imaging signal is independent of the tissue depth, therefore ideally suited for imaging deep-sitting tumors *in vivo*.

Indeed, PET imaging is used routinely in clinical practice for monitoring cellular/molecular events *de novo* in cancer cells, as well as during pharmacological and radiation therapy [Muehllehner and Karp, 2006]. The ideal PET probe for cancer detection should selectively target tumor cells *in vivo*, bypassing normal organs. In the realm of molecular imaging, the most popular PET probe used clinically is 18-fluorodeoxyglucose (FDG), which demonstrates greater accumulation in tumors owing to the elevated metabolic rates of neoplastic cells [Muehllehner and Karp, 2006]. However, since all cells metabolize glucose, FDG is not specific for cancer cells. Therefore, in order to enhance tumor targeting, newer PET probes containing ligands specific for tumor-specific receptors have been developed, an example of which includes ^{64}Cu-octreotide (^{64}Cu-OC) for targeting somatostatin-receptor-positive tumors [Wadas et al., 2008]. However, these newer-generation PET probes also suffer from slow blood clearance and poor bioavailability, and they can lead to images with high background signals.

Incorporating PET probes into a nanocarrier improves tumor-specific accumulation of such probes. As compared to the molecular PET probes, the nanoparticle-based PET probes allows for the delivery of multiple PET probes per nanoparticle, which might make it possible to detect target cells at the single-nanoparticle level.

In nanoparticle formulation, we have used ^{124}I isotopes (with half-life of 4.2 days) containing nanoparticles for PET imaging [Kumar et al., 2010a]. These isotopes have been incorporated in an ORMOSIL nanostructure, discussed in Chapters 3 and 4. For radiolabeling, the Bolton–Hunter method-based strategy is shown in Figure 5.12. In this method, the free amino groups on the nanoparticle surface form a covalent linkage with the *N*-hydroxysuccinimide (NSH) ester group linked with ^{124}I.

SPECT imaging also uses a radioactive nucleus, but the one that emits gamma radiation as opposed to a radionucleotide for PET, which emits posi-

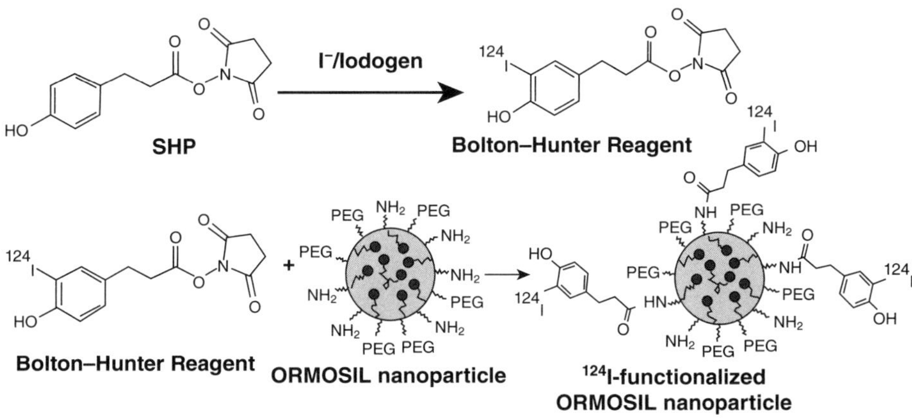

Figure 5.12. Synthesis of ^{124}I-labeled DY776-conjugated ORMOSIL nanoparticles via the Bolton–Hunter method (Reproduced with permission from Kumar et al. [2010a].)

trons [Costa, 1990]. The imaging is acquired from multiple angles (multiple projections) using a γ-ray sensitive camera. Then a tomographic reconstruction software is used to produce a three-dimensional image. However, the image resolution obtained by SPECT (typically ~1 cm) is significantly lower than that obtained by PET, which provides significantly localized information. Nevertheless, the cost of a SPECT scan is significantly lower than that for a PET scan, partly because SPECT utilizes a radionucleotide (e.g., ^{125}I) with significantly longer half-life. Again, an ORMOSIL nanoparticle formulation strategy, similar to that for PET, shown in Figure 5.12, can be used but now incorporating ^{125}I instead of ^{124}I.

There are several limitations to SPECT diagnostic images that could be overcome by nanotechnology. First, the SPECT images are limited by the specific activity of radiolabeled agent. In CT, the washout rate of the radio-opaque medium (iodine) results in poor imaging of the vasculature and soft tissue. Hence, higher specific activity of radiolabeled agent and the radio-opaque contrast medium that provides high tissue and blood vessel contrast in CT image with slow blood clearance will be extremely useful in correlating the SPECT and CT images of the small animals. The nanoparticle technology enables the incorporation of radioactive and radio-opaque media coupled with targeted optical imaging. Furthermore, the ability to surface-modify the nanoprobes to increase circulation time and target specific sites is also potentially achievable. Also, the correlation of the SPECT/CT image with histology of the plaques is of significant interest, but is currently difficult and requires multiple steps. The high-resolution imaging technique offered by the proposed nanoprobes would provide this direct correlation with histological information.

Neutron capture therapy (NCT) is based on irradiation of the nonradioactive species with epithermal/thermal neutrons, which, in turn, decays into a more stable nuclei with the emission of high-energy species such as alpha particles, Auger electrons, long-range prompt gamma rays, internal conversion electrons, and X rays [Coderre and Morris, 1999; Salt et al., 2004]. This therapy is discussed in Chapter 13 for cancer treatment. The ability of such high-energy species to cause potent localized damage to cells/tissues in close proximity forms the basis of NCT of cancer and other diseases. Boron and gadolinium atoms, owing to their high neutron capture cross section, have been traditionally used as agents for NCT [Salt et al., 2004]. Therefore, multimodal nanoparticles containing a payload of boron or gadolinium atoms, which can be specifically targeted to diseased sites, are expected to play a critical role in NCT applications, where the resulting toxicity is confined to the diseased site with minimal damage to surrounding healthy cells/tissues. Finally, it is also possible to simultaneously monitor the NCT action in real time using whole body PET or SPECT/CT scans focusing on the diseased site (e.g., cancer).

5.7. BIOLOGICAL FUNCTIONALITY

The necessary biological functionalities for a nanoplatform are: (i) biocompatibility to make the nanoparticles safe, (ii) biodistribution controlled by enhanced circulation, (iii) targeting to enhance preferential accumulation at the biological site of interest, and (iv) clearance of nanoparticles from the body, eventually. These functionalities can be introduced by appropriate choice of materials composition, surface coating, and surface functionalization. Introduction of biocompatibility requires features such as a composition that does not produce toxic species and does not illicit any immune response. Some discussions of strategies to produce nanoplatforms with this functionality have already been discussed in Chapter 3. They involve choice of material composition, surface coating with biologically compatible layers, and control of surface charge.

Polyethylene Glycol (PEG). PEG is a classical polymer with uncharged, hydrophilic, and nonimmunogenic properties, which can be physically adsorbed or covalently attached to the surface of nanoparticles. Furthermore, PEG has been shown to substantially reduce nonspecific interactions with proteins through its hydrophilicity and steric repulsion effects, thereby reducing opsonization [Alexis et al., 2008]. The chain length, shape, and density of PEG on the nanoparticle surface are the main parameters affecting the nanoparticle's surface hydrophilicity and phagocytosis and, thus, the biodistribution of the nanoparticles in the systemic circulation. Introduction of PEG on the surface also prolongs the circulation of the nanoparticles through the bloodstream when the nanoparticles are introduced by a systemic injection that can accelerate reaching of the nanoparticles to the target site.

Active targeting (discussed in Chapter 6) requires incorporation of targeting groups that recognize the biomatter expression (e.g., antigens) at the targeted biological sites. This is often done by coupling a targeting group (such as an appropriate antibody) to the surface of the nanoparticle.

Excretion of the nanoparticle may require it to be biodegradable either intrinsically or by the enzymatic degradation at the targeted site, or to be able to be excreted in urine or feces. Chemical makeup of the nanoparticle and size provide important control parameters.

5.8. MULTIFUNCTIONALITY

There is a growing interest in the development of "fusion technologies," in which two or more medical imaging techniques are combined to provide complementary information. This theme of multimodal imaging is further covered in Chapter 8, entitled "Multimodal Biomedical Imaging." For example, a CT image can provide information about tissue structure, while PET imaging can complement that data with metabolic process information. Molecular imaging is another emerging field, in which the modern tools of molecular and cell biology are integrated into advanced imaging tools that allow noninvasive imaging for early detection of disease onset, or real-time monitoring of disease treatment progress. Modern drug discovery is also increasingly reliant on molecular imaging in small animals and even in humans to advance and streamline the drug development process. The use of PET or MRI *in vivo* imaging enables easier demonstration of drug efficacy, or the lack thereof. This area also can benefit from the development of custom delivery platforms for drugs and imaging contrast agents, using molecular markers for targeting.

To help address this emerging need, we as well as other laboratories are developing nanoparticle-based probes that contain multiple contrasting agents for simultaneous use in different imaging modalities (e.g., optical, MRI, PET, SPECT, and CT imaging). Correlation of these images using a single nanoprobe will accelerate the diagnostic process while reducing stress on the patient. Another advantage of these "multimodal" nanoprobes is the ability to target specific molecular markers. This capability is increasingly important, as biochemical research becomes more successful in characterizing specific molecular factors that accompany a certain disease. These nanoparticle-based contrast agents can be surface-modified to carry various biotargeting agents (e.g., antibodies) to target specific disease markers to achieve earlier detection of diseases. Finally, these nanoprobes have the potential of providing real-time imaging, thus providing the clinician with invaluable information that can be used in designing an individually tailored therapeutic regimen for the patient.

It is anticipated that developing these targeted multimodal nanoparticle platforms will advance the sensitivity and selectivity of medical imaging technology, thus allowing for early detection of cancers or other diseases. The nanoplatforms provide the next generation of contrast agents for diagnostic systems, thereby increasing the abilities to do the following:

- Perform a direct comparison of histological analysis with whole-body imaging information, by the addition of optical imaging probes along with contrast agents for current diagnostic imaging.
- Improve early diagnosis capability for each imaging system by functionalizing nanoprobe surfaces with biotargeting agents.
- Obtain more detailed information, concerning the anatomical and physiological state of a patient at a given moment in time (real time information), as opposed to sequential information (the current state of the art in diagnostic imaging).
- Provide tailor-made treatment protocol for an individual at clinical level, by assessing different therapeutic approaches using improved real-time imaging with the nanoparticle platforms.
- Accelerate drug discovery through the use of these multimodal nanoplatforms for studying the efficacy of treatment protocol.
- Reduce patient stress related to the multiple steps necessary for a complete diagnostic workup using the current imaging technology.

Multimodal nanoparticles of several nanometers in dimension circulate freely throughout the circulatory systems of the body. Furthermore, in the case of model tumors in animals, nanoparticles have shown preferential localization in the tumor, probably due to the increased vascularity of a tumor tissue. This use of specially designed nanoprobes will facilitate enhanced contrast medical imaging of selective biological sites. These nanoprobes will shape the future of medical imaging—not only as diagnostic tools, but also for real-time monitoring of drug action coupled with therapeutic action. These are theranostic nanoparticles.

Figure 5.13 shows a schematic illustration of an approach used by us to produce multifunctional nanoparticles. As can be seen from the scheme, this platform is based on the use of precursor molecules containing covalently linked imaging probes (designated as **X** in the scheme). ORMOSIL nanoparticles incorporating these imaging probes (**X**) are synthesized by polymerization of these precursors within the nonaqueous core of microemulsion using the surfactant Tween-80. This microemulsion based sol–gel synthesis of ORMOSIL nanoparticles has been described in Chapter 4. The fluorophores, either organic dye molecules or quantum dots, can be encapsulated at the core of these optically transparent nanoparticles. It is important to stress that precursors containing multiple imaging probes (different **X** components in the scheme) for different modalities can be used simultaneously, as co-precursors to prepare multimodal nanoparticles. Optimization of a protocol for preparation of monodispersed nanoparticles using multiple co-precursors is an important step in the synthetic route.

As illustrated in Figure 5.13, the multimodal ORMOSIL nanoparticles can be synthesized with free amino groups on the surface. These free groups can also be used to attach circulation-enhancing polymer polyethyleneglycol

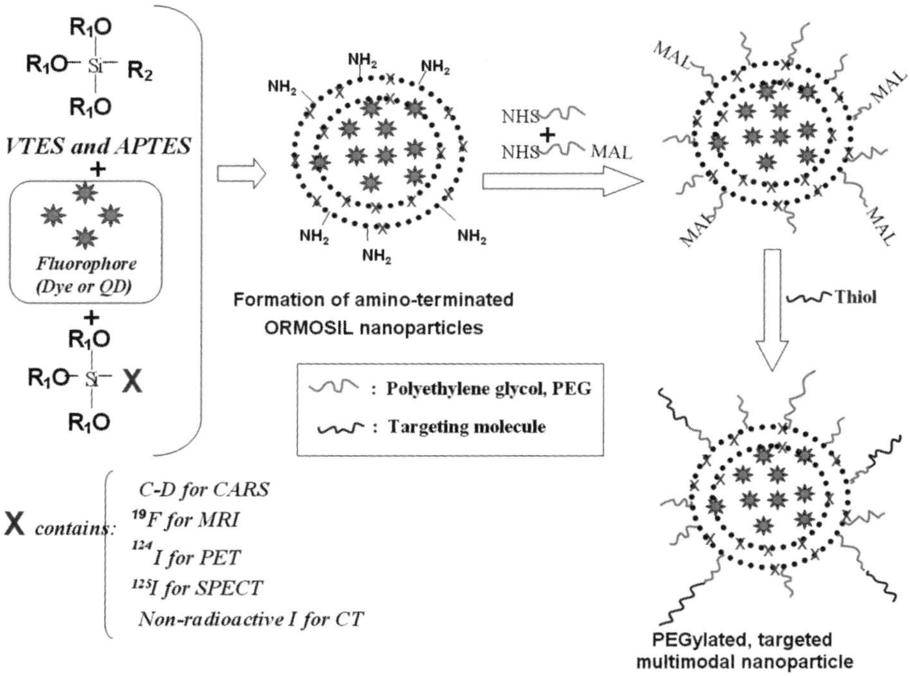

Figure 5.13. Design strategy of targeted ORMOSIL particles containing various contrast agents.

(PEG), as well as various biotargeting molecules, by a simple scheme as shown in Figure 5.13. Briefly, upon reacting these amino-terminated nanoparticles with a combination of *N*-hydroxysulfosuccinimide-PEG (NHS-PEG, 80%) and NHS-PEG-maleimide (NHS-PEG-MAL, 20%), the high-affinity binding of NHS with the primary amines results in the nanoparticle surfaces being saturated with PEG molecules (PEGylation), with a fraction (20%) of the PEGs having the reactive maleimide (MAL) groups. These maleimide groups can then be used to covalently bind thiolated targeting molecules. Biotargeting enables preferential accumulation of the nanoparticles to desired sites. Tumor targeting *in vivo* can help in enhanced accumulation of these diagnostic nanoparticles at neoplastic sites by several orders of magnitude, resulting in generation of detectable signal from very small tumor masses. This can enable a detection of the disease at very early and incipient stage. These nanoparticles can also be attached with the apoptotic marker Annexin V, for the detection of apoptotic cells at tumor sites as well as within atherosclerotic plaques.

Nanoparticle-based, multifunctional nanoprobe technology offers many advantages for medical imaging. These include:

- *Enhanced Specificity.* Nanoprobes can provide selectivity of imaging for a specific organ or a specific diseased site (e.g., tumor location). Specific targeting groups can be attached to the surface of nanoprobes in order to increase their concentration at the desired site. This technology makes it possible to create an image based on biorecognition of a molecular signature, which can facilitate the detection of disease at an early (or even pre-onset) stage. Again, the nanoparticle surface can be functionalized to attach specific biorecognition molecules that detect specific sets of enzymes or proteins produced even at an early stage.

- *Enhanced Sensitivity.* Nanoparticles can be loaded with specific molecular probes that will increase the differential signal, thereby increasing the signal to noise. In addition, new probe molecules that are not intrinsic to the body can be introduced, in which case imaging against a zero background can be obtained. An example is the radioactive isotopes for PET imaging.
- *Multifunctional Capabilities.* The nature of nanoparticles allows for synthesis of multifunctionality into a single nanoplatform. This includes combinations of multimodal imaging, drug delivery, diagnostics, and therapeutic capabilities. When coupled with mechanisms of external activation (e.g., magnetic and optical fields), these multimodal nanoparticles can themselves be used in novel therapeutic approaches with the necessary probes to track their progress in real time.

HIGHLIGHTS OF THE CHAPTER

- The various functionalities for nanomedicine applications are based on physical (optical, magnetic, thermal, radioactive) phenomena and chemical properties of biomolecules and include optical imaging, sensing, light-activated therapy, MRI imaging, magnetic separation, magnetic therapy, radio imaging, PET, neutron capture therapy, targeting, enzymatic release, and so on.
- The optical functionalities utilize either light scattering or optical transitions between two discrete energy states of a nanostructure and involve the phenomena of single-photon absorption and emission, fluorescence resonance energy transfer (FRET), Raman scattering, and multiphoton optical processes (multiphoton absorption, up-conversion) including nonlinear optical processes (second-harmonic generation, coherent anti-Stokes Raman scattering, etc.).
- Various optical nanoprobes can be designed to provide more effective optical functions for nanomedicine. Among them are organic dye-containing nanoparticles, quantum dots and rods, up-converting nanophosphors, and metallic nanoparticles.
- Typical examples of organic dye-containing nanoparticles are ORMOSIL and dendritic structures containing chemical conjugated dyes. A major

problem encountered with organic dyes is concentration quenching of fluorescence at higher level of loading and bleaching.

- Quantum dot emissions are considerably narrower, typically 20–30 nm, compared to organic fluorophores that exhibit broad emissions (>100 nm) and could be very efficient in application for multichannel detection. The quantum dots do not readily photobleach.
- Due to surface plasmon resonance (SPR), metallic nanostructures are utilized for multiplexed dark-field imaging and for SPR biosensors.
- Magnetic functionality, especially magnetic resonance imaging (MRI), plays a major role in modern diagnostics. A nanoparticle formulation for MRI containing paramagnetic ions provides some advantages, like (i) targeted delivery to the biological site to maximize contrast, (ii) enhanced local concentration of the paramagnetic ions to increase the relaxivity and thus enhance sensitivity, and (iii) reduced toxicity of the paramagnetic ion.
- Drug and gene delivery utilize the magnetic nanoparticles as the nanocarriers where targeting to a biologic site is achieved by using a magnetic field.
- Thermal functionality can be utilized both for diagnostics and therapy. For diagnostics, thermal mapping of suspected diseased sites (such as tumor) can provide useful information on the physiological activities. This functionality has also been utilized for thermal therapy. This can be direct thermal therapy such as the thermal destruction of tumor. Alternatively, it can be used to sensitize the cells for another therapeutic mode.
- Nanoparticles can be designed to contain radioactive functionality whereby they contain either a radioactive nucleus (radionucleotide) or a nucleus that can capture higher-energy radiation or an elementary particle (such as a neutron) to become radioactive. These nanoparticles again provide the advantage of increased local concentration by using targeting and high-density loading. A major function performed by these radioactive nanoparticles is in medical imaging [positron emission tomography (PET) single-photon emission computed tomography (SPECT)], although new modalities of cancer therapy using neutron capture (NCT) are a promising direction.
- The necessary biological functionalities for a nanoplatform are: (i) biocompatibility to make the nanoparticles safe, (ii) biodistribution controlled by enhanced circulation, (iii) targeting to enhance preferential accumulation at the biological site of interest, and (iv) clearance of nanoparticles from the body, eventually. These functionalities can be introduced by appropriate choice of materials composition (e.g., PEG), surface coating, and surface functionalization. Active targeting requires incorporation of targeting groups that recognize the biomatter expression (e.g., antigens) at the targeted biological sites.

- Active targeting (discussed in Chapter 6) requires incorporation of targeting groups that recognize the biomatter expression (e.g., antigens) at the targeted biological sites.
- To help address the emerging need for "fusion technologies," in which two or more medical imaging techniques are combined to provide complementary information, the development of nanoparticle-based probes that contain multiple contrasting agents for simultaneous use in different imaging modalities (e.g., optical, MRI, PET, SPECT, and CT imaging) is proposed. Correlation of these images using a single multifunctional nanoprobe will accelerate the diagnostic process while reducing stress on the patient. Another advantage of these "multimodal" nanoprobes is the ability to target specific molecular markers.

EXERCISES

True or False

1. Biological materials possess strongest absorption in the wavelength range of 800–1000 nm.
2. Gadolinium ion (Gd^{3+}) containing nanoparticles can be used as a contrast agent in MRI because the Gd^{3+} in the particle core affects the relaxation time of protons.
3. Heat can be generated in a magnetic nanoparticle when it is coupled to an AC magnetic field.
4. The hydrophilicity of a nanoparticle can be affected by the length, shape, and density of polyethylene glycol on the surface.
5. In neutron capture therapy (NCT), two coincident gamma photons are produced upon annihilation of the emitted positron by a nearby electron.

Multiple Choice

6. What kind of physical phenomena are not used for nanomedicine multi-functionalities?
   (a) Optical
   (b) Magnetic
   (c) Radioactive
   (d) Electric
7. Which of the following phenomena is not utilized in the optical functionalities?

   (a) Light scattering
   (b) Optical transitions between two discrete energy states
   (c) Phosphorescence
   (d) Ultrasound resonance

8. Which of the following problems are encountered with organic dyes in optical functionalities?
   (a) Concentration quenching of fluorescence at higher level of loading
   (b) High level of optical signal
   (c) Bleaching
   (d) Cell/tissue contamination

9. A nanoparticle formulation for MRI containing paramagnetic ions does not provide which of the following advantages:
   (a) Targeted delivery to the biological site to maximize contrast
   (b) Faster delivery
   (c) Enhanced sensitivity (due to enhanced local concentration of the paramagnetic ions)
   (d) Reduced toxicity of the paramagnetic ion

10. Which of the following techniques utilizes radioactive functionality:
   (a) PDT
   (b) FRET
   (c) PET
   (d) MRI

Answers: 1. F; 2. F; 3. T; 4. T; 5. F; 6. d; 7. d; 8. a,c; 9. b; 10. c

REFERENCES

Alexis, F., Pridgen, E., Molnar, L. K., Farokhzad, O. C., Factors affecting the clearance and biodistribution of polymeric nanoparticles, *Mol. Pharm.* **5**, 505–515 (2008).

Anand, S. S., Singh, H., Dash, A. K., Clinical applications of PET and PET-CT, *MJAFI* **65**, 353–358 (2009).

Arrueboa, M., Fernández-Pachecoa, R., Ibarra, R. M., Santamaría, J., Magnetic nanoparticles for drug delivery, *Nanotoday* **2**, 22–32 (2007).

Bailey, D. L., Townsend, D. W., Valk, P. E., Maisey, M. N., *Positron Emission Tomography: Basic Sciences*, Springer-Verlag, Secaucus, NJ, (2005).

Berry, C. C., Progress in functionalization of magnetic nanoparticles for applications in biomedicine, *J. Phys. D: Appl. Phys.* **42**, 224003 (2009).

Berry, C. C., Curtis, A. S. G., Functionalization of magnetic nanoparticles for applications in biomedicine, *J. Phys. D: Appl. Phys.* **36**, R198–R206 (2003).

Chalfie, M., Kain, S., *GFP, Green Fluorescent Protein—Properties, Applications and Protocols*, Wiley-Liss, New York (1998).

Coderre, J. A., Morris, G. M., The radiation biology of boron neutron capture therapy, *Radiat. Res.* **151**, 1–18 (1999).

Costa, D. C., Single photon emission tomography (SPET) with $^{99}Tc^{m}$-hexamethylpropycleneamineoxime (HMPAO) in research and clinical practice—a useful tool, *Vascular Med. Rev.* **1**, 179–201 (1990).

El-Sayed, I. H., Nanotechnology in head and neck cancer: The race is on, *Curr. Oncol. Rep.* **12**, 121–128 (2010).

Geraldes, C. F., Laurent, S., Classification and basic properties of contrast agents for magnetic resonance imaging, *Contrast Media Mol. Imaging* **4**, 1–23 (2009).

Furlani, E. P., Magnetic biotransport: Analysis and applications, *Materials* **3**, 2412–2446 (2010).

Huang, X. H., El-Sayed, I. H., Qian, W., El-Sayed, M. A., Cancer cell imaging and photothermal therapy in the near-infrared region by using gold nanorods, *J. Am. Chem. Soc.* **128**, 2115–2120 (2006).

Kachynski, A. V., Kuzmin, A. N., Pudavar, H. E., Prasad, P. N., Time-dimensional confocal thermal imaging using anti-Stokes luminescence, *Appl. Phys. Lett.* **87**, 023901–023903 (2005).

Kumar, R., Roy, I., Ohulchanskky, T. Y., Vathy, L. A., Bergey, E. J., Sajjad, M., Prasad, P. N., *In vivo* biodistribution and clearance studies using multimodal organically modified silica nanoparticles, *ACS Nano* **4**, 699–708 (2010a).

Kumar, R., Ohulchanskyy, T. Y., Turowski, S. G., Thompson, M. E., Seshadri, M., Prasad, P. N., Combined magnetic resonance and optical imaging of head and neck tumor xenografts using gadolinium-labelled phosphorescent polymeric nanomicelles, *Head Neck Oncol.* **2**, 35 (2010b).

Lakowicz, J. R., *Principles of Fluroescene Spectroscopy*, 3rd edition, Springer, New York (2006).

Lim, H. W., Soter, N. A., *Clinical Photomedicine*, Marcel Dekker, New York (1993).

Majewski, P., Thierry, B., Functionalized magnetite nanoparticles—Synthesis, properties, and bio-applications, *Crit. Rev. Solid State Mater. Sci.* **32**, 203–215 (2007).

Muehllehner, G., Karp, J. S, Positron emission tomography, *Phys. Med. Biol.* **51**, R117–R137 (2006).

Pankhurst, Q. A., Connolly, J., Jones, S. K., Dobson, J., Applications of magnetic nanoparticles in biomedicine, *J. Phys. D: Appl. Phys.* **36**, R167–R181 (2003).

Pankhurst, Q. A., Thanh, N. K. T., Jones, S. K., Dobson, J, Progress in applications of magnetic nanoparticles in biomedicine, *J. Phys. D: Appl. Phys.* **42**, 224001 (2009).

Periasamy, A., ed., *Methods in Cellular Imaging*, Oxford University Press, Hong Kong, 2001.

Periasamy, A., Day, R. N., *Molecular Imaging: FRET Microscopy and Spectroscopy*, Oxford University Press, New York (2005).

Periasamy, A., Clegg, R. M., *FLIM Microscopy in Biology and Medicine*, CRC Press (Taylor & Francis Group), New York (2010).

Prasad, P. N., Williams, D. J., *Introduction to Nonlinear Optical Effects in Molecules and Polymers*, Wiley, Hoboken, NJ (1999).

Prasad, P. N., *Introduction to Biophotonics*, Wiley, Hoboken, NJ (2003).

Prasad, P. N., *Nanophotonics*, Wiley, Hoboken, NJ (2004).

dos Remedios, C. G., Moens P. D. J., Fluorescence resonance energy transfer spectroscopy is a reliable "ruler" for measuring structural changes in proteins: Dispelling the problem of the unknown orientation factor, *Journal of Structural Biology* **115**, 175–185 (1995).

Safarik, I., Safarikova, M., Magnetic nanoparticles and biosciences, *Monatsh. Chem.* **133**, 737–759 (2002).

Salt, C., Lennox, A. J., Takagaki, M., Maguire, J. A. and Hosmane, N. S., Boron and gadolinium neutron capture therapy, *Russian Chem. Bull.* **53**, 1871–1888 (2004).

Shaner, N. C., Lin, M. Z., McKeown, M. R., Steinbach, P. A., Hazelwood, K. L., Davidson, M. W., Tsien R. Y., Improving the photostability of bright monomeric orange and red fluorescent proteins, *Nat. Methods* **5**, 545–551 (2008).

Wadas, T. J., Eiblmaier, M., Zheleznyak, A., Sherman, C. D., Ferdani, R., Liang, K., Achilefu, S., Anderson, C. J., Preparation and biological evaluation of ^{64}Cu-CB-TE2A-sst$_2$-ANT, a somatostatin antagonist for PET imaging of somatostatin receptor-positive tumors, *J. Nucl. Med.* **49**,1819–1827 (2008).

West, J., Halas, N. J., Applications of nanotechnology to biotechnology, *Curr. Opin. Biotechnol.* **5**, 215–217 (2000).

Zheng, Q., Ohulchanskyy, T. Y., Sahoo, Y., Prasad, P. N., Water-dispersible polymeric structure co-encapsulating a novel hexa-peri-hexabenzocoronene core containing chromophore with enhanced two-photon absorption and magnetic nanoparticles for magnetically guided two-photon cellular imaging, *J. Phys. Chem. C* **111**, 16846–16851 (2007).

CHAPTER 6

CROSSING THE BIOLOGICAL BARRIERS

Nanocarriers, covered in Chapter 3, can be introduced in the body using a number of pathways. However, to reach the biological target site (such as a specific organ or tumor), the nanoparticles must cross various biological barriers that restrict migration of foreign particles through our body. These biological barriers are designed to protect our body from invasion of harmful foreign substances and/or to minimize any damage caused by harsh environmental conditions.

Thus, nanocarriers must be imparted with "stealth" action to be able to cross these barriers in order to reach the target site. This chapter discusses this topic. Section 6.1 covers the various pathways, which can be utilized for nanoparticle delivery. Section 6.2 discusses various biological barriers. Section 6.3 introduces the characteristics of the nanoparticles, which make them able to cross the biological barriers. Section 6.4 presents some useful *in vitro* models of these biological barriers, which can be utilized to test the ability of new nanoformulations to cross a specific biological barrier.

In order for this chapter to be useful for multidisciplinary readership, any detailed description of biological principles involved is avoided.

6.1. VARIOUS DELIVERY PATHWAYS

Various delivery pathways for a nanocarrier formulation are shown in Figure 6.1. The oral delivery, perhaps most user-friendly, can be in the form of a tablet,

Introduction to Nanomedicine and Nanobioengineering, First Edition. Paras N. Prasad.

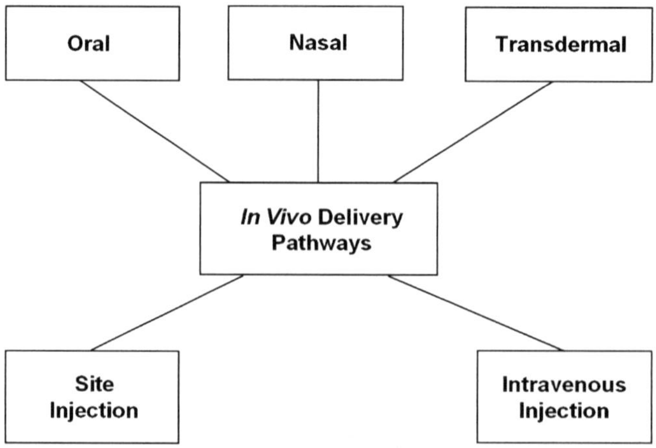

Figure 6.1. Various delivery pathways for nanocarriers.

a liquid, or a spray formulation. The tablet formulation must ensure that the nanocarriers retain their identities when compressed, generally with a "binder" matrix. The nanocarriers pass through the oral cavity into the digestive track where they can be absorbed and then passed into the bloodstream.

Nasal delivery involves inhalation or instillation of nanocarriers through the upper respiratory tract. However, most inhalational therapies require inhalers that are inserted into the mouth in order to reach the deeper lower respiratory tract airways of the lungs. For this, the formulation may be in the form of a nasal spray, a nasal liquid drop, or an inhaler similar to that which is used for bronchodilation therapy in asthma. Transdermal delivery involves delivery through the skin. Here the nanocarriers may be deposited on, or embedded in, a patch to be placed in contact with the skin. Alternatively, a cream or lotion formulation of a nanocarrier can be used to administer it. The advantage offered by this delivery is that it can be applied topically in a localized area, with an easy access to the target site.

Site injection involves administration of nanocarrier dispersion in a biocompatible fluid, directly into a biological site/organ/tissue of interest. An example is gene delivery using nanocarriers where we used a small incision in the skull to directly inject the nanocarriers containing genetic material into the brain.

Intravenous delivery, which involves direct injection of nanocarriers into the blood stream, is the most desirable form of delivery in nanomedicine. However, this is also the most challenging, because for this delivery mode to be effective, the nanocarriers must cross certain biological barriers to be discussed in the next section, escape enzymatic degradation in their pathways, and then eventually reach their target site.

6.2. VARIOUS BIOLOGICAL BARRIERS

This section discusses some important biological barriers from the nanomedicine perspective. These barriers severely restrict nanocarrier migration and thus require a rational design of pharmaceutical nanoformulations for a successful therapeutic strategy.

RES Barrier. A first line of body defense against foreign substance is provided by the reticulo endothelial system, abbreviated as RES (already briefly introduced in the Chapter 3), which consists of phagocytic cells namely monocytes and macrophages (described in Chapter 2). Monocytes are formed in the bone marrow and then transported by bloodstream throughout the body. In tissues, monocytes are differentiated to tissue-specific macrophages. In particular, the macrophages could be associated with liver (Kupffer cells), spleen, and brain (microglia). The major function of phagocytic cells is, as indicated by their name, phagocytosis, whereby monocytes and macrophages engulf cell debris as well as various pathogens including bacteria and solid foreign substances, thus eliminating them through several mechanisms. A severe problem in delivery of nanocarriers by the blood flow is posed by active phagocytosis of nanoparticles by macrophages. As a result, nanoparticles cannot cross the RES barrier, and they accumulate at the site of their capture in liver, spleen, and lymph nodes until they can be excreted by the body's elimination system. Because of this sequestration by macrophages, nanocarriers are unable to reach their target site (e.g., tumor), even if they carry targeting function (to be discussed in Chapter 7).

Although macrophages do not necessarily require molecular signaling for internalization of foreign nanoparticles, phagocytosis can be enhanced by a process called opsonization, whereby the particles nonspecifically bind to proteins called opsonins (i.e., antibodies or complement), which then are more strongly recognized by the phagocytic cells and are engulfed. Within the phagocytic cells, the nanoparticle may be chemically broken down by enzymes or other oxidative chemical species such as superoxide or nitric oxide or, if inert, either excreted or walled off by forming granuloma, depending on the size of the particle.

Skin Barrier. The skin barrier protects the body against environmental insults and microbial invasions. It is a layered structure consisting of many kinds of cells, which provide a relatively impenetrable physical barrier to entry of foreign substances of large size, as well as the ones generating harmful or immune response.

The skin is composed of three basic layers: epidermis, dermis, and subcutaneous tissue, as shown in Figure 6.2. The outer layer of skin, called epidermis, is approximately 100–150 μm thick. The outermost layer of epidermis, called the stratum corneum (SC), is comprised of a 10- to 15 μm-thick matrix of dehydrated and dead keratinocytes that are embedded in highly ordered lipid

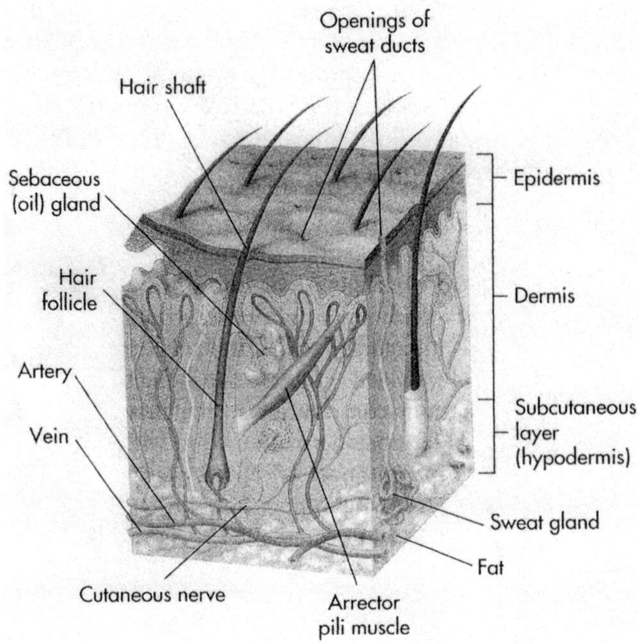

Figure 6.2. Structure of skin. (Reproduced with permission from Thibodeau and Patton [2003].)

layers, which serve as a cover [Bouwstra et al., 2003] protecting our body from entry of external materials into the skin. Below the SC lies the viable epidermis composed mostly of the keratinocytes, but containing also specialized cells: melanocytes, Langerhans cells, and Merkel cells. The melanocytes produce a pigment, melanin, which is responsible for skin color and provides protection from ultraviolet light. The Langerhans cells are similar to macrophages with respect to morphology and function. They are among first cells of immune systems that interact with pathogens entering the body through the skin, providing for a cutaneous immune system. The function of these cells is to detect and capture foreign pathogens and then transport them to the lymph nodes (introduced in Chapter 2 and further discussed in Chapter 12), where the pathogens are presented to lymphocytes in order to trigger an adaptive immune response. Finally, the Merkel cells associate with nerve fibers participating in tactile function of the skin.

The next (inner) layer of the skin is dermis, which is thick and fibrous and has an elastic fibrous structure, being composed mostly of collagen, elastin, and fibrillin. Its cellular components are the fibroblasts and macrophages, and it also contains nerves and blood vessels that are densely woven within the tissue. The dermal nerves conduit the senses of touch, pressure, pain, and temperature. The dermal blood vessels supply epidermis with nutrients. Moreover, dermis embeds the sweat glands, oil glands, and hair follicles. The base of each hair is associated with small muscles named *arrectores pilorum* (singular *arrector pili*), which connect the hair follicle to the dermis. Contraction of these muscles (e.g., in response to cold temperature or intense emotions) causes the hairs to stand on end, a condition known as "goose bumps."

At last, below the dermis lies the hypodermis, a lowermost and thickest layer of the skin. Hypodermis contains adipocytes, cells specialized in accumulating fats and serving for insulation of the body from thermal surges as well as for energy storage. It is also important to recognize that the skin plays an important role in excretion of waste from the organism.

A transdermal delivery involves the crossing of all these layers by specially engineered nanoparticles to get into the bloodstream. However, special enhancement techniques can make the skin barrier more permeable to provide easy passes for the nanocarriers. These methods are discussed in Section 6.3. Furthermore, Chapter 19 reviews various interactions of nanocarriers with the skin barrier.

Mucosal Barrier. Mucosal surfaces are present in mouth, gastrointestinal (GI), respiratory, and urogenital tracts, and conjunctiva of the eye. These surfaces are lined by epithelial cells that secrete a protective layer of thick viscous substance called mucus. Mucus involves mucins, which are glycoproteins organized primarily in cross-linked, bundled, and entangled fibers. Hence, the mucosal barrier functions are performed by both extracellular and cellular components. The main extracellular components are mucins, which provide antimicrobial protection. The barrier function of mucus is performed by steric obstruction and/or adhesion, whereby the trapped particles can be removed via mucociliary clearance. In addition, paracellular pathways between the cells are created by the apical junction complexes composed of the tight junction and subjacent adherens junction. In case of adhesion, hydrophobic interactions with mucins play an important role. The cellular components of a mucosal barrier are simply layers of epithelial cells. The epithelial cell plasma membranes are impermeable to most hydrophilic molecules.

The mucosal barrier plays an important role in the body's defense to prevent the entry of foreign substances through oral, nasal, or vaginal passage, as well as through eyes or any other moist areas containing mucus. For each organ, the biological function of mucus may be different.

Nanomedicine delivery in the context of a mucosal barrier can be broadly divided into two categories: (i) Mucoadhesive delivery in which the nanoparticles are designed to be adhesive to the mucosal surface by various interactions such as hydrogen bonding, van der Waals interactions, polymer chain penetration, hydrophobic forces, and electrostatic interactions [Lai et al., 2009]. An example is our work on mucoadhesive polyacrylic acid (PAA) nanoparticles containing ophthalmic drugs for ocular delivery [Rodman et al., 2002]. These mucoadhesive nanoparticles adhere to the mucus in the eye to provide a sustained time release of ophthalmic drugs. (ii) Transmucosal delivery in which specially tailored nanoparticles can cross the mucosal layer and deliver the drug either by transepithelial transport or by a paracellular pathway permeating through the junction complex. Section 6.3 describes the approaches, which impart a nanocarrier with the ability to either be mucoadhesive, maximizing their retention time, or be able to cross the mucosal layer for transepithelial or paracellular delivery.

The Blood–Brain Barrier. The blood–brain barrier (BBB) performs the vital function of protecting the central nervous system (CNS) in the brain from the entry and exit of endogenous as well as exogenous compounds. This barrier separates the brain from the blood supply.

The BBB is formed by junctions between endothelial cells and is further reinforced by surrounding cells: pericytes and astrocytes. The endothelial cells are connected by a complex network of adherent junctions, tight junction, and gap junctions, which permit selectivity in the entry of substances to the central nervous system. BBB thus acts as a physiological checkpoint, precluding free diffusion of circulating nanoparticles from the blood into the CNS. The BBB is impermeable to most nanocarriers, primarily because of their physical size. However, as discussed in the next section, nanoparticles can be designed to cross the BBB.

6.3. STEALTH NANOPARTICLES

A considerable effort has recently been placed in understanding the roles of physical attributes and interfaces of the nanoparticles and their interactions with these biological barriers, so that design principles can be developed for crossing them. Some basic understanding has already been developed; and based on that, some design criteria for "stealth" nanoparticles sneaking past a specific biological barrier have already emerged.

For evading capture by the RES barrier, the size and the surface charge of the nanoparticles play an important role. The smaller-sized (<30 nm) nanoparticles that have close to neutral or slightly negative charge and that are PEGylated to prolong circulation tend to evade RES capture. Placement of active targeting groups on the nanoparticle, a subject discussed in Chapter 7, also facilitates their clearance from the liver and spleen and thus assists in evading the RES capture of the nanoparticles.

Similar considerations hold for particles crossing a mucosal barrier. Smaller-sized nanoparticles are less sterically impeded and better suited to cross a mucosal barrier than the larger-sized ones. Hydrophilicity on the surface and PEGylation also enhances the transport of nanoparticles across a mucosal barrier. Another strategy involves the use of a mucolytic agent, such as a hydrolyzing enzyme, which disrupts the mucosal network to facilitate direct nanoparticle deposition on the surface of mucosal cells.

Nanoparticle characteristics can be tailored to cross the skin barrier to provide intradermal and transdermal deliveries for cosmetology and dermatology. These applications are presented in Chapter 19. As discussed in Chapter 19, nanoparticles are being developed which not only enhance the absorption of a drug into the skin, but also release the drug in a controlled manner for a prolonged period of time.

Nanoparticle-based formulations have several advantages that can enable potent drug delivery across the BBB while avoiding any damage to the barrier. These include ultra-low size, biocompatibility, nonantigenicity, an ability for targeted and controlled drug delivery, multimodality, and an ability to monitor BBB permeation in real time [Mintzer and Simanek, 2009]. In general, transport mechanisms across the BBB can be broadly divided into three types: (i) passive, (ii) carrier-mediated, and (iii) vesicular transport [Xu et al., 2008]. Lipid-soluble, nonpolar substances can enter the brain by passive diffusion across the BBB. The approach, using liposomal carriers, has been associated with several problems such as large size, toxicity, immunogenicity, as well as the instability of "carrier vectors" in biological media. As a result, these efforts have only met with limited success; often crossing the BBB was accompanied by severe, irreversible damage to the integrity of the BBB. Aqueous substance and small peptides can be transported across the endothelium by carrier-mediated influx. The vesicular transport is by either receptor-mediated or absorptive-mediated transcytosis via cationic proteins. Our group has demonstrated a successful transport of bioconjugated quantum rods (QRs) across an *in vitro* blood–brain barrier (BBB) model via a receptor-mediated transport. We used transferrin because the trafficking agent due to the transferrin receptor (TfR) is a kind of specific BBB transporter.

6.4. THE VARIOUS *IN VITRO* BARRIER MODELS

For testing the ability of a nanoparticle to cross a specific biological barrier, it is often very useful to utilize a laboratory model. These *in vitro* models are designed to include essentials of the active structural components, which provide the barrier function. In this section, some of these models are described.

Structurally, the BBB is composed of brain microvascular endothelial cells (BMVEC) connected by tight junctions (TJ) and TJ proteins, which together ensure the BBB's structural integrity. Additional components of the BBB are the surrounding capillary basement membrane and astrocytes. The astrocytic end-feet reinforce the abluminal (outer) surface of the brain endothelium and assist in the barrier function by coordinating the functional activities of BMVEC. Our *in vitro* BBB model is a well-validated co-culture technique and reflects several characteristics of the *in vivo* BBB. This is described here. The *in vitro* BBB model [Bonoiu et al., 2009] uses primary cultures of both BMVECs and normal human astrocytes, obtainable from the Applied Cell Biology Research Institute, Kirkland, WA. The BBB model, shown in

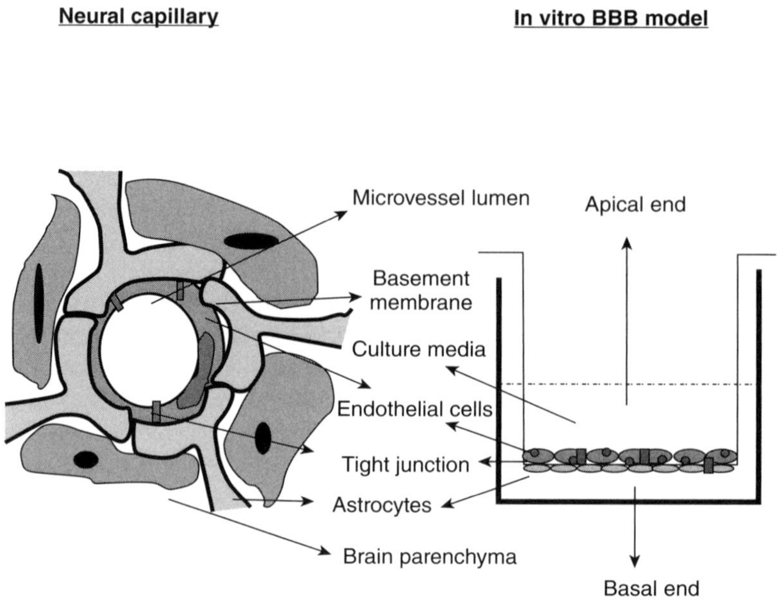

Figure 6.3. Blood–brain barrier and it modeling *in vitro*. (Partially reproduced with permission from Pardridge [1999].)

Figure 6.3, consists of two-compartment wells in a six-well culture plate, with the upper compartment separated from the lower by a 3 μM polyethylene terephthalate (PET) insert (surface area = 4.67 cm^2). The BMVECs were grown to confluency on the upper side of the insert, while a confluent layer of normal human astrocytes (NHAs) were grown on the underside. The formation of a functional and intact BBB functional model takes a minimum of 5 days, which can be confirmed by determining the transendothelial electrical resistance (TEER) value. Using the *in vitro* BBB model, we examined BBB permeability, transendothelial migration, and the efficacy of drug delivery of the QR-Tf-Saquinavir nanoplex, along with the various control formulations [Mahajan et al., 2010].

The skin barrier can be also constructed under laboratory conditions. Our *in vitro* skin barrier model is shown in Figure 6.4. It consists of skin-derived normal, human epidermal keratinocytes (NHEK), representing cells of epidermis, which is the outer layer of the skin, and normal, human dermal fibroblasts (NHFB), representing cellular constituents of dermis (see Figure 6.2). In our approach, the NHEK and NHFB cells are co-cultured, using the cell culture inserts (as shown in Figure 6.4) and allowed to differentiate to form a multilayered model of the human dermis and epidermis. The standard characterization of human epidermis includes screening of epidermis-specific dif-

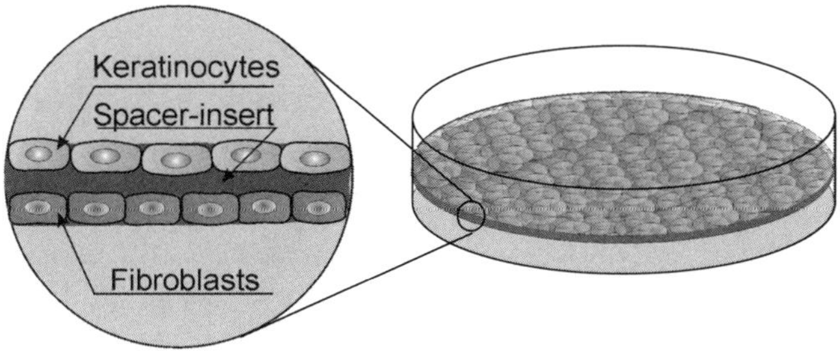

Figure 6.4. *In vitro* skin model.

ferentiation markers such as the K1/K10 cytokeratin pair, involucrin, type I epidermal transglutaminase, Filaggrin, Desmoglein1, and Claudin1 by using immunocytochemical techniques. Consistently, we observed that the expression of these markers in our model closely parallels the human skin. Our *in vitro* skin model is mitotically and metabolically active and, therefore, is useful to evaluate percutaneous absorption, transdermal drug delivery, and other studies relating to the barrier function of human skin.

The following biological endpoints for the skin barrier function can be assessed using our *in vitro* skin model: (1) evaluation of the skin barrier efficacy by measuring the trans-epidermal electrical resistance (TEER), (2) study of penetration kinetics of nanoparticle complexes or nanocarriers across the human epidermis by using bioimaging tools, (3) evaluation of expression profile of key genes involved in the above-mentioned biological processes by using RT-PCR as well as DNA microarrays, and (4) changes in histology and cytokine release by immunocytochemical and biochemical analysis.

Similarly, a diffusion chamber used by Norris and Sinko provides a useful model for mucosal barrier. Using this chamber, they studied the diffusion of variously sized polystyrene particles in a reconstituted porcine gastric mucin gel, sandwiched between the two chambers of a Transwell–Snapwell diffusion chamber (Figure 6.5), and observed a sharp decrease in translocation permeability when particle sizes approached 300 nm.

Sanders et al. [2000] reported that an extremely low percentage of carboxylated polystyrene nanospheres moved through a 220-μm-thick CF sputum layer after 150 min, with the largest nanospheres studied (560 nm) almost completely blocked by the sputum.

Finally, the most common model for the RES barrier utilizes *in vitro* cell culturing of macrophages and investigating uptake of nanoparticles in these cells via confocal microscopy [Thiele et al., 2001]. Nanoparticles with hydrophobic and/or cationic surface were found to be uptaken at a much higher rate, when compared to nanoparticles with hydrophilic and/or anionic/neutral surface.

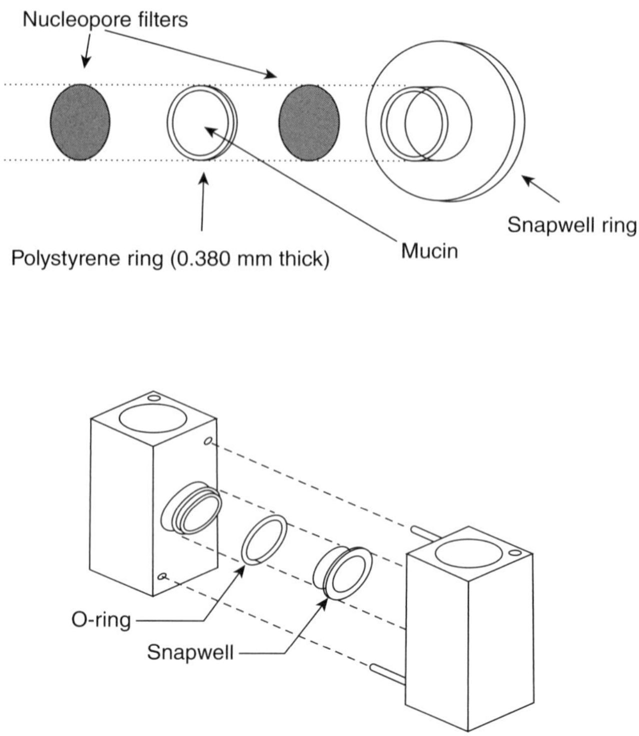

Figure 6.5. Schematic of a Transwell–Snapwell diffusion chamber. The 0.4-μm filter supplied with Snap well rings is removed and mucin is placed between two polycarbonate filters. Snap well fits within the diffusion chamber and is sealed in place with rubber O-rings. (Reproduced with permission from Norris and Sinko [1997].)

HIGHLIGHTS OF THE CHAPTER

- A nanoformulation can be introduced into the body by oral, nasal, and transdermal pathways, as well as by a site or intravenous injection. On the way to their destination, nanocarriers may have to cross several biological barriers.
- A reticuloendothelial system is a major biological barrier, which consists primarily of monocytes and macrophages. These cells can intercept and engulf nanoparticles, by a process called phagocytosis, thus reducing the efficacy of the therapeutic approaches.
- The skin barrier is a multilayer structure that prevents dehydration and protects the body against environmental insults and microbial invasions. The skin layers are called *epidermis, dermis*, and *hypodermis*. Epidermis is composed primarily of keratinocytes, but also contains melanocytes,

Langerhans cells, and Merkel cells. Dermis is mostly composed of collagen, elastin, and fibrillin fibers. Hypodermis is composed of fat-storing cells called *adipocytes*. A transdermal delivery involves the crossing of all of these layers by specifically designed nanoparticles.

- A mucosal barrier involves extended surfaces in gastrointestinal, respiratory and urogenital tracts as well as conjunctiva of the eye, which is lined by epithelial cells that secrete extracellular layer of mucus.
- Mucus involves mucins, which are glycoproteins organized primarily in cross-linked, bundled, and entangled fibers.
- With regard to the mucosal barrier, the nanomedicine formulations can be designed for either mucoadhesive or transmucosal delivery.
- The blood–brain barrier is formed by tightly bound endothelial cells and is reinforced by perivascular astrocytes and pericytes. The blood–brain barrier function is to protect the central nervous system from infections as well as from foreign substances.
- Properties of nanoparticles (size, surface coating, and charge) can be tuned to facilitate efficient crossing of a particular biological barrier.
- The smaller-sized (<30 nm) nanoparticles that have close to neutral or slightly negative charge and that are PEGylated to prolong circulation tend to evade RES capture.
- Smaller-sized nanoparticles are less sterically impeded and better suited to cross a mucosal barrier than the larger size ones. Hydrophilicity on the surface and PEGylation also enhances the transport of nanoparticles across a mucosal barrier. A mucolytic agent, such as a hydrolyzing enzyme, which disrupts the mucosal network facilitates direct nanoparticle deposition on the surface of mucosal cells.
- Nanoparticle characteristics can be tailored to cross the skin barrier to provide intradermal and transdermal deliveries for cosmetology and dermatology.
- Various *in vitro* models have been developed for modeling of the skin, mucosal, RES, and blood–brain barriers. The *in vitro* models are useful for studying how effective nanocarriers are in crossing a specific biological barrier.

EXERCISES

True or False

1. Brain microvascular endothelial cells (BMVECs) are not involved in the formation of the blood brain barrier.
2. It is possible to engineer mucoadhesive nanoparticles in order to efficiently deliver them across the mucosal barrier.

3. Kupffer cells in the liver play a critical role in the reticuloendothelial system (RES).
4. Epidermis is the outer layer of skin.

Multiple Choice

5. Intravenous delivery of drugs involves
   (a) Injection into bloodstream
   (b) Injection into target site or organ
   (c) Inhalation
   (d) Crossing the skin barrier
6. Name cells responsible for cutaneous immune response.
   (a) Melanocytes
   (b) Merkel cells
   (c) Langerhans cells
   (d) Keratinocytes
7. Name major cellular constituents of epidermis.
   (a) Keratinocytes
   (b) Adipocytes
   (c) Dermal fibroblasts
   (d) Dendritic cells

Match statements:

Statements A	Statements B
The dermis	lines the surface in mouth, gastrointestinal, (GI), respiratory, and urogenital tracts and conjunctiva of the eye.
Mucosal barrier	is formed by endothelial cells and reinforced by astrocytic end-feet.
Blood–brain barrier	includes tissue-specific phagocytic cells.
RES barrier	is mostly composed of elastic fibers and also contains sweat glands, oil glands, hair follicles, and blood vessels.

Answers: 1. T; 2. F; 3. T; 4. T; 5. a; 6. c; 7. c

REFERENCES

Bonoiu, A., Mahajan, S. D., Ye, L., Kumar, R., Ding, H., Yong, K. T., Roy, I., Aalinkeel, R., Nair, B., Reynolds, J. L., Sykes, D. E., Imperiale, M. A., Bergey, E. J., Schwartz, S. A., Prasad, P. N., MMP-9 gene silencing by a quantum dot-siRNA nanoplex delivery to maintain the integrity of the blood brain barrier, *Brain Res* **1282**, 142–155 (2009).

Bouwstra, J. A., Honeywell-Nguyen, P. L., Gooris, G. S., Ponec, M., Structure of the skin barrier and its modulation by vesicular formulations, *Prog. Lipid Res.* **42**, 1–36 (2003).

Lai, S. K., Wang, Y.Y., Hanes, J., Mucus-penetrating nanoparticles for drug and gene delivery to mucosal tissues, *Adv. Drug Deliv. Rev.* **61**, 158–171 (2009).

Mahajan, S.D., Roy, I., Xu, G. X., Yong, K. T., Ding, H., Aalinkeel, R., Reynolds, J. L., Sykes, D. E., Nair, B. B., Lin, E. Y., Prasad, P. N., Schwartz, S. A., Enhancing the delivery of anti retroviral drug "Saquinavir" across the blood brain barrier using nanoparticles, *Curr. HIV Res.* **8**, 396–404 (2010).

Mintzer, M. A., Simanek, E. E., Nonviral vectors for gene delivery, *Chem Rev*, **109**, 259–302 (2009).

Norris, D. A., Sinko, P. J., Effect of size, surface charge, and hydrophobicity on the translocation of polystyrene microspheres through gastrointestinal mucin, *J. Appl. Polym. Sci.*, **63**, 1481–1492 (1997).

Pardridge, W. M., Blood–brain barrier biology and methodology, *J. Neuro Virology*, **5**, 556–569 (1999).

Rodman, D. J., Bergey, E. J., Leibow, C., Prasad, P. N., Biophotonics perspectives for 21st century, in H. Sasabe, ed., *Nanotechnology Toward the the Organic Photonics*, pp. 29–40, Gootech Ltd., Chitose, Japan (2002).

Sanders, N. N., De Smedt, S. C., Van Rompaey, E., Simones, P., De Baets, F., Demeester, J., cystic fibrosis sputum: A barrier to the transport of nanospheres, *Am. J. Respir. Crit. Care Med.* **162**, 1905–1911 (2000).

Thibodeau, G. A., Patton, K. T., *Anatomy & Physiology*, 5th edition, Mosby, St. Louis (2003).

Thiele, L., Rothen-Rutishauser, B., Jilek, S., Wunderli-Allenspach, H., Merkle, H. P., Walter, E., Evaluation of particle uptake in human blood monocyte-derived cells *in vitro*. Does phagocytosis activity of dendritic cells measure up with macrophages? *J Control Release* **76**, 59–71 (2001).

Xu, G. X., Yong, K. T., Roy, I., Mahajan, S. D., Ding, D., Schwartz, S. A., Prasad, P. N., Bioconjugated quantum rods as targeted probes for efficient transmigration across an in vitro blood–brain barrier, *Bioconj. Chem.*, **19**, 1179–1185 (2008).

CHAPTER 7

BIOTARGETING

This chapter discusses the biotargeting direction in nanomedicine, which is of significant value for theranostics. Biotargeting provides nanocarriers with the ability to deliver a diagnostic and therapeutic payload to an intended biological site, whether a specific organ, tissue, cell, or organelle, for enhancing the sensitivity and specificity of diagnosis as well as for improving the effectiveness of a therapy. In Section 7.1, the need for targeting is discussed. Section 7.2 describes the various biological sites to be targeted. Section 7.3 contains a description of intracellular uptake pathway in order to understand the various strategies for targeting. Section 7.4 provides an overview of the various targeting strategies. Section 7.5 includes a list of targeting groups used for active cellular targeting.

7.1. BIOTARGETING: WHY WE NEED IT

The biotargeting ability crafted on a nanocarrier provides a major advancement in the effectiveness of theranostics. Some of the major benefits achieved by biotargeting are listed here:

- *Delivery of Diagnostic and Therapeutic Payloads to Specific Biological Sites.* This is a major gain whereby any toxic side effect arising from off-target accumulation of active agents is significantly reduced or eliminated. Thus, any damage to healthy cells/tissues is minimized.

Introduction to Nanomedicine and Nanobioengineering, First Edition. Paras N. Prasad.

- *Increase of Local Concentration.* By accumulation at the desired biological site, the nanocarriers increase the local concentration of a diagnostic or therapeutic agent. In the case of diagnosis, this will translate into enhanced sensitivity and specificity of the method. In the case of therapy, this increased local concentration will provide a more effective treatment at a lower dosage.
- *Effective Intracellular Therapy.* The ability of a nanocarrier to enter a specific cell type (by receptor-mediated endocytosis discussed in Section 7.3) and accumulate in a particular organelle, where a particular drug is most effective, enhances the therapeutic action. For example, photodynamic cancer therapy (discussed in Chapter 13) is expected to be more effective when the drug is localized in mitochondria. Similarly, for gene therapy to be effective, the genetic payload should enter the nucleus.
- *Enhanced Ability to Cross a Biological Barrier.* The targeting ability of a nanoparticle also assists this nanocarrier to cross a specific biological barrier (discussed in Chapter 6).

7.2. TARGETED BIOLOGICAL SITES

The various biological sites that can be targeted are represented in Figure 7.1. Organ-specific targeting relies on its structure and function distinct from its surrounding. For example, proteins expressed on membranes of some types of cells, or proteins found in extracellular matrices, can be targeted to promote a selective accumulation of nanocarriers in the specific tissues or organs. Furthermore, the biochemical and various enzymatic activities of a specific organ can also be used for biorecognition that can be used as the targeting modality. Tissue targeting implies targeting a specific tissue structure that may be abnormal; an example is a tumor tissue that is distinct from the normal tissue structure. Targeting of a tumor tissue by nanocarriers is a major area of nanomedicine

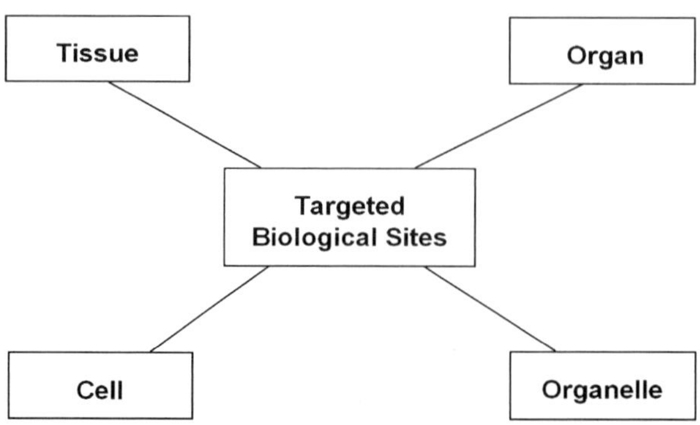

Figure 7.1. The various biological sites to be targeted by nanocarrier delivery.

for providing a more effective cancer therapy. This aspect is discussed below. Organ/tissue targeting is often labeled as first-order targeting. Targeting of a specific type of cell such as hepatic cells in the liver is called second-order targeting. This targeting involves biorecognition elements such as structure of the membrane and the various antigen receptors expressed on the targeted cell surfaces. Third-order targeting involves targeting of a particular subcellular organelle, such as the nucleus in a cell. This utilizes targeting groups with an affinity for the organelle structure/molecular constituent. Finally, nanocarriers can also be designed to contain multiple modes of targeting to separately address each order of targeting to accumulate in a specific organelle of a particular cell type in a specific organ/tissue structure.

7.3. INTRACELLULAR UPTAKE

An important aspect of targeting is the ability to utilize disease profiling, therapy, and real-time monitoring of treatment from within the cell. In order to understand cell-based targeting mechanisms, it is important to look at the mechanism by which a nanocarrier can penetrate the cell membrane and enter a specific cell type that is targeted. The general process by which a foreign substance enters a cell is called endocytosis. The general process of endocytosis is schematically represented in Figure 7.2.

There are several mechanisms of endocytosis that include clathrin-mediated endocytosis, caveolae-mediated endocytosis, macropinocytosis, and phagocytosis [Hillaireau and Couvreur, 2009]. Clathrin-coated endocytosis is the primary mechanism of entry of nanocarriers, which is mostly, but not exclusively, "receptor-mediated" (discussed in Section 7.4). This mechanism involves either (a) the interaction of bioconjugated nanocarriers with specific cell-surface receptors or (b) electrostatic interaction between positively charged nanocarriers with negatively charged receptors like heparin sulfate proteoglycans (HSPs) present on the surface of most cell types. Once endocytosed, the nanocarriers are transferred from the clathrin-coated endocytic vesicles to the endosomes and lysosomes, where the local acidic pH and presence of enzymes such as nucleases and proteases usually cause the degradation of the nanocarriers, providing the release of theranostic payloads. If one wishes to design nanocarriers that escape endolysosomal degradation, it is essential that such endocytic vesicles are disrupted for intracellular release of the intact nanocarriers. Certain cationic polymers like polyethylenimine (PEI) and polyamidoamine dendrimers have nonprotonated amino groups, which can buffer the acidic pH by attracting free protons. A direct consequence of this "proton-sponge" effect is the influx of excess chloride ions and water within the vesicles, leading to endosomal or lysosomal bust. The "helper" phospholipid components of many liposomes, such as 1,2-dioleoyl-3-phosphatidylethanolamine (DOPE), are also responsible for fusion with the endosomal membranes, leading to membrane destabilization and consequent release of the entrapped

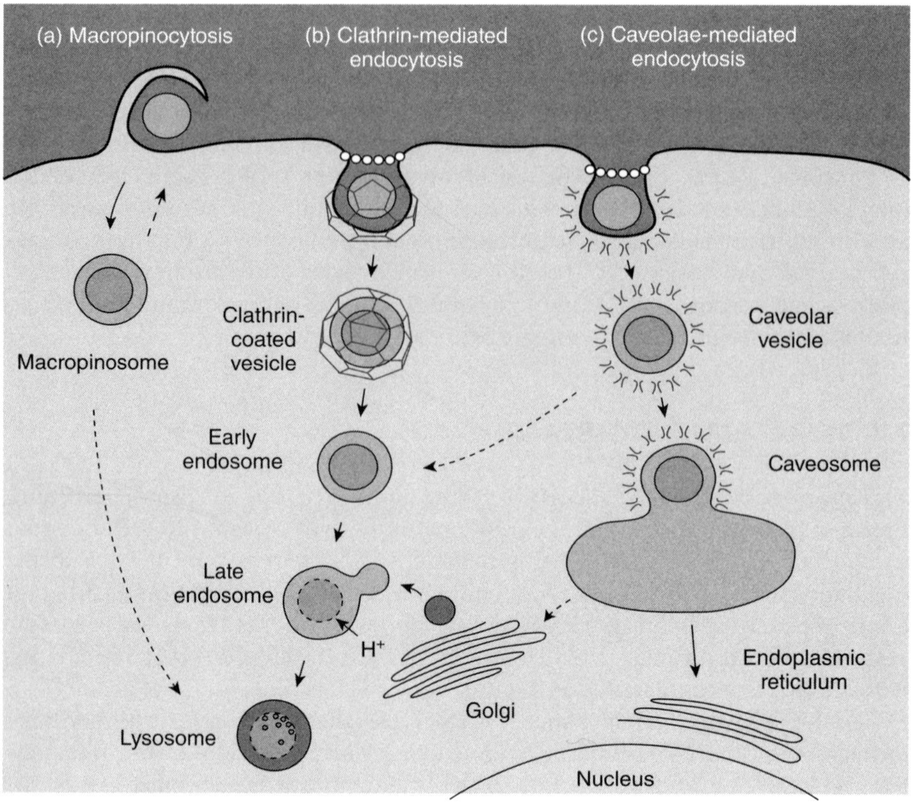

Figure 7.2. Schematic representation of the process of endocytosis. (Reproduced with permission from Hillaireau and Couvreur [2009].)

DNA [Nishikawa and Huang, 2001]. Polyethylene glyconoic acid (PLGA) polymeric nanocarriers are also shown to possess endo-osmolytic properties owing to their selective reversal of the surface charge (from anion to cationic) in the acidic endolysosomal compartment. In other cases like polylysine, where the nanocarrier does not have any intrinsic endosome-rupturing capacity, additional "endosomolytic" agents such as viral peptides have been used to aid the endosome disruption and release of the nanoparticles [Curiel, 1994].

In addition to clathrin-mediated endocytosis, caveolae-mediated endocytosis has also been shown to be a major mechanism of receptor-mediated nanoparticle uptake, particularly in endothelial cells [Hillaireau and Couvreur, 2009]. Here, after binding to the cell surface, the nanoparticles are coated with dimeric proteins, called caveolins, and are taken intracellularly by "flask-shaped" caveolar vesicles. Because these vesicles do not contain nucleases and proteases, this is a preferred route for the delivery of drug-loaded nanocarriers that are prone to endolysosomal degradation. Also, for effective cellular delivery, this mechanism does not rely on the inclusion of potentially toxic lysosomolytic factors with the nanocarrier formulation. From the perspective of pulmonary delivery, this is a major uptake mechanism, since caveolin is also abundantly expressed on lung capillaries and type I alveolar cells.

Macropinocytosis, or fluid-phase endocytosis, is another mode of endocytotic uptake of nanocarriers by cells, but occurs nonspecifically [Ragnarsson et al., 2008]. Finally, phagocytosis (also discussed in Chapter 6) is the nonendocytotic uptake process that requires the coating of nanoparticles by opsonin proteins that are subsequently captured by specialized cells, called phagocytes, and are carried to phagocyte-rich organs such as liver and spleen. These cellular uptake processes and their implications in drug and gene delivery are reviewed in detail by Hillaireau and Couvreur [2009].

7.4. TARGETING STRATEGIES

The various targeting strategies used are summarized in Figure 7.3. Field-directed targeting utilizes an external field gradient (such as a magnetic field) to enhance the cellular uptake of the nanocarriers. For magnetic-field-directed cellular uptake, the nanocarriers contain magnetic materials that allow the nanocarriers to move in a magnetic field (also see Chapter 5, Section 5.4). We have utilized this approach for selective delivery of silica nanoparticles, which encapsulate iron-oxide nanoparticles and quantum dots, to cancer cells [Law et al., 2008]. Figure 7.4 shows the enhanced cellular uptake of magnetic particle encapsulated in a larger silica nanoparticle. The magnetic field enhanced cellular uptake has also been used for drug and gene delivery (Furlani, 2010; Schillinger et al., 2005].

Passive targeting is achieved by exploiting the inherent cell/tissue morphology that favors enhanced nanoparticle uptake. Passive targeting has been mainly used to target cancer tissues. Tumor tissues have high endothelial

Figure 7.3. The various targeting strategies used for nanomedicine.

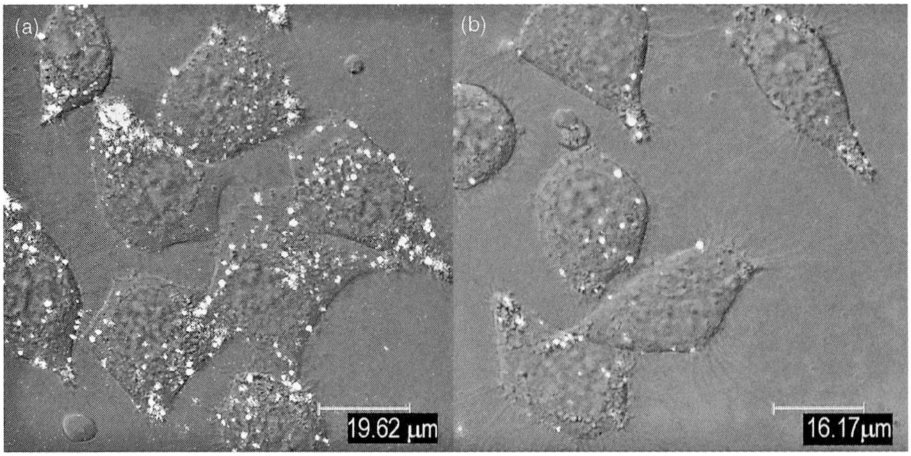

Figure 7.4. Magnetically enhanced cellular uptake of silica nanoparticles. Two-photon excited fluorescence microscopic images of Panc-1 cells treated with Tf-conjugated hybrid nanoparticle, (a) with and (b) without the presence of external magnetic field. (Reproduced with permission from Law et al. [2008].)

leakage and poor lymphatic drainage. As a result, any agents such as nanocarriers circulating in the blood have higher chances of accumulating into the cancerous areas and residing there for a long time, as compared to the surrounding normal areas. This effect is commonly known as the "enhanced permeability and retentivity" (EPR) effect. The "stealth" nanoparticles described in Chapter 6, which escape the RES capture and have prolonged circulation time in the blood, can permeate through the gaps or fenestraes of the tumor-associated endothelial barrier and invade the tumor sites. Several types of nanoparticles, including polymeric nanoparticles, gold nanospheres/rods, quantum dots, iron-oxide nanoparticles, dendrimers, and liposomes, have been targeted to tumor sites using this passive targeting. Both Doxil® and Abraxane®, the only two clinically approved cancer-drug-doped nanoparticulate carriers to date, rely on this targeting mode to deliver drugs to a cancer site. Other than cancer, passive targeting has also been exploited in drug delivery to the macrophages in the liver which have an inherent tendency to capture nanoparticles circulating in the systematic circulation. This liver-specific drug delivery has been utilized for the treatment of several hepatic diseases.

Intrinsic targeting refers to the property of a molecular probe or a drug to selectively accumulate at a specific biological site. Examples are mitochondrial and nuclear staining dyes used for optical imaging which can also be utilized as an attachment to nanocarriers to direct them to a specific organelle. Another example is a class of therapeutic agents, the photosensitizers used for photodynamic therapy (PDT), which show inherent ability to preferentially target and be retained by cancer cells. Our group has used this feature of a photosensitizer, also often called PDT drug, incorporated in a nanoparticle for targeting of cancer cells (see Chapter 13).

Active targeting can be achieved by providing nanoparticles with a molecular "address," which allows them to home in specifically at target sites, often by permeating across a biological barrier that restricts the free access of nanoparticles to underlying organs/cells/subcellular organelles. This targeting mode encompasses all kinds of targeted delivery, such as delivery to tumor, lungs, liver, brain, and so on. Different biological sites within the body overexpress "signature" biomolecules, called biomarkers. By conjugating a nanocarrier with targeting groups which specifically interact with such site-specific biomarkers, targeted delivery to the sites of interest can be achieved. Examples of such delivery modes include (a) conjugating polysaccharides for liver-targeted delivery and (b) monoclonal antibodies for tumor-targeted delivery.

For active targeting, the cell surface biomarkers should have some intrinsic properties that make them a suitable tumor-specific target. The most important factor is the exclusive expression of the biomarkers (surface antigens and receptors) on the tumor cells only, and not on the normal cells. Also, these biomarkers should not be released into blood circulation, which otherwise may be lead to blocking of the actively targeted moieties in the blood itself. One major impact area of active targeting is in cancer therapy. Because of tumor-associated metabolic and other abnormalities, cancer cells overexpress a number of receptor biomolecules (antigens) on their surface, which serve as indicators of neoplastic growth. Such receptors serving as biomarker signatures can be generic for all tumor types. Examples are folic acid or transferrin receptor. Alternatively, there can be antigens that are specific to the origin of the cancer, such as prostate-specific membrane antigen (PSMA) for prostate cancer and her/2-neu receptor for breast cancer. Detailed molecular research of cancer cells has further shed light on the diversity of such biomarkers that differentially express themselves at various stages of cancerous transformation. A class example is pancreatic cancer, where the various stages of transformation, spanning from normal cells to highly invasive cancer cells, can be correlated with the overexpression of several such biomarkers. Targeting such cancer biomarker receptors with diagnostic nanoparticles conjugated with the corresponding ligands can help clinicians to accurately stratify the stage and progression of this disease. Such accurate "staging" of this disease will lead to a better and "personalized" prescription of the desired surgical/therapeutic regimen, thus maximizing clinical benefit.

7.5. TARGETING GROUPS

A variety of biomolecules have been used as conjugates to nanocarriers to introduce targeting ability on them. The cellular uptake mechanism using the targeting molecules, which includes antibodies, peptides, aptamers, and small molecules, usually exploits the receptor mediated endocytosis. First, the actively targeted nanoparticles bind specifically to the surface receptors of the tumor cells. The nanoparticle–antigen receptor complex is further internalized by folding of the plasma membrane to form early endosomes. These endosomes aid in trafficking of the nanoparticles by pH change and activating the lysozymes, thus releasing the free drug or the nanoaprticles to the cytoplasm. From here the nanoparticles are transported to their target organelle, depending on the physicochemical properties of the nanoparticles. The released receptors are replenished back to the cell surface, once the bound nanoparticle–antigen is released from the complex. This section provides a brief description of some important classes of targeting agents.

Targeting Small Molecules. Folic acid is an example of relatively small molecules which has extensively been utilized to target cancer cells. The advantage is their small size compared to antibodies (discussed below), which facilitates their intracellular entry. Folate receptors are extensively present on the tumor cells, and they bind very specifically with high affinity to the folic acid conjugates. The cellular internalization mechanism of these folate-bound nanoparticles occurs via the receptor-mediated endocytosis [Leamon et al., 2004]. However, they tend to be non-specific in differentiating one type of cancer from another.

Antibodies. These are large proteins that selectively bind with an antigen (expression on the surface of the cell being targeted) because of their geometric (site) compatibility. Very often an antibody–antigen pair's selective association in terms of their conformational compatibility is represented as a lock (antibody) and key (antigen) combination, as shown in Figure 7.5. The use of antibodies for active targeting was initially attempted for direct conjugation with the drug molecules. However, due to the nonselective conjugation of a

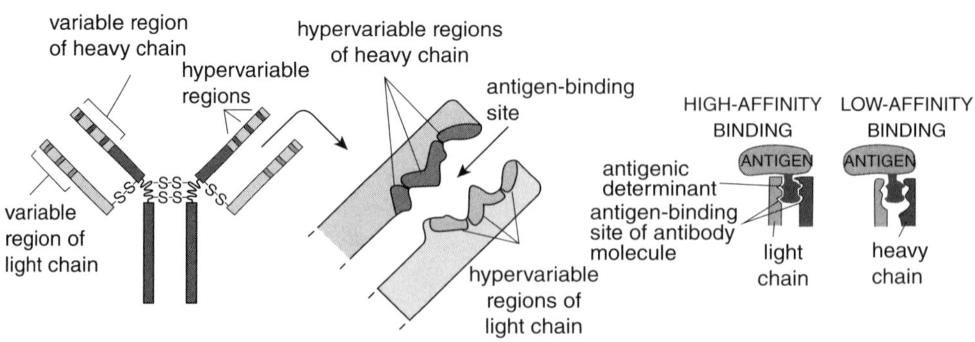

Figure 7.5. Schematic representation of antibody–antigen selective recognition. (Reproduced with permission from Alberts et al. [2002].)

large number of drug molecules to antibody active sites, the immune specificity of the antibody was compromised and this system was discouraged after clinical trials due to lower specificity. In this regard, nanoparticles containing the drug payloads presented very attractive agents for active targeting, because the surface of these nanoparticles can be easily functionalized with different antibodies without compromising their targeting affinity.

Aptamers. These are small-chain oligomers of nucleic acid (oligonucleotides) or amino acids (peptides) which are designed to selectively bind with biomarkers of a specific biological site. The oligonucleotide aptamers consist of short strands of DNA or RNA, engineered through repeated rounds of *in vitro* selection. They offer specificity in biorecognition equivalent to that of antibodies, but with the advantage that they can be produced in a test tube by chemical synthesis, have long shelf storage properties, and elicit very little immunogenicity. Peptide aptamers consist of a variable short-length peptide attached at both ends to a protein scaffold. The use of this double structure strategy provides constraints in binding, leading to increase in the binding affinity of a peptide aptamer that is comparable to an antibody. The peptide loop is typically of 10–20 amino acids, and the scaffold is of a soluble protein. In recent years, aptamer targeting has received considerable attention because of the flexibility of its structural manipulation.

One of the successful examples of aptamer targeting is anti-PSMA aptamers. These aptamers have been shown previously to bind to the prostate cancer cells with nanomolar affinity. Langer's group has coupled the PSMA aptamers to the PLGA-based nanoparticles and have shown efficient targeting capabilities via a receptor mediated internalization mechanism [Farokhzad et al., 2004; Chu et al., 2006]. The design of a self-assembled aptamer–micelle nanostructure by Tan's group exhibits selective and strong binding to the mimicking tumor site. In this design, a lipid tail is attached onto the end of an aptamer inserted with a PEG linker. Under the critical micelle condition, this amphiliphilic unit self-assembles into a spherical micelle with a hydrophobic core (Figure 7.6) [Wu et al., 2010]. Our group has also conjugated PSMA aptamer to the ORMOSIL nanoparticles to target the LNCaP cell line [unpublished results].

The only limitation associated with the aptamer arises from the nucleic acid charge (due to the negatively charged phosphate backbone), which leads to electrostatic repulsion from negatively charged cell surface. The Tan group has recently developed the cell-SELEX method to generate a panel of aptamers successfully [Xiao et al., 2008]. This method enables the generation of multiple aptamers for molecular recognition of target cells and can be used as excellent molecular recognition probes. They have further worked extensively with the sgc8 aptamer to show its enhanced targeting capabilities.

Integrin Receptor Targeting Peptides. Another mode of active targeting of tumors takes advantage of the phenomenon of angiogenesis of cancer, which is the formation of new blood vessels (neovasculature) around growing

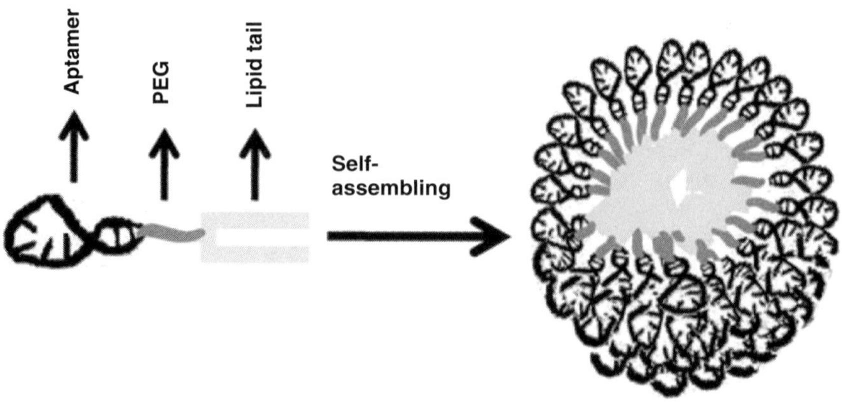

Figure 7.6. DNA aptamer–micelle for efficient delivery to cancer cancers developed by Tan's group. (Reproduced with permission from Wu et al. [2010].)

tumors in order to meet their demand of high nutrient supply. During angiogenesis, the endothelial cells surrounding tumor tissue reorganize themselves into capillaries that are attached to the extracellular matrix (ECM). Integrins such as αvβ3 are cell-surface receptors that anchor these endothelial cells to the ECM and transducer signals from the extracellular environment to the cell's interior. During angiogenesis, the αvβ3 integrins are known to be overexpressed on the surface of the endothelial cells surrounding tumor tissues. Therefore, the αvβ3 integrin serves as an extremely potent target from the point of view of the development of novel diagnostic and therapeutic probes that would specifically "home in" on tumor sites *in vivo*. Of special interest is the triplet peptide Arg-Gly-Asp (RGD), which is known to specifically bind to the αvβ3 integrins overexpressed on the tumor endothelium, thus serving as an antagonist against vascular endothelial cells and tumor growth. In addition to their potential role as a therapeutic molecule, RGD peptides-conjugated nanoparticles loaded with imaging probe(s) can also be used as a diagnostic tool to image αvβ3 integrin expression on developing tumors. Several groups, including our group, have demonstrated robust tumor targeting using nanoparticles conjugated with the RGD peptide. Specifically, our group has demonstrated early diagnosis of pancreatic cancer using several types of quantum dots, such as CdSe/CdS/ZnS, InP/ZnS, Silicon, and so on, using noninvasive optical imaging in small animals. Besides diagnostics, RGD peptide-mediated delivery of drug-loaded nanoparticles has resulted in significant tumor regression *in vivo*.

The conjugation of targeting agents to the nanoparticles surface can be achieved by the two most common approaches, either by direct covalent conjugation or by noncovalent interactions between nanoparticles and targeting molecules [De et al., 2008; Rosi and Mirkin, 2005; Pellegrino et al., 2005]. The covalent conjugation can be carried out either by direct coupling between the functional groups present on the biomolecules and nanoparticles surface or by using heterobifunctional cross-linkers. Direct covalent conjugation can be done through a variety of well-known chemical reactions such as carbodiimide-mediated amidation or esterification, Traut's reagent, and so on [Wang, 2003]. Noncovalent assembly of nanoparticles and biomolecules is usually based on electrostatic interactions, intercalation, complementary binding, and so on.

HIGHLIGHTS OF THE CHAPTER

- By applying different targeting strategies, nanocarriers can be delivered to a specific tissue or organ (first-order targeting), a specific type of cell (second-order targeting), or even specific subcellular structures (third-order targeting).
- Targeted delivery increases the therapeutic drug load in the diseased parts of the organism, thereby enhancing drug efficiency, while reducing the off-target accumulation of the drugs and keeping the unwanted side effects low.
- Biotargeting of drugs to specific cellular organelles enhances molecular selectivity of the drug action and can reduce the side effects.
- The cellular uptake of pharmaceutical nanocarriers occurs in a process of endocytosis. The process of endocytosis can be subdivided into four categories, namely, clathrin-mediated endocytosis, caveolae-mediated endocytosis, macropinocytosis, and phagocytosis.
- In particular, the clathrin-mediated endocytosis occurs via small vesicles coated with a protein named clathrin; caveolae-mediated endocytosis occurs through numerous invaginations (caveolae) on the plasma membrane; pinocytosis utilizes uptake of small particles through numerous small vesicles; and phagocytosis is a process in which specialized cells (such as macrophages) internalize relatively large particles of 0.7 μ or larger. All types of endocytosis are utilized to design the mechanism of regulated uptake of nanocarriers.
- Biotargeting can be either active or passive. Passive targeting utilizes specific features in the structure of a targeted site that would preferentially retain nanocarriers. Example of passive targeting is an accumulation of nanoparticles into malignant tissues, due to impaired circulation common for most solid tumors.
- Active biotargeting involve conjugating a nanocarrier with targeting groups, which specifically interact with site-specific biomarkers. Examples of active targeting include conjugating nanocarriers with polysaccharides for liver-targeted delivery or antibodies for tumor-targeted delivery.
- Functionalization of nanocarriers for biotargeting can be achieved by using various targeting groups including antibodies, peptides, aptamers, and other small molecules. Antibodies (immunoglobulins) are large

macromolecules capable of selective binding to specific molecules, such as receptors expressed on the surface of diseased cells. Aptamers are small nucleic acid or amino acids ligands designed for selective binding with biomarkers expressed at a specific biological site. An example of a small molecule as a targeting group is a folic acid or folate. Because folate receptors are frequently overexpressed in cancer cells, folate-functionalized nanoparticles are used for recognition and therapies of tumors. Similarly, integrin receptor targeting peptides are used for targeting of cell-surface integrin receptors in endothelial cells, which are overexpressed in many types of tumors.

EXERCISES

True or False

1. Targeting of a specific type of cell is called first-order targeting.
2. Passive targeting refers to the property of a molecular probe to selectively accumulate at a specific biological site.
3. Nanoparticle can be decorated with folic acid for active tumor targeting.
4. Drug can be specifically delivered to the targeted site by the direct conjugation of drug with antibodies.
5. Tumor can be recognized through the targeting of integrin receptor using nanoparticles.

Multiple Choice

6. Which of the following targeting is known as third-order targeting?
   (a) Tumor
   (b) Mitochondria
   (c) Macrophage
   (d) Hippocampus
7. The process in which a nanocarrier penetrates the cell membrane is called
   (a) Hydrolysis
   (b) Angiogenesis
   (c) Pathogenesis
   (d) Endocytosis
8. Which of the following polymer has the "proton-sponge" effect in the endosome?
   (a) Polyethylene glycol
   (b) Polyethylenimine

   (c) Polylysine
   (d) Liposome

9. Enhanced permeability and retentivity effect is an example of
   (a) Intrinsic targeting
   (b) Receptor-mediated targeting
   (c) Passive targeting
   (d) Field-directed targeting

10. Which of the following is not a targeting molecule?
   (a) PSMA
   (b) Transferrin
   (c) Doxil
   (d) Aptamer

Answers: 1. F; 2. F; 3. T; 4. F; 5. T; 6. b; 7. b; 8. b; 9. c; 10. c

REFERENCES

Alberts, B., Johnson, A., Lewis, J., Raff, M., Roberts, K., Walter, P., *Molecular Biology of the Cell*, Garland Publishing, New York (2002).

Chu, T. C., Twu, K. Y., Ellington, A. D., Levy, M., Aptamer mediated siRNA delivery, *Nucleic Acids Res.*, **34**, e73 (2006).

Curiel, D. T., High-efficiency gene transfer employing adenovirus–polylysine–DNA complexes, *Nat. Immun.*, **13**, 141–164 (1994).

De, M., Ghosh, P. S., Rotello, V. M., Applications of nanoparticles in biology, *Adv. Mater.* **20**, 4225–4241 (2008).

Farokhzad, O. C., Jon, S., Khademhosseini, A., Tran, T. T., LaVan, D. A., Langer, R., Nanoparticle–aptamer bioconjugates: A new approach for targeting prostate cancer cells, *Cancer Res.* **64**, 7668–7672 (2004).

Furlani, E., Magnetic biotransport: Analysis and applications, *Materials* **3**, 2412–2446 (2010).

Hillaireau, H., Couvreur, P., Nanocarriers' entry into the cell: Relevance to drug delivery, *Cell Mol. Life Sci.* **66**, 2873–2896 (2009).

Law, W.-C., Yong, K.-T., Roy, I., Xu, G., Ding, H., Bergey, E. J., Zeng, H., Prasad, P. N., Optically and magnetically doped organically modified silica nanoparticles as efficient magnetically guided biomarkers for two-photon imaging of live cancer cells. *J. Phys. Chem. C* **112**, 7972–7977 (2008).

Leamon, C. P., Reddy, J. A., Folate-targeted chemotherapy, *Adv. Drug Deliv. Rev.* **56**, 1127–1141 (2004).

Nishikawa, M., Huang, L., Nonviral vectors in the new millennium: Delivery barriers in gene transfer, *Hum. Gene Ther.* **12**, 861–870 (2001).

Pellegrino, T., Kudera, S., Liedl, T., Javier, A. M., Manna, L., Parak, W. J., On the development of colloidal nanoparticles towards multifunctional structures and their possible use for biological applications, *Small* **1**, 48–63 (2005).

Rosi, N. L., Mirkin, C. A., Nanostructures in biodiagnostics, *Chem. Rev.* **105**, 1547–1562 (2005).

Ragnarsson, E. G., Schoultz, I., Gullberg, E., *Yersinia pseudotuberculosis* induces transcytosis of nanoparticles across human intestinal villue epithelium via invasin-dependent macropinocytosis, *Lab. Invest.* **88**, 1215–1226 (2008).

Schillinger, U., Brill, T., Rudolph, C., Huth, S., Gersting, S., Krotz, F., Hirschberger, J., Bergemann, C., Plank, C., Advances in magnetofection—Magnetically guided nucleic acid delivery, *J. Magn. Magn. Mater.* **293**, 501–508 (2005).

Wang, J., Nanoparticle-based electrochemical DNA detection, *Anal. Chim. Acta* **500**, 247–257 (2003).

Wu, Y. R., Sefaha, K., Liu, H. P., Wang, R. W., Tan, W. H., DNA aptamer–micelle as an efficient detection/ delivery vehicle toward cancer cells, *Proc. Natl. Acad. Sci. USA* **107**, 5–10 (2010).

Xiao, Z., Shangguan, D., Cao, Z., Fang, X., Tan, W., Cell-specific internalization study of an aptamer from whole cell selection, *Chemistry* **14**, 1769–1775 (2008).

CHAPTER 8

MULTIMODAL BIOMEDICAL IMAGING

Biomedical imaging has become one of the most relied-upon tools in healthcare for diagnosis and treatment of human diseases. The evolution of medical imaging from plain radiography (radioisotope imaging), to X-ray imaging, to computer-assisted tomography (CT scans), to ultrasound imaging, to magnetic resonance imaging (MRI), and to optical imaging has led to revolutionary improvements in the quality of healthcare available today to our society. A nanoparticle platform allows for simultaneous use of various imaging techniques as well as to utilize novel imaging methods that can lead to early detection, screening, diagnosis, and image-guided treatment of life-threatening diseases and cancer. There is a clear need for extending imaging to the cellular and molecular levels that can lead to the detection of early stages of the formation of a disease or cancer or early molecular changes during intervention or therapy.

The necessary multifunctionalities for multimodal imaging have already been discussed in Chapter 5. This chapter details the different types of medical imaging techniques which can be applied for *in vitro* and *in vivo* diagnosis. The various imaging techniques are introduced, and their relative merits are discussed in Section 8.1. Section 8.2 provides a detailed discussion of various optical imaging techniques and gives examples of both *in vitro* and *in vivo* imaging. Section 8.3 discusses magnetic resonance imaging and the applications of nanoparticle technology to enhance the contrast as well as to provide

Introduction to Nanomedicine and Nanobioengineering, First Edition. Paras N. Prasad.

for another modality of MRI. Section 8.4 focuses on CT scans, Section 8.5 discusses radionuclide imaging, and Section 8.6 covers ultrasound imaging. Electron microscopy, a routine technique for bioimaging, is not discussed in detail, because very little has been reported on nanoparticle enhancement for this imaging modality. Section 8.7 provides details of the relatively new photoacoustic tomography and how nanoparticle technology has advanced the capability of this imaging technique. Terahertz imaging so far has provided only limited biological utility and thus is not discussed in detail. Finally, Section 8.8 discusses examples of a nanoparticle platform which can provide integrated multimodal imaging capabilities.

8.1. BIOMEDICAL IMAGING TECHNIQUES

A number of biomedical imaging techniques such as MRI, X-ray, radio imaging, and ultrasound imaging are now routinely used in disease diagnosis. However, there are other imaging techniques that can also provide valuable information on diseases at their early stages and their progression as well as on treatment-induced digression. Figure 8.1 lists these imaging techniques. Some of them may be more suitable for *in vitro* and others more useful for *in vivo*.

A brief description of each imaging modality is presented here.

Optical Imaging. Optical bioimaging utilizes an optical contrast provided by a difference in the optical properties such as light transmission, reflection, and fluorescence as well as by multiphoton processes, between the region to be imaged and the surrounding background region. For detailed coverage of optical bioimaging the reader is referred to the book *Introduction to Biophotonics* [Prasad, 2003]. The various optical functionalities utilized have already

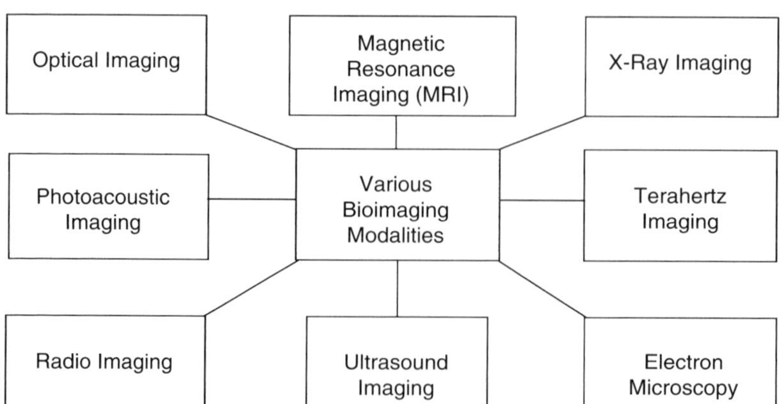

Figure 8.1. The various biomedical imaging techniques.

been discussed in Chapter 5. The nanoparticle technology has made a significant impact on enhancing the capabilities of optical imaging. It provides new nanoemitters such as quantum dots, quantum rods, upconverting nanoparticles, and aggregation-enhanced two-photon emitters introduced in Chapter 5. Second, nanomanipulation of optical excitation energy can be used to selectively highlight a specific biological site as well as to study macromolecular interactions (such as protein–protein interactions). Third, combining spectroscopy of various kinds in a localized domain with imaging leads to multispectral imaging that can provide information on distribution, interactions, and redistribution of various molecular contents at a biological site (cells or tissues). This detailed information can be used to probe molecular events that are early signatures of a disease, or which indicate a cell response to specific stimuli (e.g., drug response). Also, a nanoparticle platform allows one to use multiple targets with a different color emitter to simultaneously target and identify many biological signatures (biomarkers) of a specific disease by multiplexed imaging. The advantage of the optical technique is a more detailed molecular imaging with very high special resolution. In addition, by using nanoprobes and nanoemitters that are sensitive to pH, temperature, activities of a specific ion, and the electrical potential on the cell surface, one can also perform functional imaging. Optical imaging can be performed on *in vitro*, *ex vitro*, and *in vivo* specimen that lead to a highly integrated approach for diagnostics. Section 8.2 presents a detailed description of nanoparticle-enabled multimodal and multiplexed *in vitro* and *in vivo* optical imaging.

Magnetic Resonance Imaging (MRI). This imaging modality utilizes spin transition between the spin states of a nucleus, induced by an rf field. As discussed in Chapter 5, the nuclei utilized routinely for MRI are protons present in water, which is the most abundant chemical constituent of the body. Here the imaging contrast used to distinguish a biological region of interest from the surrounding background involves the relaxation times of the excited spin state. The concept of two relaxation times T_1 and T_2 and the imaging modalities have been discussed in Chapter 5. A magnetic nanoparticle with the ability to selectively accumulate at a specific biological site (e.g., by active targeting discussed in Chapter 7) changes one of these relaxation times and thus provides contrast from the surrounding region [Varadan et al., 2008]. Examples of nanoparticle-enabled MRI are presented in Section 8.3. MRI provides good spatial resolution (25–100 μm) and can be used for both morphological and functional imaging.

A new-generation MRI being developed in many laboratories, including ours, utilizes spin transitions of another nucleus ^{19}F that can easily be incorporated in a nanoparticle formulation to enhance the local concentration of ^{19}F at a desired biological site and thus increase the sensitivity. Thus, the considerably lower magnetic resonance signal of ^{19}F compared to a proton can be compensated for by significantly enriching its concentration at a biological site to be imaged. Furthermore, since ^{19}F is not intrinsic to the body, targeting of a specific biological site can produce MRI images with great spatial selectivity against a zero background. In addition, incorporation of ^{19}F imaging capability requires only a minor modification of the instruments currently used for proton based MRI.

X-Ray Imaging. X-ray imaging is routinely used in medical diagnosis. Here the property utilized is attenuation of X ray in tissues, because different compounds inside the body exhibit differences in absorption/scattering and thus attenuation of X-ray. A popular form of X ray imaging, called computed tomography (CT), scans the X ray over desired sites to be imaged and utilizes an inverse reconstruction algorithm to generate a three-dimensional image. Nanoparticle contrast enhancers are designed to contain iodine, barium salts, and so on, which strongly absorb X rays and thus enhance the contrast from the surrounding regions. X-ray-based CT imaging provides a good anatomical visualization of body structures and detailed map of various organs.

Radio Imaging. This modality uses a radioactive nucleus (radionuclide) that distributes and/or accumulates at the site of biological interest to provide a radioactive signal that is detected by a scintillator to reconstruct the image [Wagner et al., 1995]. The two major techniques of radionuclide imaging are positron emission tomography (PET) and single photon emission computed tomography (SPECT). The physical principles of these modalities have already been discussed in Chapter 5. The various radioisotopes used for PET imaging are ^{16}F, ^{64}Cu, ^{68}Ga, and ^{124}I which are incorporated in a nanoparticle formulation either by chemical conjugation or anchored onto the nanoparticle using chelating groups. For SPECT imaging, the radioisotopes commonly used are ^{125}I, ^{111}In, and ^{99m}Tc. The advantages offered by radionuclide imaging are unlimited tissue penetration, high sensitivity, and the ability to provide quantitative analogues that are very useful for preclinical and translational applications.

Ultrasound Imaging. Ultrasound imaging (often called sonogram) is one of the most widely used medical imaging techniques that is low cost, can be performed by a portable imaging system, and is highly safe [Bloch et al., 2004]. For this modality of imaging, the contrast agents used can resonate in an applied ultrasound beam, undergo shell oscillations, or expand and contract in response to acoustic pressure changes, which can be used for enhancement of contrast.

The three types of particle formulations as contrast enhancers for ultrasound imaging that have been extensively studied are microbubbles, liposomal agents, and perfluorocarbon-emulsion nanoparticles. The microbubble contrast agents are not really representative of nanotechnology because their sizes are in micrometers, but they have been extensively used. These contrast enhancers increase the reflected ultrasound echoamplitude from the blood pool to overcome a relatively small acoustic impedance difference between the red blood cells and plasma, thus allowing for detection of blood vessels and capillaries.

The microbubble agents consist of a gas bubble contained in an outershell. An example is an air-filled microbubble with a denatured albumin shell that is an FDA-approved contrast agent. Liposomal agents consist of liposomes that are less than 1 μm in diameter. Perfluorocarbon emulsion nanoparticles developed by Lanza et al. [1996] are composed of a lipid-encapsulated perfluorocarbon liquid that reflects acoustic wave. These particles have a mean diameter of about 250 nm.

Electron Microscopy. Electron microscopy has been widely used for imaging of fixed cells and tissue specimens. Electron microscopy can be used in two modes: (a) scanning electron microscopy (SEM), which provides a 3D image with some sacrifice in spatial resolution (≥10 nm), and (b) transmission electron microscopy (TEM), which utilizes the transmission of an electron beam through the sample. It is TEM which has been more extensively used for biomedical imaging. It provides the highest spatial resolution (≥1 nm) derived from the very short wavelength of an electron wave compared to optical waves (for a detailed discussion on this topic, the reader is referred to the monograph *Nanophotonics* by this author [Prasad, 2004]). This high resolution provides high fine structural details at the organelle level. Because intrinsic cellular constituents do not strongly scatter electrons, nanoparticles of heavy elements can be used to increase scattering and thus enhance contrast. Electron microscopic imaging of cellular structures of a targeted pancreatic cancer cell was performed in our laboratory using targeting gold nanoparticles. This method has limited medical imaging utility because it cannot be used for *in vivo* imaging.

Photoacoustic Imaging. This is a relatively new biomedical imaging method. Excellent comprehensive reviews of this method and its potential application are provided by Kim et al. [2010] and Xu et al. [2006]. It utilizes the photoacoustic effect, which is the generation of acoustic waves by the absorption of electromagnetic energy such as that from light or from an rf field. When light is used to generate acoustic signal, photoacoustic imaging combines ultrasonic imaging with high contrast provided by light absorption in the biological region to be imaged. In this method, a biological target is irradiated with a short-pulsed laser beam which is absorbed either by the intrinsic components of the biological regions or by targeting nanoparticles that have accumulated in this region. The excitation energy deposited in this region by optical excitation is partially converted into heat that generates a pressure rise through thermoelastic expansion. This pressure rise leads to their propagation as an acoustic wave that is detected by ultrasonic transducers. Figure 8.2 provides a simplified schematic of the photoacoustic (PA) imaging principle. PA images are obtained by using the arrival times of the ultrasonic waves. In the more common version of PA imaging, computed tomographic image processing is used; thus the method is also referred to as photoacoustic computed tomography (PACT). This method combines the merit of both optical imaging (higher

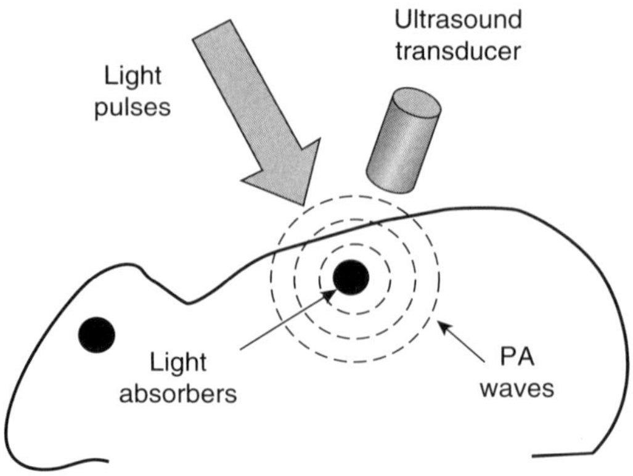

Figure 8.2. A simplified schematic of photoacoustic imaging principles. (Reproduced with permission from Kim et al. [2010].)

contrast compared to ultrasound imaging) and ultrasound imaging (longer penetration depth compared to optical imaging).

Gold nanostructures have been extensively used to enhance the contrast as well as to provide target imaging [Kim et al., 2010]. Section 8.7 provides a detailed description of PACT and gives examples of nanoparticle enhanced bioimaging.

Terahertz Imaging. This technique of imaging has found a broad range of applications ranging from materials analysis to security [Chan et al., 2007]. It utilizes attenuation of a terahertz beam that is in the frequency range of 1–3 terahertz (far-infrared region). Terahertz imaging can be performed in time domain using a pulse source or in frequency domain using a CW source. It utilizes both the absorption and refraction properties of a domain to generate the imaging contrast. Since the wavelength of a terahertz radiation is significantly longer than a tissue structure, scattering of these waves is very small. Water strongly absorbs terahertz radiation, thus making this technique not suitable for biomedical imaging. Generally, the technique has been utilized for imaging *ex vivo* dried biological specimens [Woodward et al., 2003]. Although the nanoparticle contrast enhancer has been explored for terahertz bioimaging, practical utility of this method for biomedical imaging is still limited. Thus, this method will not be discussed any further.

Finally, a comparison of the features of these different imaging modalities is presented in Table 8.1. As can be seen, different imaging modalities produce complementary information that can be advantageously used to establish a diagnosis, to assist the clinician in developing an individualized therapeutic approach, or to allow early detection of onset of a disease.

As seen from the table, optical imaging has the highest sensitivity and spatial resolution but lacks the ability to obtain anatomical and physiological

Table 8.1. Comparison of Various Biomedical Imaging Technologies

Imaging Technology	Imaging Modality	Spatial Resolution and Sensitivity	Applications	Contrast Agents	Requirement for Contrast Agents	Resolution Goal of Imaging Technology			
						Anatomy	Physiology	Metabolism	Molecular
Optical	Fluorescence, vibrational intensity, bioluminescence	Molecular-μm-mm Single cell to 10^5 cells	Metabolism, receptor, tumor progression, gene expression, therapy assessment	Fluorescent dyes, quantum dots	Can be used with or without contrast agents but mostly required				
MRI	Atomic nucleus	0.1 mm 10^{-4} moles	Metabolism, tumor progression gene expression, therapy assessment	Paramagnetic ions (Gd, Mn and Fe_2O_3)	Can be used with or without contrast agents but mostly required				
PET	Radioactive tracer	5–10 mm 10^{-12} moles	Blood flow, metabolism, volume profusion	Radioactive nucleotides (^{11}C, ^{18}F, ^{124}I)	Required				
SPECT	Radioactive tracer	5–10 mm 10^{-10} moles	Blood flow, netabolism, volume profusion	Radioactive nucleotides (^{39}Tc, ^{123}I, ^{111}In, ^{125}I)	Required				
CT	X ray	50–150 mm mg	Anatomy, reveal contrast, contrast movement	Radio-opaque ions (I, Br, Ba)	Can be used with or without contrast agents but mostly required				
Ultra-sound	Acoustic	5 mm	Anatomy	Micro-bubbies	Can be used with or without contrast agents				

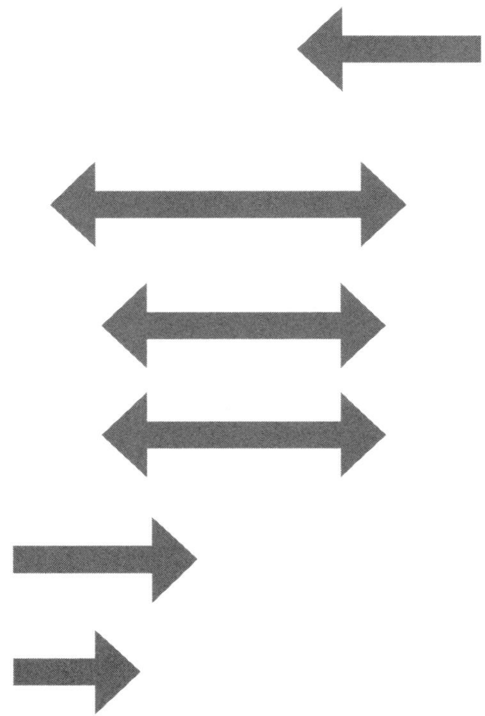

in vivo imaging. At the same time, CT and ultrasound imaging can be used to produce anatomical information but lack the capacity to yield molecular level targeted imaging capability. Until recently, imaging technologies have primarily focused on a single aspect of a disease process—for example, the anatomical structure or a functional change. Some technologies, such as MRI, can go further and image both anatomy (proton density) and other physiological aspects of tissue, such as nuclear spin relaxation times. However, combining an anatomical technique, such as CT, with a nuclear medicine-based functional technique (SPECT or PET), has traditionally required advanced image analysis algorithms to align the images from the two modalities. Therefore, there is an increased interest in developing new ways of combining these modalities to obtain complementary information. The nanoparticle enabled multimodal biomedical imaging is the topic of Section 8.8.

8.2. OPTICAL BIOIMAGING

8.2.1. Fluorescence Microscopy

Optical bioimaging covers a broad scope as shown in Figure 8.3. It utilizes a number of spectroscopic techniques with a broad range of microscopic methods

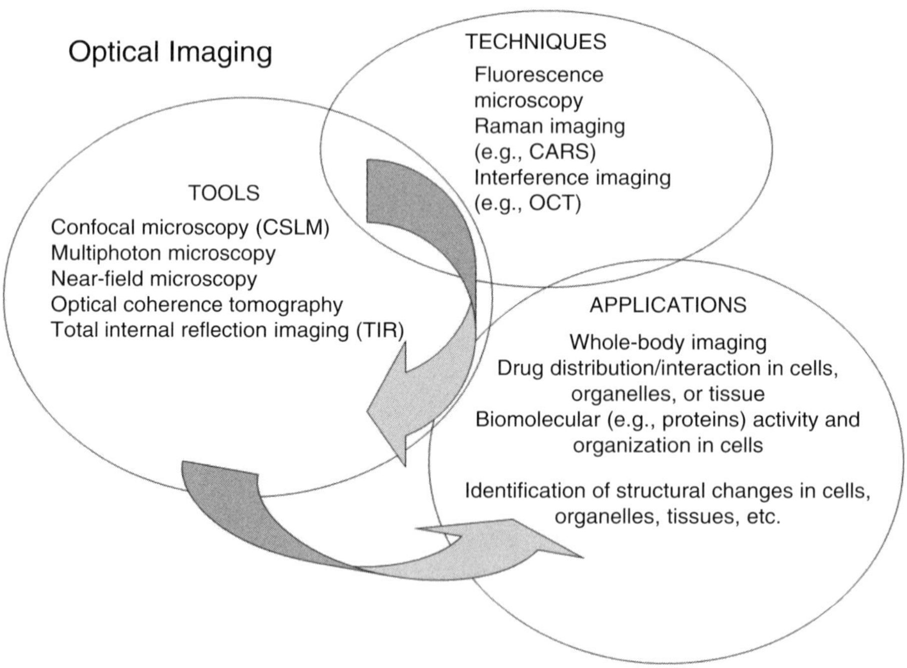

Figure 8.3. Optical bioimaging: Techniques, microscopic methods, and applications.

and can provide information on a wide length scale, from molecular level to whole body. In this section, we cover the various nanoemitters used for fluorescence imaging which is the most widely used optical method. Then scattering based methods are briefly discussed followed by examples of nonlinear optical techniques. Multiplexed cellular imaging for *in vitro* study is presented. Finally, examples are provided for *in vivo* optical imaging.

To obtain high spatial resolution in fluorescence-based three-dimensional imaging, confocal and multiphoton microscopies are commonly employed, allowing to obtain detailed information on cellular interactions and dynamics [Prasad, 2003]. Even though confocal microscopy provides a high-resolution optical sectioning capability, it has some inherent problems. Since the confocal aperture reduces the fluorescence signal level, one needs a higher excitation power, which increases the possibility of photobleaching of dyes. Multiphoton microscopy offers the following advantages over confocal microscopy, (a) significantly reduced fluorescence background due to the relatively low two-photon cross section of most biomolecules responsible for autofluorescence in biological systems, (b) reduction of photobleaching by selective excitation of the focal volume, and (c) improved depth penetration in scattering samples due to the use of excitation light in a near-infrared (NIR) spectral range. These advantages are especially important for imaging thick samples or for ultra-low concentration of analytes (approaching single molecular imaging). However, a major problem in two-photon excited fluorescence microscopy is the low two-photon absorption cross section of most of the commonly used fluorophores.

Photoluminescent nanoparticles for confocal and multiphoton microscopy offer a number of advantages over traditional fluorescent dye molecules, such as increased signal/noise ratio, higher stability against photobleaching, and facile chemistry of conjugation with targeting biomolecules due to availability of numerous reactive groups on the nanoparticle surface. Some important classes of photoluminescent nanoparticles have already been discussed in Section 5.3 of Chapter 5. These nanoparticles also show strong two-photon excitation to be useful for two-photon microscopy. Here we briefly highlight their applications in optical bioimaging.

An example of organic dye nanoaggregates that show aggregation enhanced fluorescence (AEF) under two-photon excitation is 9,10-bis[4′-(4″-aminostyryl) styryl]anthracene (BDSA) [Kim et al., 2006]. We have succeeded in loading ORMOSIL nanoparticles with high concentration of BDSA (about 30–40% of BDSA with respect to nanoparticle matrix), and no quenching was observed. A nanoparticle formulation with 25% loading of the BDSA dye (~2500 molecules per nanoparticle) has a fluorescence QY of 0.2 per dye molecule. Thus, BDSA-ORMOSIL nanoprobe provides a 500-fold enhancement over a single fluorescent molecule with a fluorescent QY of unity. Figure 8.4 shows a schematic of the BDSA-ORMOSIL nanoparticles and two-photon image of cultured cancer cells internalized with them [Kim et al., 2007a]. Gold nanoparticles have also been used for two-photon luminescence imaging. The article by Tong

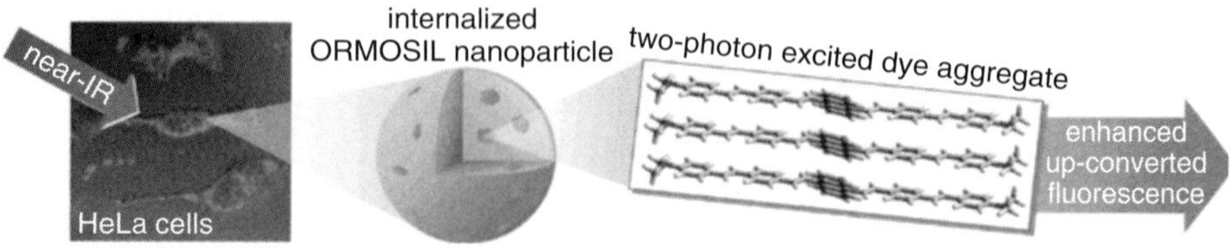

Figure 8.4. ORMOSIL nanoparticles containing nanoaggregates with enhanced two-photon excited emission as high-signal-output nanoprobes for two-photon fluorescence bioimaging. (Reproduced with permission from Kim et al. [2007a].)

et al. [2009] has reviewed the developments in the application of nanorods as multifunctional agents for imaging and therapy, but mostly focusing on the therapeutic modality with imaging capability.

Quantum dots and rods, discussed in Chapter 5, are widely used nanoemitters because of their advantageous size-tunable optical properties, high quantum efficiency, high molar extinction coefficients, broad excitation and narrow emission spectra, high resistance to photobleaching, and distinctive emission lifetimes [Prasad, 2004]. Two general approaches are being commonly used to transfer organically dispersible QDs to aqueous phase for bioimaging. The first one involves replacing hydrophobic moieties from the QDs surface with hydrophilic ligands, such as mercapto acids. The second method involves encapsulating the QDs within an amphiphilic nanostructure such as a polymeric micelle or an ORMOSIL shell (see Chapter 4) which can further be conjugated to targeting biomolecules.

In our group, a number of different QDs have been prepared. The example presented here (see Figure 8.5) is a core/shell/shell-type CdSe/CdS/ZnS quantum rod nanostructures (QRs) for multiplexed *in vitro* imaging. To demonstrate this, CdSe/CdS/ZnS QRs with two different aspect ratios (3.5, orange-emitting; 4, red-emitting) were prepared, and the orange-emitting and the red-emitting QRs were conjugated with transferrin (Tf) and folic acid (FA), respectively. *In vitro*, we have treated KB cells, which are known to overexpress the corresponding receptors for both transferrin (TfR) and folic acid (FAR), with both of these QR bioconjugates. Both of these QRs (orange and red) were internalized within the same cell, with some areas of emission overlap. This result clearly suggests that it is possible to study the simultaneous uptake of multiple QR-bioconjugates by using *in vitro* confocal microscopy.

8.2.2. Quantitative FRET Microscopy

The process of Forster resonance energy transfer (FRET) has been described in Chapter 5. Quantitative FRET microscopy is a powerful tool used to study protein–protein interaction in their native environment. Over the last decade, FRET methods have been developed for microscopic analysis of molecular associations and conformations within living organisms [Periasamy and Day,

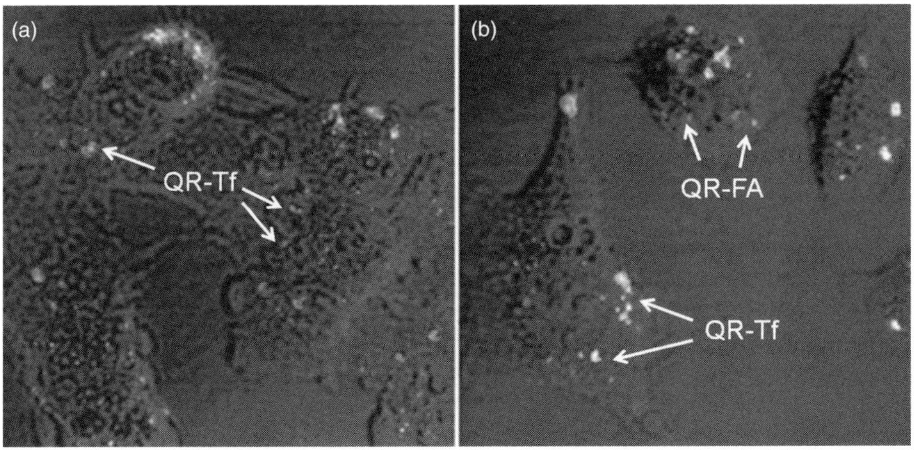

Figure 8.5. (a) Confocal microscopy image of KB cells treated with Tf-conjugated orange- and red-emitting CdSe/CdS/ZnS QRs. (b) Confocal microscopy image of KB cells treated with Tf-conjugated orange-emitting and FA-conjugated red-emitting QRs Color version of this figure can be found in Yong et al. [2008]. (Reproduced with permission from Yong et al. [2008].)

2005; Periasamy and Clegg, 2010]. The development of the spectral variants of green fluorescent protein (GFP) for tagging of proteins by genetic manipulation (see the literature at this URL: http://www.kcci.virginia.edu/Literature/index.php), together with quantitative FRET microscopy, promises to provide unprecedented new insights into understanding the dynamics and localization of molecular interactions inside living cells.

Since FRET depopulates the excited state of the donor (D), it results in a decreased probability of photon emission from D and a shortening of D's lifetime in the excited state; meanwhile, the probability of photon emission from the acceptor (A) increases. Measurement of each event can provide direct proof of energy transfer. Thus FRET can provide two- or three-dimensional spatial and temporal information on protein–protein interactions in single living cells under physiological conditions. In a two-color FRET, two different fluorophores are used to label two proteins. Figure 8.6 provides an example of a two-color FRET that utilizes lifetime measurements to study localization of homo-dimerization of CCAAT/enhancer-binding protein alpha (C/EBPα) in living cell nucleus microscopy. The decrease in fluorescence lifetime decay of cerulean indicates the dimerization of the C/EBPα in a living cell nucleus.

Employing a traditional two-color FRET method to determine the interrelationships of three cellular components (A, B, C) would require sequential imaging and analysis of the FRET-pair combinations of AB, BC, and AC. In a three-color FRET, three different fluorophores are used to label three

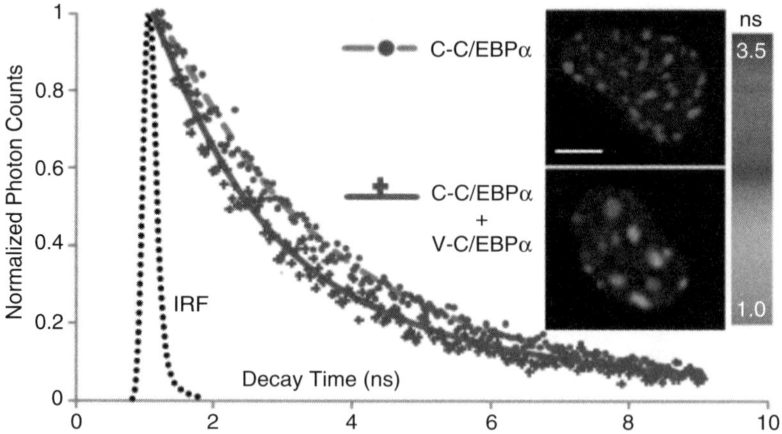

Figure 8.6. Localization of homo-dimerization of CCAAT/enhancer-binding protein alpha (C/EBPα) in living cell nucleus using two-color. C/EBPα was tagged with either cerulean (C) or Venus (V). The fluorescent lifetime decay kinetics for the C-C/EBPα (FRET donor) in the absence and the presence of V-C/EBPα (FRET acceptor) were determined by fitting the measured decay data into a single or double exponential decay model, respectively, with the measured instrument function (IRF). (Reproduced with permission from Sun et al. [2011].)

proteins. This three-color FRET analysis system is of great benefit to determine protein complex assemblies during signaling events, trafficking dynamics, or cytokinesis, particularly in the context of treatments, which may alter the relationship of the three components of interest. Recently, a three-color FRET model based on the detection of the sensitized emissions from acceptors through steady-state imaging has been reported. This model has three symbolic fluorophores—F1 (donor), F2 (acceptor to F1 and donor to F3), and F3 (acceptor)—and assumes that spectral overlap (donor emission–acceptor absorption) of each pair (F1–F2, F1–F3, F2–F3) is sufficient for FRET to occur [Sun et al., 2010]. Sun et al. used a three-color FRET model and applied it to monitor the interaction of dimerized proteins C/EBPα with Hp1α [Sun et al., 2010].

An important application of FRET using the dye aggregates exhibiting aggregation enhancement discussed above is in optical highlighting. We demonstrated this concept by co-encapsulating within ORMOSIL the BDSA (donor) with another fluorescent dye, Nile Red (acceptor), whose absorption spectrum overlaps with the BDSA emission to obtain an effective intraparticle FRET [Kim et al., 2007b]. Following the treatment of HeLa cells using ORMOSIL nanoparticles co-encapsulating BDSA and Nile Red, bright red fluorescent cellular staining (from Nile Red) was observed upon two-photon excitation, with minimum concurrent green cellular staining (from BDSA), owing to an efficient intraparticle FRET between the BDSA and Nile Red.

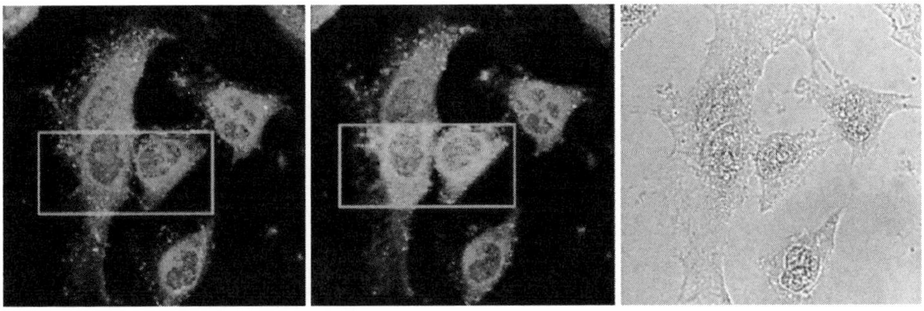

Figure 8.7. Donor–acceptor channel-merged two-photon fluorescence images of HeLa cells treated with NP-AD for 1 h, before (left panel) and after (central panel) acceptor bleaching in the areas indicated by boxes. Right panel shows transmission image of cells. (Reproduced with permission from Kim et al. [2007b].)

Upon selectively photobleaching the acceptor molecules at a specific intracellular region of interest (ROI), the donor emission was found to increase significantly at the ROI. Thus, with the use of this acceptor bleaching technique, either a whole cell or part of a cell can be optically highlighted, as shown in Figure 8.7. This technique allows the study of intracellular particle dynamics by following the temporal fluorescence change in the ROI. Owing to the intracellular movement of the nanoprobes, following selective photobleaching, the emission in the acceptor channel increases by the entry of acceptor-unbleached intact nanoprobes (denoted as NP-AD), while the emission in the donor channel decreases by the exit of acceptor-bleached ones (denoted as NP-D).

8.2.3. Technical Challenges for *In Vitro* Imaging

Technical challenges for *in vitro* imaging and some of the approaches used by our group are highlighted in Figure 8.8. A major focus is on designs of nanoemitters that emit in the near IR in order to utilize the biological window of maximum optical transparency. This topic has also been discussed in detail in Chapter 5, Section 5.2. Thus, recent efforts have been concentrated in developing near-IR nanoemitters [Jaffer et al., 2009]. In our own lab, we have developed CdTe-based quantum dots and rods for this purpose [Law et al., 2009; Yong et al., 2011]. Another important class has been rare-earth-doped up-converting nanophosphors discussed in Section 5.3 of Chapter 5. Of particular interest is our formulation of nanophosphors that absorb at ~974 nm and emit at ~800 nm.

A major concern with Cd-based quantum dots and rods is their potential toxicity due to release of the Cd ions. A strategy used with some success is surface functionalization and biocompatible coating as illustrated in Figure 8.8. Another approach leading to a growing area of research activities is use

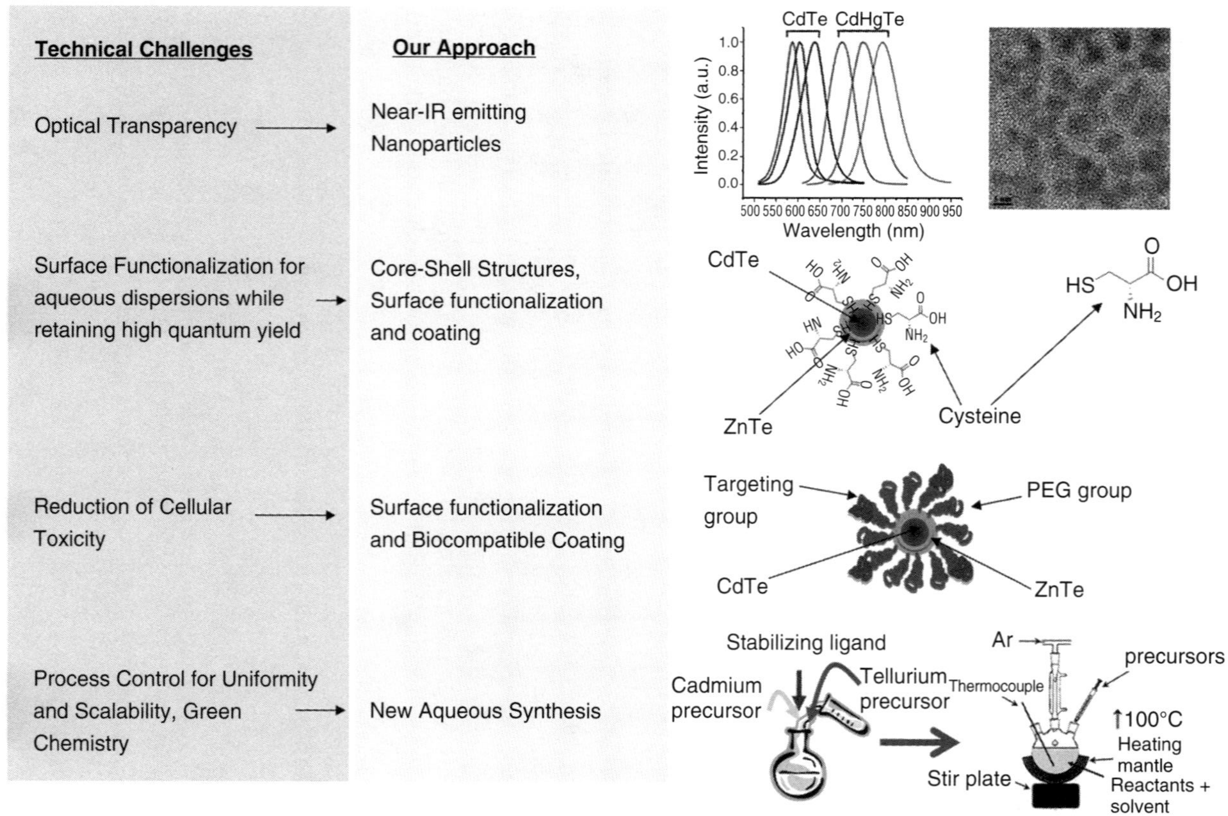

Figure 8.8. Technical challenges for *in vitro* optical imaging and representative strategies to overcome them.

of Cd-free quantum dots and nanoemitters. In our own laboratory, we have successfully used InP/ZnS and $CuInS_2$/ZnS core–shell-type quantum dots. A nanoemitter showing great promises for both *in vitro* and *in vivo* optical imaging is silicon quantum dots, discussed in Chapter 4 as well as in Chapter 5. Below we provide an example of their application for *in vivo* imaging using near-IR emitting Si quantum dots.

8.2.4. *In Vivo* Optical Imaging

In our group, *in vivo* tumor targeting was demonstrated by encapsulating NIR Si QDs in phospholipids micelle and conjugating to RGD peptides that bind to $\alpha_v\beta_3$ integrins in the tumor vasculature. *In vivo* optical imaging was carried out using a Maestro GNIR FLEX fluorescence imaging system (CRi, Cambridge, MA). Excitation of the photoluminescence is provided by the light from a Xe lamp, introduced with fiber bundle into imaging chamber using appropriate excitation band-pass filter to select suitable wavelength range of excitation. An additional long pass filter is located in front of input objective of the optical head, to cut off the excitation light. A tunable filter is automatically stepped in 5- or 10-nm increments from 500 to 950 nm, while a CCD camera captures images at each wavelength with constant exposure. The resulting images are loaded into a single data structure in memory, forming a spectral stack with a spectrum at every pixel. With the software, small differences in spectra can be rapidly detected and analyzed. The Si QDs injected into the animals are tracked by subtracting the background, which originate from autofluorescence and scattered excitation light passing through the cut-off imaging filters. A spectral unmixing algorithm (Maestro software) is then applied to create the unmixed images. Figure 8.9 shows a schematic of the *in vivo* optical imaging approach and an example of the obtained targeted tumor optical imaging.

8.2.5. Optical Coherence Tomography

Optical coherence tomography (often abbreviated as OCT) is a bioimaging technique that is rapidly growing in its applications [Fercher, 2010]. It is a reflection imaging technique similar to ultrasound imaging, except that a light wave (usually in the near-IR to IR range) scattered from a specific biological/tissue site is used to image. The sensitivity of the scattered light, as well as its selectivity from a specific backscattering site, is achieved by using the interference between the backscattered light and a reference beam. The OCT method of imaging is particularly suited for a highly scattering medium, such as a hard tissue (teeth) [Xu et al., 1999]. The interference between the propagating wavefronts of two light sources occurs when both wavefronts have well-defined coherence (phase relation) within the overlapping region. This well-defined coherence of a wavefront from a source is maintained within a distance called *coherence length*. Therefore, if both the reference beam and

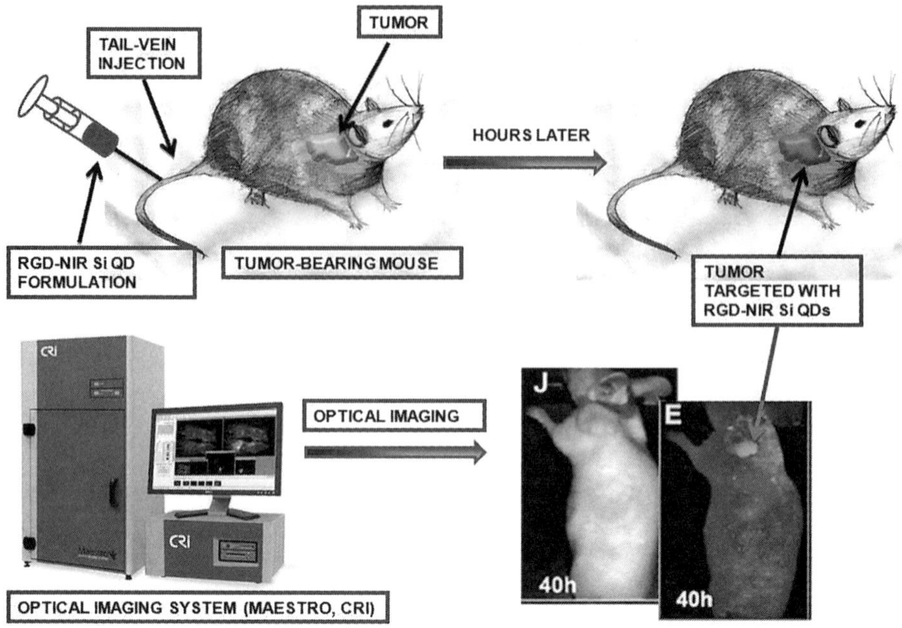

Figure 8.9. Schematic of *in vivo* optical imaging and bright field/photoluminescence images of Panc-1 tumor-bearing mouse injected with NIR Si QDs conjugated with RGD peptide targeting tumor. (Reproduced with permission from Erogbogbo et al. [2011].)

the beam back-reflected from a scattering site are derived from the same light source, a well-defined interference pattern will be produced only if their path-length difference is within the coherence length. This behavior is shown in Figure 8.10 [Prasad, 2003].

A displacement of the reference beam produces the path-length difference between the light reflected from the reference mirror and the backscattered photons from the scattering sample. In the case of a fully coherent source (such as a high-coherence laser source), the interference between the reference beam and the backscattered beam can be maintained over a large path-length difference induced by reference mirror displacement. Thus no selectivity to backscattering from a specific depth in the sample can be achieved in this case. The case on the right-hand side is for a low-coherence source (with a short coherence length). In this case the interference pattern between the reference beam and the backscattered beam is produced only when their path difference is within the coherence length. In a three-dimensional scattering medium, at any given location of the reference beam mirror, only a given depth range (defined by the coherence length) of backscattered light will interfere. Therefore, by scanning the reference mirror, one can achieve the depth discrimination. This is called time-domain OCT (TD OCT). The interference patterns

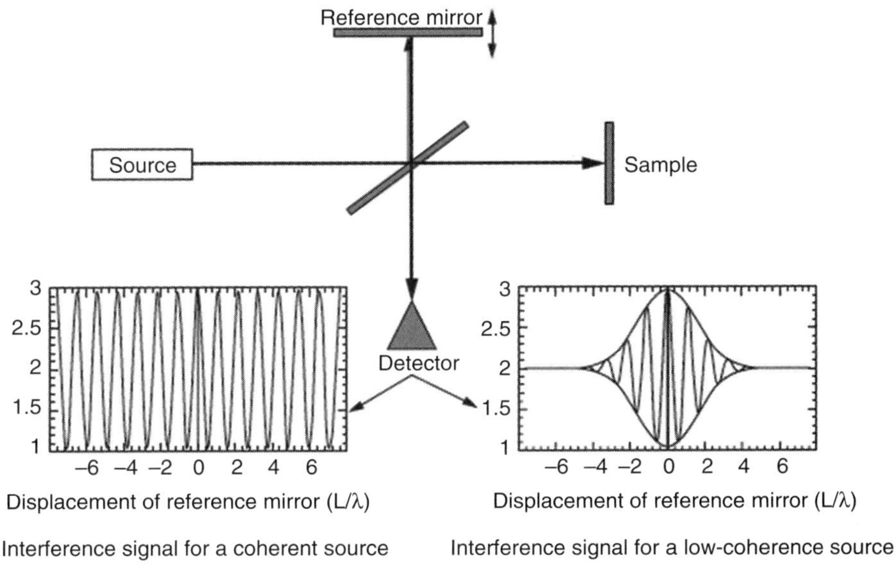

Figure 8.10. The interference signal as a function of the reference mirror displacement in case of a coherent source (e.g., laser) and a low-coherence source (e.g., SLD) are shown here. (Reproduced with permission from Prasad [2003].)

contain the information about the refractive index variation of the sample (tissue) which thus provides an optical image.

In OCT, the axial (depth) resolution is defined by the coherence length of the light source. The shorter the coherence length, the better the depth resolution. A broad-band light source has a short coherence length. Therefore, one often uses an incoherent, but bright, light source such as a superluminescent diode [Prasad, 2003] or a laser with a low coherence (such as a femtosecond laser source with a broad-band width associated with it). Fujimoto and coworkers [Swanson et al., 1993] developed a compact optical-fiber-based OCT setup as shown in Figure 8.11. In this arrangement, a dual-core fiber is used. One core of the fiber transmits the broad-band light source and splits it into the two arms: the sample probe and the reference. The other core of the fiber collects the backscattered signal and reflected reference beam and combines to produce the interference.

Another variation of a fiber-based OCT has been used by Colston and others [Colston et al., 1998] for dental applications.

Besides the initially used time-domain technique, using a path-length-modulated reference beam, Fourier-domain (FD)-based OCT has been developed [Smith et al., 1989; Fercher, 2010]. Employing the channeled spectra technique, FD OCT can be implemented based either on a spectrometer ("spectral" FD OCT) or on a tunable laser ("frequency tuning" or "frequency swept" FD OCT). In FD OCT, depth information can be obtained by a Fourier

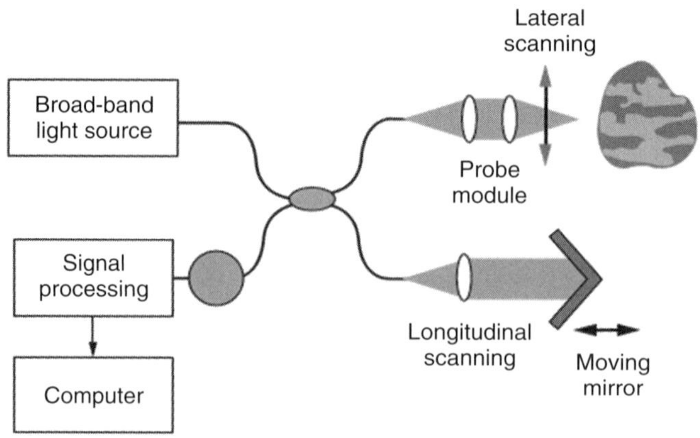

Figure 8.11. A fiber-based OCT. (Reproduced with permission from Swanson et al. [1993].)

transform of the light spectrum at the output of an interferometer. According to the Wiener–Khintchine theorem, which states the relation between the auto correlation function and the spectral power density, the depth scan can be directly calculated from the acquired spectra, without movement of the reference mirror. This feature dramatically increases scanning speed. In comparison with TD OCT, where rate of scan limited by the movement of the mechanical parts, in FD OCT the acquisition rate is limited only by the detector frame transfer rate and the computer calculation time to perform the Fourier transform of the acquired raw data. As a consequence, high scanning speed results in lower data variability due to the patient's movements. At the same time the reduced losses during a single (2D) scan enhance the signal-to-noise ratio of raw data. All these features enhance the quality of imaging and diagnostics reliability.

Modern tunable lasers based on developments in the telecommunication fields have recently replaced imperfect external-cavity tunable laser diodes, by cavity-tuning "swept laser sources," where the spectral filtering is performed inside the laser cavity. Now this direction becomes the most popular among the other OCT techniques.

Advantages provided by OCT optical imaging are:

(i) Resolutions of 4–20 μm compared to 110 μm for high-frequency ultrasound.

(ii) At or near real-time imaging.

(iii) Relatively straightforward integration with small catheter/endoscopes for the fiber-based design.

(iv) Fiber-based OCT are compact, portable, and noncontact and can be combined with laser spectroscopy and Doppler velocimetry to monitor blood flow (discussed in Chapter 12).

Compared to two-photon multiphoton microscopy, the advantages offered by OCT are:

(i) Higher capability of imaging turbid medium (e.g., biological tissue).
(ii) High depth resolution, even when the depth resolution of the objective is low.
(iii) No need to use fluorescent stains or tags as in the case of confocal fluorescence or TPLSM.
(iv) Compact design compared to TPLSM setup (using SLD instead of Ti:sapphire laser as light source).

However, the depth resolution provided by OCT is lower than that for two-photon fluorescence or confocal microscopy with high magnification objective.

8.2.6. Super-Resolution Fluorescence Microscopy

Super-resolution imaging is a rapidly developing area which can achieve resolution previously only achievable using electron microscopy. It is based on fluorescence imaging (Section 8.2.1) and is designed to go beyond the diffraction limited resolution (Abbe limit), thus allowing to probe biological structures with nanometer resolution. These techniques can be broadly classified into three groups: illumination sculpting (4-Pi, STED), single molecule localization (PALM, STORM), and patterned or structured illumination (SIM, I^5M).

Illumination sculpting techniques can further be divided into evanescent wave techniques (such as near-field scanning optical microscopy (NSOM), total internal reflection fluorescence (TIRF) microscopy), interference-based (4Pi, I^5M) techniques [Hell and Stelzer, 1992; Gugel et al., 2004], and techniques based on photophysical sharpening of point spread function (STED). The evanescent wave-based techniques can improve spatial resolution in selected dimension to the range of <50 nm. For example, NSOM can obtain ~50-nm XY resolution, but can only be used to probe surfaces. Similarly, TIRF imaging can obtain <50 nm axial resolution (with diffraction-limited XY resolution), but again the applications are limited to the interface of glass–water interface (e.g., membrane structures of adherent cells). These specialized techniques are described in *Introduction to Biophotonics* by Prasad [Prasad, 2003].

Interference-based techniques, on the other hand, can probe in three dimensions, but the improvement in resolution can only be in one dimension. For example, the 4-Pi technique utilizes two high NA objectives to focus light in opposing direction, and it relies on the interference between these two beams to shrink the focus spot size in the axial direction. This method can achieve <100-nm axial resolution, however, retaining diffraction-limited resolution in XY direction.

Stimulated Emission Depletion (STED). Photophysical sculpting of the point spread function (PSF), on the other hand, can improve the XYZ resolution and be used to obtain super resolution images in 3D. Stimulated emission depletion (STED) was the first concrete and feasible concept showing that, in fluorescence microscopy, the diffraction barrier can be broken using such manipulation of the PSF [Hell and Wichmann, 1994]. The basic concept behind point scanning STED microscopy is to confine the fluorescence emission of fluorescent markers to a region that is much smaller than that covered by the diffraction-limited excitation spot. To achieve this feature, a second red-shifted "depletion" beam, shaped into doughnut profile, is overlaid with the excitation beam. This high-power doughnut beam (typically at a wavelength matching the tail end of the emission spectra) creates a condition for stimulated emission, thereby de-exciting the molecules in its region to prevent them from producing their natural broadband fluorescence (Figure 8.12). Thus the resulting fluorescence signal is confined to a spot much smaller than the spot size of the excitation beam. This sculpted emission PSF can be achieved in all directions (X,Y, and Z) with a properly a shaped depletion beam [Harke et al., 2008]. Though most STED imaging studies have been performed using selected organic fluorophores, recently there have been reports of the use of various inorganic materials for STED imaging [Han et al., 2009].

Super-Resolution Imaging with Localization of Single Molecules. It has been known for some time that emission from single molecules can be localized with a precision much better than the limit imposed by diffraction [Ambrose and Moerner, 1991; Thompson et al., 2002; Yildiz et al., 2003]. The localization accuracy in this case is photon-limited and can be estimated as $s/(N)^{0.5}$, where s is the standard deviation of the PSF and N is the number of photons detected. Based on this concept, a densely populated fluorophore distribution can be divided into a sequence of sparser population and individual fluorophore location can be sequentially localized to reconstruct a full image (Figure 8.13a, b). Several different methods have been proposed for this division of fluorophore distribution into multiple single molecule events or a sequence of sparse fluorophore distributions, each of which has minimal overlap in the images of the individual emitters. Some examples of using transitions between (metastable) on and off states, for the creation of localizable subpopulation of single molecules, are photoactivated localization microscopy (PALM) [Betzig et al., 2006; Hess et al., 2006], stochastic optical reconstruction microscopy (STORM) [Rust et al., 2006; Bates et al., 2007], ground-state depletion microscopy followed by individual molecule return (GSDIM) [Folling et al., 2008], and direct stochastic optical reconstruction microscopy (dSTORM) [Heilemann et al., 2008]. Figure 8.13c shows a wide-field image of microtubules stained with rhodamine 6G, while Figure 8.13b shows the GSDIM images of the same area. Some of these localization techniques can be extended for improving axial resolution as well, for obtaining super-resolved 3D images [Huang et al., 2008].

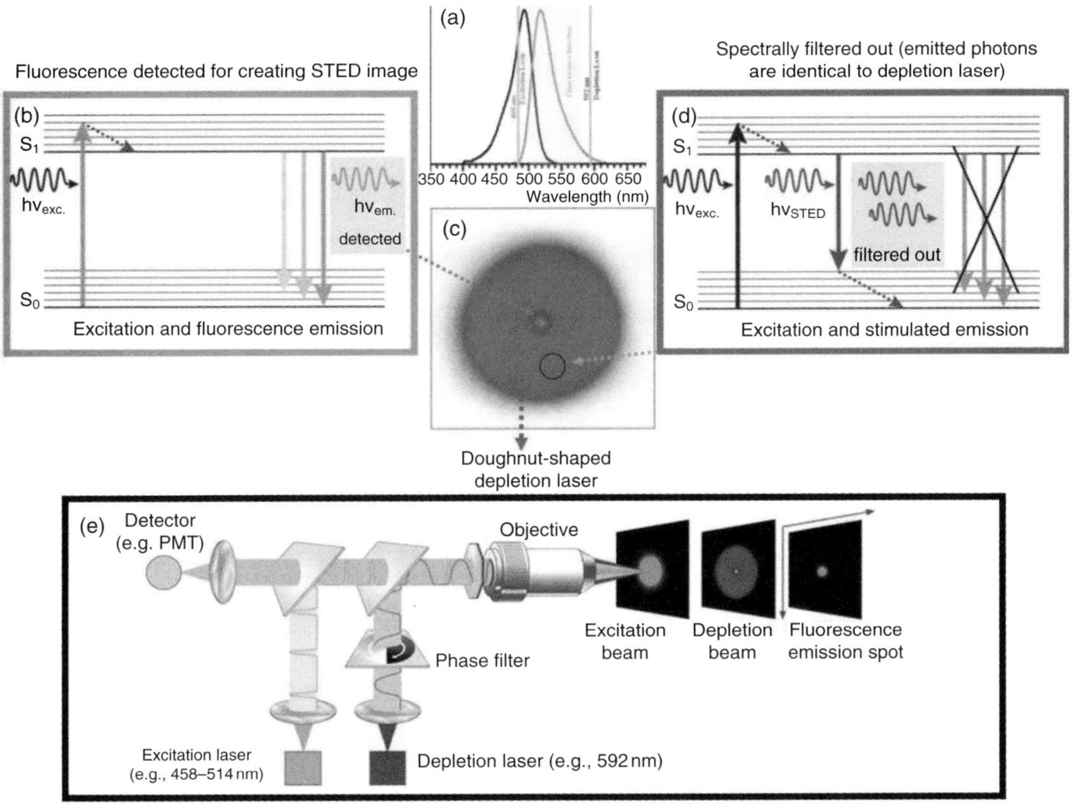

Figure 8.12. Principle of STED imaging. (a) A typical condition for selecting dyes for STED imaging. The depletion laser is chosen close to the tail of the emission spectra. (b) Jablonski diagram showing the emission spectra of typical STED dyes. (c) Doughnut-shaped depletion laser is overlaid with the excitation beam with a Gaussian profile. (d) Jablonki diagram showing the process of stimulated emission condition created by the depletion laser. (e) Illustration of a typical STED setup. A depletion laser with a phase plate in its path to create a doughnut-shaped beam is merged with the excitation laser. (Image provided by Haridas Pudavar, Ph.D., courtesy of Leica Microsystems. Copyright Leica Microsystems, 2012, all rights reserved.)

Super-Resolution Imaging by Using Patterned Illumination. In another approach, a super-resolution image is extracted from a series of images acquired with a patterned illumination. In this approach, commonly referred to as structured illumination (SIM), resolution improvement is achieved by illuminating the sample with a sequence of periodic patterns of high spatial frequencies [Gustafsson, 1999, 2000]. From the Moiré patterns appearing in these sequential images (typically obtained by rotating the illumination pattern), information about still higher spatial frequencies can be computationally extracted. This technique can be extended for obtaining super

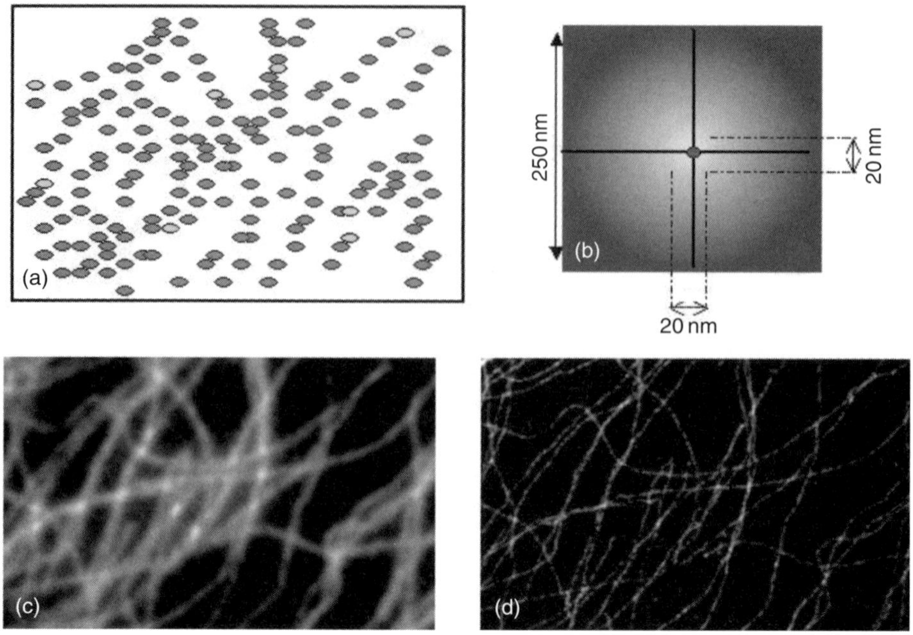

Figure 8.13. Principle of localization techniques. (a) All localization-based super-resolution imaging techniques (PALM, STORM, GSDIM, etc.) involve sequential localization of single-molecule switching and readout. Here gray circles indicate the molecular state before switching (nonfluorescent or fluorescing in different color), while the light gray circles indicate the switched state. (b) Each switched molecule image on the camera is localized by calculating the centroid with much higher precision than the diffraction-limited spot size. (c) Wide-field image of microtubules stained with rhodamine 6G antibody. (d) Ground-state depletion image of the same sample obtained with a commercial GSD system. (Image provided by Haridas Pudavar, Ph.D., courtesy of Leica Microsystems.)

resolution in the axial direction, as well for creating super-resolution images in 3D [Gustafsson et al., 2008; Pielage et al., 2008; Schermelleh et al., 2008).

Nanotechnology and Super Resolution. Among these techniques, STED and SIM can be used for imaging nanoparticles [Kolesov et al., 2011; Tzeng et al., 2011]. Tzeng et al. were able to demonstrate the use of fluorescent nano-diamond (FND) particles, described in Chapter 3, for super-resolution imaging using the STED technique (Figure 8.14). The high photostability of nanoparticles, such as FND particles, can be extremely important for super-resolution techniques. Though localization-based super-resolution imaging is usually achieved by single-molecule event detection, there are reports of using the same technique for single-particle detection such as quantum dot blinking

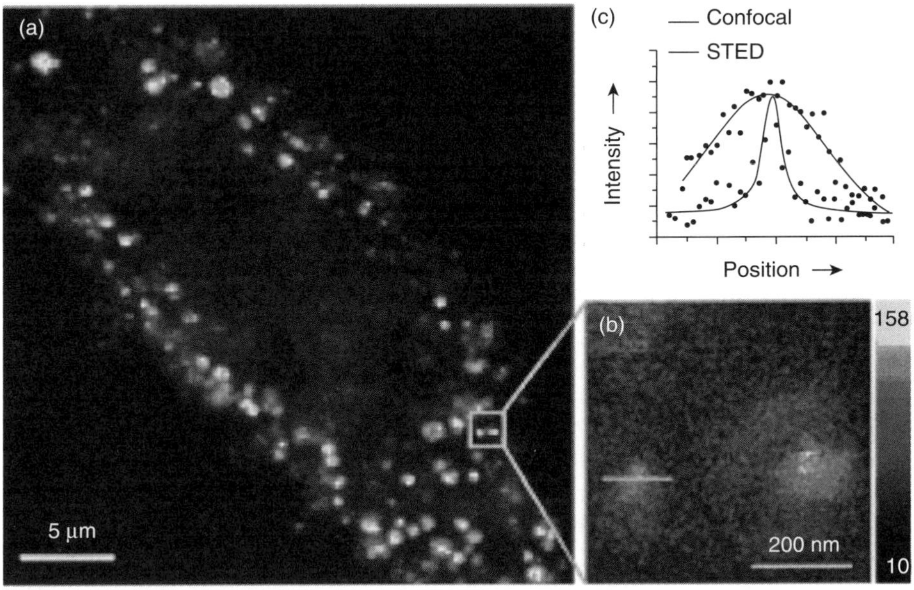

Figure 8.14. Confocal and STED imaging of HeLa cells labeled with BSA-conjugated fluorescent nano-diamond particle (FNDs) by endocytosis. (a) Confocal image acquired by raster scanning of an FND-labeled cell. The fluorescence image of the entire cell is shown and demonstrates fairly uniform cell labeling by BSA-conjugated FNDs. (b) STED image of single BSA-conjugated FND particles enclosed within the box in (a). (c) Confocal and STED fluorescence intensity profiles of the particle indicated in (b) with a line. Solid curves are best fits to one-dimensional Gaussian (confocal) or Lorentzian (STED) functions. (Reproduced with permission from Tzeng et al. [2011].)

[Lidke et al., 2005]. Hence potentially some of these nanoparticles can be combined with localization techniques for achieving improved resolution.

Wang et al. [2012] have demonstrated the potential of exploiting caveolae, which are membrane invaginations of 50–100 nm in diameter that bud from the membrane to form vesicles, to efficiently deliver therapeutic nanoparticles across endothelial barrier using super-resolution microscopy [Wang et al., 2012]. To determine the caveolar size and dynamics in live endothelial cells, they developed the super-resolution microscopy technique using a dual-color nanoparticle pair.

8.3. MAGNETIC RESONANCE IMAGING

Two types of magnetic resonance imaging (MRI) will be discussed in this section. The first one is proton-based MRI, which is currently in practice for medical imaging. The second one is ^{19}F-based MRI, which holds significant promises for next-generation MRI. In both cases, a nanoparticle platform provides significant advantages. The parameters that are used in MRI are as follows:

- Spin density. It is the concentration of the spin nucleus (proton or ^{19}F).
- T_1-relaxation time, which is the spin-lattice relaxation describing the decay of excited spin population.
- T_2-relaxation time, also known as spin–spin relaxation that describes the energy transfer between adjacent spin nuclei.

A change in the value of these parameters from the tissue of interest to the surrounding tissue provides the contrast for MRI. However, the intrinsic contrast for proton MRI derived from protons in water is not large; hence for contrast enhancement, a contrast agent is introduced which accumulates at the biological site (tissue region of interest). This aspect has been described in Chapter 5 and is also briefly mentioned in the overview Section 8.1 of this chapter.

The contrast agents used for proton MRI are materials with superparamagnetic or paramagnetic properties as discussed in Chapter 5, Section 5.4. Here two examples will be discussed in detail. One is Gd^{III}, which is the best contrasting agent due to its large magnetic moment. It decreases the T_1-relaxation time because of the through-space dipole–dipole interaction. This produces increased signal intensity in the MRI signal compared to the intensity in the MRI signal without Gd^{III}. The most widely used Gd contrast agents are chelates of Gd with the diethyltriamine pentaacetic acid (DTPA) or tetraazacyclododecane tetraacetic acid (DOTA). However, there is a major concern about the release of Gd ions which are known to be toxic. Thus, nanoparticle formulations utilizing a Gd form which is strongly bound, so that no free Gd ions are released, are being explored [Kumar et al., 2010a; Chen et al., 2011]. Another factor to be considered is that only water molecules in the near vicinity of Gd^{III} produce a significant change in T_1. Thus a nanoparticle formulation should produce an increase in the number of water molecules in the first coordination sphere to increase the sensitivity.

We have utilized a Gd^{III}-containing nanoparticle formulation that also includes NIR-emitting (phosphorescent) molecules for dual-mode optical and MRI imaging [Kumar et al., 2010a]. Figure 8.15 shows the T_1-weighted MRI image of mouse injected with this Gd^{III}-containing nanoparticle formulation. The other widely used and FDA-approved MRI contrast enhancer is the superparamagnetic iron oxide (Fe_3O_4) nanoparticles, also discussed in Chapter 5. In contrast to the Gd^{III}-based contrast enhancer, the Fe_3O_4 nanoparticles reduce the T_2-relaxation time of water and thus reduce the MRI signal leading to a dark contrast. These nanoparticles typically of $\leq$30nm can be surface-functionalized with a biotargeting group to accumulate in a tissue region of interest (such as a tumor). In many laboratories including ours, these Fe_3O_4

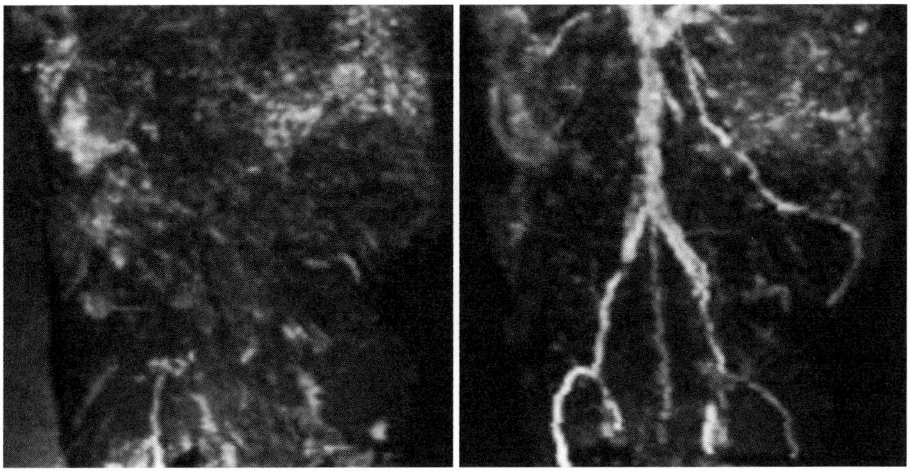

Figure 8.15. *In vivo* imaging with Gd^{III}-containing nanoparticles. Maximum intensity projections generated from T_1-weighted 3D MRI images showed visible contrast enhancement within tumor, liver, and blood vessels after intravenous injection (right) in comparison with the preinjected animal image (left). (Reproduced with permission from Kumar et al. [2010a].)

nanoparticles have been conjugated to or co-encapsulated with an optical nanoemitter (such as a quantum dot) for combined MRI and optical bimodal imaging [Law et al., 2008; Erogbogbo et al., 2010; Lai et al., 2008; Mi et al., 2010].

In vivo ^{19}F MRI offers several advantages compared to proton-based MRI. First, ^{19}F-containing compounds can be directly imaged by MR without background contamination from other molecules or anatomical structures. Secondly, ^{19}F MRI acquisitions yield images of the three-dimensional distribution of ^{19}F-containing molecules and, therefore, enable direct quantitative measurements of the biodistribution, pharmacokinetics and pharmacodynamics of administrated agents [Maxwell et al., 1991]. Thirdly, for high-resolution localization of ^{19}F signals, images can subsequently be registered with high-resolution/proton MRI and/or acquired directly with arbitrarily high spatial resolution dependent only upon signal-to-noise (S/N) per unit time considerations (approximately 17% lower ^{19}F S/N compared to proton S/N per molar concentration). Lastly, ^{19}F MRI T_1 relaxation rates of many perfluorocarbon emulsions have been shown to correlate to pO_2 concentrations in solution and in preliminary *in vivo* studies [Kim et al., 2003]. This ability could allow for noninvasive measurement of tissue oxygenation before, during, and after therapeutic intervention for assessing delivery of radiation, chemotherapy, and/or photodynamic therapy (PDT) to yield improved patient outcome. However, the main limitation of ^{19}F MRI is the paucity of available fluorine-containing

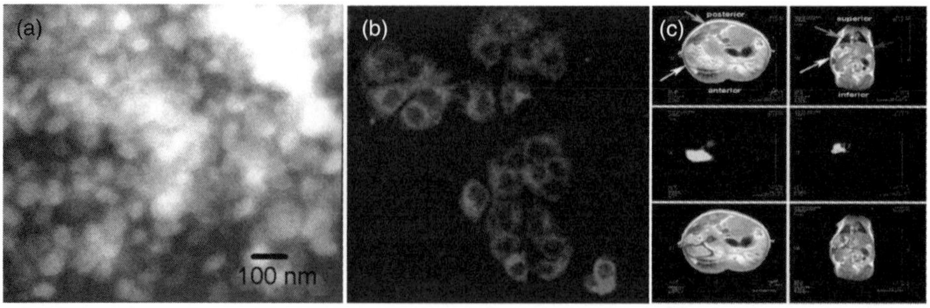

Figure 8.16. (a)TEM image of HPPH and fluorine-19-doped ORMOSIL nanoparticles. (b) *In vitro* confocal microscopic (optical) image of cancer cells. (c) *In vivo* F-19 MR image (stomach) of mice. (Reproduced with permission from Prasad et al., US patent application.)

compounds, which can be administered in sufficient quantities for *in vivo* imaging while remaining nontoxic.

We have designed and synthesized ORMOSIL nanoparticles (diameter < 100 nm) encapsulating fluorophores and incorporating an abundance of ^{19}F atoms in the hydrocarbon arm using an ORMOSIL precursor 3,3,3-trifluoropropyl-trimethoxysilane (TFPTMS). We have demonstrated the feasibility of using these particles under *in vivo* conditions for ^{19}F MRI and *in vitro* fluorescence imaging (Figure 8.16).

8.4. X-RAY CT IMAGING

CT imaging utilizes attenuation of X rays in tissues to obtain morphological and vascular information within the body. Utilizing differences in X-ray absorption and attenuation by different components in the body, one can achieve anatomical visualization of body structures and obtain high-contrast images of blood vessels, stomach, and gastrointestinal organs. Commonly used contrast agents are iodine, barium salt, and gastrografin (see the information in the website of the American College of Radiology Imaging Network, http://www.acrin.org). These contrast agents have been used in the past in molecular or salt forms. In this traditional approach, these contrast agents are relatively short-lived and nonspecific, with no targeting ability. Various nanoparticle contrast enhancers for CT have been introduced which overcome the limitations. In this section, these nanoparticle-based contrast enhancers are described.

One group is the various nanoparticle formulations containing an iodinated structure. One example is a low-density lipoprotein (LDL) nanoparticle that is loaded with poly-iodinated triglyceride (ITG) [Hill et al., 2010]. The LDL

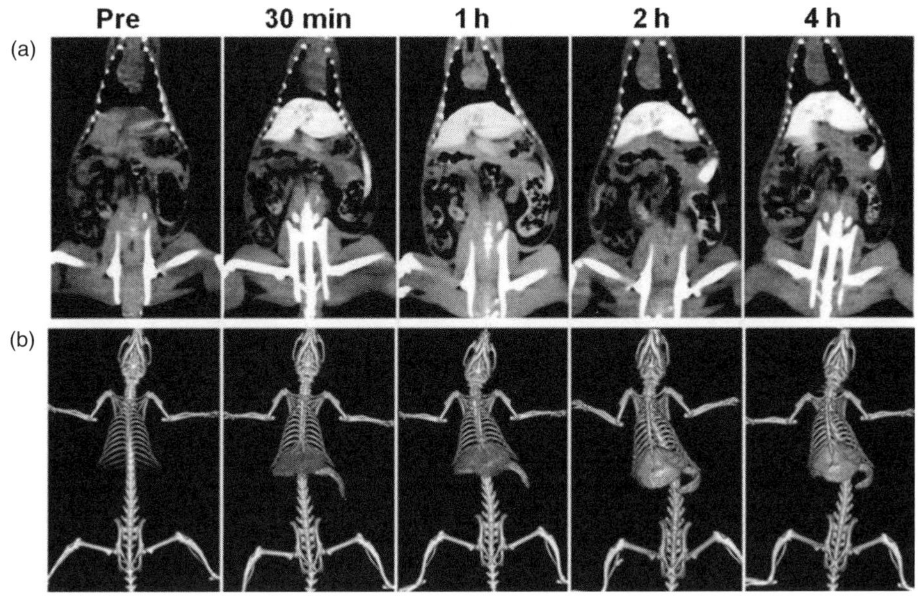

Figure 8.17. *In vivo* X-ray CT imaging. (a) serial CT coronal views of a rat following intravenous injection of 1 mL of PVP-coated Bi_2S_3 nanoparticles. (b) Corresponding 3D renderings of *in vivo* CT images above. (Reproduced with permission from Ai et al. [2011].)

particles are formed by a phospholipid monolayer that has a hydrophobic core into which an ITG, 2-oleoylglycine 1,3-bis(iodopanoate) with a high affinity for LDL, was incorporated. The use of LDL imparts these nanoparticle contrast enhancers with the ability to target the LDL cell surface receptor, LDLR, which is overexpressed in several types of tumors.

Bismuth-containing nanoparticles have also been proposed. An example is provided by bismuth sulfide, Bi_2S_3, nanoparticles [Ai et al., 2011]. As an injectable CT imaging agent, these nanoparticles exhibit excellent stability at high Bi concentration. To make these nanocrystals more stable with long resistance time (avoid fast RES clearance), their surface was coated with a biocompatible polymer, polyvinyl pyrrolidone (PVP). Figure 8.17 shows the CT image of a live mouse obtained with these nanoparticles.

A very promising contrast enhancer for CT is gold nanoparticles [Hainfeld et al., 2006]. These gold nanoparticles are ~1.9 nm in diameter and thus clear through the kidneys. They absorb X rays 3 times more than an equal weight of iodine, thus providing higher contrast than the standard iodine agents. A commercial formulation of these gold nanoparticles as X-ray contrast agents for *in vivo* CT imaging is available from Nanoprobes (NY) under the trade name AuroVist™. They can be applied for imaging using standard micro CT, clinical CT, planar X ray, or mammography.

8.5. RADIO IMAGING

Various nanoparticle formulations for both radionuclide imaging methods, PET and SPECT, have been reported. The flexibility of a nanoparticle platform lends itself to the use of a variety of chemistry to attach various radionuclides, listed in Section 8.1, for imaging to provide quantitation.

For PET imaging, both organic and inorganic nanostructures have been utilized [Welch et al., 2009]. Multiarmed polymers, comprised of a methoxy-terminated poly(ethylene glycol) (MPEG) chain and another segment functionalized with 1,4,7,10-tetraazacyclododecanetetraacetic acid (DOTA) for chelation of the radionuclide, ^{64}Cu, is an example of an organic nanostructure used for PET. These nanoscale star polymers, ranging in size from 25 to 70 nm, provide a long circulation time. Another advantage of using a nanoparticle approach is the ability to attach increased number of radionuclides to each nanoparticle so that PET imaging can be performed at a much lower dose of nanoparticles. Click chemistry has been used to attach a large number of ^{18}F atoms, another nucleotide for PET imaging [Devaraj et al., 2009].

A number of nanoparticle formulations incorporating the various radionuclides suitable for SPECT imaging, as discussed in Chapter 5, have been reported. ^{125}I has received a great deal of attention for incorporating into a nanoparticle. An example is a silver nanoparticle that is capped with poly(*N*-vinyl-2-pyrrolidone) (PVP), with an average size of 12 nm which was radiolabeled by chemisorption of ^{125}I, with a >80% yield [Chrastina and Schnitzer, 2010]. In this work, they intravenously injected these nanoparticles in a BALB/c mouse and evaluated the *in vivo* distribution pattern by noninvasive whole-body SPECT imaging. Another example is ^{125}Cd-labeled CdTe/ZnS core–shell nanoparticles for *in vivo* SPECT/CT imaging [Woodward et al., 2007]. In our own group, an ORMOSIL-based nanoclinic platform incorporating ^{124}I for PET and ^{125}I for SPECT has been used. This ORMOSIL nanoclinic has been used for multimodal imaging, which is discussed in Section 8.8.

8.6. ULTRASOUND IMAGING

The use of external contrast agent provides gain in sensitivity of ultrasound imaging [Bloch et al., 2004]. Furthermore, the ability to introduce targeting leads to selective imaging a specific biological site, with significantly improved contrast. Because of the long wavelength of these very low frequency ultrasound waves, the particles used as contrast enhancers have to be of larger size (≥250 nm) to efficiently reflect them. Thus, the contrast enhancing agents for ultrasound do not in strict sense involve nanotechnology which is generally considered to cover the length scale 1–100 nm. Some important classes of contrast enhancers have already been described in Section 8.1.

The most widely used contrast agent for ultrasound imaging in clinical use and research are microbubbles that have a gaseous bubble enclosed in a shell.

Such microbubbles are commercially available. An example is Optison, commercialized by Mallinckrodt Inc., St. Louis, Missouri, which has perfluoropropane within a denatured albumin shell. Another example is Definity by Bristol–Myers Squibb Medical Imaging, Inc. of Billerica, Massachusetts, which consists of a perfluoropropane-containing agent within a phospholipid shell.

Lanza et al. [1996] have developed perfluoroemulsion nanoparticles, 250 nm in size, consisting of lipid encapsulated perfluorocarbon liquid. Liposomes of diameters less than 1 μm have also been used as ultrasound contrast agents. In the clinical usage, a small amount of the contrast agent is injected intravenously, which enhances the echo signal from the blood pool.

The microbubbles used in the size range 1–8 μm are still small enough to pass through the microcirculation, with substantial persistence in the bloodstream, producing a contrast effect that lasts 3–10 min. One major area of ultrasound imaging where the contrast agents have made a major impact is echocardiography, in which the use of the contrast agents containing gaseous bubbles allows to look at defects in the heart [Soliman et al., 2007]. An abnormal communication will produce the microbubbles to pass into the left side of the heart; in contrast, they will exit a normal heart through the pulmonary artery.

A major thrust in this area is the development of targeted contrast agents, such as targeted microbubbles, which are linked to ligands that bind to receptors on the cells of interest such as inflamed cells or cancer cells. These targeting ligands can be readily attached with the carbodiimide chemistry discussed in Chapter 4. A major issue is that the current targeting strategy produces low adhesion efficiency, which means that only a small fraction of injected microbubbles bind to the biomarkers on the targeted cell.

The microbubble contrast agent also can serve as a drug delivery vehicle by ecapsulating the drug or its nanoformulation, which can be released by breaking the microbubble at higher ultrasonic power.

8.7. PHOTOACOUSTIC IMAGING

A number of nanoparticle formulations have been used to enhance contrast and to image targeted biological sites [Kim et al., 2010; Song et al., 2009]. An important class of contrast enhancers for photoacoustic imaging is gold nanostructures which exhibit strong optical absorption in the near IR, derived from their localized surface plasmon resonance (see Chapter 5, Section 5.3). An example is provided by gold nanocages used by Song et al. [2009] for photoacoustic sentinel lymph node mapping on a rat model. These nanocages can be produced in a broad size range from 35 to 100 nm, providing tunability of their localized surface plasmon resonances to cover the near-IR region which allow deep optical penetration. In addition, gold is also biocompatible. Song et al. were able to achieve sentinel lymph node mapping as deep as ~33 nm. *In vitro* targeted imaging has also been demonstrated [Kim et al., 2010]. Other gold nanostructures, such as gold nanoshells and nanorods, have also been used which provide surface plasmon resonances in the near-IR. Gambhir and colleagues from Stanford University [de la Zerda et al., 2010] have used indocyanine green dye-enhanced single walled carbon nanotubes, conjugated with cyclic Arg-Gly-Asp (also called RGD) peptides as targeted contrast agents, for photoacoustic imaging of tumors. They showed that the photoacoustic signal from a tumor in a mouse was 300 times higher than that from mice injected with nontargeted nanotubes. The good contrast ability of the carbon nanotubes is again derived from its strong absorption of light which gets efficiently converted into heat by rapid nonradiative relaxation processes.

8.8. MULTIMODAL IMAGING

There is a growing interest in the development of "fusion technologies," in which two or more diagnostic approaches are combined to provide complementary information. For example, a CT image can provide information about tissue structure, while PET imaging can complement that data with metabolic process information. Molecular imaging is another emerging field, in which the modern tools of molecular and cell biology are integrated into advanced imaging tools that allow noninvasive imaging for early detection of disease onset, or real-time monitoring of disease treatment progress. Modern drug discovery is also increasingly reliant on molecular imaging in small animals and even in humans to advance and streamline the drug development process. The use of PET or MRI *in vivo* imaging enables easier demonstration of drug efficacy, or the lack thereof. This area also can benefit from the development of custom delivery platforms for drugs and imaging contrast agents, using molecular markers for targeting.

To help address this emerging need, nanoparticle-based probes that contain multiple contrasting agents for simultaneous use in different imaging modalities (e.g., optical, MRI, PET, SPECT, and CT imaging) are being developed in many laboratories [Louie, 2010; Liu et al., 2010; Sharma et al., 2006; Kim et al., 2009; Jin et al., 2010; Frullano and Meade, 2007]. Correlation of these images using a single nanoprobe will accelerate the diagnostic process while reducing stress on the patient. Another advantage of these "multimodal" nanoprobes is the ability to target specific molecular markers. This capability is increasingly important, because biochemical research becomes more successful in characterizing specific molecular factors that accompany a certain disease. These nanoparticle-based contrast agents can be surface-modified to carry various biotargeting agents (e.g., antibodies) for targeting-specific disease markers to achieve earlier detection of diseases. The nanoplatforms provide the next generation of contrast agents for diagnostic systems, thereby increasing the abilities to do the following:

- Perform a direct comparison of histological analysis with whole-body imaging information, by the addition of optical imaging probes along with contrast agents for current diagnostic imaging.

- Improve early diagnosis capability for each imaging system by functionalizing nanoprobe surfaces with biotargeting agents.
- Obtain more detailed information, concerning the anatomical and physiological state of a patient at a given moment in time (real-time information), as opposed to sequential information (the current state of the art in diagnostic imaging).
- Provide tailor-made treatment protocol for an individual at clinical level, by assessing different therapeutic approaches using improved real-time imaging with the nanoparticle platforms.
- Accelerate drug discovery through the use of these multimodal nanoplatforms for studying the efficacy of treatment protocol.
- Reduce patient stress related to the multiple steps necessary for a complete diagnostic workup using the current imaging technology.

The references cited above in this section provide many examples of multimodal imaging that have been demonstrated. Many of them are examples of bimodal like combining PET with CT or PET with MRI [Louie, 2010]. Dual modal imaging combining optical with MRI have also been extensively investigated, and various nanoparticle formulations including core–shell structures have been proposed [Louie, 2010; Frullano and Meade, 2007; Kim et al., 2009]. Our earlier work on nanoclinics, discussed in Chapter 3, also reported incorporation of both optical and magnetic probes providing dual modal imaging [Levy et al., 2002].

We recently have developed multifunctional ORMOSIL-based nanoclinics for multimodal imaging (Kumar et al., 2010b). The design of this silica-based nanoclinic has been presented in Figure 4.13 in Chapter 4. We have used this multifunctional nanoclinic platform to obtain optical, ^{19}F MRI, PET, and SPECT imaging as shown in Figure 8.18.

HIGHLIGHTS OF THE CHAPTER

- The evolution of medical imaging from plain radiography (radioisotope imaging), to X-ray imaging, to computer-assisted tomography (CT scans), to ultrasound imaging, to magnetic resonance imaging (MRI), and to optical imaging has led to revolutionary improvements in the quality of healthcare.
- A nanoparticle platform allows for simultaneous use of various imaging techniques and also allows us to utilize novel imaging methods that can lead to early detection, screening, diagnosis, and image-guided treatment of life-threatening diseases. It can extend imaging to the cellular and molecular levels enabling detection of molecular changes at early stages of a disease during intervention or therapy.
- Optical bioimaging utilizes an optical contrast provided by a difference in the optical properties such as light transmission, reflection, and

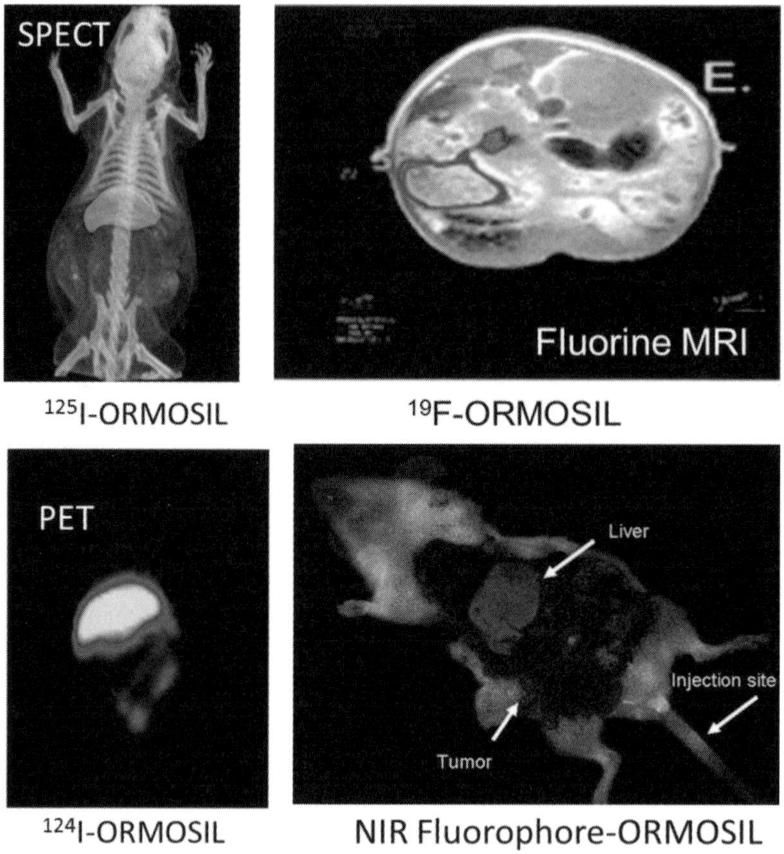

Figure 8.18. Illustrative multimodal images obtained using a multifunctional ORMOSIL nanoclinic. Accumulation of nanoparticles in liver is clearly seen in all modes. (Reproduced with permission from Kumar et al. [2010b].)

fluorescence, as well as by multiphoton processes, between the region to be imaged and the surrounding background region. It can be used on *in vitro*, *ex vivo*, and *in vivo* specimens, providing detailed molecular imaging with very high special resolution.

- Nanoparticle technology enhances the capabilities of optical imaging by (1) providing new optical contrast agents such as nanoemitters (quantum dots and rods, upconverting nanoparticles, aggregation enhanced two-photon emitters, etc.) or nanoscatters (e.g., plasmonic metallic nanoparticles) and (2) using energy transfer to study macromolecular interactions (such as protein–protein interactions). Combining localized spectroscopy with imaging leads to multispectral imaging to provide information on distribution, interactions, and redistribution of various molecular contents at a biological site (cells or tissues).

- Confocal and multiphoton microscopies produce high-spatial-resolution fluorescence-based three-dimensional images, allowing us to obtain detailed information on cellular interactions and dynamics. Multiphoton microscopy offers the following advantages over confocal microscopy: (a) significantly reduced fluorescence background due to the relatively low two-photon cross section of most biomolecules responsible for autofluorescence in biological systems, (b) reduction of photobleaching by selective excitation of the focal volume, and (c) improved depth penetration in scattering samples due to the use of excitation light in a NIR spectral range.
- Photoluminescent nanoparticles for confocal and multiphoton microscopy offer a number of advantages over traditional fluorescent dye molecules, such as increased signal-to-noise ratio, higher stability against photobleaching, and facile chemistry of conjugation with targeting biomolecules due to availability of numerous reactive groups on the nanoparticle surface.
- Silica-based nanoparticles loaded with the organic dye nanoaggregates, showing aggregation enhanced fluorescence under two-photon excitation, can be employed as optical imaging probes that demonstrate manifold enhancement over a single fluorescent-molecule-based probe.
- Quantum dots and rods are widely used nanoemitters because of their advantageous size-tunable optical properties, high quantum efficiency, high molar extinction coefficients, broad excitation and narrow emission spectra, high resistance to photobleaching, and distinctive emission lifetimes.
- A major focus is on designs of nanoemitters that emit in the near IR in order to utilize the biological window of maximum optical transparency. CdTe-based quantum dots and rods can be used for this purpose. Up-converting nanoparticles also show promise for near-IR imaging.
- A major concern with the most commonly used Cd-based quantum dots and rods is their potential toxicity due to release of the Cd ions. A strategy used with some success is surface functionalization and biocompatible coating. Another approach is use of Cd-free quantum dots, such as InP/ZnS and $CuInS_2$/ZnS or Si nanocrystals.
- NIR emitting Si quantum dots were shown to be useful as efficient optical probes, allowing photoluminescence imaging *in vitro* and *in vivo*.
- Quantitative Forster resonance energy transfer (FRET) microscopy utilizing energy transfer between a donor and an acceptor, monitored either by the intensity or the lifetime change of the donor fluorescence, is a powerful tool used to study protein–protein interaction in their native environment.
- Three-color FRET provides information on the interrelationships of three cellular components (A, B, C) which allows protein complex

assemblies during signaling events, trafficking dynamics, or cytokinesis, particularly in the context of treatments.

- An important application of FRET using dye-aggregate-filled nanoparticles is in optical highlighting, where intraparticle energy transfer between nanoaggreagates as a donor and co-doped with nanoparticles containing another organic dye as an acceptor is realized. Upon selectively photobleaching of the acceptor molecules, the donor emission increases significantly, optically highlighting the region of interest (e.g., a cell or part of a cell).
- Optical coherence tomography (OCT) utilizes scattering of the light (usually in NIR to IR range) from a specific biological/tissue site to image. The sensitivity of the scattered light, as well as its selectivity from a specific backscattering site, is achieved by using the interference between the backscattered light and a reference beam. The OCT method of imaging is particularly suited for a highly scattering medium, such as a hard tissue (e.g., teeth).
- Advantages provided by OCT optical imaging in comparison with confocal or multiphoton microscopies are: (i) higher capability of imaging a turbid medium (e.g., biological tissue); (ii) high depth resolution; (iii) no need to use fluorescent stains or tags; (iv) fiber-based OCT being compact, portable, and noncontact. However, the depth resolution provided by OCT is lower than that for two-photon fluorescence or confocal microscopy with a high magnification objective.
- Super-resolution optical imaging is based on fluorescence imaging to go beyond the diffraction-limited resolution (Abbe limit), thus allowing to probe biological structures with nanometer resolution. These techniques can be broadly classified into three groups: illumination sculpting, single molecule localization, and patterned or structured illumination.
- Illumination sculpting techniques can be divided into evanescent wave techniques [such as near-field scanning optical microscopy (NSOM), total internal reflection fluorescence (TIRF) microscopy], interference-based (4Pi, I^5M) techniques, and techniques based on photophysical sharpening of point spread function (STED). The evanescent wave-based techniques can improve spatial resolution in selected dimension to the range of <50 nm.
- Interference-based techniques can probe in three dimensions, but the improvement in resolution can be only in one dimension. The 4-Pi technique utilizes two high NA objectives focusing light in opposing direction and relies on the interference between these two beams to shrink the focus spot size in axial direction. This method can achieve <100-nm axial resolution, retaining diffraction-limited resolution in XY direction.
- Photophysical sculpting of the point spread function (PSF) can improve XYZ resolution and be used to obtain super-resolution images in 3D. Stimulated emission depletion (STED) was the first concrete and feasible

concept showing that the diffraction barrier in fluorescence microscopy can be broken using such manipulation of PSF.

- STED microscopy utilizes confinement of fluorescence to a region that is much smaller than that covered by the diffraction-limited excitation spot, by using a second red-shifted "depletion" beam, shaped into doughnut profile, overlaid with the excitation beam. This high-power doughnut beam creates a condition for stimulated emission, thereby de-exciting the molecules to prevent them from generating its natural broadband fluorescence. Thus the resulting fluorescence signal is confined to a spot much smaller than the spot size of the excitation beam.
- Super-resolution optical imaging can be achieved with single molecule localization, since the emission from single molecules can be localized with a precision much better than the limit imposed by diffraction. A densely populated fluorophore distribution can be divided into a sequence of sparser population, and individual fluorophore location can be sequentially localized to reconstruct a full image. Some examples of creation of localizable subpopulation of single molecules are photoactivated localization microscopy (PALM), stochastic optical reconstruction microscopy (STORM), ground-state depletion microscopy followed by individual molecule return (GSDIM), and direct stochastic optical reconstruction microscopy (dSTORM).
- In structured illumination microscopy (SIM), the resolution improvement is achieved by illuminating the sample with a sequence of periodic patterns of high-spatial frequencies. From the patterns appearing in these sequential images (typically obtained by rotating the illumination pattern), information about still higher spatial frequencies can be computationally extracted.
- Nanotechnology can be used in super-resolution imaging, since higher photostability of nanoemitters in comparison with single molecules is extremely important for super-resolution techniques. STED and SIM with photoluminescent nanoparticles have already been demonstrated. Single-particle event, such as quantum dot blinking, has been utilized for localization-based super resolution.
- Technical challenges for optical imaging are (a) increasing optical transparency in the biological site of imaging, (b) retaining high-fluorescence quantum yield when nanoemitters are functionlized to make them water-dispersible, (c) reducing cellular toxicity, and (d) having process control for uniformity, scalability, and green chemistry.
- *In vivo* optical imaging can be used to optically track the administered nanoparticles to obtain information on biodistibution, targeting, and therapeutic action.
- Magnetic resonance imaging is an important bioimaging modality that utilizes spin transition between the spin states of a nucleus, induced by an rf field. The nucleus utilized routinely for MRI is proton in water, which

is the most abundant chemical constituent of the body. MRI provides good spatial resolution (25–100 μm) and can be used for both morphological and functional imaging.

- A new-generation MRI utilizes spin transitions of another nucleus, ^{19}F, which can easily be incorporated in a nanoparticle formulation to enhance the local concentration of ^{19}F at a desired biological site and thus increase the sensitivity. Thus, the considerably lower magnetic-resonance signal of ^{19}F compared to a proton can be compensated for by significantly enriching its concentration at a biological site to be imaged.
- Since ^{19}F is not intrinsic to the body, targeting of a specific biological site can produce MRI images with great spatial selectivity against a zero background. Incorporation of ^{19}F imaging capability requires only a minor modification of the instruments currently used for proton-based MRI.
- The parameters that are used in MRI are: (i) spin density; it is the concentration of the spin nucleus (proton or ^{19}F); (ii) T_1-relaxation time, which is the spin-lattice relaxation, describing the decay of excited spin population; (iii) T_2-relaxation time, also known as spin–spin relaxation, which describes the energy transfer between adjacent spin nuclei.
- One of the best contrasting agents in MRI is Gd^{III}, due to its large magnetic moment. It decreases the T_1-relaxation time because of the through-space dipole–dipole interaction. This produces increased signal intensity in the MRI signal compared to the intensity in the MRI signal without Gd^{III}. Nanoparticles containing Gd^{III} ions are demonstrated to be used as T_1 contrast agents.
- The superparamagnetic iron oxide (Fe_3O_4) nanoparticles, another MRI contrast enhancer, reduce the T_2-relaxation time of water and thus reduce the MRI signal, leading to a dark contrast.
- X-ray imaging routinely used in medical diagnosis exploits attenuation of X ray in tissues to obtain morphological and vascular information within the body. X-ray-based computed tomography (CT) imaging generates a three-dimensional image to provide a good anatomical visualization of body structures and detailed map of various organs. The nanoparticle contrast enhancers for CT contain iodine, barium salts, and so on, which strongly absorb X ray.
- Radio imaging uses a radionuclide that distributes and/or accumulates at the site of biological interest to provide a radioactive signal that is detected by a scintillator to reconstruct the image. The advantages offered by radionuclide imaging are unlimited tissue penetration, high sensitivity, and the ability to provide quantitative analogues that are very useful for preclinical and translational applications.
- The two major techniques of radionuclide imaging are positron emission tomography (PET) and single-photon emission computer tomography (SPECT). Various nanoparticle formulations for both PET and SPECT

include such radioisotopes as ^{16}F, ^{64}Cu, ^{68}Ga, and ^{124}I for PET and ^{125}I, ^{111}In, and ^{99m}Tc for SPECT. Radioisotopes are incorporated in a nanoparticle formulation either by chemical conjugation or anchored onto the nanoparticle using chelating groups.

- For ultrasound imaging, the contrast agents can resonate in an applied ultrasound beam, undergo shell oscillations, or expand and contract in response to acoustic pressure changes, which can be used for enhancement of contrast. Because of the long wavelength of these very low frequency ultrasound waves, the particles used as contrast enhancers have to be of larger size (≥250 nm) to efficiently reflect them.
- The three types of particle formulations as contrast enhancers have been used for ultrasound imaging: microbubbles, liposomal agents, and perfluorocarbon-emulsion nanoparticles. A major thrust in this area is the development of targeted contrast agents, such as targeted microbubbles, which are linked to ligands that bind to receptors on the cells of interest such as inflamed cells or cancer cells.
- The microbubble contrast agent also can serve as a drug delivery vehicle by encapsulating the drug or its nanoformulation, which can be released by breaking the microbubble at higher ultrasonic power.
- Electron microscopy, widely used for imaging of fixed cells and tissue specimens, can be used in two modes: (a) scanning electron microscopy (SEM), which images with some sacrifice in spatial resolution (≥10 nm), and (b) transmission electron microscopy (TEM), which utilizes the transmission of an electron beam through the sample, providing the highest spatial resolution (≥1 nm). Because intrinsic cellular constituents do not strongly scatter electrons, nanoparticles of heavy elements can be used to increase scattering and thus enhance contrast.
- In photoacoustic (PA) imaging, a biological target is irradiated with a short-pulsed laser beam that is absorbed either by the intrinsic components of the biological regions or by targeting nanoparticles accumulated in this region. The absorbed energy is partially converted into heat, resulting in generation of acoustic waves that are detected by ultrasonic transducers.
- PA imaging combines the merit of both optical imaging (higher contrast compared to ultrasound imaging) and ultrasound imaging (longer penetration depth compared to optical imaging). Photoacoustic computed tomography (PACT) utilizes computed tomographic image processing.
- The nanoparticle platforms can provide capabilities for combining all imaging modalities, yielding complementary informations that can be used to establish a diagnosis to assist the clinician in developing an individualized therapeutic approach, or to allow early detection of onset of a disease.
- The nanoparticle-based multimodal imaging probes can be surface modified to carry various biotargeting agents (e.g., antibodies) to target specific

disease markers for achieving earlier detection of diseases. These nanoprobes have the potential of providing real-time imaging, thus providing the clinician with invaluable information that can be used in designing an individually tailored therapeutic regimen for the patient.

EXERCISES

True or False

1. Optical imaging is applicable only to an *in vitro* specimen.

2. Confocal and multiphoton microscopies are employed to obtain high spatial resolution 3D images.

3. A major concern with the most commonly used Cd-based quantum dots and rods is their potential toxicity.

4. Super-resolution optical imaging can go beyond the diffraction limit for optical imaging.

Multiple Choice

5. Magnetic resonance imaging provides good spatial resolution of
- **(a)** 200–300 nm
- **(b)** 1–10 μm
- **(c)** 25–100 μm
- **(d)** 200–300 μm

6. Computed tomography (CT) imaging is
- **(a)** X-ray-based
- **(b)** Ultrasound-based
- **(c)** radionuclide-based
- **(d)** Optical-based

7. To be used as contrast enhancers in ultrasound imaging, particles have to be of size
- **(a)** ≥250 nm
- **(b)** ≥30 nm
- **(c)** ≥1 μm
- **(d)** ≥10 μm

8. To be efficient contrast agents in photoacoustic imaging, particles have to have
- **(a)** An intense absorption
- **(b)** An intense scattering

(c) An intense fluorescence

(d) An intense magnetization

Answer: 1. F; 2. F; 3. T; 4. T; 5. c; 6. c; 7. a; 8. a

REFERENCES

Ai, K., Liu, Y., Liu, J., Yuan, Q., He, Y., Lu, L., Large-scale synthesis of Bi253 nanodots as a contrast agent for *in vivo* X-ray computed tomography imaging, *Adv. Mater.* **23**, 4886–4891 (2011).

Ambrose, W. P., Moerner, W. E., Fluorescence spectroscopy and spectral diffusion of single impurity molecules in a crystal, *Nature* **349**, 225–227 (1991).

Bates, M., Huang, B., Dempsey, G. T., Zhuang, X., Multicolor super-resolution imaging with photoswitchable fluorescent probes, *Science* **317**, 1749–1753 (2007).

Betzig, E., Patterson, G. H., Sougrat, R., Lindwasser, O. W., Olenych, S., Bonifacino, J. S., Davidson, M. W., Lippincott-Schwartz, J., Hess, H. F., Imaging intracellular fluorescent proteins at nanometer resolution, *Science* **313**, 1642–1645 (2006).

Bloch, S. H., Dayton, P. A., Ferrara, K. W., Targeted imaging using ultrasound contrast agents, *IEEE Eng. Med. Biol. Mag.* **23**, 18–27 (2004).

Chan, W. L., Deibel, J., Mittleman, D. M., Imaging with terahertz radiation, *Rep. Prog. Phys.* **70**, 1325–1379 (2007).

Chen, G., Ohulchanskyy, T. Y., Law, W. C., Ågren, H., Prasad, P. N., Monodisperse NaYbF4:Tm3+/NaGdF4 core/shell nanocrystals with near-infrared to near-infrared upconversion photoluminescence and magnetic resonance properties, *Nanoscale* **3**, 2003–2008 (2011).

Chrastina, A., Schnitzer, J. E., Iodine-125 radiolabeling of silver nanoparticles for *in vivo* SPECT imaging, *Int. J. Nanomed.* **5**, 653–659 (2010).

Colston, B. W., Jr., Everett, J. M., Da Silva, L. B., Otis, L. L., Stroeve, P., Nathel, H., Imaging of hard- and soft-tissue structure in the oral cavity by optical coherence tomography, *Appl. Opt.* **37**, 3582–3585 (1998).

De la Zerda, A., Liu, Z., Bodapati, S., Teed, R., Vaithilingam, S., Khuri-Yakub, B. T., Chen, X., Dai, H., Gambhir, S. S., Ultrahigh sensitivity carbon nanotube agents for photoacoustic molecular imaging in living mice, *Nano Lett.* **10**, 2168–2172 (2010).

Devaraj, N. K., Keliher, E. J., Thurber, G. M., Nahrendorf, M., Weissleder, R., 18F-Labeled nanoparticles for *in vivo* PET-CT imaging, *Bioconjug. Chem.* **20**, 397–401 (2009).

Erogbogbo, F., Yong, K.-T., Hu, R., Law, W.-C., Ding, H., Chang, C.-W., Prasad, P. N., Swihart, M. T., Biocompatible magnetofluorescent probes: Luminescent silicon quantum dots coupled with superparamagnetic iron(III) oxide, *ACS Nano* **4**, 5131–5138 (2010).

Erogbogbo, F., Yong, K.-T., Roy, I., Hu, R., Law, W.-C., Zhao, W., Ding, H., Wu, F., Kumar, R., Swihart, M. T., Prasad, P. N., *In vivo* targeted cancer imaging, sentinel lymph node mapping and multi-channel imaging with biocompatible silicon nanocrystals, *ACS Nano* **5**, 413–423 (2011).

Fercher, A. F., Optical coherence tomography—Development, principles, applications. *Z. Med. Phys.* **20**, 251–276 (2010).

Folling, J., Bossi, M., Hell, S. W., Fluorescence nanoscopy by ground-state depletion and single-molecule return, *Nat. Methods* **5**, 943–945 (2008).

Frullano, L., Meade, T. J., Multimodal MRI contrast agents, *J. Biol. Inorg. Chem.* **12**, 939–949 (2007).

Gugel, H., Bewersdorf, J., Jakobs, S., Engelhardt, J., Storz, R., Hell, S. W., Cooperative 4Pi excitation and detection yields sevenfold sharper optical sections in live-cell microscopy, *Biophys. J.* **87**, 4146–4152 (2004).

Gustafsson, M. G. L., Extended resolution fluorescence microscopy, *Curr. Opin. Struct. Biol.* **9**, 627–634 (1999).

Gustafsson, M. G. L., Surpassing the lateral resolution limit by a factor of two using structured illumination microscopy, *J. Microsc.* **198**, 82–87 (2000).

Gustafsson, M. G. L., Shao, L., Carlton, P. M., Wang, C. J. R., Golubovskaya, I. N., Cande, W. Z., Agard, D. A., Sedat, J. W., Three-dimensional resolution doubling in wide-field fluorescence microscopy by structured illumination, *Biophys. J.* **94**, 4957–4970 (2008).

Han, K. Y., Willig, K. I., Rittweger, E., Jelezko, F., Eggeling, C., Hell, S. W., Three-dimensional stimulated emission depletion microscopy of nitrogen-vacancy centers in diamond using continuous-wave light, *Nano Lett.* **9**, 3323–332 (2009).

Hainfeld, J. F., Slatkin, D. N., Focella, T. M., Smilowitz, H. M., Gold nanoparticles, A new X-ray contrast agent, *Br. J. Radiol.* **79**, 248–253 (2006).

Harke, B., Keller, J., Ullal, C. K., Westphal, V., Schönle, A., Hell, S. W., Resolution scaling in STED microscopy, *Opt. Express* **16**, 4154–4162 (2008).

Heilemann, M., van de Linde, S., Schüttpelz, M., Kasper, R., Seefeldt, B., Mukherjee, A., Tinnefeld, P., Sauer, M., Subdiffractionresolution fluorescence imaging with conventional fluorescent probes, *Angew. Chem. Int. Ed. Engl.* **47**, 6172–6176 (2008).

Hell, S. W., Stelzer, E. H. K., Properties of a 4Pi confocal fluorescence microscope, *J. Opt. Soc. Am. A* **9**, 2159–2166 (1992).

Hell, S. W., Wichmann, J., Breaking the diffraction resolution limit by stimulated emission: Stimulated-emission-depletion fluorescence microscopy, *Opt. Lett.* **19**, 780–782 (1994).

Hess, S. T., Girirajan, T. P. K., Mason, M. D., Ultra-high resolution imaging by fluorescence photoactivation localization microscopy, *Biophys. J.* **91**, 4258–4272 (2006).

Hill, M. L., Corbin, I. R., Levitin, R. B., Cao, W., Mainprize, J. G., Yaffe, M. J., Zheng, G., *In vitro* assessment of poly-iodinated triglyceride reconstituted low-density lipoprotein, *Acad. Radiol.* **17**, 1359–1365 (2010).

Huang, B., Wang, W., Bates, M., Zhuang, X., Three-dimensional super-resolution imaging by stochastic optical reconstruction microscopy, *Science* **319**, 810–813 (2008).

Jaffer, F. A., Libby, P., Weissleder, R., Optical and multimodality molecular imaging. Insights into atherosclerosis, *Arterioscler. Thromb. Vasc. Biol.* **29**, 1–7 (2009).

Jin, Y., Jia, C., Huang, S.-W., O'Donnell, M., Gao, X., Multifunctional nanoparticles as coupled contrast agents, *Nat. Commun.* **1**, 1–8 (2010).

Kim, S., Zheng, Q., He, G., Bharali, D., Pudavar, H., Baev, A., Prasad, P. N., Aggregation-enhanced fluorescence and two-photon absorption in nanoaggregates of a 9,10-bis

[4′-(4″-aminostyryl)styryl]anthracene derivative, *Adv. Funct. Mater.* **16**, 2317–2323 (2006).

Kim, S., Pudavar, H., Bonoiu, A., Prasad, P. N., Aggregation-enhanced fluorescence in organically modified silica nanoparticles: A novel approach toward high-signal-output nanoprobes for two-photon fluorescence bioimaging, *Adv. Mater.* **19**, 3791–3795 (2007a).

Kim, S., Huang, H., Pudavar, H. E., Cui, Y., Prasad, P. N., Intraparticle energy transfer and fluorescence photoconversion in nanoparticles, an optical highlighter nanoprobe for two-photon bioimaging, *Chem. Mater.* **19**, 5650–5656 (2007b).

Kim, C., Favazza, C., Wong, L. V., *In vivo* photoacoustic tomography of chemicals: High-resolution functional and molecular optical imaging at new depths, *Chem. Rev.* **110**, 2758–2782 (2010).

Kim, J., Piao, Y., Hyeon, T., Multifunctional nanostructural materials for multimodal imaging, and simultaneous imaging and therapy, *Chem. Soc. Rev.* **38**, 372–390 (2009).

Kim, J. G., Zhao, D. W., Song, Y. L., Constantinescu, A., Mason, R. P., Liu, H. L., Interplay of tumor vascular oxygenation and tumor pO_2 observed using near-infrared spectroscopy, an oxygen needle electrode, and F-19 MR pO_2 mapping, *J. Biomed. Opt.* **8**, 53–62 (2003).

Kolesov, R., Reuter, R., Xia, K., Stöhr, R., Zappe, A., Wrachtrup, J., Super-resolution upconversion microscopy of praseodymium-doped yttrium aluminum garnet nanoparticles, *Phys. Rev. B* **84**, 153413 (2011).

Kumar, R., Ohulchanskyy, T. Y., Turowski, S. G., Thompson, M. E., Seshadri, M., Prasad, P. N., Combined magnetic resonance and optical imaging of head and neck tumor xenografts using gadolinium-labelled phosphorescent polymeric nanomicelles, *Head Neck Oncol.* **2**, 35 (2010a).

Kumar, R., Roy, I., Ohulchanskky, T. Y., Vathy, L. A., Bergey, E. J., Sajjad, M., Prasad, P. N., *In vivo* biodistribution and clearance studies using multimodal organically modified silica nanoparticles, *ACS Nano* **4**, 699–708 (2010b).

Lai, C.-W., Wang, Y.-H., Lai, C.-H., Yang, M.-J., Chen, C.-Y., Chou, P.-T., Chan, C.-S., Chi, Y., Chen, Y.-C., Hsiao, J.-K., Iridium-complex-functionalized Fe_3O_4/SiO_2 core/shell nanoparticles: A facile three-in-one system in magnetic resonance imaging, luminescence imaging, and photodynamic therapy, *Small* **4**, 218–224 (2008).

Lanza, G. M., Wallace, K. D., Scott, M. J., Cacheries, W. P., Abendschein, D. R., Chistg, D. H., Scharkely, A. M., Miller, J. G., Gaffney, P. J., Wickline, S. A., A novel site-targeted ultrasonic contrast agent with broad biomedical application, *Circulation* **94**, 3334–3340 (1996).

Law, W.-C., Yong, K.-T., Roy, I., Xu, G., Ding, H., Bergey, E. J., Zeng, H., Prasad, P. N., Optically and magnetically doped organically modified silica nanoparticles as efficient magnetically guided biomarkers for two-photon imaging of live cancer cells, *J. Phys. Chem. C* **112**, 7972–7977 (2008).

Law, W.-C., Yong, K.-T., Roy, I., Ding, H., Hu, R., Zhao, W., Prasad, P. N., Aqueous-phase synthesis of highly luminescent CdTe/ZnTe core/shell quantum dots optimized for targeted bioimaging, *Small* **5**, 1302–1310 (2009).

Levy, L., Sahoo, Y., Kim, K. S., He, G. S., Bergey, E. J., Prasad, P. N., Nanochemistry: Synthesis and characterization of multifunctional nanoclinics for biological applications, *Chem. Mater.* **14**, 3715–3721 (2002).

Lidke, K., Rieger, B., Jovin, T., Heintzmann, R., Superresolution by localization of quantum dots using blinking statistics, *Opt. Express* **13**, 7052–7062 (2005).

Liu, Z., Kiessling, F., Gätjens, J., Advanced nanomaterials in multimodal imaging: Design, functionalization, and biomedical applications, *J. Nanomater.* **2010**, 894303 (2010).

Louie, A., Multimodality imaging probes: Design and challenges, *Chem. Rev.* **110**, 3146–3195 (2010).

Maxwell, R. J., Frenkel, T. A., Newell, D. R., Bauer, C., Griffiths, J. R., F-19 Nuclear-magnetic-resonance imaging of drug distribution *in-vovo*—The disposition of an antifolate anticancer in mice, *Magn. Reson. Med.* **17**, 189–196 (1991).

Mi, C., Zhang, J., Gao, H., Wu, X., Wang, M., Wu, Y., Di, Y., Xu, Z., Mao, C., Xu, S., Multifunctional nanocomposites of superparamagnetic (Fe_3O_4) and NIR-responsive rare earth-doped up-conversion fluorescent (NaYF4:Yb,Er) nanoparticles and their applications in biolabeling and fluorescent imaging of cancer cells, *Nanoscale* **2**, 1141–1148 (2010).

Periasamy, A., Day, R. N., *Molecular Imaging: FRET Microscopy and Spectroscopy*, Oxford University Press, New York (2005).

Periasamy, A., Clegg, R. M., *FLIM Microscopy in Biology and Medicine*, CRC Press (Taylor & Francis Group), New York (2010).

Pielage, J., Cheng, L., Fetter, R. D., Carlton, P. M., Sedat, J. W., Davis, G. W., A presynaptic giant ankyrin stabilizes the NMJ through regulation of presynaptic microtubules and transsynaptic cell adhesion, *Neuron* **58**, 195–209 (2008).

Prasad, P. N., *Introduction to Biophotonics*, Wiley, Hoboken, NJ (2003).

Prasad, P. N., *Nanophotonics*, Wiley, Hoboken, NJ (2004).

Prasad, P. N., Mazurchuk, R. V., Bharali, D. J., Sukumaran, D. K., Bergey, E. J., Spernyak, J. A., Seshadri, M., Hybrid inorganic nanoparticles, methods of using and methods of making, US Patent Application 20060222587.

Rust, M. J., Bates, M., Zhuang, X., Sub-diffraction-limit imaging by stochastic optical reconstruction microscopy (STORM), *Nat. Methods* **3**, 793–795 (2006).

Schermelleh, L., Carlton, P. M., Haase, S., Shao, L., Winoto, L., Kner, P., Burke, B., Cardoso, M. C., Agard, D. A., Gustafsson, M. G. L., Leonhardt, H., Sedat, J. W., Subdiffraction multicolor imaging of the nuclear periphery with 3D structured illumination microscopy, *Science* **320**, 1332–1336 (2008).

Sharma, P., Brown, S., Walter, G., Santra, S., Moudgil, B., Nanoparticles for bioimaging, *Adv. Colloid Int. Sci.* **123–126**, 471–485 (2006).

Smith, L. M., Dobson, C., Absolute displacement measurements using modulation of the spectrum of white light in a Michelson interferometer. *Appl. Opt.* **28**, 3339–3342 (1989).

Soliman, O. I. I., Geleijnse, M. L., Meijboom, F. J., Nemes, A., Kamp, O., Nihoyannopoulos, P., Masani, N., Feinstein, S. B., Ten Cate, F. J., The use of contrast echocardiography for the detection of cardiac shunts, *Eur. J. Echocardiogr.* **8**, S2–S12 (2007).

Song, K. H., Kim, C., Cobley, C. M., Xia, Y., Wang, L.V., Near-infrared gold nanocages as a new class of tracers for photoacoustic sentinel lymph node mapping on a rat model. *Nano Lett.* **9**, 183–188 (2009).

Sun, Y., Wallrabe, H., Booker, C., Day, R. N., Periasamy, A., Three-color spectral FRET microscopy localizes three interacting proteins in living cells, *Biophys. J.* **99**, 1274–1283 (2010).

Sun, Y., Day, R. N., Periasamy, A., FLIM–FRET microscopy imaging to study protein–protein interactions in living cells, *Nat. Protocols* **6**, 1324–1340 (2011).

Swanson, E. A., Izatt, J. A., Hee, M. R., Huang, D., Lin, C. P., Schuman, J. S., Puliafito, C. A., Fujimoto, J. G., *In vivo* retinal imaging by optical coherence tomography, *Opt. Lett.* **18**, 1864–1866 (1993).

Thompson, R. E., Larson, D. R., Webb, W. W., Precise nanometer localization analysis for individual fluorescent probes, *Biophys. J.* **82**, 2775–2783 (2002).

Tong, L., Wei, Q., Wei, A., Cheng, J. X., Gold nanorods as contrast agents for biological imaging: Optical properties, surface conjugation and photothermal effects, *Photochem. Photobiol.* **85**, 21–32 (2009).

Tzeng, Y.-K., Faklaris, O., Chang, B.-M., Kuo, Y., Hsu, J.-H., Chang, H.-C., Superresolution imaging of albumin-conjugated fluorescent nanodiamonds in cells by stimulated emission depletion, *Angew. Chem. Int. Ed.* **50**, 2262–2265 (2011).

Varadan, V. K., Chen, L., Xie, J., *Nanomedicine: Design and Applications of Magnetic Nanomaterials, Nanosensors and Nanosystems*, Wiley, London (2008).

Wagner, H. N., Szabo, Z., Buchanan, J. W., *Principles of Nuclear Medicine*, W. B. Saunders, Philadelphia, 2nd edition (1995).

Wang, Zh., Tiruppathi, C., Minshall, R., Malik, A. B., Imaging caveolae-mediated transport of nanoparticles using superresolution microscopy, *Proc. SPIE* **8231**, 8231-1 (2012).

Welch, M. J., Hawker, C. J., Wooley, K. L., The advantages of nanoparticles for PET, *J. Nucl. Med.* **50**, 1743–1746 (2009).

Woodward, J. D., Kennel, S. J., Mirzadeh, S., Dai, S., Wall, J. S., Richey, T., Avenell, J., Rondinone, A. J., *In vivo* SPECT/CT imaging and biodistribution using radioactive $Cd^{125m}Te/ZnS$ nanoparticles, *Nanotechnology* **18**, 175103 (5 pp) (2007).

Woodward, R. W., Wallace, V. P., Pye, R. J., Cole, B. E., Armone, D. P., Linfield, E. H., Pepper, M., Terahertz pulse imaging of *ex vivo* basal cell carcinoma, *J. Invest. Dermatol.* **120**, 72–78 (2003).

Xu, F. M., Pudavar, H. E., Prasad, P. N., Dickensheets, D., Confocal enhanced optical coherence tomography for nondestructive evaluation of paints and coatings, *Opt. Lett.* **24**, 1808–1810 (1999).

Xu, M., Wang, L. V., Photoacoustic imaging in biomedicine, *Rev. Sci. Instrum.* **77**, 041101 (2006).

Yildiz, A., Forkey, J. N., McKinney, S. A., Ha, T., Goldman, Y. E., Selvin, P. R., Myosin V walks hand-over-hand: Single fluorophore imaging with 1.5-nm localization, *Science* **300**, 2061–2065 (2003).

Yong, K. T., Roy, I., Pudavar, H. E., Bergey, E. J., Tramposch, K. M., Swihart, M. T., Prasad, P. N., Multiplex imaging of pancreatic cancer cells by using functionalized quantum rods, *Adv. Mater.* **20**, 1412–1417 (2008).

Yong, K.-T., Law, W.-C., Roy, I., Jing, Z., Huang, H., Swihart, M. T., Prasad, P. N., Aqueous phase synthesis of CdTe quantum dots for biophotonics, *J. Biophoton.* **4**, 9–20 (2011).

CHAPTER 9

BIOSENSING

Biosensing forms a major component of medical diagnostics, a topic that will be revisited again in Chapter 10. *In vitro* biosensing using biopsied cells, blood, urine, sweat, sputum, bronchialveolar davage (BAL), and other body fluid samples collected from humans, provide detailed analysis of a rich variety of secreted biomolecules, which carry significant information about the health and well-being of the body. These samples are also potent indicators of any malfunctioning of the body and can play a critical role in the diagnosis and staging of several, if not all, human diseases. In addition, biosensing provides detection methods for new strands of microbial organisms and the spread of infectious diseases. This chapter is focused on biosensing only and its role in medical diagnostics. Other methods of diagnosis supplementing biosensing will be discussed in Chapter 10.

In this chapter, the principles of biosensing will be discussed in Section 9.1, where an overview of the role of nanoparticles and nanostructures in enhancing the capabilities and scopes of biosensing is also presented. Section 9.2 discusses nanoparticle-based optical sensing. Section 9.3 discusses nanostructure-based magnetic sensing. Section 9.4 describes nanoparticle-based electric sensing. Section 9.5 presents nanoparticle-based electrochemical biosensing. Section 9.6 covers another variation of electrochemical biosensing using electrochemiluminescence for detection. Finally, Section 9.7 discusses implantable *in vivo* bioelectronic sensors where nanotechnology also promises to make a significant impact.

Introduction to Nanomedicine and Nanobioengineering, First Edition. Paras N. Prasad.

9.1. PRINCIPLES OF BIOSENSING

Biosensors are analytical devices that can detect chemical or biological species or microorganisms. They can be used to monitor the changes in the *in vivo* concentrations of an endogenous species as a function of a physiological change induced internally by disease manifestation or by the invasion of microbes, bacteria, viruses, and toxins. Biosensors thus have a wide range of applications including:

- Clinical diagnostics
- Drug development
- Environmental monitoring (air, water, and soil)
- Food quality control

A biosensor in general utilizes a biological recognition element that senses the presence of an analyte (the species to be detected) and creates a physical or chemical response that is converted by a transducer into a signal. The important components of a biosensor are: (i) a sampling unit that introduces an analyte into the detector and can be as simple as a circulator; (ii) a recognition element that binds or reacts with a specific analyte, providing biodetection specificity (enzymes, antibodies, or even cells, such as yeast or bacteria, have been used as biorecognition elements; the principles of biorecognition have already been discussed in Chapter 7); (iii) stimulation that, in general, can be provided by optical, electric, magnetic, or other kinds of force fields that extract a response as a result of biorecognition; and (iv) a transduction process that transforms the physical or chemical response of the biorecognition in the presence of an external stimulation, into an optical, electrical, or magnetic signal that is detected by the detection unit. The detection unit may include pattern recognition for identification of the analyte.

Based on the transduction method that provides for detection, biosensors can be classified in different categories as depicted in Figure 9.1.

Optical Biosensing. Optical detection methods for biosensing are very diverse and can also provide quantitative molecular information. It utilizes a number of spectroscopic methods such as absorption, fluorescence, and Raman and surface plasmon resonance [Prasad, 2004]. In addition, the optical methods of detection can utilize the change in a number of optical properties such as intensity, frequency, phase, polarization, and lifetime of fluorescence. For quantitative analysis of biospecies, a change in the intensity of incident light by absorption or a change in the fluorescence or Raman intensity upon binding with an analyte is utilized.

In the fluorescence detection mode, a change in the fluorescence intensity, a shift in the peak wavelength of fluorescence, a change in the fluorescence lifetime, or a change in the polarization property of fluorescence upon binding

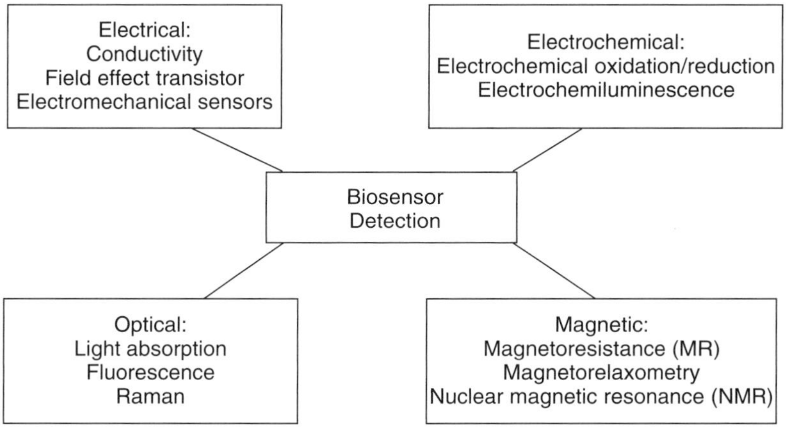

Figure 9.1. The various detection methods and the parameters monitored for biosensing.

of the biorecognition element with the biospecies to be detected is monitored. The fluorescence may be an intrinsic feature of the biorecognition element, but, more commonly, the biorecognition element is chemically conjugated with a fluorescence species (such as a dye or a quantum dot).

In the Raman detection mode, a change in the vibrational frequency of a Raman active vibrational mode or a change in the intensity is monitored. A more powerful approach, which provides orders-of-magnitude enhancement in the Raman signal and thus a significant gain in the sensitivity, is the use of surface-enhanced Raman scattering (SERS). In this case, a Raman reporter (a molecule the vibrational frequency of which is used for detection) is attached to the surface of a metallic nanoparticle or a metallic nanostructure. The enhanced local electric field due to the surface plasmon resonance (discussed in Chapter 5) leads to this enhancement. The Raman reporters coupled to the metallic (e.g., silver or gold) nanoparticles are thus used for SERS-based biodetection.

LSPR sensors utilize the unique properties of the free oscillating electrons (plasmons) on metallic nanostructures as presented in Chapter 5. Because the oscillation of surface plasmon is very sensitive to the surrounding dielectric environment, it can be used as a sensitive probe for biodetection by monitoring the shift of plasmonic peak.

Another type of biosensor utilizes the delocalization of surface plasmon resonance at the interface of a thin (~50 nm) metallic film (also discussed in Chapter 5). It is often called an SPR biosensor. Here, a change in the angle (coupling) at which the incident light generates the SPR excitation is generally used for biodetection.

In a whispering gallery mode (WGM) sensor, light is trapped in a circular orbit in a microsphere by the total internal reflections. The light can circulate in the orbit for a million times. The dielectric change associated with the analyte capture on the microsphere surface causes a shift in the resonance frequency. This resonance frequency shift can be used for biosensing.

In a photonic crystals sensor the change in the Bragg diffraction wavelength of a periodic dielectric array, occurring upon analyte capture, is used as the sensing mechanism.

All these optical biosensing techniques are discussed in more detail in Section 9.2.

Magnetic Biosensing. Magnetic biosensing principally utilizes two different approaches. In one approach, the stray (fringe) magnetic field of magnetically tagged nanoparticles is detected by a magnetic field sensor using a solid substrate platform. The magnetic field sensors have been used based on magnetoresistance effect, spin valve action, or Hall effect. These methods are described in detail in Section 9.3. The sensing mechanism utilizes the change in the stray field, as measured by the magnetic field sensors, when an analyte binds to the magnetic nanoparticles that are passed over an array of patterned structures on the magnetic sensor, where these arrays have complementary probes to capture the magnetic nanoparticles.

Magnetorelaxometry utilizes the measurement of the change of relaxation times upon analyte binding with the nanoparticles. The two types of relaxation phenomena used are: (i) the relaxation of magnetization of the nanoparticle when the magnetizing field is turned off and (ii) the spin–spin relaxation time, primarily of protons in water. Both of these relaxation times change upon binding of the magnetic nanoparticles with an analyte biomolecule.

Electrical Biosensing. A major effort in this mode of sensing has been the use of a field-effect transistor (FET). It consists of a substrate called gate and two electrodes called source and drain. The biosensing element is the channel between the source and the drain which has the biorecognition element immobilized onto its surface. Capture of an analyte by the biorecognition element produces a change in the current flowing through the devices and can thus be detected.

Quartz crystal microbalance is another example of an electrical sensor, an electromechanical sensor. The quartz crystal microbalance consists of a piezoelectric crystal that vibrates at a specific frequency when a specific electrical signal is applied. The surface of the quartz crystal is functionalized with a biorecognition capture agent, which, upon binding to an analyte, increases the mass of the crystal and subsequently causes a change in the oscillation frequency, allowing for the generation of an electrical response.

These biosensing methods are further discussed in Section 9.4.

Electrochemical Biosensing. Electrochemical biosensing has been widely used in the past; and many types of commercial sensors, such as glucose sensors, use electrochemical transduction. This method utilizes analyte detection by inducing an enzyme-catalyzed redox reaction that can either consume (oxidation) or produce (reduction) electrons to afford an electrical signal (current or voltage). This method is discussed in Section 9.5.

Another approach gaining more interest is the use of electrochemiluminescence (ECL) for detection. In this case, the electron transfer process induced by the detection of an analyte produces a specie in the excited electronic state which then luminesces to return to the ground electronic state. This luminescence is then used for detection. This method of sensing has received considerable attention recently and is discussed in detail in Section 9.6.

9.2. OPTICAL BIOSENSORS

Optical biosensing offers a broad range of transduction mechanisms utilizing changes in the frequency, phase, polarization and intensity of light. In addition, various geometries of light propagation and light–matter interaction can be used. Accordingly, vast amounts of literature, in the form of excellent recent reviews as well as books, exist on this topic. For review of earlier work, a good source is the monograph *Introduction to Biophotonics* [Prasad, 2003]. Herein, only selected examples of optical biosensing using nanostructures are presented, because a true comprehensive coverage of this topic is beyond the scope of this introductory monograph. The various optical biosensing methods utilizing nanostructure platforms, covered in this section, are summarized in Figure 9.2.

9.2.1. Fluorescencs Sensors

Fluorescence-based biosensing provides a powerful approach to monitor the physiological activities of cells that leads to a variety of intracellular diagnostics for disease profiling on cellular level. Figure 9.3 illustrates the applications of fluorescence biosensors using a nanostructure platform.

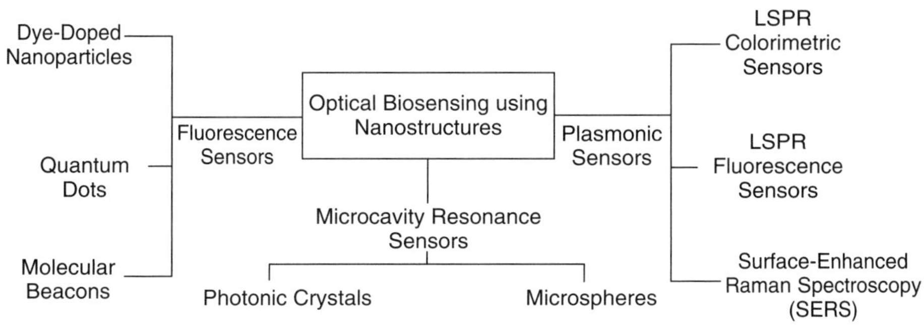

Figure 9.2. Various optical biosensing methods utilizing nanostructures. LSPR is the acronym for localized surface plasmon resonance discussed in Chapter 5.

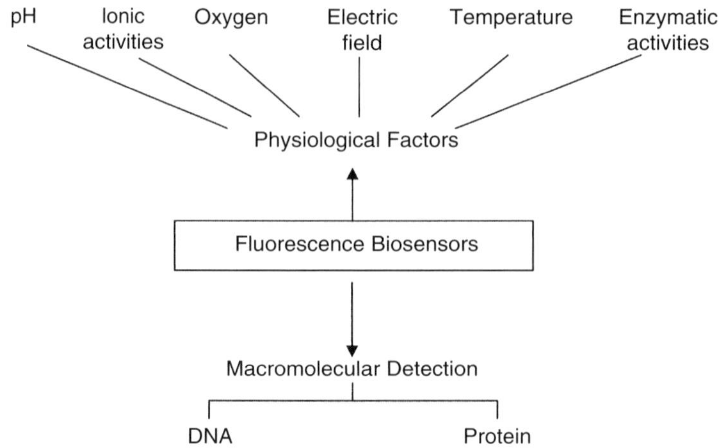

Figure 9.3. Various applications of fluorescence-based biosensors.

A major advancement in the intracellular sensing of physiological activities was produced by Kopelman's group through their design of PEBBLE (probe encapsulated by biologically localized embedding), nanosensors [Clark et al., 1999; Lee et al., 2009a]. PEBBLEs, introduced in Chapter 3, are nanoscale spherical devices consisting of sensor molecules entrapped in a chemically inert matrix. Figure 9.4 shows a schematic diagram of a PEBBLE nanosensor that can provide many functions to monitor various physiological parameters. Examples of matrix media used for the PEBBLE technology, also shown in the figure, are polyacrylamide hydrogel, sol–gel silica, and cross-linked decyl methacrylate. These matrices have been used by Kopelman's group to fabricate sensors for H^+, Ca^{2+}, Na^+, Mg^{2+}, Zn^{2+}, Cl^-, NO_2^-, O^2, NO, and glucose. The PEBBLE ranges in size from 30 to 600 nm. The matrix porosity allows for the entrapment and sensing of the analyte. The encapsulation protects the cells from any toxicity associated with the sensing dye as well as insulates the indicator dyes from cellular interferences such as protein binding. The principle behind sensing involves a change in the fluorescence intensity of an appropriately encapsulated dye in the presence of the analyte to be detected.

An example of a PEBBLE nanosensor is the calcium PEBBLE that utilizes calcium Green-1 and sulforhodamine dyes as sensing components encapsulated in a hydrophilic polyacrylamide (PAA) nanoparticles. The calcium green fluorescence intensity increases with increasing calcium concentrations, while the sulforhodamine fluorescence intensity is unaffected. The ratio of the calcium green intensity to the sulforhodamine intensity can thus be used to measure cellular calcium levels.

Similar principles have been used for sensing of other ions such as K^+. A large number of dyes exist that show strong dependence of their fluorescence intensity on the pH [Han and Burgess, 2010]. Kopelman and co-workers chem-

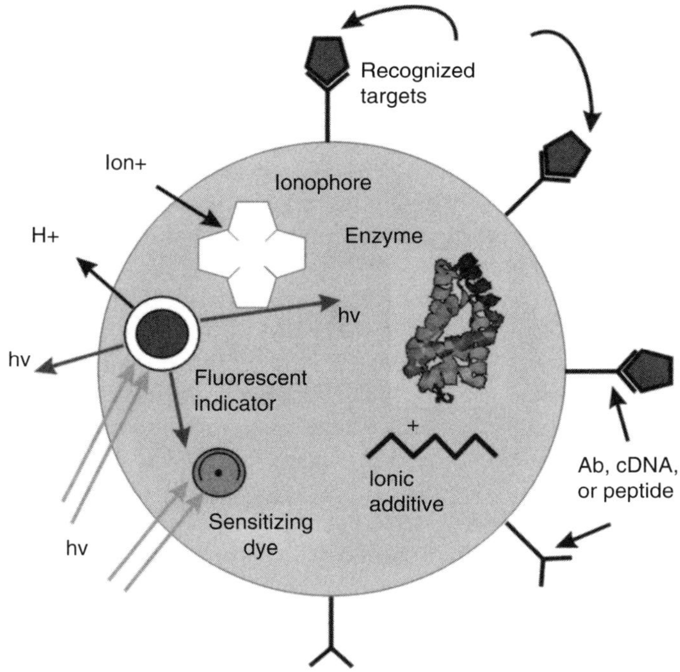

Figure 9.4. Schematics of a PEBBLE nanosensor. (Reproduced with permission from R. Kopelman, http://www.umich.edu/~koplab/research2/analytical/EnterPEBBLEs.html.

ically linked the pH sensitive dyes to a PAA nanoparticle PEBBLE for intracellular pH sensing.

We proposed a novel concept of nanoparticle-based fluorescent pH indicator for near-physiological use by matching the pK_a to the pH of the experimental system [Kim et al., 2006]. The designed particulate indicator is characterized by the nano-assembly of a pH-probing hydrophobic molecule—that is, an organically modified silica (ORMOSIL) nanoparticle that is concentrated with a ratiometric pH-responsive organic dye. ORMOSIL nanoparticles are known as water-dispersible and biocompatible carriers for hydrophobic dyes in aqueous media (Chapter 3). As for the hydrophobic dye, a naphthalenylvinylpyridine derivative (NVP) was used, with the pyridyl unit as the proton receptor. Like other pyridine-based dyes, NVP responds to protons, with fluorescence exhibiting red shift from high energy (blue) to low energy (yellow). In the NVP-concentrated ORMOSIL nanoparticle, the dyes are densely dispersed everywhere through the nanoparticle, whereas the proton-sensing event takes place mainly near the surface. Given this localized event, the high-energy neutral and the low-energy protonated states will closely coexist in the inner and surface parts of the nanoparticle, respectively, enabling an inner-to-surface energy transfer. By virtue of this, the proton sensitivity, in terms of fluorescence, is significantly improved by the energy transfer which amplifies the energy-accepting protonated surface signal; the apparent pK_a value of the nanoparticle is thus modified from that of the dye component. Pyridine-based dyes have potential for large two-photon activity as well. Nevertheless, their use in pH probing, in general, is restricted to rather acidic environments, because they are less sensitive to protons, due to low pK_a values. Our strategy, therefore, offers an easy way to increase their low pK_a values to a physiological level.

An example of an oxygen PEBBLE is a hydrophobic matrix such as ORMOSIL or polydecylmethacrylate (PDMA) encapsulating an oxygen-sensitive platinum-based dye (platinum II octaethylporphyrine ketone) as well as a reference dye for ratiometric determination of the oxygen content. The use of a hydrophobic matrix provides higher solubility for oxygen. Since this dye has fluorescence in the infrared region, the PEBBLE sensors can even be operated in human plasma samples, which do not absorb in this spectral region; scattering is also significantly reduced at this longer wavelength. In addition to the change in intensity of fluorescence, other fluorescence parameters, such as fluorescence anisotropy and fluorescence lifetime, can also be utilized for oxygen measurements [Lee et al., 2009b].

Electric potential of cell membranes as well as intracellular electric fields are also important indicators of cellular processes. Kopelman and co-workers have developed E-PEBBLES, which measure intracellular electric fields. They utilize a fast response voltage-sensitive dye (di-4-ANEPPS) inside silane-capped (polymerized) micelles [Tyner et al., 2007]. A fast reorientation of the voltage-sensitive dye, inside this hydrophobic core, provides a shorter response time. A shift in the peak position of the fluorescence spectrum, due to the electrochromic effect of the dye, provides the measurement of the electric field, again using the ratiometric method. An enzymatic PEBBLE provides an example of a glucose sensor in which the enzyme glucose oxidase (GO_x), an oxygen-sensitive ruthenium-based dye, and a reference dye are co-encapsulated within PAA nanoparticles. The enzymatic oxidation of glucose to gluconic acid depletes the local oxygen concentration, leading to an increase in the fluorescence of the oxygen-sensitive dye [Lee et al., 2009a].

Another factor that provides information on the physiological status of a cell or a subcellular unit is temperature. These are also indicators of overactive malignant cells that may have a higher temperature than the corresponding normal cell. The enzymatic activities in mitochondria can be monitored by its temperature. We introduced a fluorescence-based method to monitor local temperature [Kachynski et al., 2005]. This method utilizes the ratio of the fluorescence intensity of anti-Stokes to Stokes as a probe for the temperature. Conventional fluorescence, or Stokes emission, occurs from the optical excitation of molecules in the zero point vibrational level of the ground electronic state to the first excited electronic state, along with a subsequent radiative decay to the ground state. In the case of anti-Stokes luminescence (ASL), the optical excitation occurs from thermally excited vibrational energy states of the ground state. Since the population of the upper vibrational levels of the ground state is temperature-dependent (based on Boltzman distribution), the ASL intensity is extremely temperature-dependent (exponential dependence) as represented by the following equation:

$$I(T) \sim A\exp(-\Delta E/kT) \tag{9.1}$$

where $I(T)$ is the ASL intensity of the dye solution at temperature T, A is a proportionality coefficient, k is the Boltzmann constant, and ΔE is the activation energy to the thermally populated level.

We demonstrated the usage of the ASL phenomenon for 3D thermal mapping of cells by using a commercial dye, rhodamine 640, excited at 488 nm (argon laser line) for conventional fluorescence but excited at 632.8 nm (He–Ne laser line) for anti-Stokes luminescence. Figure 9.5a presents the ASL spectrum of a rhodamine 640 (Rh640) dye-encapsulated 35-nm silica nanoparticle, with 632.8-nm excitation. A high-intensity Nd-YAG laser at 1064 nm provided direct two-photon excitation of the Rh640 Stokes fluorescence which was used as a reference to normalize the ASL intensity. The normalization compensates for any concentration fluctuation in the dye solution, occurring during the measurement. Figure 9.5b shows the temperature dependence of the normalized ASL intensities.

This method was used to monitor the temperature distribution in a cell.

Fluorescence biosensing has also been used extensively for DNA, RNA, and protein detection [Rosi and Mirkin, 2005]. A dye-doped fluorescence biosensing example for DNA detection is provided by the work of Tan and co-workers [Zhao et al., 2003]. It utilizes silica nanoparticles containing dyes that are highly luminescent and very resistant to photobleaching by the environment. They used these nanoparticles for DNA/mRNA analysis, as well as for single bacterium detection. Sensitive DNA detection is extremely important in clinical diagnostics, gene therapy, and a variety of biomedical studies. The dye-doped silica nanoparticles were prepared using the reverse micelle reaction method described in Chapter 4. The ultrasensitive DNA/mRNA analysis was realized using a sandwich assay, shown in Figure 9.6. There are three DNA strands in this assay: capture DNA1 (5′TAA CAA TAA TCC T-biotin3′), probe DNA3 (5′ biotin-T ATC CTT ATC AAT ATT 3′) labeled with a NP (60 nm) to form a NP-DNA3 conjugate, and target DNA2 (5′GGA TTA TTG TTA AAT ATT GAT AAG GAT 3′). The combined sequences of DNA1 and DNA3 are complementary to that of the target DNA2. The biotinylated DNA1 is first immobilized onto an avidin-coated glass substrate. DNA2 and NP-DNA3 are then added for hybridization. Detection of the captured DNA2, which hybridized to the DNA1 bound to the glass surface, is accomplished by monitoring the fluorescence signals of the NP-DNA3 conjugates left on the

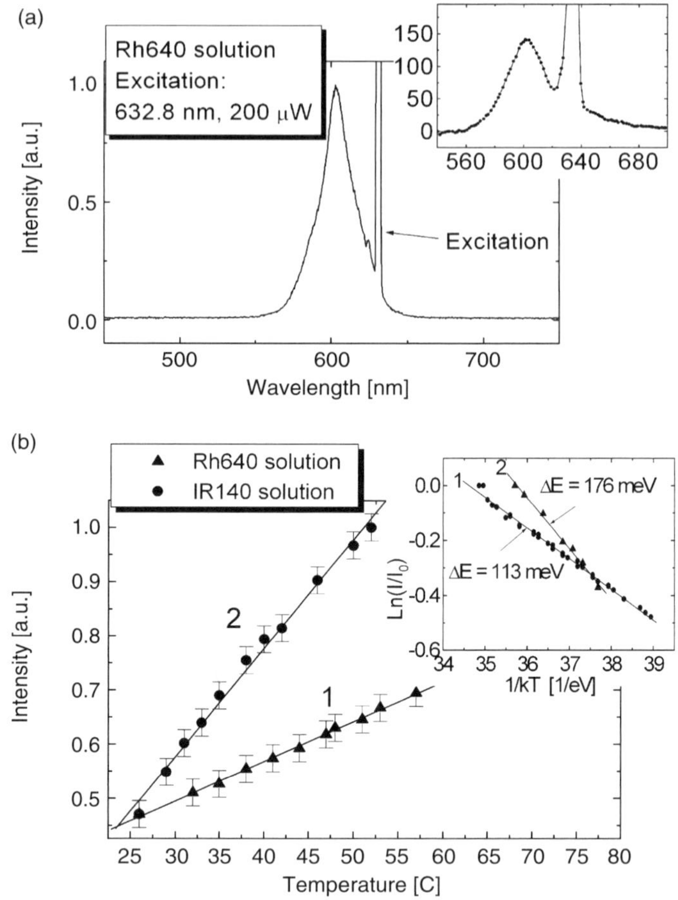

Figure 9.5. (a) ASL spectrum of Rhodamine 640 dye encapsulated in 35-nm silica shells (inset). (b) Temperature dependence of ASL for Rh640. (Reproduced with permission from Kachynski et al. [2005].)

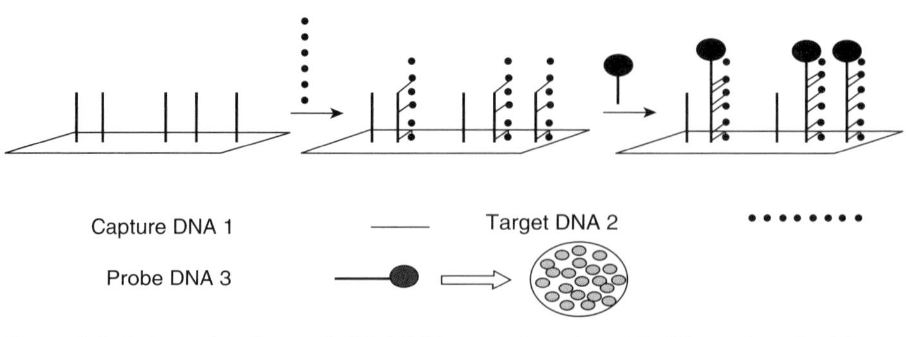

Figure 9.6. Schematic of a sandwich DNA assay based on NP. (Reproduced with permission from Zhao et al. [2003].)

glass surface after thorough washing steps and with proper excitation. The nanoparticles exhibit an excellent signaling ability in the presence of trace amounts of DNA targets. With effective surface modification, nonspecific binding and nanoparticle aggregation are minimized. In addition, the nanoparticle-based DNA bioanalysis assay can effectively discriminate even one base of mismatched DNA sequence.

Quantum Dots. Due to their narrow emission band, tunability of emission, photostability against photobleaching, and ease of surface functionalization, quantum dots and rods are now being extensively investigated for biosensing and replacing dyes for a variety of applications [Rosi and Mirkin, 2005; Han and Burgess, 2010; Zhong, 2009].

CdSe/ZnSe/ZnS core–shell–shell QDs capped with mercaptoacetic acid have been used for intracellular pH sensing in human ovarian cancer cells [Liu et al., 2007]. The ZnSe/ZnS shells around the CdSe provide protection against oxidation, while the mercaptoacetic acid makes them water-dispersible. The emission intensity of the quantum dots increases monotonically with increasing pH; they are 10-fold less fluorescent at pH 4 than at pH 10. Deng et al., [2007] used green and orange CdTe QDs as pH sensors to monitor proton flux diversity ATP synthesis for the detection of viruses.

Nie and co-workers used quantum dots as tags for the multiplexed detection of DNA [Han et al., 2001]. They labeled a DNA target with a fluorescent dye and an oligonucleotide microbead with quantum dots emitting at different wavelengths. After capturing the DNA target, a single microbead fluorescence spectrum revealed both the presence and the identity of the DNA target. Varying the number and ratios of the multiple color quantum dots can yield a unique fluorescent spectral signature (emission at different wavelengths with variable intensities) for each individual target. Alivisatos and co-workers reported the use of CdSe/ZnS core–shell QDs in a chip-based assay to detect single base-pair mutations in DNA [Gerion et al., 2003].

Molecular Beacon. The molecular beacon approach utilizes an electronic energy transfer scheme between a fluorescent unit, a dye or a QD (F1), and a fluorescent quencher (Q), which can be a metallic nanoparticle such as Au [Tan et al., 2000]. A molecular beacon consists of a loop, involving a single-stranded oligonucleotide in a specific sequence and a stem, usually consisting of five to seven complementary base pairs. The two ends of the stem consist of an emitter (F1) and a quencher (Q). In the absence of the analyte, the stem is intact, keeping the emitter and the quencher in close proximity, which quenches the emission. In the presence of the analyte, the binding or biorecognition process forces the stem apart, thus increasing the distance between the emitter and the quencher sufficiently to inhibit quenching. The result is restoration of the fluorescence of the emitter. This process is illustrated in Figure 9.7.

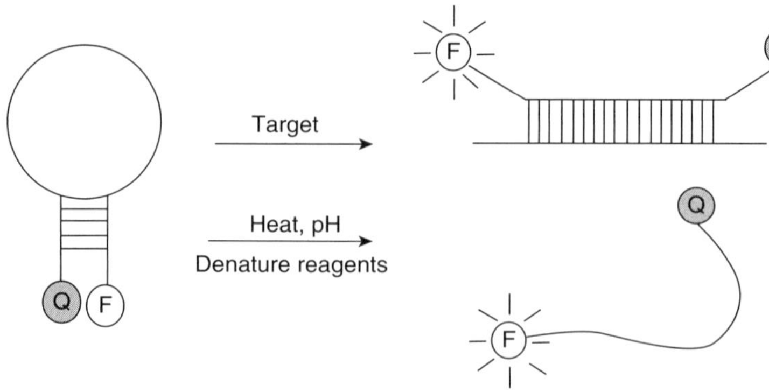

Figure 9.7. Molecular beacon approach for biosensing. Hybridization with the DNA target molecules of a complementary sequence or unwinding of the DNA with an increase in temperature, change of pH, or presence of a denaturing agent produces an increase in emission. (Reproduced with permission from Tan et al. [2000].)

The molecular beacon technique has emerged as a new class of DNA, RNA, or PNA probe. Molecular beacons with a selected sequence of bases in the loop can be synthesized to detect the complementary DNA strand by hybridization (pairing up of the complementary strands by hydrogen bonding), forcing the stem to open and restore the emission.

Recent efforts in the molecular beacon approach have focused on the utilization of quantum dots rather than organic dyes, as the emitters. This substitution provides the added photostability of the QDs to avoid the photobleaching problem of dyes. Gold nanoparticles have been similarly utilized as the emission quencher that involves the dipolar quenching mechanism when an emitter is in contact with a gold surface. Kim et al., [2004] used CdSe/ZnS core–shell QDs covered with a mercaptoacetic acid monolayer, together with an organic quencher, 4-(4′-dimethylaminophenylazo)benzoic acid (DABCYL), for complementary DNA target sequence identification. Cady et al., [2007] used a QD and a gold nanoparticle combination for sequence-specific DNA detection.

9.2.2. Plasmonic Sensors

Plasmonic biosensors are another class of widely investigated and utilized optical biosensors. Both localized surface plasmon resonances (LSPR) in metallic nanoparticles and delocalized surface plasmon resonances (SPR) in metallic nanofilms, already discussed in Chapter 5, have been utilized for biosensing. This section below details these approaches that have already been outlined in Section 9.1 and summarized in Figure 9.2.

LSPR Colorimetric Sensors. These sensors utilize a shift in the peak position of the localized surface plasmon resonance as the transduction mecha-

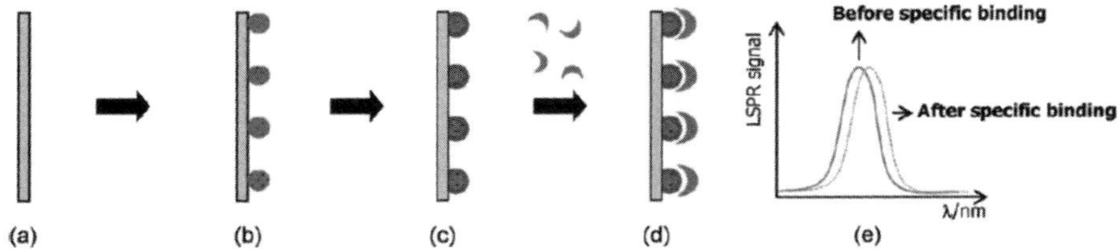

Figure 9.8. Biosensing scheme using refractive index sensing with LSPR of plasmonic nanostructures. (a) Substrate. (b) Metal nanoparticles are attached to substrate by the chemical linkers or nanolithography. (c) Metal particles are modified with the sensor moiety. (d) Analyte attaches specifically onto the recognition layer and (e) causes a change in the refractive index around the particles resulting in an LSPR shift. (Reproduced with permission from Sepúlveda et al. [2009].)

nism for analyte capture. This shift can simply produce a change in color, hence the term colorometric sensing. In the simplest scheme, the surface of a metallic nanoparticle is functionalized with a biorecognition element to capture the analyte to be detected. This binding changes the dielectric environment around the nanoparticle to shift the LSPR peak used for detection. Using the sensitivity of LSPR to the surrounding dielectric environment, also called refractive index sensing, it utilizes metallic nanoparticles which could be of different shapes, core–shell structures, and individually dispersed or immobilized on a substrate [Sepúlveda et al., 2009]. A demonstrative scheme is shown in Figure 9.8. Here the gold or silver nanoparticles can be chemically tethered to a glass surface or electrostatically bound by creating opposite charges on the metal nanoparticles and the substrate. Another method which provides control of the shape and size of the nanoparticle arrays on a solid substrate is nanosphere lithography (NSL) [Haynes et al., 2001]. In this method, a metal is evaporated onto a self-assembled, close-packed monolayer of monodispersed polystyrene nanospheres. The metal nanostructures are deposited in the interspatial regions between the polystyrene nanospheres, which are subsequently removed to leave truncated tetrahedral-shaped metallic nanoparticle arrays with a hexagonal symmetry [Prasad, 2004]. Other variations of NSL using different types of masks have been proposed to produce metallic nanostructures of different shapes, such as nanodisks, nanorings, nanoshells, ellipsoids, nanorods, or any combinations of these.

In one of the earlier works, the widely used biotin–streptavidin model was used to demonstrate LSPR biosensing. Gold nanospheres immobilized on a glass substrate were coupled through a linker [Nath and Chilkoti, 2002], and the binding with streptavidin was detected either by monitoring a change in the transmission or simply by monitoring the color change.

Another type of LSPR sensor is an aggregation sensor, where aggregation of metallic nanoparticles caused by analyte binding induces coupling of the plasmonic fields of these nanoparticles to produce a red shift of the LSPR peak. This shift is dependent on the interparticle separation, and the shift can be visually directed by a change in color. With gold nanospheres, for example, aggregation produces a shift from red to purple-blue.

Storhoff and Mirkin [1999] linked a single-stranded DNA, modified with a thiol group at one terminal, to a gold nanoparticle of ~15 nm in diameter via strong gold–sulfur interactions. The 15-nm-diameter separated gold particles exhibited a well-defined surface plasmon resonance, even when attached to DNA, which produces a burgundy-red color. When the DNA attached to the gold particle hybridizes with its complementary DNA in the test sample, the duplex formation leads to aggregation of the nanoparticles, shifting the surface plasmon resonance and thus changing the color to blue black.

Halas, West, and co-workers [Hirsch et al., 2003] used gold nanoshells with nanoshell-conjugated antibodies to demonstrate a rapid immunoassay, capable of detecting analytes within a complex biological media such as blood. The aggregation of these multiple metallic nanoparticles due to an analyte produces a red shift in the plasmon resonance, along with a decrease in the extinction coefficient. Hirsch et al., [2003] demonstrated the detection of immunoglobulins. Figure 9.9 shows a decrease in the extinction coefficient of the plasmon band at 720 nm upon aggregation of the gold nanoshells, as a result of binding to the analyte (immunoglobulin).

An interesting concept used for monitoring DNA conformational changes is that of a "plasmonic molecular ruler" introduced by Alivisatos' group

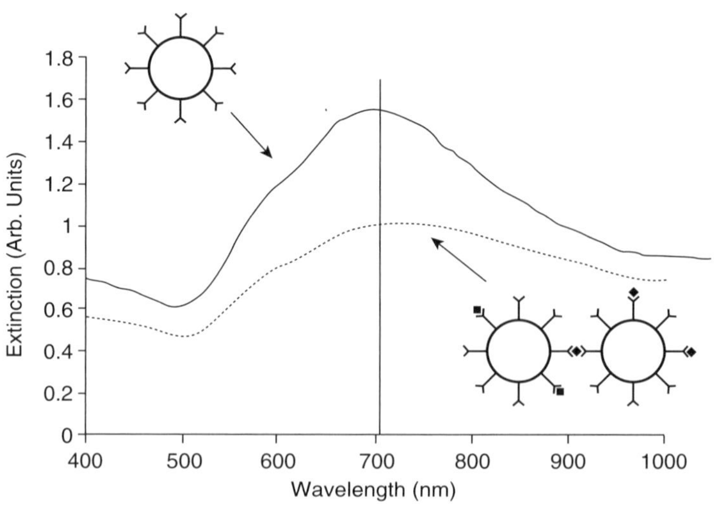

Figure 9.9. Change in the extinction coefficient of the plasmon resonance of gold nanoshells resulting from aggregation, caused by analyte binding. (Reproduced with permission from Hirsch et al. [2003].)

[Sonnichsen et al., 2005]. In a plasmonic ruler, the change in separation between two metallic nanoparticles, due to DNA hybridization, produced a color change as well as a change in the plasmonic scattering intensity.

LSPR Fluorescence Sensors. This method utilizes an enhancement in the fluorescence intensity of an emitter, when it is separated from a metallic nanoparticle at an optimum distance but is not in direct contact with the metal. When the emitter is in direct contact, the metal surface acts as a dipole-induced quencher of fluorescence as discussed earlier under the molecular beacon approach. When they are separated, however, the enhanced plasmonic field at the metal surface enhances the fluorescence and the enhancement is maximum at an optimum distance [Prasad, 2004].

Lakowicz and co-workers [Malicka et al., 2003] used metal-enhanced fluorescence for the detection of DNA hybridization. They used thiolated oligonucleotides to bind silver nanoparticles on a glass surface. The addition of a complementary fluorescein-labeled oligonucleotide produced a dramatic time-dependent 12-fold increase in the fluorescent intensity, as the hybridization process proceeded, bringing the fluorescein close to the silver nanoparticles for surface plasmon enhancement of fluoresceins intensity. This type of approach can thus be used to increase the sensitivity of DNA detection.

Mirkin's group has developed an aptamer nano-flare, a gold nanoparticle core functionalized with a dense monolayer of nucleic acid aptamers, that can directly quantify an intracellular analyte in a living cell (Figure 9.10) [Zheng et al., 2009]. In the initial state, the fluorescence of the flare strand is quenched by the gold nanoparticle. When the ATP molecules bind to the aptamer, it causes a new folded secondary structure and also causes the release of fluorescence reporters. They demonstrated that these nanoconjugates are readily taken up by cells where their signal intensity can be used to quantify intracellular ATP concentration.

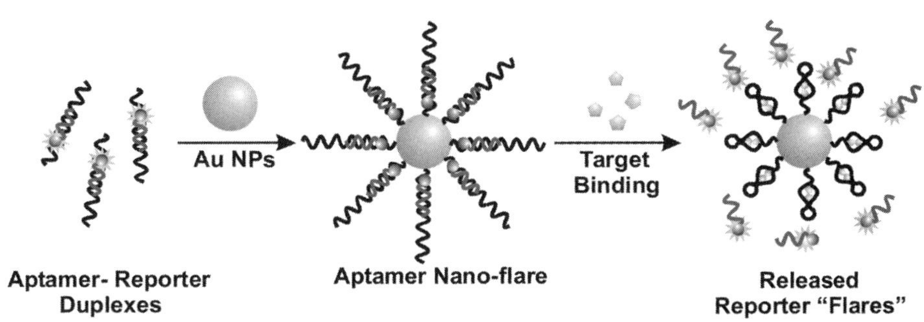

Figure 9.10. Schematic diagram of aptamer nano-flares; the reporter is capable of being released by the target molecule. (Reproduced with permission from Zheng et al. [2009].)

Surface-Enhanced Raman Scattering (SERS) Sensors. SERS is emerging as a powerful biodetection method that provides detailed information about the chemical structure and the conformation of an analyte, with the sensitivity approaching an ultimate single molecule detection [Anker et al., 2008; Qian et al., 2008]. The tremendous intensity enhancement provided by SERS enables detailed analysis of the vibrational fingerprints of proteins and has led to significant advances in SERS-based immunoassays [Han et al., 2009]. SERS-based analysis has the advantage over fluorescence-based detection, because the problem of photobleaching is not encountered. The SERS-based method can be label-free, with direct detection of a protein by adsorption onto a SERS platform. This method has been used successfully for hemoproteins such as hemoglobin. A SERS probe can be alternatively used where the SERS-enhanced Raman spectra of a molecular specie is linked to a metallic nanoparticle for the detection of a target protein. These SERS probes can thus act as Raman tags for a target analyte. For multi-protein detection, Han et al., [2009] developed a new label-free protein detection "Western SERS." This combines SERS and Western blot by using colloidal silver staining, while the method allows for multiprotein detection on a single nitrocellulose membrane. The silver nanoparticles are adsorbed along the net of the membrane, to which proteins have already adhered, to produce SERS effect.

Although the earlier work used silver nanoparticles and their aggregates for SERS, the emphasis has recently shifted to the use of gold nanoparticles, due to their environmental stability, even though the SERS enhancement produced by gold nanostructures is not as large. Another area of increased activity utilizes (a) anisotropic metallic nanostructures, such as nanoprisms, which exhibit a significantly increased electromagnetic enhancement factor when compared to spherical nanoparticles [Anker et al., 2008], and (b) aggregates such as dimers which show a significantly large plasmonic field, called hot spots, in the regions in between nanoparticles. The SERS signals obtained from these hot-spot regions are thus many orders of magnitude higher, providing greatly improved sensitivity.

Another example is our work that demonstrated the application of a pH-dependent SERS nanoprobe to living biological systems, utilizing 2-aminothiophenol (2-aminobienzenethiol, 2-ABT)-labeled silver nanoparticles [Wang et al., 2008]. Through the variations in the SERS spectra of the attached 2-ABT molecules, the probe is capable of measuring pH in its vicinity over the range of 3.0–8.0. The vibrational band of the amino group of 2-ABT was chosen as the Raman probe because it is sensitive to pH and the amino group is a common functional moiety of many biomolecules. We showed that these nanoprobes are robust and that their pH sensitivity can be used when the nanoparticles are incorporated inside living cells.

Nie and co-workers [Qian et al., 2008] used a Raman molecular beacon for DNA detection. The principle of the molecular beacon has been described above, except that a dramatic change in SERS is used as a probe for DNA hybridization.

Surface Plasmon Resonance Biosensors. Surface plasmon resonance (often abbreviated as SPR) sensors are, perhaps, the most extensively utilized optical biosensors that are also commercially available [Prasad, 2003]. The main reason for SPR biosensors to become a powerful tool in biological and chemical sensing application is their capability of real-time monitoring and label-free sensing. Quantitative measurements such as the binding kinetics and the binding affinity between two biomolecules can be readily obtained from the SPR sensorgram. General SPR instruments are constructed using the Kretschmann configuration [Wallis and Stegeman, 1986] as shown in Figure 9.11.

A microscope slide coated with a thin film of metal (usually a 40 to 50-nm-thick gold or silver film by vacuum deposition) serves as a nanoplasmonic platform, while a biosensing layer containing an immobilized biorecognition element can be coated onto the metal surface. The microscope slide is then coupled to a prism through an index-matching fluid or a polymer layer. A p-polarized laser beam (or light from a light-emitting diode) enters the prism. The reflection of the laser beam is monitored. At a certain angle, θ_{sp}, the electromagnetic wave couples to the interface as a surface plasmon. At the same time, an evanescent field propagates away from the interface that extends to about 100 nm above and below the metal surface. At this angle, the reflected intensity signal drops (Figure 9.12). This angle is determined by the relationship

$$k_{sp} = kn_p \sin\theta_{sp}$$

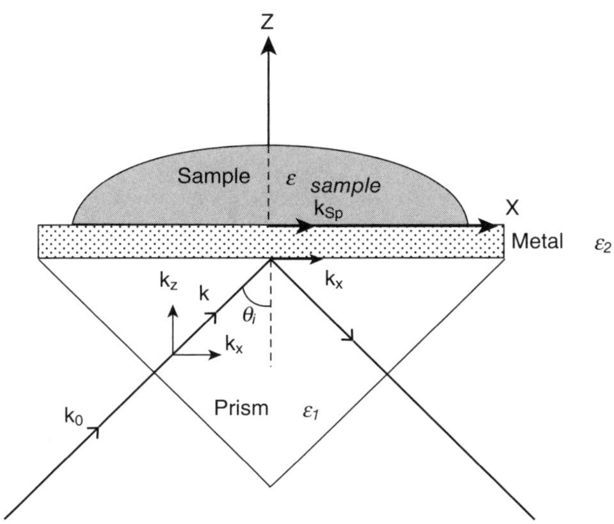

Figure 9.11. Schematics of the Kretschmann geometry for SPR biosensing. (Courtesy of W.-C. Law, SUNY at Buffalo.)

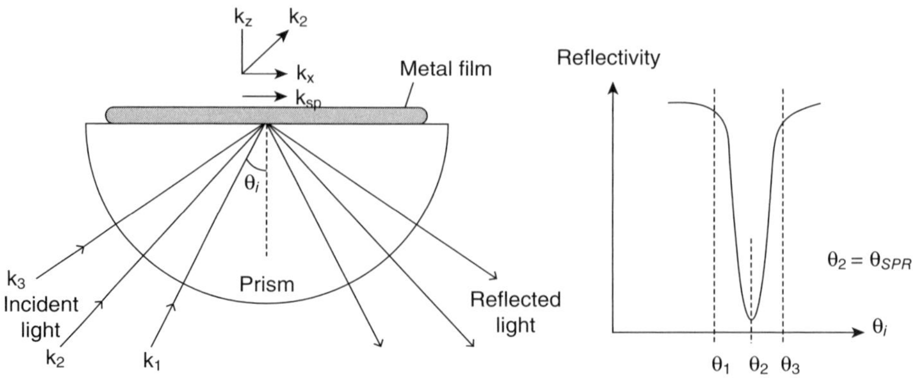

Figure 9.12. The momentum of the incident light can be resolved into k_x and k_z. SPR occur when $k_x = k_{sp}$. (Courtesy of W.-C. Law, SUNY at Buffalo.)

where k_{sp} is the wave vector of the surface plasmon, k is the wave vector of the bulk electromagnetic wave, and n_p is the refractive index of the prism.

In this experiment, one can measure the angle for the minimum value of reflectivity. The sensing response is a change in the refractive index of the sensing layer containing ligands (e.g., antibodies) upon analyte binding which is measured as a change, $\delta\theta$, in the coupling angle. In commercial SPR sensors, the change in the coupling angle is measured by a charge-coupled device (CCD) or photodiode array using a convergent light beam. This arrangement, as shown in Figure 9.13, permits real-time monitoring of the ligand–analyte binding to obtain kinetics of association and dissociation. To obtain this information, the sample solution containing the analyte flows over the sensor chip containing the ligand. During the association phase, the analyte binds to the ligand immobilized on the sensor chip, generating an increase in the response (amount of shift of the coupling angle). The magnitude of the response ($\delta\theta$) levels off over time, as an equilibrium condition between the free and the bound analyte is reached. To monitor dissociation, the flow switches to that of a running buffer that washes out the analyte (leading to dissociation of it from the ligand). During the dissociation, the magnitude of the response decreases. The generated response curve for the association and dissociation cycle is often called a *sensorgram*. The association process is also shown in Figure 9.13 as an inset on the right-hand side.

In this case, for the angular measurement as shown in Figure 9.13, the detection limit is about 10^5 in terms of the refractive index change (RIU), which corresponds to the 1 pg/mm^2 of biomaterial accumulating at the biosensor surface. Several recent studies have demonstrated that the sensitivity of SPR could be improved by at least two orders of magnitude by monitoring the phase characteristics of light instead of the intensity [Kabashin and Nikitin, 1998; Wu et al., 2004]. Our group has demonstrated (a) a phase-sensitive SPR

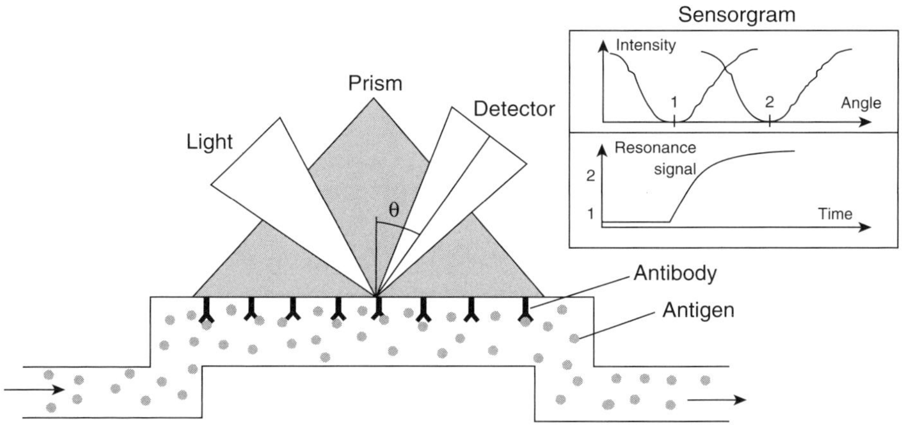

Figure 9.13. Surface plasmon resonance sensor schematic utilizing a CCD or photodiode array (*left*). The inset on the right-hand side shows the sensorgram. The top in the inset shows a shift in the SPR curve from 1 to 2 upon binding with the analyte. The bottom curve is obtained by monitoring the position of SPR resonant angle as a function of time when the analyte is introduced. (Reproduced with permission from Prasad [2003].)

biosensor based on temporal modulation of a pumping beam by a photoelastic modulator and (b) subsequent extraction of phase information at the second and the third harmonics of the modulation frequency [Law et al., 2007]. By utilizing these two characteristics simultaneously, the system can therefore detect a very small change of refractive index (or a tiny amount of biomoelcular interaction) over a wide dynamic range. Based on this setup, our group has further demonstrated a nanoparticle-enhanced SPR biosensor. Specifically, the use of nanoparticle tags/labels results in a dramatic enhancement of sensitivity due to three main factors [Hutter et al., 2001]: (i) an increase of the absolute mass in each binding event, (ii) an increase in the bulk refractive index of the analyte due to the complex refractive index of nanoparticle, and (iii) coupling between the localized surface plasmon resonance (LSPR) of metallic nanoparticles and surface plasmon resonance (SPR) of the sensing film. In our studies, gold nanorods (GNRs) with different aspect ratios (AR) were chosen as the amplification labels [Law et al., 2009]. There are indications from our work that the enhancement factor of nanoparticle-based SPR biosensor is due to the presence of plasmonic coupling between the nanoparticles and the sensing film. Maximum enhancement effect can be achieved when the LSPR peak wavelength of nanoparticles matches the oscillation frequency of effective plasmon on the SPR film (Figure 9.14).

The use of metallic nanoparticles induced an increase in the binding mass, an increase in refractive index, and a perturbation of the evanescent field that led to an enhanced sensitivity. Typically, two strategies were employed for the

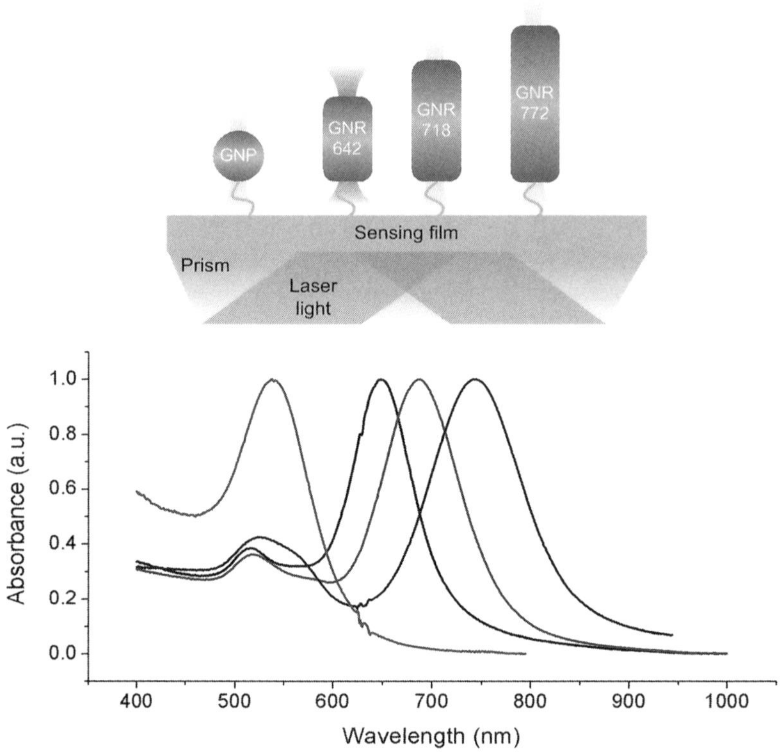

Figure 9.14. Absorption of gold nanorods with different aspect ratios and their plasmonic coupling efficiency on a SPR sensing film. GNR642 has the highest coupling efficiency when laser source operating at 785 nm is used.

Au-amplified SPR biosensors: (i) The targeted analyte is directly conjugated to the gold nanoparticle and the whole conjugate is flowed through the sensing surface; and (ii) the targeted analyte is flowing through the sensing surface, followed by a secondary nanoparticle conjugate that binds to the captured analyte. In practical terms, the second approach will be more preferable, because the pre-modification of the analyte could be omitted. Therefore, in our recent studies, we employed the superior performance offered by GNR to develop a sandwich nano-immunoassay sensor of tumor necrosis factor-alpha (TNF-α) antigen detection for early detection of (Figure 9.15) [Law et al., 2011]. TNF-α antigens were first captured by the antibodies decorating sensing film. The enhancement was accomplished by the secondary binding of GNR-antibody (GNR-Ab) to the captured TNF-α antigen (targeted analyte) that led to a larger perturbation of the SPR coupling condition. TNF-α has been recognized as a tumor promoter that may be involved in tumor initiation process [Szlosarek et al., 2006]. An ultrasensitive monitoring of the concentra-

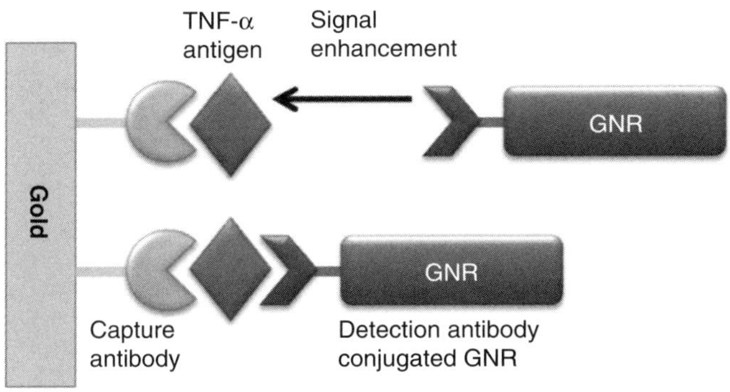

Figure 9.15. Schematic diagram of an ultrasensitive immunoassay for early detection of cancer, based on using GNR as an amplification label. (Reproduced with permission from Law et al. [2011].)

tion of this biomarker can potentially serve as a reliable diagnostic tool for early detection of cancer.

Whispering Gallery Mode Biosensor (WGM-Biosensor). In this technique, light is trapped by total internal reflection in a dielectric microcavity such as a microsphere, micro-disk, or micro-ring. This light executes a circular orbit, accompanied by an exponentially decaying field beyond the dielectric interface (~100 nm from the microcavities' perimeter). Any change in the dielectric properties at the surface of the microsphere, caused by the binding of an analyte, produces a change in the resonance frequency of the microcavity which can be used for biosensing. This sensing scheme is presented by Figure 9.16. The device gains a huge label-free sensitivity advantage over SPR because of the very narrow frequency line width associated with the microcavity resonance, in comparison to the shift in resonance frequency accompanying adsorption on an SPR sensor. Recently, Arnold and co-workers [2009] showed that even a virus could be sensitively discriminated by this device. They also found that they could easily detect the adsorption of a single Influenza A virus (~100 attograms) and quantify its size/mass. Later, Vollmer et al., [2008] designed an enhancement scheme using plasmonic nano-antenna planted on the surface of the dielectric sphere that is expected to reach single protein sensitivity (~0.1 attogram). A commercial device for clinical use is manufactured by Genalyte in San Diego.

9.2.3. Photonic Crystal Sensors

Photonic crystals are periodic dielectric arrays that diffract light at wavelengths determined by the lattice spacing between the arrays and the average refractive index of the structure [Prasad, 2004]. Colloidal synthesis of photonic

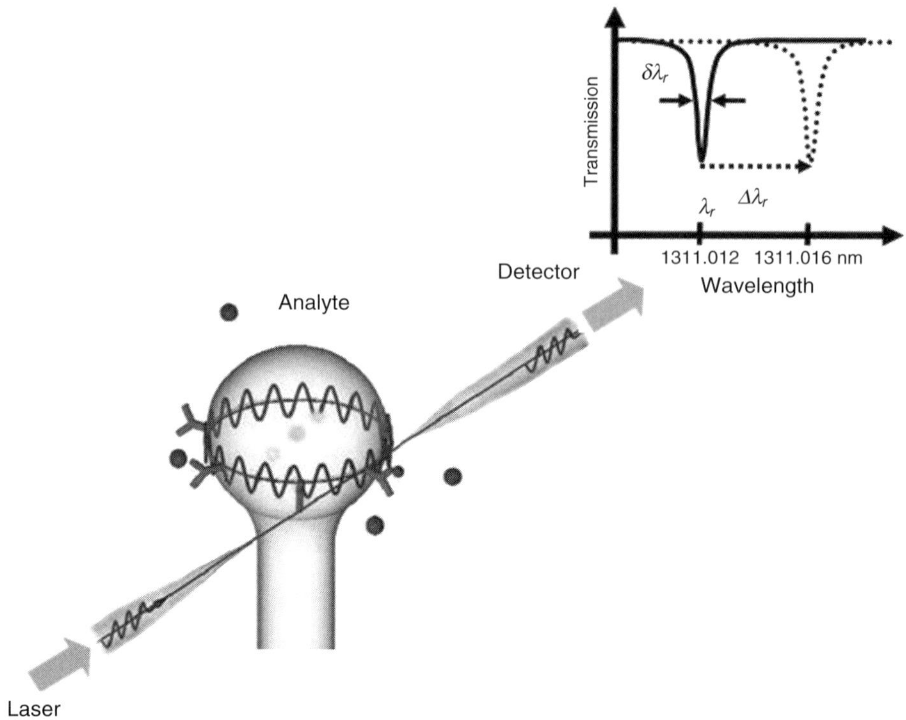

Figure. 9.16. Concept of the Whispering Gallery Mode (WGM) biosensor. WGM in a dielectric sphere driven by evanescent coupling to a tapered optical fiber. The binding of analyte molecules to the immobilized antibodies is detected from a shift of the resonance wavelength. (Courtesy of Professor Stephen Arnold.)

crystal has drawn a wide research attention because of the simplicity of the preparation procedures and the relatively lower cost. A typical example is a biosensor utilizing colloidal photonic crystal made of DNA-responsive hydrogel beads developed by Zhao et al., [2010]. A quantum dot tagged DNA-responsive hydrogel photonic beads suspension array is shown in Figure 9.17a. Any specific hybridization of the target DNA and the cross-linked single-stranded DNA in the hydrogel grid will cause a shrinking of the hydrogel and lead to a blue shift in the Bragg diffraction peak position. Figure 9.17b shows a concentration-dependent curve of the concentration of the target DNA and the corresponding peak shifts.

9.3. MAGNETIC BIOSENSORS

Magnetic biosensors have received a great deal of attention during recent years [Varadan et al., 2008; Tamanatra et al., 2008]. In the general scheme of

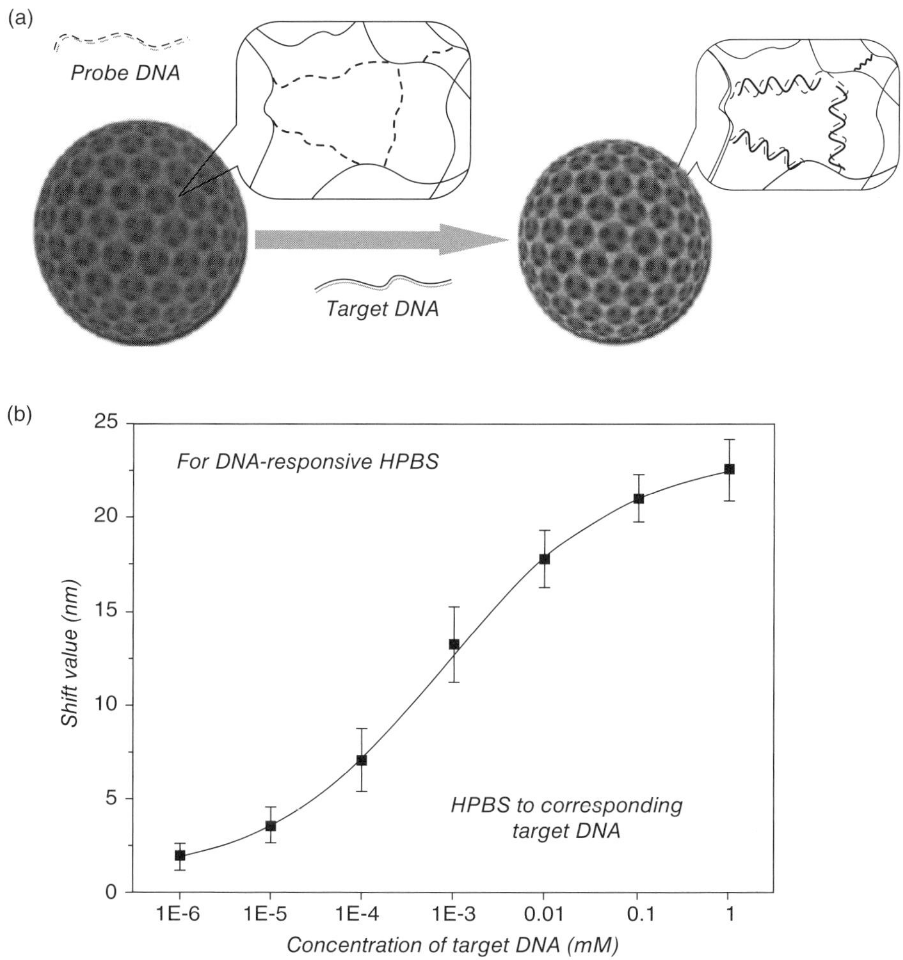

Figure 9.17. (a) Schematic diagram of the DNA-responsive hydrogel photonic beads biosensor. (b) The diffraction peak shifts when different concentrations of target DNA are exposed to the biosensor. (Reproduced with permission from Zhao et al. [2010].)

magnetic sensing, magnetic nanoparticles, discussed in Chapter 5, which are coated with specific biorecognition elements, bind to the target biospecies, the analyte to be detected. The binding is detected either by using the stray (fringe) magnetic field of these magnetic nanoparticles through a magnetic field sensor substrate platform or by a change in the relaxation time in a ferrofluid environment. These principles and techniques utilized are listed in Figure 9.18.

A magnetic biosensor offers a number of advantages [Varaden et al., 2008]:

- Very stable magnetic properties of the nanoparticles, because they do not face the problem of photobleaching encountered in optical sensing.

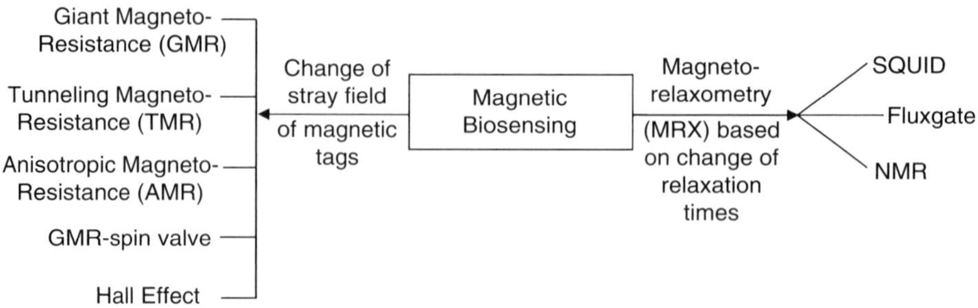

Figure 9.18. The magnetic biosensing principles and the various techniques utilizing them.

- No screening of the magnetic field by aqueous reagents or biospecies.
- Very little background interference, because most biological systems exhibit very small magnetic susceptibilities.
- Remote manipulation of magnetic nanoparticles using a field gradient, even when in biological media.
- Highly sensitive and multiple methods of magnetic transduction for biosensing.

Earlier work utilized micron-sized magnetic microbeads, but recent efforts have shifted to the use of smaller (~20 nm) magnetic nanoparticles (such as Fe_3O_4 discussed in Chapter 5) that can be readily prepared with controlled monodispersity and are thus more suited for reproducibility. The smaller-sized particles, with large surface area for analyte detection, provide increased label binding density across the sensor surface. Too-small-sized particles can also pose problems in providing high sensitivity, however, due to the low magnetic moments of a small volume. An optimum size is thus required. The various magnetic sensing schemes are described below.

Substrate-Based Stray Field Sensing. This method describes the detection of the stray fields of magnetic nanoparticle tags by their interaction (binding) to a solid-state substrate, known as a magnetic field sensor. A schematic representation of this type of sensor is shown in Figure 9.19 [Graham et al., 2004]. Here, the target biomolecular analytes [B] are captured by the array in the solid-state magnetic sensor substrate and the unbound biomolecules are washed away. The magnetic nanoparticles functionalized with a biorecognition biomolecule [A] is then passed over the array, whereby biomolecule A interacts with biomolecule B upon biorecognition and thus provides a signal in the magnetic field sensor, generated by the stray field of the nanoparticle label. The target analyte B thus acts as a linker between the nanoparticles and the magnetic field sensor. This scheme is also often called a Direct Detection Method [Varadan et al., 2008]. In a "Secondary Detection Scheme," two steps

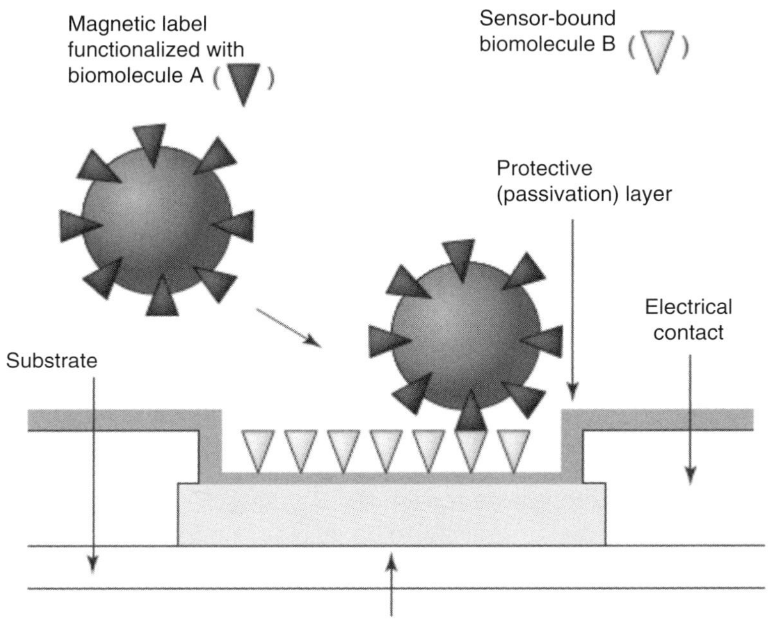

Figure 9.19. Schematics of magnetic biodetection through the use of a solid-state magnetic field sensor, the binding of magnetic nanoparticle labels to the sensor, and the measurement of their stray magnetic field. (Reproduced with permission from Graham et al. [2004].)

are involved [Graham et al., 2004]. The first step involves the tagging of a target molecule (e.g., target DNA) with a biomolecular label (e.g., biotin) that can then be captured by the complementary biorecognition probe molecule (e.g., probe DNA by hybridization) immobilized on the magnetic field sensor substrate. The second step involves the functionalization of the magnetic nanoparticles with streptavidin which is then passed over the substrate where the streptavidin–biotin binding results in the generation of magnetic transduction by the stray field of the nanoparticles. A number of different magnetic field sensors, as listed previously in Figure 9.19, have been used for sensing. They are briefly described below.

Giant Magnetoresistance (GMR) Sensors. This sensing technique utilizes the GMR effect of ferromagnetic layers separated by nonmagnetic metallic spacer layers where the electronic transport through the metallic layer (hence its resistance) becomes dependent on the spin and hence the relative orientation of the magnetization directions of the adjacent ferromagnetic layers which can be changed by an external field. Thus, the stray magnetic field of the captured magnetic labels creates a change in the resistance which acts as a transduction mechanism to produce an electrical response.

The sensitivity of a magnetoresistive material is expressed as a ratio of the change in resistance and the minimum resistance in a saturation field. The ratio comparisons are often used to assess the relative sensitivities of the sensors.

Tunneling Magnetoresistance (TMR) Sensors. In this technique, the ferromagnetic layers are separated from each other by a very thin insulating layer (e.g., Al_2O_3) and the tunneling current (resistance) between them is measured, which is strongly dependent on the magnetic field. Although there are fabrication issues such as placing top electrical contact as one measures current perpendicularly through the device, as well as dielectric breakdown of the insulating layers, the TMR devices have the appeal of significant gain in sensitivity over to the GMR devices [Reiss et al., 2005].

Anisotropic Magnetoresistance (AMR) Sensors. An AMR sensor utilizes the variation in the electron mean free path (hence resistance) as a function of the angle between the electron velocity and the conductive material's (e.g., iron) magnetization direction. This method in a stacked geometry provides a low MR ratio and, thus, a significantly lower sensitivity when compared to the XMR sensors described above. To increase the sensitivity, Miller et al., [2002] introduced a ring sensor geometry.

Hall Effect Sensors. The Hall effect involves the action of the transverse force of an externally applied magnetic field to push charge carriers to one side of a current-carrying conductor. The resulting charge carrier buildup generates an electric field called Hall voltage in a direction perpendicular to both the applied magnetic field and the flowing current. Microarray technology has been applied in the design of micron-scale Hall sensors for immunological assay [Aytur et al., 2006]. Protable microarrays can thus be fabricated for multiplexed analyte detection.

Spin-Valve Sensors. Spin valves also use the magnetoresistance effect, but they use a different magnetoresistive stack arrangement. A schematic arrangement is shown in Figure 9.20 [Tamanatra et al., 2008]. The antiferromagnetic pinning layer of MnIr forces the magnetization of the ferromagnetic CoFe layer in a particular direction, thus preventing the rotation (pinning) of the CoFe layer magnetization. But it allows the bottom NiFe layer magnetization to freely rotate in the presence of an external magnetic field because of the thick nonmagnetic conducting layer of Cu. This rotation details the presence of external magnetic fields. The net result is an exchange bias that produces a high MR ratio of 6–15%, exceeding that of other magnetoresistive devices.

Magnetorelaxometry Sensors. This type of sensing utilizes magnetic relaxation times. The two types of relaxations used in sensing are:

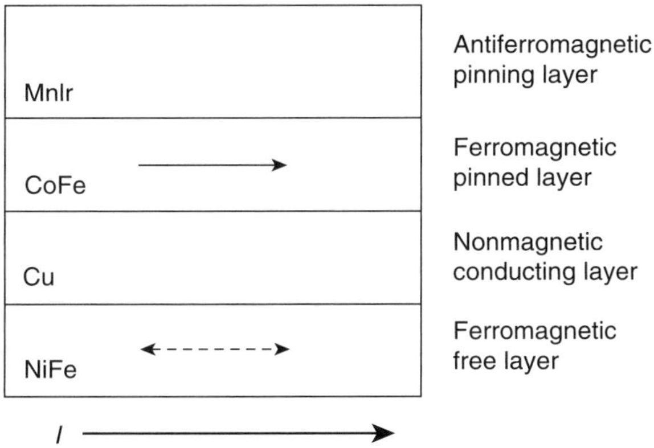

Figure 9.20. The basic stack arrangements of a spin-valve sensor. (Reproduced with permission from Tamanatra et al. [2008].)

(i) *Relaxation of the Magnetization of a Magnetic Nanoparticle after the Aligning Magnetic Field Is Turned Off.* The unbound (mobile) magnetic nanoparticles relax under Brownian rotation on a time scale, typically in microseconds. If the nanoparticles are bound to a large biomolecule, they relax via the Neel mechanism with a relaxation time on the scale of seconds. Thus on the basis of this difference in relaxation behavior (hence relaxation times), bound and unbound magnetic nanoparticles can be distinguished, even when dispersed in a liquid environment (ferroliquid state).

SQUID (superconducting quantum interference device) is used to measure the magnetization and its decay (relaxation). An example is the detection of a bacterial pathogen, *Listeria monocytogenes* [Grossman et al., 2004]. The magnetic nanoparticles are bound to the bacteria through a polyclonal antibody biorecognition and placed in a SQUID, where a pulsed magnetic field is applied to align the magnetic dipole moments and create a net magnetization. When the magnetic field is off, relaxation of magnetization by randomization of the magnetic dipole is observed. The unbound nanoparticles relax by Brownian rotation within 50 μs, too short to be measured by SQUID, but the bound particles, because of a relatively large-sized bacteria, undergoes Neel relaxation within 1 s. Although this method does not require immobilization on a solid platform or washing away of unbound particles, the operation requires low temperature.

Fluxgate magnetometers have also been used for magnetorelaxometry [Ludwig et al., 2005]. The fluxgate magnetometers are much simpler in design when compared to SQUID, because it involves a ferromagnetic core that is periodically driven by a primary coil into saturation. It can measure magnetization at room temperature, whereas SQUID, offering significantly higher sensitivity, requires operation at low temperature.

(ii) *Spin–Spin Relaxation (T_2-Relaxation) Times in a Nuclear Magnetic Resonance (NMR) Device.* The principle has been discussed in Chapter 5 and revisited in Chapter 8 under magnetic resonance imaging. Here, the change in the T_2-relaxation time of protons in water, produced by a magnetic nanoparticle bound to the analyte biomolecule by ligand–receptor interaction, is used for sensing. The basis for sensing results from the fact that the changes in T_2-relaxation times produced by bound and unbound magnetic nanoparticles are different. The advantage of this method is that even the raw samples such as a turbid biological medium or a whole-cell lysate can be analyzed, because the T_2-relaxation time is not dependent on the surrounding matrix or nonspecifically bound reagents [Grimm et al., 2004].

9.4. ELECTRICAL BIOSENSING

This method requires electrical transduction where the biosensor detection is in the form of a signal as a current or a voltage. This is the case with a field effect transistor (FET) device. In another modality discussed here, using electromechanical transduction with a quartz crystal microbalance (QCM), the signal is in the form of a charge in the frequency from the QCM resonant frequency.

FET is a semiconductor device consisting of two *n*-doped (excess electrons) regions called source and drain. These are separated by a channel consisting of a *p*-doped (excess hole) region and a metal gate (as shown in Figure 9.21). The flow of current between the source and the drain (i.e., conductance) is dependent on the voltage applied to the gate. FET-based sensors have been investigated since mid-1970s. The basic principle of biosensing utilizing an FET is also shown in Figure 9.21. The region forming the channel is functionalized

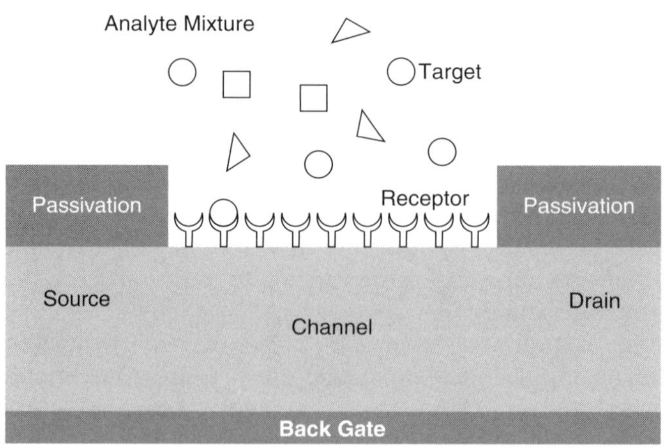

Figure 9.21. A sensing scheme utilizing a FET biosensor. (Reproduced with permission from Lee et al. [2009b].)

with receptor molecules that bind to the target molecules when exposed to the analyte solution. This binding changes the electrical properties of the channel region, resulting in a change in the conductance between the source and the drain. The sensitivity of the device is clearly limited by the diffusion characteristics of the target analyte. In addition, the electrostatic effects due to the presence of other ions (counterions) in the solution also have to be taken into account in evaluation of the device characteristics. The source and the drain regions are isolated from the analyte solution by passivation.

Recent activities in the field of FET biosensors have been the utilization of channels comprised of nanowries and nanotubes [Patolsky et al., 2006; Allen et al., 2007]. Their large localized surface areas provide enhanced nanoscale contact with the biomolecules and provide a significant gain in the sensitivity. Lieber's group [Patolsky, 2006] used silicon nanowire FETs for detecting proteins and DNA hybrids as well as cancer biomarkers. The principle is the depletion or accumulation of charge carriers in the channel by binding of the charged biomolecules to the surface of these nanowires.

Carbon nanotube based FETs have also been investigated. The electrical properties of these carbon nanotubes are strongly dependent on their physical characteristics (single-walled vs. multiwalled, diameter, length, etc.). These carbon nanotubes have been introduced in Chapter 3. The single-walled carbon nanotubes (SWCNT) can be made metallic or semiconducting. Semiconducting SWCNTs have been used to fabricate the channel region in FET sensors. The first design used a single carbon nanotube as a channel between the source and the drain [Tans et al., 1998; Martel et al., 1998]. In another approach, a network of carbon nanotubes served as a collective channel [Snow et al., 2003]. The ease of functionalization of the nanotube surface provides an opportunity for tailoring their characteristics to recognize and bind with specific analytes. As discussed in the review by Allen et al., [2007], carbon nanotube FETs have been used for the detection of proteins, antibody–antigen assays, DNA hybridization, and enzymatic reactions for glucose detection.

The quartz crystal microbalance (QCM) has gained popularity because of its simplicity and very low cost. It is a mass sensing device in which the piezoelectric oscillation frequency of a quartz crystal changes when an adsorbate deposits on the surface of the quartz crystal to change its mass. To enhance the sensitivity of QCM, various nanoparticle-based amplification schemes have been used. Mao et al., [2006] used a scheme shown in Figure 9.22 in which a thiolated single-strand DNA probe specific to *E. coli* 0157:H7 eaeA gene was immobilized onto the QCM sensor surface. After hybridization with the complementary biotinylated target DNA, streptavidin-conjugated Fe_3O_4 nanoparticles were used as "mass enhancers" to amplify the frequency change.

Chu et al., [2006] used another amplification for immunoassay in which antibody-functionalized Au nanoparticles were used as the primary amplifying probe while a dendritic-type immunocomplex of protein A and antibody-modified Au nanoparticles was used as the secondary amplifying probe.

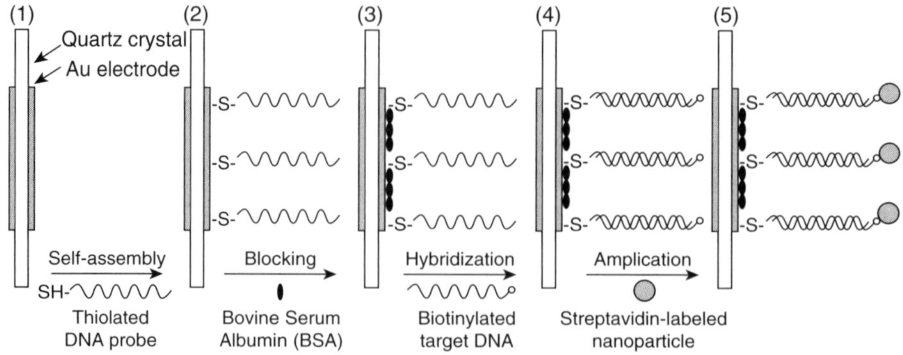

Figure 9.22. A mass-enhancing scheme used with QCM to lower the detection limit of *E. coli.* (Reproduced with permission from Mao et al. [2006].)

9.5. ELECTROCHEMICAL BIOSENSING

In electrochemical sensing, a chemical electron transfer reaction is used as the transduction mechanism. It is one of the oldest biosensing principles and has been utilized in many commercial sensors, such as glucose sensors. Two different types of electrochemical sensors described here utilize the electron transfer in two different ways, which are described in detail in this section.

Electrochemical Cell Sensors. In this case, the electron transfer transduction takes place on an electrode of an electrochemical cell to produce a voltage or a current charge. One major appeal is the low cost of these types of sensors, with the addition of good sensitivity.

Nanomaterials have made a significant impact on electrochemical biosensing. Semiconductor and metallic nanoparticles have been utilized in electrochemical biosensing. The functions of these nanoparticles have been widespread, some of which are summarized below [Luo et al., 2006; Wang and Hu, 2009].

Immobilization of Biomolecules. This function makes use of the large surface area and high surface energy of nanoparticles, which allows for the adsorption or immobilization of biomolecular analytes. The metallic or semiconductor nanoparticles can be immobilized on the working electrode of an electrochemical cell first, by using a variety of methods such as physical adsorption, chemical bonding, electrodeposition, or electropolymerization with a redox polymer. By using cross-linking elements, nanoparticles multilayers can provide a porous film coating with increased surface area on the electrode. A variety of methods can be used to immobilize the biomolecules onto the surface of nanoparticles. The nanoparticles may carry charges that can be used for electrostatic binding to a charged biomolecule. With proteins, the amine group can be used to attach to the gold nanoparticle surface. In yet another scheme, DNA strands (oligonucleotides) can also be modified with special functional groups such as thiol (-SH) that can bind to a metallic nanoparticle surface. Binding of proteins to nanoparticles have also led to preserving the bioactivity of the proteins during protein detection.

Catalysis of Electrochemical Reactions. Many nanoparticles, such as metallic nanoparticles, exhibit catalytic properties that can decrease the overpotential for many analytically important electrochemical reactions. A nitric oxide (NO) electrochemical sensor, for example, uses the catalytic activity of gold nanoparticles on a platinum microelectrode to lower the overpotential for NO oxidation by about 250 mV [Zhu et al., 2002]. Platinum nanoparticles also exhibit good catalytic properties and they have been used in H_2O_2 sensors where a carbon film electrode is modified with platinum nanoparticles [You et al., 2003].

Enhancement of Electron Transfer. Electrochemical sensing is based on the principle of electron transfer. Redox-enzyme biosensing utilizes the electron transfer between the active site of an enzyme and the electrode. Direct electrical communication between the enzyme, when the active site is surrounded by a thick insulating protein layer, is difficult and thus the electrical signal is not often detectable. Nanoparticles have been used to facilitate electron transfer between the active site of an enzyme and an electrode by acting as an electron transfer mediator. For example, hemoglobin, immobilized with ZrO_2 nanoparticles on a pyrolytic graphite electrode, has exhibited improved electrochemical biosensing [Liu et al., 2004].

Labeling Biomolecules. A nanoparticle-labeled biomolecule with the retention of its activity interacts with its counterpart (biorecognition); a quantitative assessment can be made through the dissolution of the nanoparticle labels—for example, metallic nanoparticles with the use of methods like stripping voltammetry. Stripping voltammetry is a general electroanalytical technique for trace metal measurements [Katz et al., 2004].

Simultaneous detection of three different DNA targets [Wang et al., 2003] have also been made. Three different nucleic acids are immobilized onto three different magnetic nanoparticles, which then capture three different target DNA strands labeled by different nanoparticles such as ZnS, CdS, and PbS by complementary hybridization. Stripping voltammetry was used to detect the concentrations of Zn, Cd, and Pb ions enabling simultaneous analysis of the three DNA targets.

Nanoparticles as Reactants. Nanoparticles themselves can act as electrochemical reactants (as opposed to just being catalysts as described above). An example is provided by the MnO_2 naoparticles, which, in one case, were introduced on the surface of a glucose biosensor. This caused electrochemical oxidation of ascorbic acid during glucose detection [Xu et al., 2004].

In recent years, conducting polymers have emerged as another class of nanostructured materials for electrochemical biosensing. Both electrochemically formed films and nanoparticle formulations of conducting polymers have been used [Luo et al., 2006]. Another area of development has been the use of nanostructured electrodes. A patterned carbon nanofiber nanoelectrode array was recently used by Meyyappan and co-workers for multiplexed ultrasensitive detection [Arumugam et al., 2009]. They used electron beam patterened vertically aligned carbon nanofibers as the sensing element. They used this electrode configuration for the detection of DNA targets from *E. coli* 0157:H7.

9.6. ELECTROCHEMILUMINESCENCE BIOSENSING

Electrochemiluminescence (ECL) sensing involves the principle of a species at the working electrode surface undergoing electron transfer to form an excited-state species that emits light. This electroluminescence is used as the transduction pathway for biodetection. The major advantage of ECL over fluorescence detection is that no light source is needed for the excitation, thus eliminating the background light scattering and impurity emissions.

ECL biosensing is a growing area of interest, with numerous types of nanomaterials such as nanoparticles and nanotubes of metals, semiconductors, carbon, or polymers being used in this type of biosensing [Qi et al., 2009; Bertoncello and Forster, 2009]. Described below are the two pathways used for electrochemiluminescence in biodetection. In one pathway, electrochemically generated intermediate species interact with the electrodes by an annihilation process to form both ground-state and electronically excited species, whereby the excited species emit light. An example is the ECL of 9,10-diphenylanthracene (DPA) as follows [Santhanam and Bard, 1965]:

$$\text{DPA} - e^- \rightarrow \text{DPA}^{*+} \text{ (oxidation at the anode)}$$

$$\text{DPA} + e^- \rightarrow \text{DPA}^{*-} \text{ (reduction at the cathode)}$$

$$\text{DPA}^{*-} + \text{DPA}^{*+} \rightarrow \text{DPA} + \text{DPA}^{*} \text{ (annihilation)}$$

$$\text{DPA}^{*} \rightarrow \text{DPA} + h\nu \text{ (luminescence)}$$

The second approach utilizes a co-reactant for the ECL. Here ECL is generated by applying a potential at an electrode, where a luminophore species is deposited. Through the application of an appropriate polarity (positive or negative) potential at the electrode, both the luminophore and the co-reactant (in the solution) undergo oxidation or reduction, resulting in the formation of radicals and intermediate states. The intermediate states decompose to form highly reactive oxidizing or reducing species that interact with the oxidized or reduced luminophore to create an excitation to produce luminescence. An example is a $Ru(bpy)_3^{2+}$-based luminophore in the presence of tripropylamine

(TPA) as the co-reactant. The luminophore species are regenerated at the electrode, with consumption of the co-reactant molecules.

Recently, carbon nanotubes (CNT) have attracted a great deal of attention where the luminophore has been deposited on them. Polymer/CNT/nanoparticles/$Ru(bpy)_3^+$ composites have drawn considerable attention. ORMOSIL/CNTS/$Ru(bpyr)_3^{2+}$ composites have been used for protein detections [Tao et al., 2007]. Tao et al. used this type of composite for detection of herring sperm double-stranded DNA. ECL detection using conducting polymers has also been demonstrated [Chang et al., 2008]. The electron transfer process in quantum dots has also been used for ECL biosening, with the electroluminescence generated when reduced quantum dots (e.g., CdSe) collide with the oxidized forms and produce electronically excited quantum dots by the annihilation process described earlier. Thioglycolic acid-capped CdSe QDs were recently coupled with glucose oxidase (GOX), whereby a decrease in the ECL response of GOX-CdSe QDs after the addition of glucose was measured for glucose detection.

9.7. *IN VIVO* BIOELECTRONIC SENSORS

The development of *in vivo* implantable biosensors for healthcare has shown a remarkable progress over the past decade. Continuous monitoring and management of chronic diseases (e.g., diabetes), using *in vivo* bioelectronics provide great potential to detect various body disorders and alert the patients by remotely transmitting the signal through the electronics. The implantable biosensors are miniaturized, about the size of a penny, and composed of biocompatible materials. A subcutaneous implantable glucose and lactate biochip developed by Guiseppi-Elie's group is shown in Figure 9.23 [Guiseppi-Elie et al., 2005]. The miniaturized ($2 \times 4 \times 0.5$ mm) biochip contains two platinum-based enzyme electrodes to provide dual analyte sensing and detection. The biorecognition layer of the electrodes is fabricated using a hydrogel in which glucose oxidase and lactate oxidase can be immobilized. The Ag/AgCl electrode acts as a reference to detect the amperometric change between the electrodes.

The use of electroconductive hydrogels, which combines the biocompatibility of hydrogel and the electrical properties of a conductive polymer, was first described by Guiseppi-Elie in 1995 [Guiseppi-Elie et al., 1995]. Poly(hydroxyethyl methacrylate) (poly(HEMA)) and polypyrrole (PPy) are two typical examples of electroconductive hydrogels that have shown a significant reduction of inflammatory response and can be potentially applied in making clinical diagnostic devices [Guiseppi-Elie, 2010].

Figure 9.24a shows a PPy implant with dimensions of $2 \times 3 \times 0.25$ mm developed by Langer and Sur [George et al., 2005]. The device was surgically implanted into the cerebral cortex of a rat (Figure 9.24b). After 6 weeks, the rats were scarified and the implanted site was examined using immunofluorescence.

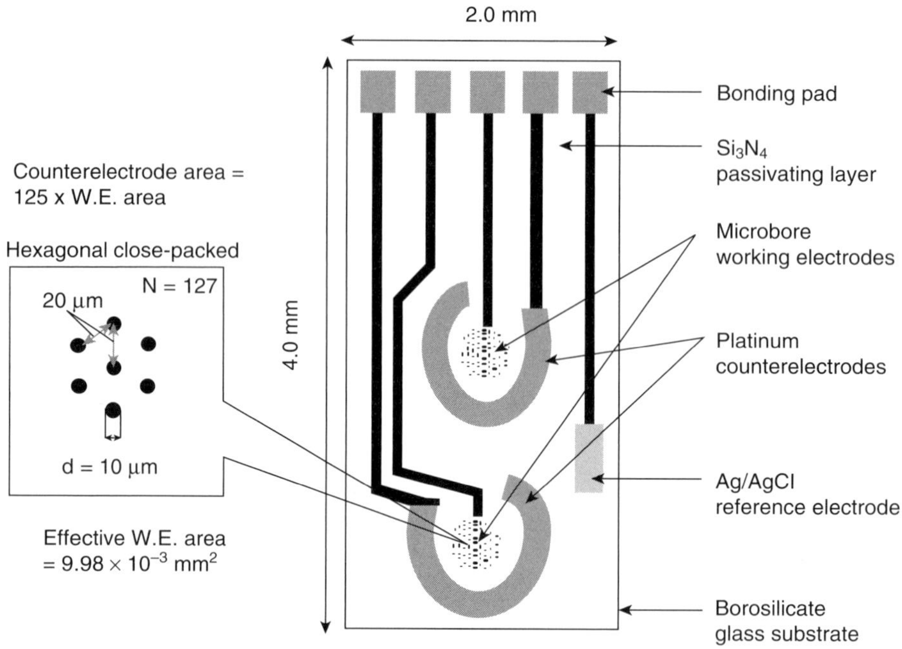

Figure 9.23. Schematic of the amperometric biochip for monitoring of glucose and lactate. (Reproduced with permission from Guiseppi-Elie et al. [2005].)

The results revealed that the implants were completely enveloped into the tissue without observing any immunoresponse which indicate that the PPy material was nontoxic. Figure 9.24c illustrates the histological slice at 6 weeks post-implantation, showing that the tissue (neural parenchyma) completely enveloped the implants.

The application of nanoparticles in the FET-based and ECL biosensor has been discussed above. The same strategy can also be employed in the *in vivo* bioelectronic sensor. A comprehensive review describing the synergetic effect of nanotechnology and implantable biosensors is available in the literature [Vaddiraju et al., 2010]. The main issues in implantable biosensors, such as biofouling, foreign body response, sensor drifting, and poor temporal resolution, can be readily addressed by using nanotechnology. For instance, carbon nanotubes can be utilized as nanoelectrodes in a glucose biosensor (Figure 9.25) [Lin et al., 2004]. Improved signal-to-noise ratio and enhanced detection limits can be accomplished by using a loosely packed nanoelectrode ensemble to avoid the generation of an overlapping signal associated with the interelectrode interference.

In addition, in an electrochemical biosensor, the combination of multi-wall carbon nanotubes (MWNTs) and gold nanoparticles (GNPs) can serve as connectors between the electrode and the enzyme to enhance the effective elec-

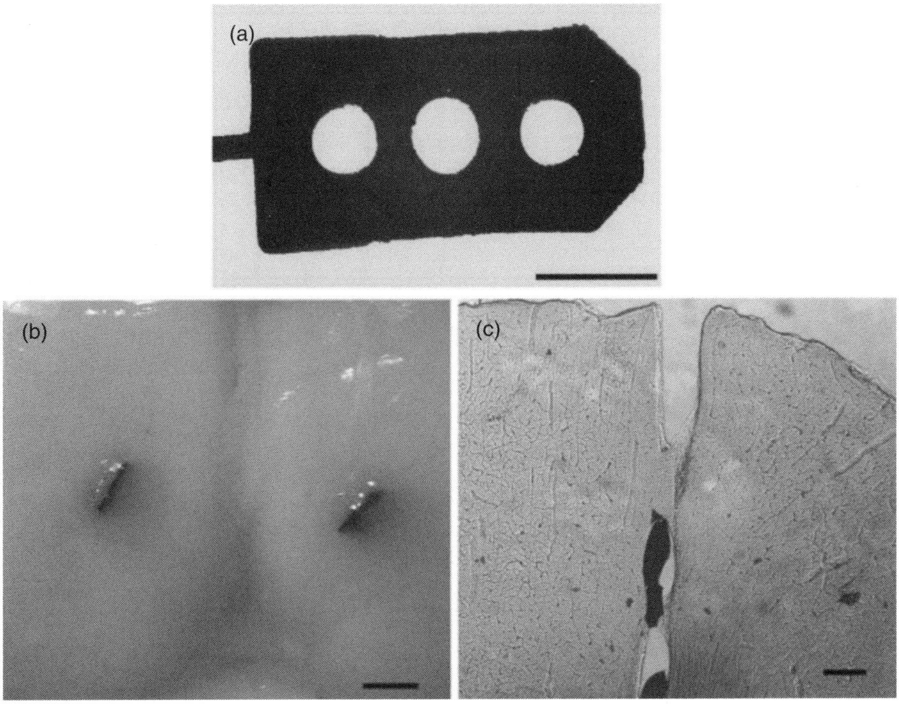

Figure 9.24. (a) Picture of a PPy implant. Scale bar: 1 mm. (b) Two PPy implants placed in the rat's cortex. Scale bar: 2 mm. (c) Histological slice at 6 weeks post-implantation. Scale bar: is 200 μm. (Reproduced with permission from George et al. [2005].)

trode surface area and facilitate fast direct electron transfer between redox proteins (hemoglobin (Hb)) and the electrode surface (Figure 9.26) [Chen et al., 2007].

Incorporation of nanomaterials into the implantable biosensors offers diverse advantages to improve the performance of biosensing capability, sensor lifetime, and reduced inflammatory response.

HIGHLIGHTS OF THE CHAPTER

- Biosensors are analytical devices that can detect chemical, species, biological species, and microorganisms.
- A biosensor utilizes a biological recognition element that senses the presence of an analyte and creates a physical or chemical response that is converted by a transducer into a signal.
- The important components of a biosensor are: (i) a sampling unit, (ii) a recognition element, (iii) a stimulation as a result of biorecognition, and (iv) a transduction.

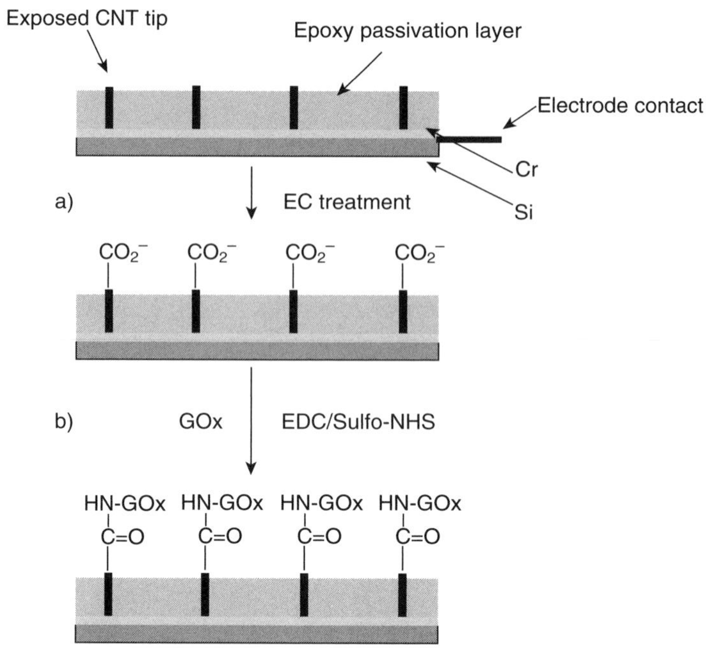

Figure 9.25. Fabrication of a glucose biosensor based on CNT nanoelectrode ensembles: (a) Electrochemical treatment of the CNT nanoelectrode ensembles for functionalization. (b) Coupling of the enzyme to the functionalized CNT. (Reproduced with permission from Lin et al. [2004].)

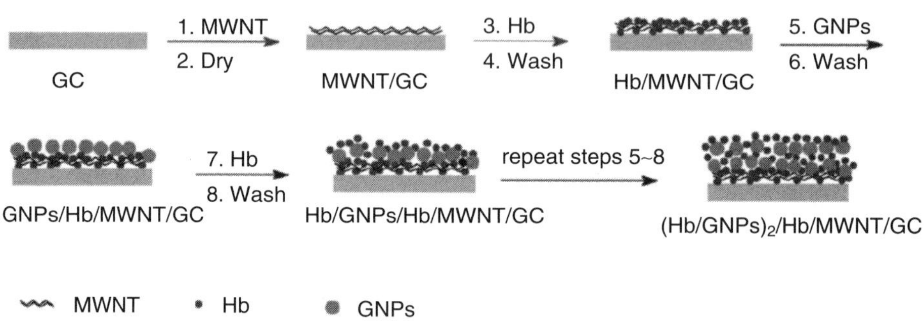

Figure 9.26. Hydrogen peroxide biosensor based on the immobilization of hemoglobin on MWNTs and GNPs. (Reproduced with permission from Chen et al. [2007].)

- Optical biosensing technique detects changes in the intensity, spectral, phase angle, lifetime, and polarization of fluorescence/scattered light upon interaction with an analyte.
- Electrochemical (ECL) biosensing technique utilizes analyte detection by the induction of an enzyme-catalyzed redox reaction.

- Fluorescence-based biosensing serves for intracelllullar mapping of macromolecules as well as for probing of variety of factors in the molecular environment, including pH, temperature, electric field, ionic content, enzymatic activities, and so on. These data are analyzed for diagnostics and disease profiling.
- PEBBLEs are photonic nanosensors, consisting of sensor molecules entrapped in a chemically inert matrix; these are an example of a nanoparticles-based multisensing device.
- The ratio of the intensities of anti-Stokes to Stoke fluorescence can be used as a probe for measuring intracellular temperature.
- Dye-doped silica nanoparticles were developed for DNA/mRNA analysis by monitoring the fluorescence signal of the conjugates extracted from the substrate.
- Due to the unique properties of semiconductor nanocrystals, quantum dots and rods are being extensively investigated for biosensing and replacing dyes for a variety of applications.
- A molecular beacon approach utilizes an electronic energy transfer scheme between a fluorescent unit and a fluorescent quencher. In the absence of the analyte, the distance between the emitter and the quencher is close and the emission is quenched. In the presence of the analyte, the biorecognition process forces them apart and the emission restores.
- The working principle of a plasmonic biosensor is based on monitoring the change of the plasmonic properties of metallic nanoparticles (localized surface plasmon resonance, LSPR) and metallic films (surface plasmon resonance, SPR).
- In an LSPR colorimetric sensor, gold or silver nanoparticles can be chemically tethered to a glass surface or electrostatically bound by creating opposite charges on the metal nanoparticles and the substrate. The shift of the LSPR peak position of metallic nanoparticle, which is associated with the transduction mechanism for analyte capture, is monitored for sensing.
- In a LSPR fluorescence sensor, the metal-enhanced fluorescence is monitored. The fluorescence intensity of an emitter can be enhanced when it is separated from a metallic nanoparticle at an optimum distance.
- Surface-enhanced Raman scattering (SERS) sensing provides detailed information about the chemical structure and the conformation of an analyte, with the sensitivity approaching the ultimate single molecule detection. The tremendous intensity enhancement provided by SERS enables a detailed analysis of the vibrational fingerprints of proteins.
- Surface plasmon resonance (SPR) sensors are the most extensively utilized and also commercially available optical biosensors. In the Kretschmann geometry using prism coupling widely used to excite surface plasmons, the sensing response is a change in the refractive index of the

sensing layer containing ligands (e.g., antibodies) upon analyte binding, which is measured as a change in the surface palsmon coupling angle.

- A phase-sensitive SPR biosensor, based on temporal modulation of the pumping beam by a photoelastic modulator, enhances the sensitivity.
- A nanoparticle-based SPR further enhances sensitivity, with efficient plasmonic coupling between the nanoparticle and the sensing film.
- A whispering gallery mode biosensor (WGNB) utilizes light trapped by internal refection in a dielectric microcavity such as a microsphere. An analyte binding on the surface of the dielectric microcavity at the interface causes a shift in the resonance frequency of the microcavity.
- Photonic crystals are ordered dielectric arrays that diffract light at wavelengths determined by the lattice spacing between the arrays and the average refractive index of the photonc structure.
- Biosenisng using photonic crystal can be accomplished by monitoring the shift in the peak wavelength position of diffraction induced by analyte interaction.
- Magnetic biosensing technique measures the stray magnetic field on a solid substrate platform based on magnetoresistance effect, spin valve action or Hall effect.
- The detection of the stray field of magnetic nanoparticle tags by their interaction (binding) to a solid substrate is known as substrate-based stray field sensing.
- Giant magnetoresistance (GMR) sensors utilize the GMR effect of ferromagnetic layers separated by nonmagnetic metallic spacer layers, where the electronic transport through the metallic layer (hence its resistance) becomes dependent on the spin.
- The stray magnetic field of the captured magnetic labels creates a change in the resistance which acts as a transduction mechanism to produce an electrical response.
- In a tunneling magnetoresistance (TMR) sensor, the ferromagnetic layers are separated from each other by a very thin insulating layer, and the tunneling current (resistance) between them is measured.
- The anisotropic magnetoresistance (AMR) sensor utilizes the variation in the electron mean free path as a function of the angle between the electron velocity and the conductive material's magnetization direction.
- A Hall Effect sensor involves the action of the transverse force of an externally applied magnetic field to push charge carriers to one side of a current-carrying conductor.
- A spin-valve sensor also uses the magnetoresistance effect with a different magnetoresistive stack arrangement, involving four layers: (i) ferromagnetic free layer, (ii) nonmagnetic conducting layer, (iii) ferromagnetic pinned layer, and (iv) antiferromagnetic pinning layer.

- A magnetorelaxometry sensor measures the relaxation time of the magnetization of a magnetic nanoparticle after the aligning magnetic field is turned off. If the magnetic nanoparticles are bound to a large biomolecule, the relaxation time will be prolonged.
- The electrical biosensing technique requires electrical transduction where the biosensor detection is in the form of a signal as a current or a voltage.
- The field effect transistor (FET) is a semiconductor device consisting of two *n*-doped (excess electrons) regions called source and drain. The flow of current between the source and the drain (i.e., conductance) is dependent on the voltage applied to the gate (a *p*-doped region with excess holes). Biomolecular binding changes the electrical properties, resulting in a change in the conductance between the source and the drain.
- Quartz crystal microbalance (QCM) is a mass sensing device in which the piezoelectric oscillation frequency of the quartz crystal changes, when an adsorbate deposits on the surface of the quartz crystal to change its mass.
- In electrochemical sensing, a chemical electron transfer reaction is used as the transduction mechanism.
- Metallic or semiconductor nanoparticles can be immobilized on the working electrode of an electrochemical cell to exhibit catalytic properties that generate an electrochemical response.
- Electroluminescence (ECL) sensing involves species at the working electrode surface undergoing electron transfer to form an excited-state species that emits light.
- *In vivo* implantable biosensor enables continuous monitoring and management of chronic disease and early detection of body malfunctions. It also contains miniaturized electronic circuitry to alert the patients by remotely transmitting the signal.
- The main issues involved in the implantable biosensor are complexity in miniaturization and the induced inflammatory response.

EXERCISES

True and False

1. In the Raman detection mode, a shift in the peak wavelength of fluorescence is monitored.

2. Magnetorelaxometry utilizes the measurement of the charge of relaxation times upon the analyte binding with the nanoparticles.

3. Field-effect transistor-based biosensor is an example of magnetic biosensor.

4. The ratio of the fluorescence intensity of anti-Stokes to Stokes can be utilized for temperature measurement.

5. In a molecular beacon, the emission can only be observed when the fluorescent unit is very close to the quencher.
6. The plasmonic peak position of metallic nanoparticle is sensitive to the change of surrounding dielectric environment.
7. The Kretschmann geometry is widely used to excite surface-enhanced Raman scattering.
8. One of the problems in the magnetic biosensor is photobleaching of nanoparticles.
9. The magnetization and relaxation of nanoparticles can be measured using SQUID.

Multiple Choice

10. Which of the following molecules is the biorecognition element for biotin?
    - **(a)** DNA
    - **(b)** Streptavidin
    - **(c)** IgG antibody
    - **(d)** silver
11. Which of the following is *not* an optical biosensor?
    - **(a)** SERS
    - **(b)** SPR
    - **(c)** PEBBLE
    - **(d)** QCM
12. Which of the following can be used for the characterization of magnetic nanoparticle?
    - **(a)** SQUID
    - **(b)** FET
    - **(c)** CNT
    - **(d)** ECL sensor
13. Which of the following is *not* true for colloidal CdSe quantum dots?
    - **(a)** It has a narrow emission band.
    - **(b)** The luminescence can be quenched by a thin layer of ZnS shell.
    - **(c)** The luminescence can be quenched by a thin layer of Au shell.
    - **(d)** The emission will be red-shifted with larger size.
14. Which of the following is true for a FET-based biosensor?
    - **(a)** It measures the vibrational frequency of biomolecule.
    - **(b)** A laser should be used as the light source.

(c) It consists of an *n*-doped "source" and a *p*-doped "drain."

(d) The electrical properties of the channel region will be changed during the detection.

Answers: 1. F; 2. T; 3. F; 4. T; 5. F; 6. T; 7. F; 8. F; 9. T; 10. b; 11. d; 12. a; 13. b; 14. d

REFERENCES

Allen, B. L., Kichambare, P. D., Star, A., Carbon nanotube field-effect-transistor-based biosensors, *Adv. Materi.* **19**, 1439–1451 (2007).

Anker, J. N., Hall, W. P., Lyandres, O., Shah, N. C., Zhao, J., van Dyne, R. P., Biosensing with plasmonic nanosensors, *Nature Mater.* **7**, 442–453 (2008).

Arnold, S., Keng, D., Shopova, S. I., Holler, S., Zurawsky, W., Vollmer, F., Whispering Gallery Mode Carousel—A photonic mechanism for enhanced nanoparticle detection in biosensing, *Opt. Express* **17**, 6230–6238 (2009).

Arumugam, P. U., Chen, H., Siddiqui, S., Weinrich, J. A. P., Jejelowo, A., Li, J., Meyyappan, M., Wafer-scale fabrication of patterned carbon nanofiber nanoelectrode arrays: A route for development of multiplexed, ultrasensitive disposable biosensors, *Biosens. Bioelectron.* **24**, 2818–2824 (2009).

Aytur, T., Foley, J., Anwar, M., Boser, B., Harris, E., Beatty, P. R., A novel magnetic bead bioassay platform using a microchip-based sensor for infectious disease diagnosis, *J. Immunol. Methods* **314**, 21–29 (2006).

Bertoncello, P., Forster, R. J., Nanostructured materials for electrochemiluminescence (ECL)-based detection methods: Recent advances and future perspectives, *Biosens. Bioelectron.* **24**, 3191–3200 (2009).

Cady, N. C., Strickland, A. D., Batt, C. A., Optimized linkage and quenching strategies for quantum dot molecular beacons, *Mol. Cell. Probes* **21**, 116–124 (2007).

Chang, Y.-L., Palacios, R. E., Fan, F.-R. F., Bard, A. J., Barbara, P. F., Electrogenerated chemiluminescence of single conjugated polymer nanoparticles, *J. Am. Chem. Soc.* **130**, 8906–8907 (2008).

Chen, S., Yuan, R., Chai, Y., Zhang, L., Wang, N., Li, X., Amperometric third-generation hydrogen peroxide biosensor based on the immobilization of hemoglobin on multiwall carbon nanotubes and gold colloidal nanoparticles, *Biosens. Bioelectron.* **22**, 1268–1274 (2007).

Chu, X., Zhao, Z. L., Shen, G. L., Yu, R. Q., Quartz crystal microbalance immunoassay with dendritic amplification using colloidal gold immunocomplex, *Sens. Actuators B* **114**, 696–704 (2006).

Clark, H. A., Hoyer, M., Philbert, M. A., Kopelman, R., Optical nanosensors for chemical analysis inside single living cells. 1. Fabrication, characterization, and methods for intracellular delivery of PEBBLE sensors, *Anal. Chem.* **71**, 4831–4836 (1999).

Deng, Z., Zhang, Y., Xue, J., Tang, F., Wei, Q., Green and orange CdTe quantum dots as effective pH-sensitive fluorescent probes for dual simultaneous and independent detection of viruses, *J. Phys. Chem. B* **111**, 12024–12031 (2007).

George, P. M., Lyckman, A. W., LaVan, D. A., Hegde, A., Leung, Y., Avasare, R., Testa, C., Alexander, P. M., Langer, R., Sur, M., Fabrication and biocompatibility of polypyrrole implants suitable for neural prosthetics, *Biomaterials* **26**, 3511–3519 (2005).

Gerion, D., Chen, F. Q., Kannan, B., Fu, A. H., Parak, W. J., Chen, D. J., Majumdar, A., Alivisatos, A. P., Room temperature single-nucleotide polymorphism and multiallele DNA detection using fluorescent nanocrystals and microarrays, *Anal. Chem.* **75**, 4766–4772 (2003).

Graham, D. L., Ferreira, H. A., Freitas, P. P., Magnetoresistive-based biosensors and biochips, *Trends in Biotechnology*, **22**, 455-462, (2004).

Grimm, J., Perez, J. M., Josephson, L., Weissleder, R., Novel nanosensors for rapid analysis of telomerase activity, *Cancer Res.* **64**, 639–643 (2004).

Grossman, H. L., Myers, W. R., Vreeland, V. J., Bruehl, R., Alper, M. D., Bertozzi, C. R., Clarke J., Detection of bacteria in suspension by using a superconducting quantum interference device, *PNAS* **101**, 129–134 (2004).

Guiseppi-Elie, A., Electroconductive hydrogels: Synthesis, characterization and biomedical applications, *Biomaterials* **31**, 2701–2716 (2010).

Guiseppi-Elie, A., Brahim, S., Slaughter, G., Ward, K. R., Design of a subcutaneous implantable biochip for monitoring of glucose and lactate, *IEEE Sens. J.* **5**, 345–355 (2005).

Guiseppi-Elie A., Sheppard N. F., Jr., ed. Conferring biospecificity to electroconductive polymer-based biosensor devices, in *ACS Northeast Regional Meeting (NERM)*: Rochester, New York (1995).

Han, J., Burgess, K., Fluorescent indicators for intracellular pH, *Chem. Rev.* **110**, 2709–2728 (2010).

Han, M., Gao, X., Su, J. Z., Nie, S., Quantum-dot-tagged microbeads for multiplexed optical coding of biomolecules, *Nat. Biotechnol.* **19**, 631–635 (2001).

Han, X. X., Zhao, B., Ozaki, Y., Surface-enhanced Raman scattering for protein detection, *Anal. Bioanal. Chem.* **394**, 1719–1727 (2009).

Haynes, C. L., Van Duyne, R. P., Nanosphere lithography: A versatile nanofabrication tool for studies of size-dependent nanoparticle optics, *J. Phys. Chem. B* **105**, 5599–5611 (2001).

Hirsch, L. R., Jackson, J. B., Lee, A., Halas, N. J., West, J. L., A whole blood immunoassay using gold nanoshells, *Anal. Chem.*, **10**, 2377–2381 (2003).

Hutter, E., et al., Role of substrate metal in gold nanoparticle enhanced surface plasmon resonance imaging, *J. Phys. Chem. B* **105**(1), 8–12 (2001).

Kabashin, A. V., Nikitin P. I., Surface plasmon resonance interferometer for bio- and chemical-sensors, *Opt. Commun.* **150**(1–6), 5–8 (1998).

Kachynski, A. V., Kuzmin, A. N., Pudavar, H. E., Prasad, P. N., Time-dimensional confocal thermal imaging using anti-Stokes luminescence, *Appl. Phys. Lett.* **87**, 023901–023903 (2005).

Katz, E., Willner, I., Wang, J., Electroanalytical and bioelectroanalytical systems based on metal and semiconductor nanoparticles, *Electroanalysis* **16**, 19–44 (2004).

Kim, J. H., Morikis, D., Ozkan, M., Adaptation of inorganic quantum dots for stable molecular beacons, *Sens. Actuators B* **102**, 315–319 (2004).

Kim, S., Pudavar, H., Prasad, P. N., Dye-concentrated organically modified silica nanoparticles as a ratiometric fluorescent pH probe by one- and two-photon excitation, *Chem. Commun.* 2071–2073 (2006).

Law, W.-C., Yong, K.-T., Baev, A., Prasad, P. N., Sensitivity improved surface plasmon resonance biosensor for cancer biomarker detection based on plasmonic enhancement, *ACS Nano* **5**, 4858–4864 (2011).

Law, W.-C., Yong, K.-T., Baev, A., Hu, R., Prasad, P. N., Nanoparticle enhanced surface plasmon resonance biosensing: Application of gold nanorods, *Opt. Express* **17**, 19041–19046 (2009).

Law, W.-C., Markowicz, P., Yong, K.-T., Roy, I., Baev, A., Patskovsky, S., Kabashin, A. V., Ho, H.-P., Prasad, P. N., Wide dynamic range phase-sensitive surface plasmon resonance biosensor based on measuring the modulation harmonics, *Biosens. Bioelectron.* **23**, 627–632 (2007).

Lee, Y.-E., Smith, R., Kopelman, R., Nanoparticle PEBBLE sensors in live cells and *in vivo*, *Annu. Rev. Anal. Chem.* **2**, 57–76 (2009a).

Lee, K., Nair, P. R., Scott, A., Alam, M. A., Janes, D. B., Device considerations for development of conductance-based biosensors, *J. Appl. Phys.* **105**, 102046–102059 (2009b).

Lin, Y., Lu, F., Tu, Y., Ren, Z., Glucose biosensors based on carbon nanotube nanoelectrode ensembles, *Nano Lett.* **4**, 191–195 (2004).

Liu, S., Dai, Z., Chen, H., Ju, H., Immobilization of hemoglobin on zirconium dioxide nanoparticles for preparation of a novel hydrogen peroxide biosensor, *Biosens. Bioelectron.* **19**, 963–969 (2004).

Liu, Y.-S., Sun, Y., Vernier, P. T., Liang, C.-H., Chong, S. Y. C., Gundersen, M. A., pH-Sensitive photoluminescence of CdSe/ZnSe/ZnS quantum dots in human ovarian cancer cells, *J. Phys. Chem. C* **111**, 2872–2878 (2007).

Ludwig, F., Mäuselein, S., Heim, E., Schilling, M., Magnetorelaxometry of magnetic nanoparticles in magnetically unshielded environment utilizing a differential fluxgate arrangement, *Rev. Scientific Instruments* **76**, 106102–106104 (2005).

Luo, X., Morrin, A., Killard, A. J., Smyth, M. R., Application of nanoparticles in electrochemical sensors and biosensors, *Electroanalysis* **18**, 319–326 (2006).

Malicka, J., Gryczynski, I., Geddes, C. D., Lakowicz, J. R., Metal-enhanced emission from indocyanine green: A new approach to in vivo imaging, *J. Biomed. Opt.* **8**, 472–478 (2003).

Mao, X., Yang, L., Su, X. L., Li, Y., A nanoparticle amplification based quartz crystal microbalance DNA sensor for detection of *Escherichia coli* O157:H7, *Biosens. Bioelectron.* **21**, 1178–1185 (2006).

Martel, R., Schmidt, T., Shea, H. R., Hertel, T., Avouris, P., Single- and multi-wall carbon nanotube field-effect transistors, *Appl. Phys. Lett.* **73**, 2447–2449 (1998).

Miller, M. M., Prinz, G. A., Cheng, S. F., Bounnak S., Detection of a micron-sized magnetic sphere using a ring-shaped anisotropic magnetoresistance-based sensor: A model for a magnetoresistance-based biosensor, *Appl. Phys. Lett.* **81**, 2211 (2002).

Nath, N., Chilkoti, A., A Colorimetric Gold Nanoparticle Sensor to interrogate biomolecular interactions in real time on a surface, *Anal. Chem.* **74**, 504–509 (2002).

Patolsky, F., Zheng, G., Lieber, C. M., Nanowire-based biosensors, *Anal. Chem.* **78**, 4260–4269 (2006).

Prasad, P. N., *Introduction to Biophotonics*, Wiley-Interscience, Hoboken, NJ (2003).

Prasad, P. N., *Nanophotonics*, Wiley-Interscience, Hoboken, NJ (2004).

Qi, H. L., Peng, Y., Gao, Q., Zhang, C. X., Applications of nanomaterials in electrogenerated chemiluminescence biosensors, *Sensors* **9**, 674–695 (2009).

Qian, X., Zhou, X., Nie S., Surface-enhanced raman nanoparticle beacons based on bioconjugated gold nanocrystals and long range plasmonic coupling, *J. Am. Chem. Soc.* **130**, 14934–14935 (2008).

Reiss, G., Brueckl, H., Huetten, A., Schotter, J., Brzeska, M., Panhorst, M., Sudield, D., Becker, A., Kamp, P. B., Puehler, A., Wojczykowski, K., Jutzi, P., Magnetoresistive sensors and magnetic nanoparticles for biotechnology, *J. Mater. Res.* **20**, 3294–3302 (2005).

Rosi, N. L., Mirkin, C. A., Nanostructures in biodiagnostics, *Chem. Rev.* **105**, 1547–1562 (2005).

Santhanam, K. S. V., Bard, A. J., Chemiluminescence of electrogenerated 9,10-diphenylanthracene anion radical, *J. Am. Chem. Soc.* **87**, 139–140 (1965).

Sepúlveda, B., Angelomé, P. C., Lechuga, L. M., Liz-Marzán, L. M., LSPR-based nanobiosensors, *Nano Today* **4**, 244–251 (2009).

Snow, E. S., Novak, J. P., Campbell, P. M., Park, D., Random networks of carbon nanotubes as an electronic material, *Appl. Phys. Lett.* **82**, 2145–2147 (2003).

Sonnichsen, C., Reinhard, B. M., Liphardt, J., Alivisatos, A. P., A molecular ruler based on plasmon coupling of single gold and silver nanoparticles, *Nat. Biotechnol.* **23**, 741–745 (2005).

Storhoff, J. J., Mirkin, C. A., Programmed materials synthesis with DNA, *Chem. Rev.* **99**, 1849–1862 (1999).

Szlosarek, P., Charles, K. A., Balkwill, F. R., Tumour necrosis factor-[alpha] as a tumour promoter, *Eur. J. Cancer* **42**(6), 745–750 (2006).

Tamanatra, C. R., Mulvaney, S. P., Rife, J. C., Whitman, L. J., Magnetic labeling, detection, and system integration, *Biosens. Bioelectron.* **24**, 1–13 (2008).

Tan, W., Fang, X., Li, J., Liu, X., Molecular beacons: A novel DNA probe for nucleic acid and protein studies, *Chem. Eur. J.* **6**, 1107–1111 (2000).

Tans, S. J., Verschueren, A. R. M., Dekker, C., Room-temperature transistor based on a single carbon nanotube, *Nature* **393**, 49–52 (1998).

Tao, Y., Lin, Z.-J., Chen, X.-M., Chen, X., Wang, X.-R., Tris(2,2′-bipyridyl)ruthenium(II) electrochemiluminescence sensor based on carbon nanotube/organically modified silicate films, *Anal. Chim. Acta* **594**, 169–174 (2007).

Tyner, K. M., Kopelman, R., Philbert, M. A., Nanosized voltometer enabled cellular-wide electric field mapping, *Biophys. J.* **93**(4), 1163–1174 (2007).

Vaddiraju, S., Tomazos, I., Burgess, D. J., Jain, F. C., Papadimitrakopoulosa, F., Emerging synergy between nanotechnology and implantable biosensors: A review, *Biosens. Bioelectron.* **25**, 1553–1565 (2010).

Varadan, V., Chen, L. F., Xie, J., *Nanomedicine: Design and Applications of Magnetic Nanomaterials, Nanosensors and Nanosystems*, John Wiley & Sons (2008).

Vollmer, F., Arnold, S., Whispering-gallery-mode biosensing: label-free detection down to single molecules, *Nature Method* **5**, 591–596 (2008).

Vollmer, F., Arnold, S., Keng, D., Single virus detection from the reactive shift of a whispering-gallery mode, *Proc. Natl. Acad. Sci. USA* **105**, 20701–20704 (2008).

Wallis, R. F., Stegeman, G. I., *Electromagnetic Surface Excitations*, Springer-Verlag, Berlin (1986).

Wang, F., Hu, S., Electrochemical sensors based on metal and semiconductor nanoparticles, *Microchim. Acta* **165**, 1–22 (2009).

Wang, J., Liu, G., Merkoçi, A., Electrochemical coding technology for simultaneous detection of multiple DNA targets, *J. Am. Chem. Soc.* **125**, 3214–3215 (2003).

Wang, Z. Y., Bonoiu, A., Samoc, M., Cui, Y. P., Prasad, P. N., Biological pH sensing based on surface enhanced Raman scattering through a 2-aminothiophenol-silver probe, *Biosens. Bioelectron.* **23**, 886–891 (2008).

Wu, S. Y., Ho, H. P., Law, W. C., Lin, C., Kong, S. K., Highly sensitive differential phase-sensitive surface plasmon resonance biosensor based on the Mach–Zehnder configuration, *Opt. Lett.* **29**, 2378–2380 (2004).

Xu, J.-J., Luo, X.-L., Du, Y., Chen, H.-Y., Application of MnO_2 nanoparticles as an eliminator of ascorbate interference to amperometric glucose biosensors, *Electrochem. Commun.* **6**, 1169–1173 (2004).

You, T., Niwa, O., Tomita, M., Hirono, S., Characterization of platinum nanoparticle-embedded carbon film electrode and its detection of hydrogen peroxide, *Anal. Chem.* **75**, 2080–2085 (2003).

Zhao, Y. J., Zhao, X. W., Tang, B. C., Xu, W. Y., Li, J., Hu, J., Gu, Z. Z., Quantum-dot-tagged bioresponsive hydrogel suspension array for multiplex label-Free DNA detection, *Adv. Funct. Mater.* **20**, 976–982 (2010).

Zhao, X., Tapec-Dytioco, R., Tan, W., Ultrasensitive DNA detection using highly fluorescent bioconjugated nanoparticles, *J. Am. Chem. Soc.* **125**, 11474–11475 (2003).

Zheng, D., Seferos, D. S., Giljohann, D. A., Patel, P. C., Mirkin. C. A., Aptamer nano-flares for molecular detection in living cells, *Nano Lett.* **9**, 3258–3261 (2009).

Zhong, W., Nanomaterials in fluorescence based biosensing, *Anal. Bioanal. Chem.* **394**, 47–59 (2009).

Zhu, M., Liu, M., Shi, G., Xu, F., Ye, X., Chen, J., Jin, L., Jin, J., Novel nitric oxide micro-sensor and its application to the study of smooth muscle cells, *Anal. Chim. Acta* **455**, 199–206 (2002).

CHAPTER 10

HIGH-THROUGHPUT MULTIPLEXED DIAGNOSTICS

Bioimaging and biosensing discussed in Chapters 8 and 9 form an important basis for modern diagnostics. In addition, methods are needed to provide the ability to simultaneously and rapidly detect a large number of biomarkers (analytes) that are indicators of a specific disease manifestation. This chapter presents some of the major biotechnologies for such high-throughput multiplexed diagnostics where nanotechnology is expected to make a major impact. The two such major biotechnologies are flow cytometry and microarray technologies. Both are briefly described here, and their details can be found in the monograph Introduction to Biophotonics by this author [Prasad, 2003].

This chapter starts with an overview of a comprehensive diagnostic strategy in Section 10.1. Section 10.2 describes the method of flow cytometry. Section 10.3 describes the approach of enzyme-linked immunosorbent assay (ELISA), which is an earlier version of microarray technology for protein analysis in a microwell plate geometry for multiplexed detection. Section 10.4 provides a description of the microarray technologies used for both DNA and protein detection. Section 10.5 presents the suspension bead array (SBA) assay utilizing microbeads in a flow system (flow cytometer) for multiplexed rapid analysis. Each section covering these specific topics also includes a discussion of the impact of nanotechnology in the respective areas.

Introduction to Nanomedicine and Nanobioengineering, First Edition. Paras N. Prasad.

10.1. COMPREHENSIVE DIAGNOSTIC STRATEGY

This section presents an overview of a comprehensive diagnostic approach that utilizes the information, from anatomical to molecular and from structural to functional, that is necessary for a complete diagnosis of a disease to select the best therapeutic approach or treatment. Chapter 8 provides a description of medical imaging approaches that can be used for disease diagnosis, based on anatomical and tissue structural information. Biosensing methods discussed in Chapter 9 can be used to monitor the functional aspects that involve the evaluation of physiological parameters and characterization of metabolic functions.

A major emphasis of modern diagnostics, particularly in relation to early-stage detection of a disease or infection, is diagnosis on cellular and molecular levels. Cellular diagnostics involves cell profiling on the basis of cell morphology and cellular functions as measured by its physiological status, as well as antigen expressions on the cell surface. Cellular imaging and biosensing discussed in Chapters 8 and 9 can provide a wealth of information. In addition, flow cytometry, discussed here in Section 10.2, is a rapid method of cell profiling, where thousands of cells per second can be analyzed, with the high precision of identifying even one specific type of cell in a heterogeneous population of millions of cells. It can readily be used for multiplexed detection utilizing simultaneous identification of multiple (as many as 30 or even more) parameters relating to various biomarkers.

Molecular diagnosis utilizes detection and quantification of various macromolecules such as DNA, RNA, and proteins, as well as small biomolecules such as hormones. A major area of considerable current interest for molecular diagnostics is genomics, which deals with genome structure and function. Genomics, derived from *genome*, is a term used to refer to an organism's complete set of genetic information. The genomics encompasses a very broad range of directions related to genome sequencing and gene mapping, gene expression, and regulation. Among the most prominent branches of genomic are (i) *structural genomics*, which deals with the high-resolution 3D studies of proteins structures as function of genetic code, and (ii) *functional genomics*, which analyzes function of individual genes as well as gene networks seeking a complete understanding of the gene functioning of a biological system.

Genomics monitors expressed genes whose translation ultimately yields the respective proteins that need to be catalogued and analyzed in order to understand biological complexity and functions. This is the field of proteomics. Analogous to genomics, structural proteomics deals with the identification, profiling, and quantification of proteins, while functional proteomics involves a study of protein structure, localization, modification, interactions, activities, and functions. Proteomics focuses on high throughput: (i) protein profiling dealing with determination of the abundance, modification, and subcellular localization of proteins in a given cell or tissue and (ii) determination of protein function.

DNA detection for molecular diagnostics utilizes identification of presence or absence as well as quantification of specific genes. In addition, function of genes, as specific gene expression in the form of under or over expression, can be utilized by analysis of the corresponding proteins.

A major diagnostic method using immunoassay with antibody is based on the detection of soluble proteins released in body fluids such as saliva, urine, blood, or sputum. This method is particularly useful for early-stage diagnosis of cancer when soluble biomarkers are released even at a precancerous stage. The method of ELISA, protein microarrays, and flow cytometry using microbeads, as discussed in subsequent sections, can be used for detecting even a trace amount of these soluble proteins. An example is ovarian cancer, in which case the use of serum proteomic patterns can accurately identify it with 100% sensitivity and 95% specificity. Another type of biomarker is the growth factor signaling proteins that promote tumor cell growth, mobility, survival, and differentiation. They can be analyzed by using body fluids.

Cytokines are a class of low-molecular-weight, soluble proteins that are mediators of many cellular activities such as migration, activation, phagocytosis, proliferation, and apoptosis (see website http://accessmedicine.com/resourceTOC.aspx?resourceID=651). They also participate in responding to infections, trauma, burns, sepsis, and other tissue injuries. Thus, the analysis and quantitation of cytokines in body fluids and circulating plasma is of great value to diagnostics. The methods described in this chapter are well-suited for cytokine assay.

10.2. FLOW CYTOMETRY

Principles and Applications. A flow cytometer is an optical diagnostic device, which currently is used in research and clinical laboratories for disease profiling by measuring the physical and/or chemical characteristics of cells. The term *cytometry* refers to the measurement of physical and/or chemical characteristics of cells or, in general, of any biological assemblies [Shapiro, 1995; Givan, 2001; Stewart et al., 2002]. In flow cytometry, such measurements are made, while the cells, the biological assemblies, or microbeads flow in a suspension in a single file, past a sensing point. The sensing is conducted by using an optical technique, where the beam from a light source interacting with each individual cell, a bioassembly, or a microbead produces scattering and fluorescence. The optical response is used to determine cellular features and organelles, providing counts and ability to distinguish different types of cells in a heterogeneous population. Generally the cells or the intracellular products are tagged with special fluorescently labeled antibodies or dyes that bind to cellular components capable of producing fluorescence. Flow cytometry yields measurement of various optical responses representing a set of properties, also called *parameters*, which provide unique characteristics of a specific type of cell. The identification and quantification of a particular type of cell or soluble biomolecule captured on a microbead can then be used to correlate with a specific pathological condition that can be used to identify a specific disease or microbial invasion.

The current market for flow cytometry is near one billion dollars worldwide, showing the wide usage of this instrument, even through the primary application has been on eukaryotic cells. The usage of flow cytometry for the study of prokaryotic cells is beginning to emerge. An instrument capable of accurately measuring the properties of microbes, whether bacterial, fungal, or viral, can be expected to significantly expand the current market. The impact will be felt in many diverse areas, such as the pharmaceutical industry, microbiology applications, and agriculture. As pharmaceutical research moves more toward target-directed intervention and biological intervention, and away from drug screening with intact animals, flow cytometry offers a cost-effective method for the development and testing of agents by pharmaceutical industries.

New-generation flow cytometers significantly impact on the ability of physicians and clinicians to quickly determine microorganism infections in humans, animals, food supplies, and water supplies with a compact high-performance cytometer. Flow cytometry also encourages agricultural research and livestock development. Selection of sex type in feed animals, or the ability to selectively sort male and female sperm cells, has an enormous economic impact on cattle industry. The ability to select desired animal gender with flow cytometric techniques, and to use artificial insemination ensures desired selection of offspring gender and physical traits. This capability dramatically reduces breeding and maintenance costs. This application is expected to grow substantially in the coming decades.

Flow cytometry has proved to be a powerful technique for cell cycle analysis, where measurements of DNA content can be used to provide a great deal of information about the cell cycle. Cell division, apoptosis, and necrosis [Darzynkiewicz et al., 1997] can be studied. Furthermore, metabolic characteristics such as calcium flux, mitochondrial activity, cellular pH, and free radical production in live cell populations can be probed and quantified in real time. Flow cytometry using fluorescent tagged protein or a reporter gene can readily be used to measure gene expression in cells transfected with recombinant DNA. Thus using flow cytometry one can readily measure the following (*source:* http://facs.stanford.edu/5minuteguide.htm): (i) expression of proteins and (ii) transfection efficiency. In addition, transfection assays can be combined with staining and sorting for other markers. Also, flow cytometry can be used to purify transfected cells for further analysis or use. Other cellular functions studied by flow cytometry are (*source:* http://cyto.purdue.edu/flowcyst): (i) phagocytosis, (ii) intracellular cytokines, (iii) oxidative burst, and (iv) membrane potential.

Basic Design. A flow cytometer incorporates a light source, an illumination optic, and a light collection optic. Figure 10.1 presents the schematic of a flow

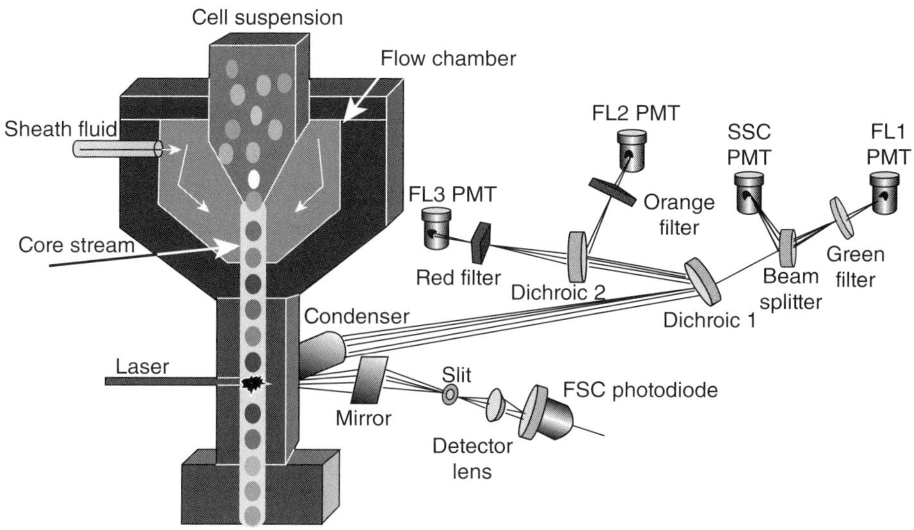

Figure 10.1. Schematic of a flow cytometer showing details. (Courtesy of C. Stewart/RPCI.)

cytometer. Although a flow cytometer measures optical response from one individual cell at a time, progress in fluidity and detection, together with rapid data acquisition and processing software, can now readily provide the ability to routinely detect and analyze 5000 cells per second with speeds of up to 75,000 cells per second in the more advanced instruments. The two significant advantages offered by a flow cytometer over a traditional microscope are (a) its rapid throughput rate at which each cell is interrogated and (b) the ability to sort selected populations while maintaining viability. Thus real-time monitoring of many biological events can be achieved.

A flow cytometer may use a single excitation wavelength or a number of excitation wavelengths from different laser sources. The laser beams can be coaxial or separated, so that one or more interrogation points occur. In the older versions of flow cytometers, the blue-green wavelength region was covered by an argon ion laser, which provides excitation wavelengths of UV (350–365 nm) as well as 488 nm and 514.5 nm. The red region was covered by a helium neon laser (632 nm). The current trend, however, is to replace these bulky and inefficient gas lasers by compact solid-state lasers. A new prospect is the use of near IR lasers (700–800 nm) to excite IR dyes, whereby the problem due to autofluorescence interference is considerably reduced.

The flow cuvette involves hydrodynamic focusing of cells onto the laser beam. A cell suspension is introduced through a bore, called the *core*, which is about 20 μm wide at the top of a conical tapered surrounding concentric larger bore (cuvette) of dimension 200 μm. This larger bore is used to flow another stream of flowing fluid (water or saline) called the *sheath*. Therefore, in this arrangement, a core stream of the cell suspension is injected into the center of the sheath stream. Sheath and cell fluids are usually under differential pressure. Therefore, during the flow, the sheath fluid produces hydrodynamic focusing of coaxial flow of the core fluid, whereby the two streams maintain their relative positions, do not mix significantly, and move at the velocity of the sheath fluid. The hydrodynamic focusing produces the flow of cells in a single file (one cell at a time).

The illumination optics, most frequently used in current flow cytometers, utilizes a laser beam's profile to ensure homogeneous irradiation of the cells in a flow. For instance, cylindrical optics serve well for this purpose, usually consisting of two crossed cylindrical lenses, and produce an elliptical-shaped beam: (a) 10–20 μm in dimension parallel to the direction of cell flow and (b) 60 μm in dimension perpendicular to the flow dimensions. This shape of the beam ensures a wider illumination field across the width of the flow, so that the optical response (fluorescence or scattering) does not fluctuate if cells stray from the center of the beam. At the same time, the smaller (20 μm) dimension of the elliptical beam parallel to the flow direction allows cells to pass in and out of the light illumination quickly and avoids simultaneous illumination of more than one cell at a time.

The collection optics consists of a train of optical ports to separate and collect various optical responses produced by illumination of each flowing cell. These optical responses are: (i) forward scattering (FSC), (ii) side scattering (SSC), and (iii) various fluorescence signals at different wavelengths collected at 90° to the laser beam. The number of optical responses detected (hence the number of photodetectors used) defines the number of measured parameters in a flow cytometer.

For collection of forward-scatter, a beam stop is placed in front of the beam passing through the flowing fluid to block the direct laser beam. The low-angle scattered light, which constitutes FSC, is generated usually at a bend angle of ~10° and is not blocked by the beam stop; it is focused by a collecting lens at a photodetector such as a photodiode. The side-scattered signal and fluorescence are collected at an angle of 90° to the illuminating laser beam and the flowing stream. For this purpose, the 90° light signal is processed by an arrangement of various dichroic mirrors. A specific dichroic mirror reflects fluorescence of a given wavelength while transmitting light of other wavelengths. As illustrated in Figure 10.1, the fluorescence signals of three different wavelengths (red, orange, and green) are separated by placing the dichroic mirrors in the path of the collected light. This arrangement also separates the side-scatter (SSC) signal. The beams, after being reflected from a specific dichroic mirror, pass through band-pass filters to further spectrally discriminate the desired fluorescence from other light. Then they are focused, through a pinhole for further spatial filtering to reduce stray light, onto the photodetector. In the case of light collection for the side scatter, the beam passes through a narrow band-pass filter (±5 nm), centered at the laser wavelength, before going to the photodetector.

The forward-scatter (FSC) signal relates to the cell size, shape, and refractive index and reflects the cross-sectional area and volume of the cell. For example, a dead cell appears smaller than the corresponding living cell, because its refractive index is more like the surrounding stream, due to the leaky outer membrane. The dead cells, therefore, bend less light into the FSC detector than the corresponding living cells.

The side-scatter (SSC) signal is produced by reflection and refraction from the variation in cell structure. Therefore, SSC is related to the cell's surface texture and internal structure. SSC is sometimes also referred to as a granularity signal, because it provides information on the granularity of the cell. For example, granulocytic blood cells, which have granular and irregular nuclei, produce a significantly more intense SSC signal compared to that from more regular lymphocytes or erythrocytes.

Fluorescence labeling allows one to selectively label (stain) a specific subpopulation of cells in order to investigate their cell structures and functions. For example, immunofluorescence, used for immunophenotyping, involves staining of cells with antibodies that are conjugated to fluorescent dyes, often called fluorochromes by the flow cytometry community. This staining is typically used to label antigens on the cell surface, but by permeabilizing the cell with a detergent or alcohol, antibodies can also be directed to cytoplasmic targets. The two approaches used to immunofluorescently stain cells are: (i) direct immunofluorescence in which an antibody is directly conjugated to a fluorochrome; thus the cells are stained in a single step; and (ii) indirect immunofluorescence in which the primary antibody to cell surface antigen is not labeled. Instead a second antibody specific to the primary antibody is conjugated to a fluorochrome.

Modern flow cytometry utilizes a large array of monoclonal antibodies that are specific for various antigen proteins on the cell surface [Givan, 2001]. The antigens defined by these antibodies are characterized by a CD number designation, where CD stands for "cluster of differentiation," which defines a particular protein differentiating one type of cells from another. The CD numbers now range from CD1 to over 350, as the number of characterized antigens has steadily grown. These CD number designations are used to specify an antibody that is specific to a particular antigen.

Another optical parameter monitored can be the polarization of the scattered light, which provides information on the birefringence produced by the cell structure such as that of eosinophil granules.

Data Display. Disease profiling involves correlations of data and their display in useful formats that allow one to obtain profiles of a certain cell population. These various formats of data display are:

Single Parameter Histograms. In this mode of display, also known as frequency histograms, one plots the relative signal of one parameter (fluorescence from a specific labeling fluorochrome or a particular scatter light signal)

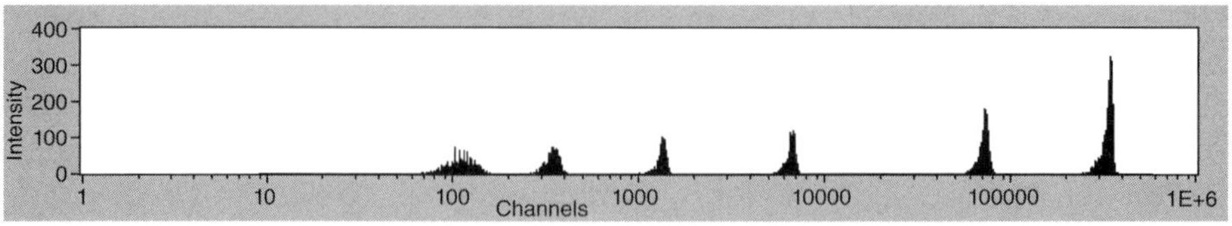

Figure 10.2. Methods for describing the histogram distribution of signal intensities from a population of intensity calibration kit of microspheres. The plots show the number of beads on the vertical axis against channel numbers (related to fluorescence intensity) on the horizontal axis. The histogram clearly demonstrates the separation of six groups of events with different fluorescence intensity.

against the number of events (number of cells registering this intensity level). A certain intensity range is assigned to a certain channel number. Therefore, the histogram plot often substitutes the intensity range by the channel number that is represented on the horizontal axis. This histogram plot permits one to look at the intensity distribution from a large number of events for a given fluorescence wavelength range. Figure 10.2 shows some typical histograms obtained for a set of polystyrene microspheres stained by a fluorescent dye for laser excitation using 488-nm wavelength (L-14814, Molecular Probes). This set is usually used for intensity calibration of flow cytometers and contains mono-sized microbead water dispersion of six microsphere subsets. Each subset differs from the others by amount of dye embedded in the beads, which results in different fluorescence intensity after the excitation event. In histogram, the *Y* axis represents the number of beads (or event counts) for each channel (defined as fluorescence signal intensity) which is represented on the *X* axis. Software programs then allow one to obtain a statistical analysis of the data, keeping in mind that histograms allow this analysis for one parameter at a time.

The quality of the data obtained on a given instrumental is represented by the coefficient of variation (CV) which is defined as the standard deviation of a series of values divided by the mean of those values. It is expressed as a percentage. For the histogram format of data display, the CV is represented by the width of the histogram curve. Often CV is used to calibrate the alignment of a flow cytometer. A narrower CV using standards implies a better alignment and a higher-quality performance for a specific flow cytometer. In an actual analysis, the CV data can be used to measure the variation in characteristics within a cell population. In the histogram display, markers can be used to define regions of intensity that are of interest.

Dot Plots. Correlation of one parameter with another can often be used to distinguish different types of cells. A frequently used display for this purpose is a two-dimensional dot plot in which the two axes represent the intensity channels for selected two parameters. Each cell then is placed as a dot on the

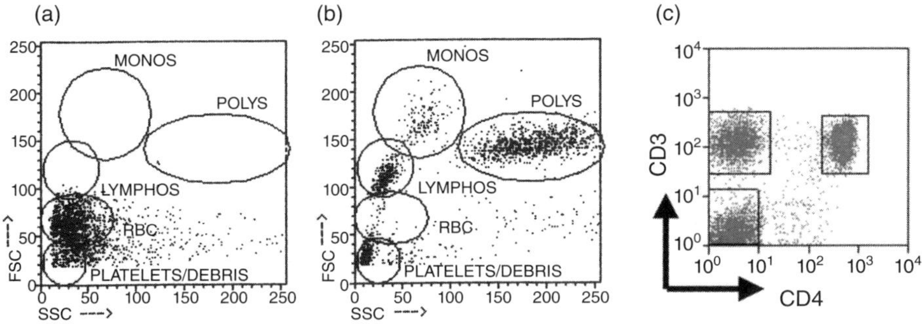

Figure 10.3. Examples of the flow cytometry results shown as dot plots. Patterns in plots (a) and (b) correspond to the blood cells identification by scattering properties (FSC versus SSC) before and after lysis, respectively. (Abbreviations: RBC, red blood cells; LYMPHOS, lymphocytes, POLYS, polymorphs; MONOS, mononuclear cells). Plot (c) demonstrates identification of T-cell subpopulations using fluorescence staining of CD3 and CD4 cell surface biomarkers. (Reproduced with permission from Givan [2001] and Toussirot et al. [2009].)

plot according to the two intensity channels it registers in for the selected two parameters. A four-parameter datum requires six two-parameter correlations, each represented by a dot plot. Dot plots are frequently used when the number of cells available is not a limiting factor and where the fluorescence intensity associated with stained cells is bright enough so that it is clearly separated from that of cells which are negative for each fluorescence marker. Figure 10.3 represents an example of dot plots. Plots (a and b) illustrate the scattering properties of the whole peripheral blood (a) and the same pattern after erythrocyte lysis (b) [Givan, 2001]. The lytic treatment of blood removes primarily the red cells, which reveals the plot clusterization on polymorphs (granular leukocytes) and mononuclear cells. The regions of five clusters representing the distinct populations of leukocytes and platelets demonstrate different level combinations of forward-scatter and side-scatter counts. Figure 10.3c demonstrates identification of three subpopulations of lymphocytes according to CD classifications of biomarker molecules on these cells surfaces [Toussirot et al., 2009].

Digital Data Processing. At the early stage, commercial flow cytometers used analogue signal processing techniques to extract simple features such as pulse height, width, and area of the signal generated by photodetectors. The drawback of this analogue data processing is the limited number of features that can be used to discriminate populations of the cells. Conventional digital flow cytometers allow us to digitize optical signals, extract novel features, and use a wide digital data array to discriminate populations of cells. A digital format leads into the next generation of flow cytometry instrumentation, enables fast and more precise cellular analysis, and performs archival or retrieval transforms, such as conversion from linear to log scales, fluorescence compensation, gating, and cell sort decision-making.

Fiber-Based New-Generation Flow Cytometer. In commonly used flow cytometers, described above, the use of multiple excitation sources and/or interrogation zones requires separate light splitting (set of mirrors and band-pass filters) and detection (photomultipliers, avalanche photodiodes, etc.) modules for each interrogation zone. This constitutes a major obstacle for instrument improvements in at least three areas: (1) miniaturization (footprint), (2) ruggedness, maintenance, and portability (complex alignment of optical elements and its stability), and (3) overall cost of the instrument.

The desire to measure more parameters by researchers and clinical laboratory scientists forces instrument designers to use more excitation sources and, consequently, more optical elements for color separation and more detectors to measure optical signals. Consequently, one needs to develop solutions to make flow cytometers more compact, portable, and less expensive. Faced with the above challenges, we developed a unique color separation technique and time-multiplexed detection scheme for use in flow cytometry. This technology takes advantage of fiber optics, which can be integrated with most of other optical and detection hardware used in a flow cytometer. Our novel optical unit uses confocal fiber-optic fluorescence pickup and a miniature multichannel spectrograph, instead of a set of mirrors and filters, to disperse emitted light into color channels. The layout of this new flow cytometer is shown in Figure 10.4. The collection optics injects the multicolor signals from the cuvette

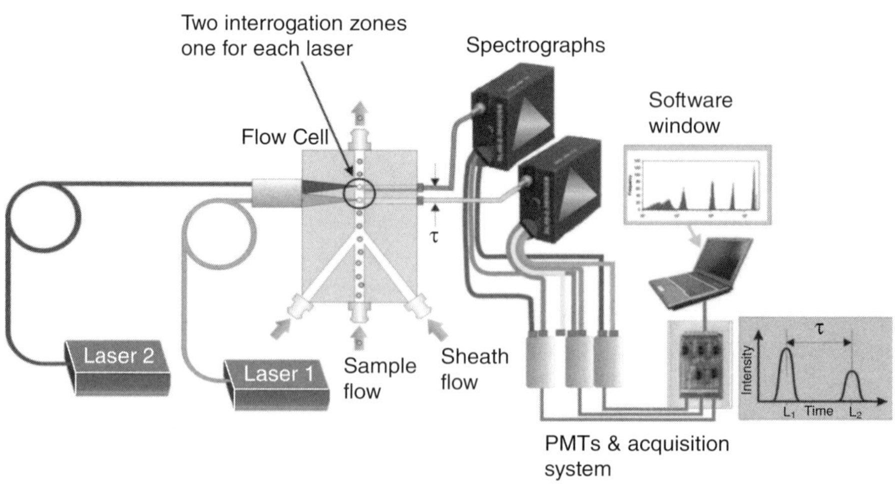

Figure 10.4. Layout of a flow cytometer with two lasers, two interrogation zone and two fiber-coupled spectrographs for color separation.

into the input fiber of a miniature spectrograph. On the other hand, the spectrograph is coupled with an array of output optical fibers, located in the spectrograph's image plane, into which color-separated light is fed. With carefully selected numerical aperture and diameter of optical fibers, each fiber of the output fiber array carries a bandwidth of about 30 nm, sufficiently narrow to distinguish between different fluorescent probes having close emission spectra but wide enough to cover the spectral width of a particular fluorescent probe. Each spectrograph's output channel coupled into an optical fiber can be used to deliver light to a detector. When more than one excitation source of different wavelengths is used, multiple spectrographs are then necessary to resolve fluorescence signals arising from different excitation sources and directed via optical fibers to a detector. Because fluorescent events excited by spatially separated laser beams are time-delayed, a single detector that is programmed to accept signals over a prescribed time interval can detect them. This is essentially a time-multiplexing detection method that allows detecting a large numbers of fluorescence parameters using a small number of detectors. For example, with three spatially separated excitation sources and three fluorescence detectors, one can measure up to nine color parameters plus scatter (scatter is detected with an additional detector, usually a fast photodiode). Figure 10.4 shows schematically the main components of a flow cytometer wherein a system with two laser and two interrogation zones is shown (the flow cytometer has two lasers and two fiber-coupled spectrographs for color separation).

This time multiplexing/demultiplexing data collection, detection, and digital acquisition represents a revolutionary, novel approach leading to cost-saving instrument design. Color separation can also be accomplished with a single multichannel spectrometer, when the input fibers corresponding to different interrogation zones are aligned in the input spectrograph's focal plane to project well-separated spectral images at the spectrograph output. This design provides more detection channels per detector and substantially lowers the overall cost.

Nanoparticle Technology for Flow Cytometry. Traditionally organic dyes, often called fluorochromes in the context of flow cytometry, have been used to label antibodies that produce fluorescence tagging of target cells by binding to the complementary antigens expressed on the cell surface. More recently, interests have shifted to the use of nanoparticles for labeling antibodies. Quantum dots (QDs) have been utilized in place of organic dyes [Ibanez-Peral et al., 2008]. Many of the advantages offered by quantum dots over organic dyes are the same as for bioimaging and biosensing discussed, respectively, in Chapters 8 and 9 earlier. To reiterate, these are: (i) photostability i.e., they do not photobleach; (ii) size-tunable narrow emission bands, allowing for multiple parameters and for multiplexed analyses using a large number of channels over a broad spectral range; and (iii) a single excitation source able to excite a number of different quantum dots emitting at different wavelengths.

Stable quantum dot formulations (by size and compositional variations) can be prepared to make them emissive in the near-IR region where biological cells exhibit maximum optical transparency. In contrast, near-IR emitting organic dyes are generally not photostable. This feature of quantum dots allows near-IR flow cytometry to be used to increase the reliability of analysis of biological cells. Because this technique is for *in vitro* diagnosis, which does not require live cells, toxicity of quantum dots is not of any concern. There are numerous demonstrated applications of QDs in flow cytometry. Here only selected examples are presented. QDs have been used in combination with organic fluorochromes for phenotyping of multiple antigen-specific T-cell populations, where 17 different fluorescence emissions were used for multiplexed analysis [Chattopadhyay et al., 2006]. Flow cytometric detection of bacteria and pathogens have also been made with the use of QDs labeling [Edgar et al., 2006; Ferrari et al., 2007].

Metallic nanoparticles can also be used as labels where their plasmonically enhanced scattering at a specific wavelength, discussed in Chapter 5, can be used for detection. Another prospect is the use of SERS Raman probes for targeting and using flow cytometry in the Raman detection mode. The SERS Raman probes have already been discussed in Chapter 9, entitled "biosensing". The fiber-based, spectrograph-coupled flow cytometer, discussed above, is well-suited for Raman detection, where the spectrograph with additional suitable filter can conveniently be used to detect Raman signal. For Raman labeling, it is important to make sure that the vibrational band of the Raman tag is not in regions of vibrational bands of the cellular constituents. An example developed in our lab is the SERS Raman tag containing molecules with a -CN group, which has the vibrational frequency of ~2236 cm^{-1}.

10.3. ENZYME-LINKED IMMUNOSORBENT ASSAY (ELISA)

Immunoassay in General. Immunoassay is the method of identifying soluble protein analytes, antigens, in body fluids such as blood and urine by using the specificity of antigen binding with immunoglobulin proteins, the antibodies. Antigens have specific binding sites called epitopes, at which an antibody can attach. The corresponding antibodies can be monoclonal; these have only one region on its 3D structure to recognize one epitope on the corresponding antigen and bind to it. In contrast, polyclonal antibodies are composed of antibodies that recognize and bind to multiple epitopes on the corresponding antigen.

In an immunoassay, the antigen–antibody complex generates the various transduction signals, whether optical or electrical, as discussed in Chapter 9. It may simply involve the binding of an analyte with a fluorescently labeled antibody (capture antibody), thus fluorescence detection is used to identify the analyte. This is the fluorescence immunoassay. We shall return to this method in the Section 10.4, which discusses protein microarray.

Commonly Used ELISA Format. A widely used immunoassay for protein analysis and quantification in body fluids in order to diagnose a wide range of diseases and infections is the enzyme-linked immunosorbent assay, ELISA. This assay involves an antibody linked to an enzyme so that an enzyme-catalyzed chemical reaction of a corresponding substrate (enzyme–substrate complexation leading to chemical transformation of the substrate) produces a product that provides an appropriate optical (or electrical in some recent approaches, see below) transduction signal to be detected.

The ELISA method is highly sensitive, reproducible, and simple; thus there are a number of commercial sources of ELISA kits as well as suitable antibodies. Generally, ELISA is performed on a solid platform in the form of a plastic microtiter plate consisting of a number of wells. The most common format is a 96-well plate, but plates with 384 or even 1536 wells are also available. The most commonly used and highly sensitive indirect ELISA technique involves a sandwich assay schematically represented in Figure 10.5, which utilizes a sandwich inclusion of the target antigen between two antibodies, a capture antibody and a detection antibody, the latter being conjugated (linked) to an enzyme. While using monoclonal antibody combinations as the matched capture and detection pairs, it is important to ensure that they recognize separate epitopes on the target antigen and thus do not hinder each other's binding to the antigen.

The steps involved in ELISA detection and quantitation of a target analyte antigen can thus be summarized as follows:

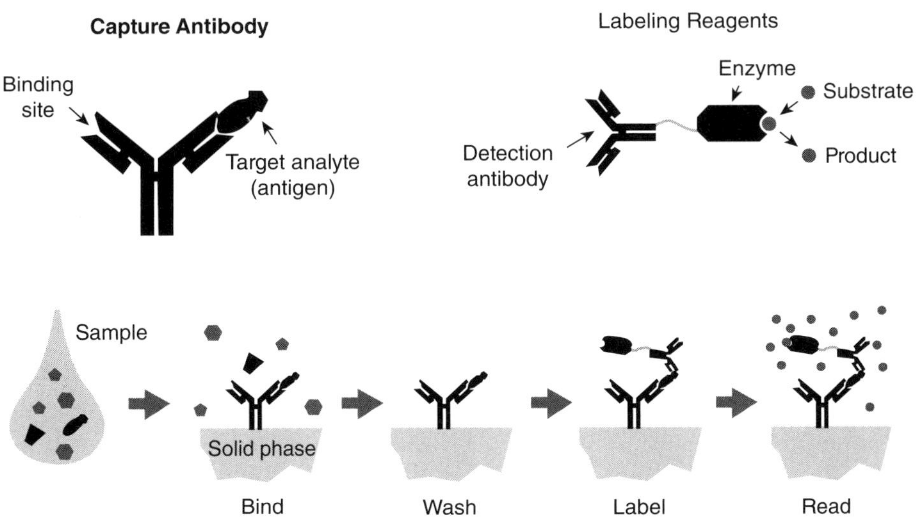

Figure 10.5. Schematics for ELISA detection using a double antibody sandwich configuration. (Reproduced with permission from http://derc.ucsd.edu/cores/mouse-phenotyping.shtml.)

- The surface of the well in the microtiter plate is coated with adhesive agents that can bind to the capture antibody.
- The wells in the microtiter plate are injected with the capture antibody, either manually or robotically. The amount of capture antibody attached to the well is adjusted by varying its concentration.
- The free binding sites in the microtiter plate are then saturated by incubating with an excess of a soluble inert protein such as bovine serum albumin (BSA).
- The target sample (or control) is added to the wells and incubated with the capture antibodies to allow for their binding to specific analytes.
- Any unbound products are removed by washing.
- The detection antibody, linked with an appropriate enzyme such as a horse radish peroxidase (HRP), is diluted in the BSA-containing buffer and is now added and incubated for a period of 1–2 h or more at room temperature and in a humid atmosphere.
- The plate is then washed with PBS.
- A substrate material specific to the enzyme which produces soluble products is now added. In the case of the enzyme, HRP, the substrates used are: (i) 2,2′-azo-bis(3-ethylbenzathiazoline-6-sulfonic acid (ABTS), (ii) *o*-phenylenediamine (OPD), and (iii) 3,3′,5,5′-tetramethylbenzidine base (TMB), which yield soluble products of green, orange, and blue colors, respectively.
- The antigen is detected optically by color change (colorimetric detection) and quantified by measuring the optical density (absorbance) of the produce of the substrate reaction at its absorption maximum, using an ELISA plate reader.

A modification of this double sandwich antibody assay is a triple antibody assay that involves the detection of an antigen by binding with a secondary antibody (detection antibodies) followed by the labeling of the secondary antibody with a third antibody that is linked to an enzyme. This modification provides an opportunity to amplify the signal as several detection antibodies can bind to a secondary antibody. The sensitivity range of the sandwich ELISA can be in picomolar to micromolar range.

Variations of Common ELISA Format. Another variation of ELISA is Dot-ELISA [Ndao, 2009; Pappas, 1988], in which a nitrocellulose or other paper membrane is used instead of a plastic microtiter plate. This solid surface allows the use of a small volume of the sample. In this case, the enzyme-catalyzed reaction of the substrate produces a color dot on the membrane which can be read visually. The dot ELISA, being rapid and cost-effective, has been extensively utilized for the detection of various human and animal parasitic diseases [Pappas, 1988; Ndao, 2009].

As described above, ELISA normally utilizes an optical transduction yielding to colorimetric detection. However, other transduction mechanisms provided by the choice of an appropriate enzyme–substrate pair have been utilized. For example, enzyme-mediated pH changes can be used to detect and quantify an antigen [Stern et al., 2010]. Stern et al. utilized this transduction mechanism with ELISA to detect an analyte using a nanowire field-effect transistor (FET) sensor. The FET biosensors have been discussed in Chapter 9. They were able to detect interleukin-2 in physiologically buffered solution at concentrations as low as 1.6 pg mL^{-1}.

Another transduction mechanism involves the use of an enzyme–substrate reaction to generate chemiluminescence, where the luminescence intensity is then used for quantitative characterization of an antigen [Pappert et al., 2010]. An example is the HRP enzyme with luminol and peroxide to generate chemiluminescence. It should be recognized that the ELISA method is a scheme to couple antibody–antigen specific binding for biorecognition with an enzyme–substrate reaction for detection. Thus, this method can be utilized in a variety of formats. As described above, it is commonly used in the format of a planar platform (microwell plates). However, as first introduced by Knight et al. [2004] for cytokine analysis, ELISA can also be used in a microarray format, described in detail in the next section (Section 10.4). Knight et al. [2004] introduced a microarray immunoassay (MI) by spotting antibodies on nitrocellulose pads affixed to glass slides. The use of nitrocellulose has already been described above for dot ELISA. They demonstrated the application of the MI for rapid and sensitive detection of 16 human cytokines.

Nanotechnology for ELISA. Nanotechnology has been applied to ELISA in a number of different ways, which are described here. Nanoparticles have been used as carriers of the detection antibody linked with an enzyme, with a goal to amplify the optical signal. Ambrosi et al. [2010] used gold nanoparticles as carriers for the anti-CA 15-3-HRP antibody–enzyme conjugate for the analysis of CA 15-3 antigen, which is an important biomarker present in blood samples and can be used for breast cancer diagnostics. The gold nanoparticle complexed with multiple anti-CA 15-3-HRP conjugates, as depicted in Figure 10.6, amplifies the optical response for colorimetric detection in the sandwich-type ELISA detection.

Another application utilizes magnetic nanoparticles for immunomagnetic separation (IMS) to provide enrichment of the analyte in complex matrices. An example is provided by the work of Pappert et al. [2010], who utilized nanoparticle-based IMS and combined it with sandwich chemiluminescence-ELISA for selective enrichment and quantification of *E. coli*. Figure 10.7 illustrates the scheme they used for separation and enrichment of the bacteria using a magnetic field, subsequent elution, and then capture to detect them by the sandwich ELISA chemiluminescence method.

Another application of nanoparticles and nanosensing is at the transduction–detection level. As discussed above, an enzyme–substrate reaction can be used

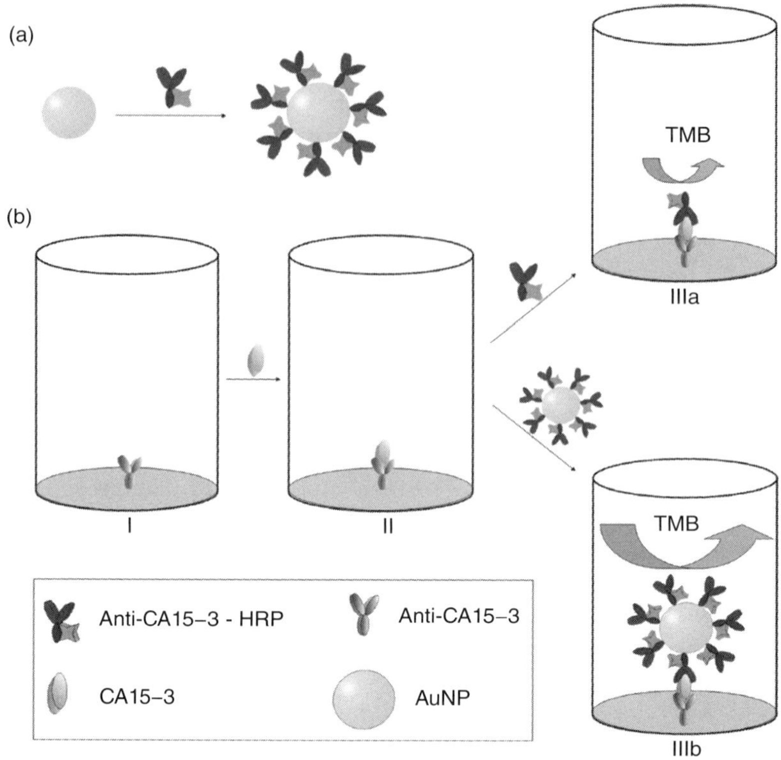

Figure 10.6. Schematic (not in scale) of (a) the preparation of the complex Au-anti-CA15-3-HRP and (b) the sandwich-type ELISA procedure without (IIIa) and with (IIIb) the application of AuNPs as the signal enhancer. (Reproduced with permission from Ambrosi et al. [2010].)

to produce a pH change that can be detected with enhanced sensitivity using nanoparticles. Nanoparticle biosensors for enhanced sensitivity to detect pH changes have already been discussed in Chapter 9. A redox reaction involving electron transfer can also utilize nanoparticles using various electrical sensing mechanisms described again in Chapter 9.

Another example is the use of ELISA utilizing field effect transistor sensing as employed by Stern et al. [2010], which has already been mentioned earlier. In their method, which they call nanoelectronic ELISA, the ELISA assay produces a pH change that is detected by using an indium oxide nanowire-based ion sensing FET. Figure 10.8 schematic shows this method. In this approach, before the addition of the substrate, urea, the indium oxide (In_2O_3) surface is protonated to produce a relatively large channel current. The addition of urea increases the pH, which deprotonates the In_2O_3 surface to produce a decrease in the positive charge on the gate, which subsequently produces a decrease in the channel current for the *n*-type device.

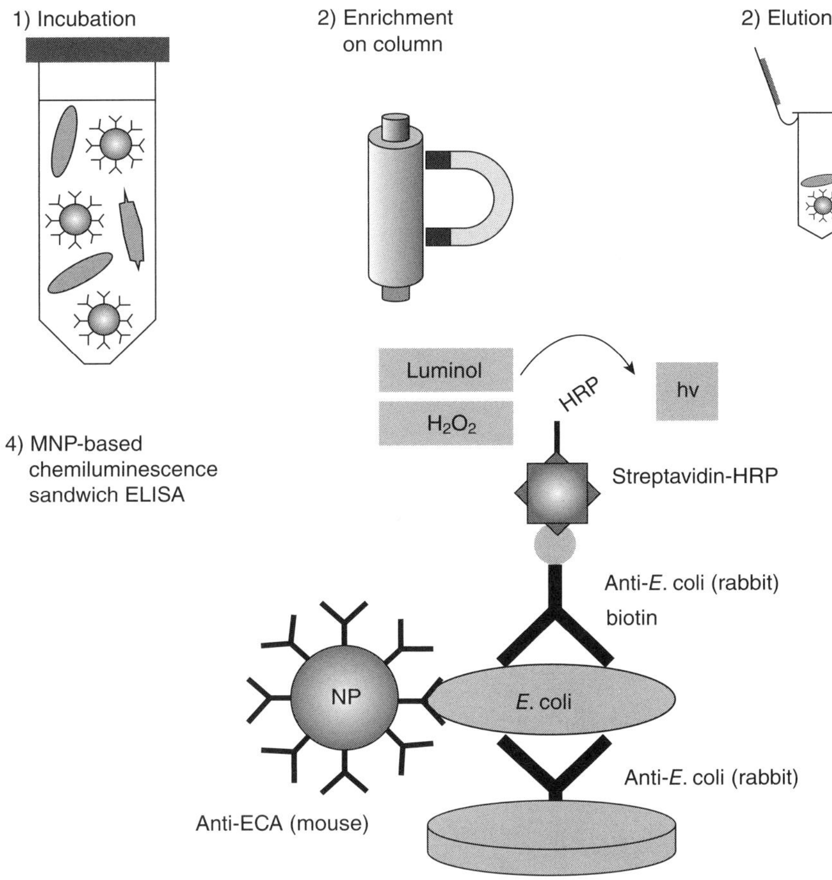

Figure 10.7. Schematic illustration of bacterial separation and of enrichment followed by their detection using sandwich ELISA. (Reproduced with permission from Pappert et al. [2010].)

10.4. MICROARRAYS TECHNOLOGY

General Perspectives. Microarray technology is a powerful, universally applicable analytical tool for rapid and simultaneous detection and analysis of a large number of cells, or molecular contents and proteins, with high sensitivity and precision. Thousands of DNA and RNA sequences can be simultaneously probed using a microarray to provide a detailed insight into the molecular profiles of a disease. Traditionally, it has been used for phenotyping and genotyping of cells. Phenotyping is the process of identifying cells with specific marker such as antigen expression on the diseased cells. When antibodies are used, the process is called immunophenotyping. When genetic constructs are used, it is called genotyping.

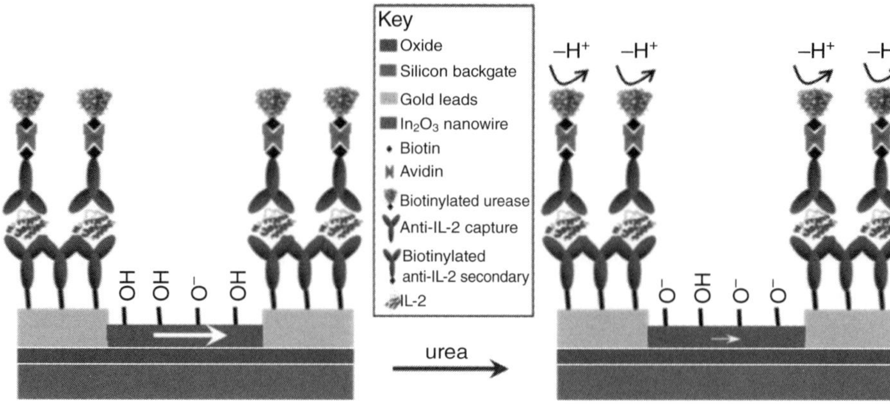

Figure 10.8. Schematic description of the nanoelectronic-ELISA approach where the use of urea as the substrate induces a pH increase that deprotonates the hydroxyl groups to decrease channel current. (Reproduced with permission from Stern et al. [2010].)

The microarray technology provides valuable inputs that aid in understanding the molecular basis of health and disease to advance molecular medicine and to accelerate development of customized drugs. The microarray technology is an extension of biosensing techniques, but now utilizing a micropatterned array of biosensors for simultaneous detection of multiple analytes. These arrays allow rapid and simultaneous probing of a large number of DNA, RNA, and proteins.

For details on the microarray technology, the reader is referred to Introduction to Biophotonics [Prasad, 2003]. This current section provides a brief description of this technology and the impact of nanotechnology on it. Using a DNA microarray, tens of thousands of genes can be screened for mutations or expression levels. The microarray technology has practically revolutionized the field of genome research, making it possible to perform studies ranging from (a) gene expression monitoring and transcription profiling for drug target identification to (b) large-scale identification of single nucleotide polymorphisms (SNP), which results from a base variation at a single nucleotide position.

DNA Microarray Technology. Traditionally, the DNA microarray represents arrays of tiny spots—measuring 5–150 μm, depending on the method of fabrication—on a solid support such as a glass strip or silicon chip.

At these sites, fragments of single-stranded DNA ranging from 20 to 1000 or even more bases are attached. These arrays identify DNA sequences of a gene in a sample by using fluorescently labeled messenger RNA (m-RNA) or c-DNA. The c-DNA (complementary DNA) is a single-stranded DNA that is produced from m-RNA by a process called reverse transcription. The fluores-

TABLE 10.1. Two Main Approaches for DNA Microarray Technology

Ex situ—Spotted Arrays Brown et al., Stanford	*In situ*—Oligonucleotide Arrays (Affymetrix Inc.)
↓	↓
DNA oligomers (probes) tethered to glass substrate using robotics and printing technology	DNA oligomers (probes) are synthesized on the surface of the chip (micropatterning using photolithograph)

↓

Labeled mRNA, cDNA, or ESTs (targets) annealed with arrays

↓

Laser scanner applied for fluorescence detection of positive pairing of probe and target (scanner usually employs confocal detection)

[a]ESTs, expressed sequence tags.

cently labeled m-RNAs or c-DNAs are then hybridized with the compleimentary DNA fragments immobilized on the chip. The chips are subsequently scanned with high-speed fluorescence detectors. The two main pieces of hardware for microarray technology are: (i) the microarray slide spotter and (ii) the microarray scanner. The two main approaches used to fabricate DNA microarrays are described in Table 10.1. Using the spotted array method, current capabilities permit printing of more than 31,000 elements on a microscope slide.

The Oligonucleotide Arrays technology, also called by a trade name Genechip® and introduced by Affymetrix (*source:* www.affymetrix.com), involves *in situ* synthesis of oligonucleotides of known sequence in a site-specific arrangement on a substrate. The process of fabrication of a Genechip® combines combinatorial DNA synthesis chemistry. This approach can produce hundreds of thousands of different oligonucleotide probes packed at an extremely high density. These oligonucleotides are up to 25 nucleotides (25mer) long. A schematic of the photolithographic process used is shown in Figure 10.9. The process of fabrication of a Genechip® combines DNA synthesis chemistry with photolithographic techniques adapted from the semiconductor industry [Barone et al., 2001]. The photolithographic process utilizes ultraviolet light projected through holes in masks to deprotect photolabile groups and

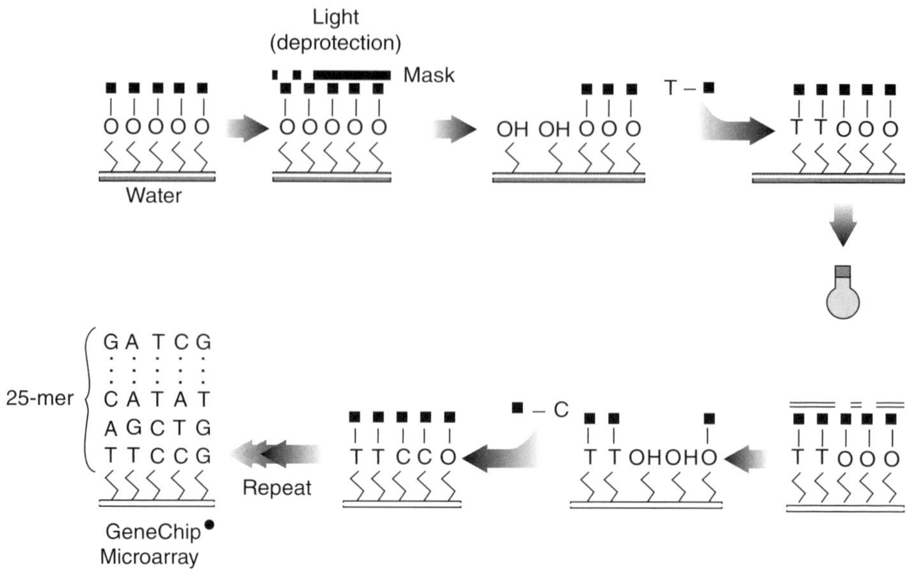

Figure 10.9. The use of a unique combination of photolithography and combinatorial chemistry to manufacture Genechip® Arrays. (Reproduced with permission from Prasad [2003]).

subsequently direct parallel and stepwise synthesis of oligonucleotides with a specific sequence. When using a fused silica or a planar glass substrate, its surface is first covalently modified using a silane reagent to provide hydroxyalkyl groups that serve as the initial synthesis sites. These sites are then extended with linker groups, protected with a photolabile-protecting group, such as 5′-(α-methyl-6-nitropiperonyloxycarbonyl), abbreviated as MeNPOC, which can be activated at specific spatial locations by UV light exposure for addition of nucleoside phosphoramidite monomers, also containing the photolabile group at the 5′ (or 3′) position. The photodeprotection is induced by the ~350-nm wavelength of the UV light from a commercial photolithographic exposure system.

Repetition of the cycle of changing the mask, deprotecting by photolithography and adding a nucleotide to ~70 times, can allow the synthesis of a complete array of thousands of 25-mer oligonucleotides in parallel. Generally, 20 pairs of oligonucleotides are arrayed to represent each gene. Each nucleotide matching the gene (perfect match, PM) is paired with a second mismatch oligonucleotide differing only by a central nucleotide (mismatch, MM). MM oligonucleotides serve to detect nonspecific and background hybridization, important for quantifying weakly expressed m-RNAs. The maximum achievable microarray density is determined by the spatial resolution provided by the photolithographic process. Typical dimensions for each array are

24 μm × 24 μm on a 1.6-cm^2 chip. This method has been used to display 65,000–400,000 DNA oligonucleotides on a 1.6-cm^2 glass surface [Lockhart et al., 1996].

Whereas the basic principles of microarray assays developed in 1980–1990 have not significantly changed, nanotechnology has enhanced the sensitivity and productivity of DNA microarray techniques. For example, Nanosys Inc. has developed a nanowire to serve as a substrate for DNA and protein microarrays (*source:* http://www.nanosysinc.com). They claim that the nanowire structure provides a 100-fold-higher binding area, as compared to conventional glass or silicone surfaces, thus increasing the sensitivity and reliability of the assay. The miniaturization of arrays can also significantly decrease the consumption of biomaterials, which is critical from the commercial standpoint. In this regard, the highest sensitivity for DNA and RNA quantitative analysis is achieved by PCR and RT-PCR; however, the cost of this assay is prohibitive for mass-scale analysis. This problem was addressed by the design of high-density chips with many thousands of nanoliter-scale PCR reaction volumes that makes assay more affordable and less time-consuming. (*Source*: http://www.appliedbiosystems.com). A number of new DNA microarray technologies are emerging. It is not possible to describe all of them within the limited scope of this monograph.

Protein Microarray Technology. In a protein microarray, a glass slide serving as a substrate for a protein chip is printed with thousands of protein probes [Mitchell, 2002; Poetz et al. 2005]. Depending on the design of the assay, the probes could be used for specific antibodies, aptamers, antigens, allergens, and so on; however, antibody microarrays are used most commonly. A biological sample is subsequently spread over the chip, and any binding is detected. The detection methods used with protein chips can be of many different types. The optical methods involving fluorescence and surface plasmon resonances have already been discussed in Section 10.2.

Although the projected market for protein chips is expected to surpass that for gene chips and already a number of companies are producing protein chips, the production of a protein chip and its successful utilization requires more stringent conditions. The fundamental differences between DNA and protein molecular structure contribute to difficulties in a simple extension of the DNA microarray technology to be used for fabrication of protein arrays. Some of the important differences between DNA and proteins from this perspective are listed below:

- While DNA involves a double-stranded rigid structure, a protein's organization is determined by a complex set of three-dimensional structures, namely primary, secondary, tertiary, and quaternary (Prasad [2003]; see also Chapter 2), which are extremely sensitive to the environment. Thus DNA molecules are robust; they can be hybridized, dried, and rehydrated without damage to their fundamental structure. Proteins, on the other hand, are unstable and easily denatured at solid–liquid and liquid–air interfaces, which can diminish their capability for specific interactions in the assay.

- The principle for biorecognition used in the DNA microarrays involves hybridization of the single-stranded c-DNA or m-RNA with a complementary strand, which is highly specific. In the case of protein microarrays, one often utilizes a wide array of immobilized antibodies. Unlike DNA capture, which is high affinity, the current antibody production capability produces low-affinity capture antibodies that severely compromise the validity of the interpretation of the result.
- While DNA microarray technology can utilize PCR methods to amplify detection, no equivalent of PCR exists for proteins.
- Antibodies, being generally glycosylated, have large surface areas for interactions and thus exhibit significant cross-reactivity between target proteins. The large surface area occupied on a small protein spot also promotes denaturation.

Just as in the case of a DNA microarray, a protein microarray is produced on a glass or a silicon substrate that has been treated with an aldehyde or other agent to immobilize the protein capturing agent such as an antibody [MacBeath and Schreiber, 2000]. The reaction of the aldehyde group with the amino group of the antibody (or protein) to form a Schiff base linkage is used for immobilization. In this scheme, following the attachment of the proteins to the substrate, the unreacted aldehyde groups are quenched to minimize nonspecific binding by immersing the substrate in a buffer solution containing bovine serum albumin (BSA). The schematic of this process is shown in Figure 10.10.

In the case of printing of peptides or very small proteins, BSA obscures the molecules of interest. In such cases, first a molecular layer of BSA is attached to a glass substrate. Then BSA is activated with *N,N′*-succinimidyl carbonate to produce active residues that can react with surface amines on the proteins. The unreacted sites on the substrate are then quenched with glycene.

A number of other approaches have been utilized to produce protein microarrays. They include the use of layers of aluminum or gold, hydrophilic polymers, and polyacrylamide gels for immobilization of capture agents.

Another approach pursued by Zyomyx of Hayward, California, utilizes photolithography to etch miniature wells on the surface of a silicon chip. The proteins or antibodies are immobilized in the flow chambers on the chip to maintain an aqueous solution environment.

Another approach used by Large Scale Biology of Vacaville, California, utilizes hundreds of thin plastic rods, each doped with a particular antibody and bundled together in a sheaf. Chips in the form of micrometer-thin slices are then produced from this sheaf by cutting them transversely.

The technique of rolling circle amplification (RCA) [Lizardi et al., 1998] has been used to increase the sensitivity for multiplexed arrays on antibody

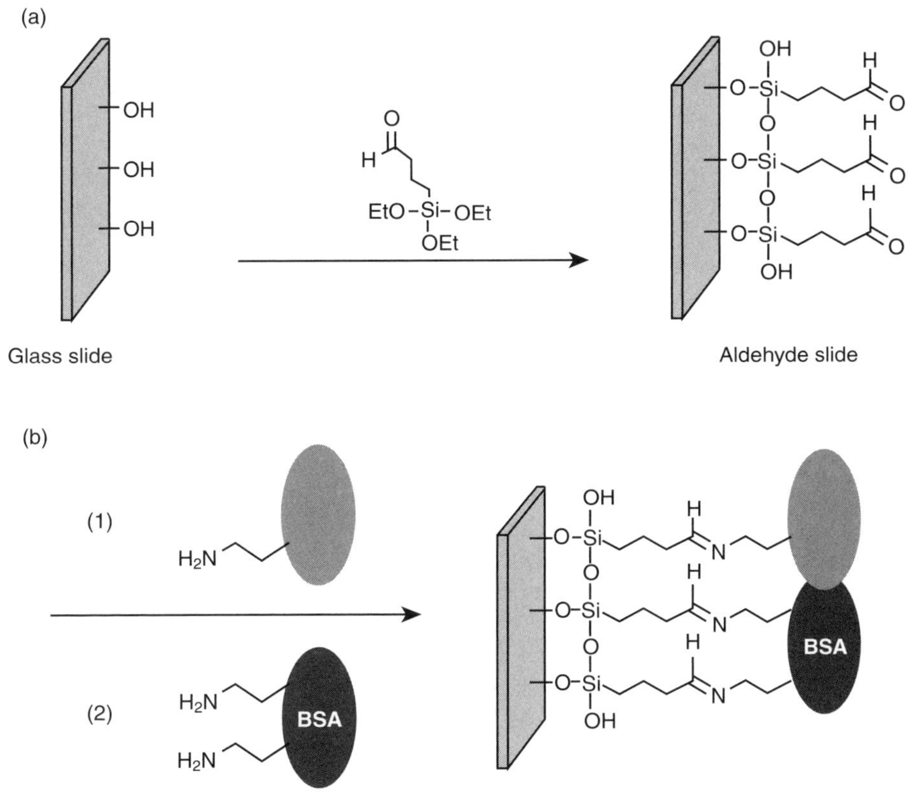

Figure 10.10. (a) Aldehyde group attachment to the substrate. (b) Immobilization of a protein at an array site and subsequent quenching of the unreacted site with BSA. (Reproduced with permission from Prasad [2003].)

microarray [Schweitzer et al., 2000]. An adaptation of RCA is used where the 5′ end of an oligonucleotide primer is attached to an antibody. Consequently, the presence of a circular DNA, DNA polymerase, and nucleotides allows rolling circle replication to generate a concatamer of circle DNA sequence copies attached to the antibody. The concatamer, which consists of tandem arrays of monomeric DNA molecules with complementary ends, can then be detected by hybridization with fluorescently labeled complementary oligonucleotide probes. This method of amplification has been reported to yield a 100 to 1000-fold improvement over detection using simply fluorescently labeled antibodies or streptavidin [Schweitzer et al., 2000].

Some Examples of Application of Microarrays.

Cancer Diagnosis. Since the DNA microarrays are in the most advanced state of development with applications in cancer diagnostics, this section discusses some examples of the microarray applications. The microarray technology is emerging as a powerful diagnostic tool in identifying molecular signatures of cancer and other diseases. It provides a rapid screening method to systematically explore changes in expression accompanying progression of cancer or a disease. It is particularly useful in the cases of tumors where various subclasses are morphologically indistinguishable. This technique has been used for many types of cancer, such as breast cancer, diffuse large B-cell lymphoma, leukemia, colon adenocarcinoma, and ovarian cancer.

The DNA array technology provides a method for rapid genotyping, facilitating the diagnosis of diseases for which a gene mutation has been identified. It also assists in the diagnosis of diseases for which known gene expression biomarkers, of a pathologic state or signature gene, exist. Golub et al. [1999] used microarray technology to distinguish between two similar diseases, acute myloid leukemia (AML) and acute lymphoblastic leukemia (ALL). They chose these cancers to validate the microarray method, because the two forms are already distinguishable by a number of methods. They applied a 6817-gene Affymetrix chip to 38 bone marrow samples and were able to identify 50 genes whose expression most distinguished AML from ALL. Golub and his colleagues are using DNA chips to investigate if gene expression patterns can predict which patients will respond to standard chemotherapy [Steinberg et al., 2000]. DNA microarrays were used by Alizadeh et al. [2000] to distinguish between two previously undifferentiated forms of diffuse large B-cell lymphoma. Welsh et al. [2001] used DNA microarray analysis to elucidate molecular markers for ovarian cancer. DNA microarrays were also used by Haviv and Campbell [2002] for assessing ovarian cancer gene expressions. Dhanasekaran et al. [2001] identified prognostic markers for prostate cancer. A great deal of activities utilizing microarray technology has also been focused on breast cancer [Moon et al., 2001; Hedenfalk et al., 2001].

Elucidation of Biochemical Signaling Pathways. DNA microarrays are also playing a significant role in the elucidation of many complex biochemical pathways [Shoemaker and Linsley, 2002]. Expression profiling has been used to monitor signaling pathways. This was accomplished by monitoring, in parallel, changes in expressions of thousands of genes affected by proximal signaling events. Ideker et al. [2001] used a combination of expression profiling and proteomic analysis to systematically analyze the metabolic galactose utilization pathway from yeast, following systematic perturbations by genetic deletion.

Microbial Interactions. DNA microarrays have been successfully used to elucidate the interaction between bacteria and their hosts. The effect of a commensal bacterium on gene expression in the intestine was reported by Hooper et al. [2001]. Microarrays were used by Huang et al. [2001] and Boldrick et al. [2002] to investigate the effects of pathogenic microorganisms on gene expression by dendritic cells and unfractionated peripheral blood mononuclear cells, respectively. They found massive and unique gene-expression changes triggered by different organisms.

Drug Development. The presence of alternate gene forms or atypical expression of a gene involved in drug action or metabolism can manifest as resistance to therapy or an atypical response to therapy. Pharmacogenomic and toxicogenomic studies correlate the genetic profile of individual patients and the individual response to a drug or toxin, respectively. The information obtained from these studies can be used to design arrays that will assist in the selection of custom and rational drug therapy. The identification of signature genes or biomarkers indicative of a disease process can identify candidate targets for therapeutic intervention.

Arrays assist in the identification of sentinel genes that demonstrate altered expression in a given cell or tissue type in response to drug or toxin exposure. Creating profiles of sentinel genes associated with drugs sharing a common mechanism allows potential new therapies to be rapidly screened for similar activities. This facilitates the selection of compounds for further investigation and may reduce the need for animal testing. An *in vitro* screen for potential toxicity has the potential to reduce drug-screening costs, prevent human suffering, and reduce product liability.

Analysis of sentinel genes can assist in determining the mechanism of action of a drug or toxin. Given that there is a multitude of events triggered by the initial action of a drug, screening thousands of genes at one time can identify multiple potential drug effectors. This allows robust hypotheses of drug mechanism to be formed and tested in subsequent investigations.

The valuable assistance provided by rapid screening using microarray technology can lead to accelerated FDA approval. FDA has established an accelerated review that allows approval based on evidence of the effect of the candidate drug on a surrogate endpoint, a disease-specific biomarker or signature gene. Once a drug has been shown to affect the expression of a surrogate endpoint in the desired manner, the FDA can grant marketing approval. Prior to this juncture, it has been difficult to identify disease biomarkers due to limited technology and individual patient variability. DNA array analysis allows multiple sequences to be searched for the presence of a suitable biomarker or a group of biomarkers. Testing the effects of a drug on a group of biomarkers may compensate for differences that reflect genetic variability.

10.5. SUSPENSION BEAD ASSAY

Basic Principle. The suspension bead assay (SBA), originally introduced by Luminex Corporation, thus also frequently referred to as the Luminex technology, utilizes a polymeric microsphere platform rather than a planar platform [Sukhanova et al., 2008; Krishnan et al. 2009]. In the original Luminex assay, color-coded microspheres were used to carry out a sandwich-type assay

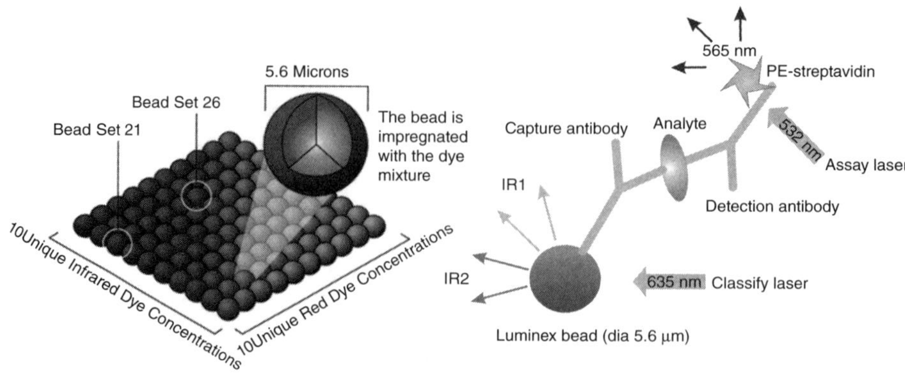

Figure 10.11. The Luminex SBA technology approach.

similar to that in ELISA. Here the capture antibody is immobilized on the surface of the microsphere that captures the target antigen. Then a detection antibody conjugated to a reporter dye through biotin–streptavidin coupling chemistry is used for quantitative characterization of the analyte. The Luminex approach utilizes 5.6-μm polystyrene microspheres that are color-coded with a red emitting dye and a near-infrared emitting dye at 10 different concentrations as shown in Figure 10.11, thus providing 100 detectable codes. The fluorescence of this IR dye can be excited by a 635-nm laser.

The dyes can be incorporated in the microsphere by simply suspending an organic solvent containing the dyes, where the microspheres swell to allow the dye molecules to diffuse in. Then the microbeads are transferred to an aqueous phase where they shrink, thus entrapping the dye molecules. The surface of these polystyrene microbeads is chemically modified to be coupled to the capture antibodies for immunoassay. A number of methods have been utilized ranging from passive adsorption to chemical conjugation. In the sandwich immunoassay, a detection antibody conjugated to a reporter dye (e.g., R-phycoerythrin, abbreviated as PE) through biotin–streptavidin coupling, is then used to quantify the captured antigen concentration. The cartoon diagram, shown in Figure 10.11, illustrates this scheme. In this scheme, the fluorescence intensities of the dyes coding the beads, and excited by 635 nm, identifies the antibody–antigen pair, while the intensity of fluorescence of the reporter dye, an orange emitter that is excited by a separate 532-nm green laser, provides a quantitative assay of the antigen.

Other suppliers of colored microbeads have used single color coding, but coupled with different microbead sizes, which can produce different intensities of low-angle forward scatter (FSC) in a flow cytometer, as discussed in Section 10.2. Furthermore, as discussed below, another modification introduced in the microbead technology is the use of quantum dots in place of organic dyes. By using different capture antibodies on distinct microspheres and in combination with different reporter fluorochromes (or quantum dots), one can simultaneously detect and quantify multiple analytes. Assay kits are commercially available to simultaneously detect as many as 42 different cytokines.

The microbead technology can be used in a planar solid support format such as with a microwell plate where a scanner can be used to detect the fluorescence signals from different microbeads. However, a suspension microbead assay utilizing a flow cytometer has become more popular because of the ease of detecting multiplexed signals with a high throughput. Moreover, recent advances in producing fiber-based flow cytometers, as described in Section 10.2, have provided further impetus to broadening the applications of quantum-dot-based microbead assay.

SBA for Protein Immunoassay. A major area of application of the microbead assay has been in protein analysis to use proteomics for disease diagnostics. Soluble protein expressions like cytokines have extensively been analyzed. A number of companies supply kits for simultaneous analysis of multiple cytokines.

Another area of protein detection for early detection of cancer is based on analysis of circulating autoantibodies [Sukhanova and Nabiev, 2008]. The circulating autoantibodies against tumor-associated antigens appear long before any pathological manifestations. Thus the identification and quantitative analysis of these autoantibody-related molecular signatures can serve as a very powerful method for detecting cancer predisposition and its early stage.

The microbead assay has also been used to diagnose infectious diseases by monitoring of appropriate antigen signatures present in blood or plasma [Nolan and Mandy, 2006], as well as for developing an understanding of molecular mechanisms of signal transduction [Leifert et al., 2005]. G-protein-coupled receptors (GPCRs) form an important class, which are involved in a number of physiological processes.

SBA for Nucleic Acid Detection. SBA is being increasingly used for nucleic acid detection [Krishnan et al., 2009; Dunbar, 2006]. In this case, a synthetically produced oligonucleotide generally acts as a capture unit (instead of capture antibody) and is conjugated to the microbead. Fragments of known sequence DNA and RNA have also been used. PCR is first used to amplify the unknown nucleic acid (see Section 10.4), labeled with biotin and hybridized with the oligonucleotide on the microbead. Then the reporter fluorochrome PE complexed with streptavidin is added to bind with the hybridized nucleic acid, and its fluorescence is used for detection. Applications so far have included analysis of single nucleotide polymorphism (SNP), disease screening utilizing genomics, gene expression profiling, microbead detection, and analysis of small-chain si-RNA and microRNA in human cancer [Krishnan et al., 2009, Nolan and Mandy, 2006].

Application of Nanotechnology. Various kinds of nanoformulations have been used for the improvement in the SBA approach. One approach [Van Blaaderen and Vrij, 1992; Lawrie et al., 2003] has incorporated dye containing nanoshells around silica microspheres. Lawrie et al. [2003] were able to prepare microspheres coated with up to six different fluorescent-dye shells alternating with nonfluorescent silica spacer shells. These types of nanostructured microspheres provide a large number of optical signatures (codes) from a combination of fluorescence wavelength, intensity, size, and refractive index. However, careful selection of the dye set to minimize fluorescence resonance energy transfer has to be made.

A major direction is the use of quantum dots instead of organic dyes. A variation of the colored microbead assay is incorporation of quantum dots (QDs) to color code the microbead as well as use their conjugate with the detection antibody as the reporter fluorescent tag. The advantages of using QDs are, as discussed previously, significantly superior multiplexing capability due to their narrow emission width, tremendous resistance against photobleaching, and higher sensitivity due to greater brightness. Furthermore, extremely broad QD's absorption profile allows the use of a single-laser excitation source to generate fluorescence from quantum dots of different sizes and compositions, emitting over a broad spectral range (see Chapter 5). The most commonly used QDs have been CdTe and CdSe in a core–shell geometry using a ZnS shell. CdTe provides the benefit of tuning their emission to the near-IR region.

QDs have been incorporated in microspheres using a variety of methods. Infiltration and entrapment of QDs into microspheres by using a swelling method described earlier for the dye case was used by Nie and co-workers [Han et al., 2001]. Another method is the introduction of QDs during the polymerization process to produce the microspheres [Sukhanova and Nabiev, 2008]. The third method incorporates the QDs in a shell around the microsphere. We have successfully used polyelectrolyte coatings around a microsphere to incorporate QDs in between two layers of oppositely charged polyelectrolytes [Unpublished results]. The QDs can conveniently be conjugated to the detection antibody by using biotin–streptavidin coupling.

A simplified approach used by us does not involve color coding of the microbeads, but employs a sandwich immunoassay technology for uncolored microbead containing the capture antibody. Here, the detection antibody can contain one-or multiple-color QDs for both identification and quantification of the antigen. Figure 10.12 provides an example of this approach for the immunoassay of tumor necrosis factor-alpha (TNF-α), which is a biomarker for many cancers. According to this protocol, 3-μm polystyrene microbeads were coupled with the capture antibody. Then, a microbead suspension was incubated with the analyte solution (TNF-α), washed, and subsequently incubated with the detection antibody conjugated to the CdTe/ZnS core–shell QDs. Finally, this bead suspension was analyzed through the fiber-based flow cytometer described in Section 10.2. The emission and scattering signals were

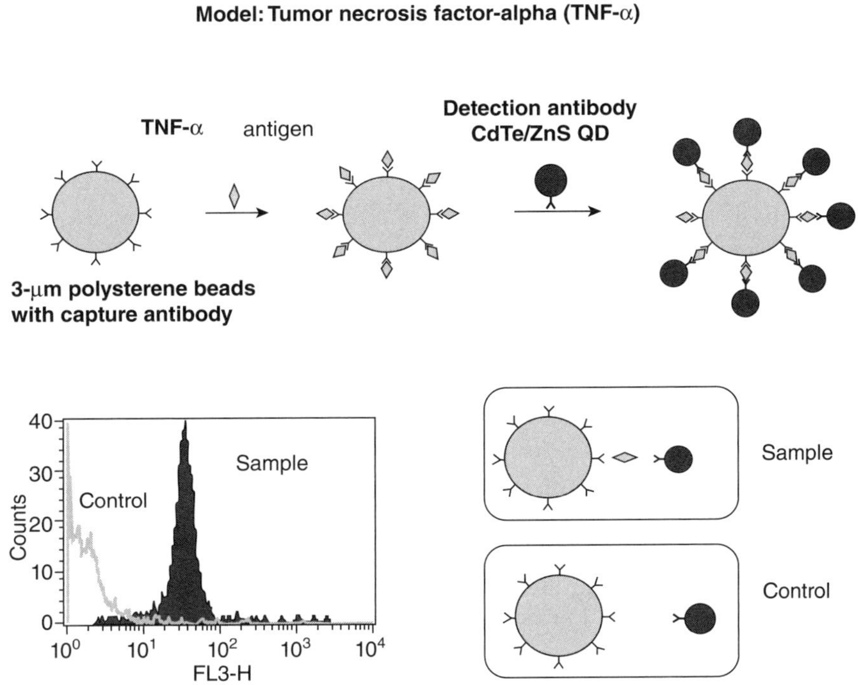

Figure 10.12. The ILPB multiple-color QDs immunoassay schematics.

collected through the optical fibers, separated by the spectrograph and processed using the data acquisition system. A control experiment with no antigen binding was also run to distinguish the background. A fluorescence signal from QDs bound to the analyte was clearly obtained, as shown on the single-parameter histogram of Figure 10.12. This QD-based color-coding technology of analytes is compatible with most commercial models of the flow cytometers.

Another type of nanoparticles that can also be used are SERS Raman reporters, already mentioned in Section 10.2 in conjunction with cell labeling for flow cytometry detection. Other SERS-based applications include a spectroscopic assay based on silver nanorods as SERS enhancers, being developed for rapid detection of trace levels of viruses with high sensitivity and specificity [Jain, 2007]. The technique measures the change in frequency of an NIR laser as it scatters viral DNA or RNA.

HIGHLIGHTS OF THE CHAPTER

- A major emphasis of modern diagnostics, particularly in relation to early-stage detection of a disease or infection, is diagnosis on cellular and molecular levels. Cellular diagnostics involves cell profiling on the basis of cell morphology and cellular functions, as measured by its physiological status, as well as antigen expression on the cell surface.

- A flow cytometer is an optical diagnostic device used for disease profiling by measuring the physical and/or chemical characteristics of cells or biological assemblies.
- The basic steps of flow cytometry are (i) fluorescence labeling of biological substances in cells, (ii) hydrodynamic focusing to produce a laminar flow of cells in a single file at the interrogation point, (iii) laser illumination and collection of optical response, and (iv) data acquisition and processing.
- The illumination optics focuses the laser beam(s) to an elliptical spot coincident with the flowing suspension to illuminate each cell, one at time, as it passes through the illumination zone.
- Cells are illuminated with one or more laser beams designed to selectively excite particular fluorochromes used to stain specific substances expressed by the cells (cell subpopulation).
- The optical response detected by the collection optic system consists of forward-scattered count (FSC), side-scattered count (SSC), and the fluorescence intensities of the various fluorochromes used to stain the substances expressed by different cells.
- FSC and SSC, together with the number of fluorescent wavelengths detected, define the total number of parameters collected by a flow cytometer. The FSC signal is often used to measure cell size, the larger the size of the cell, the greater the FSC signal. The SSC signal, generally collected at 90° from the excitation beam, provides information on the granularity of the cell (differences in refractive index).
- The fluorescence signals at different wavelengths, also detected at 90° from the excitation beam, are separated from SSC and from each other generally by using sets of dichroic mirrors and wavelength selective filters.
- Flow cytometry generates a large amount of data representing multiparameter detection. A single-parameter histogram is a plot of each channel in one mode of display which represents the relative signal height (channel number) of one parameter (FSC, SSC, or a specific fluorescence) against the event count (number of events) registering this intensity level.
- Dot plots, also referred to as bitmaps or scattergrams, are bivariate plots in which each cell is placed as a dot on the plot according to the two intensity channels it registers in for the selected two parameters represented on the x and y axes. Dot plots are useful for correlation of one parameter with others and can be used to distinguish different types of cells.
- Immunophenotyping, which refers to identification of cells using fluorochrome-conjugated antibodies as immunosensitive probes for

protein antigens expressed by cells, is an important application for cellular analysis.

- Measurements of the DNA content of cell is a major application of flow cytometry and is used to characterize disease or a tumor.
- Flow cytometer detection of soluble biomolecules captured on a microbead can be then used to correlate with a specific pathological condition that can be used to identify a specific disease or microbial invasion.
- The use of confocal fiber-optic fluorescence pickup and a miniature multichannel spectrograph, instead of a set of mirrors and filters, to disperse emitted light into color channels, make flow cytometers more compact, portable, and less expensive.
- QDs have been utilized for labeling antibodies to be used in flow cytometry due to (i) photostability, (ii) size-tunable narrow emission bands, allowing for multiparameters, multiplexed analyses using a large number of channels over a broad spectral range, and (iii) a single excitation source (one large) being able to excite a number of different quantum dots emitting at different wavelengths.
- The most commonly used and highly sensitive ELISA technique involves a sandwich assay, which utilizes a sandwich inclusion of the target antigen between two antibodies, a capture antibody and a detection antibody, the latter being conjugated (linked) to an enzyme.
- Nanotechnology has been applied to ELISA in a number of different ways: (i) gold nanoparticles have been used as carriers of the detection antibody linked with an enzyme, with a goal to amplify the optical signal; (ii) magnetic nanoparticles have been utilized for immunomagnetic separation to provide enrichment of the analyte; (iii) in nanoelectronic ELISA, the ELISA assay produces a pH change that is detected by using an indium oxide nanowire-based ion sensing field effect transistor.
- Microarray technology is used for rapid and simultaneous detection of a large number of biologics, with high throughput, sensitivity, and precision. Microarrays are micropatterned arrays of biosensing capture agent that allow rapid and simultaneous probing of a large number of DNA and proteins.
- The microarray technology has significantly accelerated the development of two exciting frontiers: (i) genomics dealing with mapping and sequencing of genomes and determining gene functions and (ii) proteomics covering the structure, localization, modifications, interactions, and functions of proteins.
- The two major types of biological microarrays are (i) DNA microarrays to identify DNA sequences of a gene and (ii) protein microarrays to capture and identify proteins of interest.
- The various steps of microarray technology are (i) patterning of microarrays, (ii) immobilization/*in situ* fabrication of biorecognition elements at

the microarray locations, (iii) biorecognition of fluorescently labeled specimens, (iv) scanning/readout of microarrays, and (v) data collection and processing.

- The two main approaches to fabricate DNA microarrays are (i) spotted arrays in which single-stranded c-DNA or finite-size nucleotide oligomers (oligonucleotides) are covalently bonded by printing (spotting) on a polylysine film on glass and (ii) oligonucleotide arrays, also called Genechip®, which utilizes photolithography to produce *in situ* synthesis of oligonucleotides of known sequence in a site-specific arrangement.
- Analytes detected by DNA microarrays are single-stranded DNA fragments, m-RNA, produced from the sample cell by the transcription process, or c-DNA (complementary DNA) produced from m-RNA by reverse transcription. The DNA microarrays use fluorescence detection of the fluorescently labeled c-DNA or m-RNA sample.
- To enhance the sensitivity of detection, a well-known chemical process called the polymerase chain reaction (PCR) is used to create multiple copies of m-RNA from a specific fragment of the parent DNA in the specimen.
- Protein microarrays are usually fabricated by using a substrate, covalently bonded to an aldehyde which then is coupled to a protein capturing agent (e.g., an antibody) to immobilize it. The capture agents used in protein microarrays are various antibodies, synthetic polypeptide ligands, aptamers, and oligonucleotides.
- A technique called rolling circle amplification (RCA) can be used to increase the sensitivity of multiplexed arrays on antibody microarrays.
- Protein microarrays have not found as wide an acceptance as have the DNA microarrays, because of the difficulties in handling them, some of which are as follows: (i) proteins are not as robust as DNA and thus readily denature, (ii) protein microarrays usually have to be kept in water to reduce denaturation, (iii) protein detection is not as specific as DNA, and (iv) no equivalent to PCR exists for proteins to enhance sensitivity.
- Microarray technology has been already utilized in cancer diagnosis, elucidation of biochemical signaling pathways, microbial interactions, and drug development.
- The suspension bead assay (SBA), utilizes a polymeric microsphere platform rather than a planar platform, where color-coded microspheres are used to carry out a sandwich-type assay similar to that in ELISA. The capture antibody is immobilized on the surface of the microsphere that captures the target antigen. Then a detection antibody conjugated to a reporter dye through biotin–streptavidin coupling chemistry is used for quantitative characterization of the analyte.
- Examples of nanoformulations used for the improvement in the SBA approach are: (i) dye containing nanoshells incorporating around silica microspheres and (ii) use of quantum dots instead of organic dyes.

EXERCISES

True or False

1. Genomics is a type of molecular diagnosis, which deals with the study of genome structure and function.

2. The most commonly used ELISA technique involves a sandwich assay, which utilizes a sandwich inclusion of the target antigen between antibody and enzyme.

3. Nanotechnology has been applied to ELISA in the way of employing the magnetic nanoparticles for immunomagnetic separation to provide enrichment of the analyte.

4. In a protein microarray, the biological assemblies or microbeads flow in a suspension in an individual manner and past a sensing point.

Multiple Choice

5. A flow cytometer is an optical diagnostic device for disease profiling
   (a) By measuring the flow rate of cells or biological assemblies
   (b) By measuring the physical and/or chemical characteristics of cells or biological assemblies
   (c) By measuring the surface charges of cells or biological assemblies
   (d) By measuring the toxicities of cells or biological assemblies

6. The sensing in the flow cytometer utilizes an optical technique where the beam from a light source interacting with each individual cell/bioassembly produces
   (a) Illumination
   (b) Scattering or fluorescence
   (c) Heating
   (d) Sound wave

7. A point where a light source interacts with a cell/bioassembly passing through the flow cell of flow cytometer is called:
   (a) Active point
   (b) Measurement point
   (c) Interrogation point
   (d) Hot spot

8. The main principle of DNA microarrays is
   (a) Auto-fluorescence of DNA molecules
   (b) The property of DNA to be dissolved in water

   (c) The property of hybridization between two DNA strands
   (d) The damage of DNA

9. To carry out a sandwich-type assay, the suspension bead assay (SBA) utilizes
   (a) A polymeric microsphere platform
   (b) A planar substrate
   (c) A spherically shaped large substrate
   (d) A magnetic filtration

Answer: 1. T; 2. F; 3. T; 4. F; 5. b; 6. b; 7. c; 8. c; 9. a

REFERENCES

Alizadeh, A. A., Eisen, M. B., Davis, R. E., Ma, C., Lossos, I. S., Rosenwald, A., Boldrick, J. G., Sabet, H., Tran, T., Yu, X., Powell, J. I., Yang, L. M., Marti, G. E., Moore, T., Hudson, J., Lu, L. S., Lewis, D. B., Tibshirani, R., Sherlock, G., Chan, W. C., Greiner, T. C., Weisenburger, D. D., Armitage, J. O., Warnke, R., Levy, R., Wilson, W., Grever, M. R., Byrd, J. C., Botstein, D., Brown, P. O., Staudt, L. M., Distinct types of diffuse large B-cell lymphoma identified by gene expression profiling, *Nature* **403**, 503–511 (2000).

Ambrosi, A., Airo, F., Merkoci, A., Enhanced gold nanoparticle based ELISA for a breast cancer biomarker, *Anal Chem.* **82**, 1151–1156 (2010).

Barone, A. D., Beecher, J. E., Bury, P. A., Chen, C., Doede, T., Fidanza, J. A., McGall, G. H., Photolithographic synthesis of high-density oligonucleotide probe arrays, *Nucleosides Nucleotides Nucl. Acids* **20**, 525–531 (2001).

Boldrick, J. C., Alizadeh, A. A., Diehn, M., Dudoit, S., Liu, C. L., Belcher, C. E., Botstein, D., Staudt, L. M., Brown, P. O., Relman, D. A., Stereotyped and specific gene expression programs in human innate immune responses to bacteria, *Proc. Natl. Acad. Sci. USA* **99**, 972–977 (2002).

Chattopadhyay, P. K., Price, D. A., Harper, T. F., Betts, M. R., Yu, J., Gostick, E., Perfetto, S. P., Goepfert, P., Koup, R. A., De Rosa, S. C., Bruchez, M. P., Roederer, M., Quantum dot semiconductor nanocrystals for immunophenotyping by polychromatic flow cytometry, *Nat. Med.* **12**, 972–977 (2006).

Darzynkiewicz, Z., Juan, G., Li, X., Gorczyca, W., Murakami, T., Traganos, F., Cytometry in cell necrobiology: Analysis of apoptosis and accidental cell death (necrosis), *Cytometry* **27**, 1–20 (1997).

Dhanasekaran, S. M., Barrette, T. R., Ghosh, D., Shah, R., Varambally, S., Kurachi, K., Pienta, K. J., Rubin, M. A., Chinnaiyan, A. M., Delineation of prognostic biomarkers in prostate cancer, *Nature* **412**, 822–828 (2001).

Dunbar, S. A., Applications of Luminex xMAP technology for rapid, high-throughput multiplexed nucleic acid detection, *Clin. Chim. Acta* **363**, 71–82 (2006).

Edgar, R., McKinstry, M., Hwang, J., Oppenheim, A. B., Fekete, R. A., Giulian, G., Merril, C., Nagashima, K., Adhya, S., High-sensitivity bacterial detection using biotin-tagged phage and quantum-dot nanocomplexes, *Proc. Natl. Acad. Sci. USA* **103**, 4841–4845 (2006).

Ferrari, B. C., Bergquist, P. L., Quantum dots as alternatives to organic fluorophores for cryptosporidium detection using conventional flow cytometry and specific monoclonal antibodies: Lessons learned, *Cytometry A* **71**, 265–271 (2007).

Givan, A. L., *Flow Cytometry: First Principles*, 2nd edition, Wiley-Liss, New York (2001).

Golub, T. R., Slonim, D. K., Tamayo, P., Huard, C., Gaasenbeek, M., Mesirov, J. P., Coller, H., Loh, M. L., Downing, J. R., Caligiuri, M. A., Bloomfield, C. D., Lander, E. S., Molecular classification of cancer: Class discovery and class prediction by gene expression monitoring, *Science* **286**, 531–537 (1999).

Han, M., Gao, X., Su, J. Z., Nie, S., Quantum-dot-tagged microbeads for multiplexed optical coding of biomolecules, *Nat. Biotechnol.* **19**, 631–635 (2001).

Haviv, I., Campbell, I. G., DNA microarrays for assessing ovarian cancer gene expression, *Mol. Cell. Endocrinol.* **191**, 121–126 (2002).

Hedenfalk, I., Duggan, D., Chen, Y. D., Radmacher, M., Bittner, M., Simon, R., Meltzer, P., Gusterson, B., Esteller, M., Kallioniemi, O. P., Wilfond, B., Borg, A.,Trent, J., Raffeld, M., Yakhini, Z., Ben-Dor, A., Dougherty, E., Kononen, J., Bubendorf, L., Fehrle, W., Pittaluga, S., Gruvberger, S., Loman, N., Johannsoson, O., Olsson, H., Sauter, G., Gene-expression profiles in hereditary breast cancer, *N. Engl. J. Med.* **344**, 539–548 (2001).

Hooper, L. V., Wong, M. H., Thelin, A., Hansson, L., Falk, P. C., Gordon, J. I., Molecular analysis of commensal host–microbial relations hips in the intestine, *Science* **291**, 881–884 (2001).

Huang, Q., Liu, D. Y., Majewski, P., Schulte, L. C., Korn, J. M., Young, R. A., Lander, E. S., Hacohen, N., The plasticity of dendritic cell responses to pathogens and their components, *Science* **294**, 870–875 (2001).

Ibanez-Peral, R., Bergquist, P. L., Walter, M. R., Gibbs, M., Goldys, E. M., Ferrari, B., Potential use of quantum dots in flow cytometry, *Int. J. Mol. Sci.* **9**, 2622–2638 (2008).

Ideker, T., Thorsson, V., Ranish, J. A., Christmas, R., Buhler, J., Eng, J. K., Bumgarner, R., Goodlett, D. R., Aebersold, R., Hood, L., Integrated genomic and proteomic analyses of a systematically perturbed metabolic network, *Science* **292**, 929–934 (2001).

Jain, K. K., Applications of nanobiotechnology in clinical diagnostics, *Clini. Chem.* **53**, 2002–2009 (2007).

Knight, P. R., Sreekumar, A., Siddiqui, J., Laxman, B., Copeland, S., Chinnaiyan, A., Remick, D. G., Development of a sensitive microarray immunoassay and comparison with standard enzyme-linked immunoassay for cytokine analysis, *Shock* **21** 26–30 (2004).

Krishnan, V. V., Khan, I. H., Luciw, P. A., Multiplexed microbead immunoassays by flow cytometry for molecular profiling: Basic concepts and proteomics applications, *Crit. Revi. Biotechnol.* **29**, 29–43 (2009).

Lawrie, G. A., Battersby, B. J., Trau, M., Synthesis of optically complex core–shell colloidal suspensions: Pathways to multiplexed biological screening, *Adv. Funct. Mater.* **13**, 887–896 (2003).

Leifert, W. R., Aloia, A. L., Bucco, O., Glatz, R. V., McMurchie, E. J., G-protein-coupled receptors in drug discovery: Nanosizing using cell-free technologies and molecular biology approaches, *J. Biomol. Screen.* **10**, 765–779 (2005).

Lizardi, P. M., Huang, X. H., Zhu, Z. R., Bray-Ward, P., Thomas, D. C., Ward, D. C., Mutation detection and single-molecule counting using isothermal rolling-circle amplification, *Nat. Genet.* **19**, 225–232 (1998).

Lockhart, D. J., Dong, H. L., Byrne, M. C., Follettie, M. T., Gallo, M. V., Chee, M. S., Mittmann, M., Wang, C. W., Kobayashi, M., Horton, H., Brown, E. L., Expression monitoring by hybridization to high-density oligonucleotide arrays, *Nat. Biotechnol.* **14**, 1675–1680 (1996).

MacBeath, G., Schreiber, S. L., Printing proteins as microarrays for high-throughput function determination, *Science* **289**, 1760–1763 (2000).

Mitchell, P., A perspective on protein microarrays, *Nat. Biotechnol.* **20**, 225–229 (2002).

Moon, D. H., Lee, S. J., Park, K. Y., Park, K. K., Ahn, S. H., Pai, M. S., Chang, H., Lee, H. K., Ahn, I. M., Correlation between Tc-99m-pertechnetate uptakes and expressions of human sodium iodide symporter gene in breast tumor tissues, *Nucl. Med. Biol.* **28**, 829–834 (2001).

Ndao, M., Diagnosis of parasitic diseases: Old and new approaches, *Interdisciplin. Perspect. Infect. Dis.*, Article ID 278246, 1–15 (2009).

Nolan, J. P., Mandy, F., Multiplexed and microparticle-based analyses: Quantitative tools for large-scale analysis of biological systems, *Cytometry Part A* **69A**, 318–325 (2006).

Pappas, M. G., Recent applications of the dot-ELISA in immunoparasitology, *Vet. Parasitol.* **29**, 105–129 (1988).

Pappert, G., Rieger, M., Niessner, R., Seidel, M., Immunomagnetic nanoparticle-based sandwich chemiluminescence-ELISA for the enrichment and quantification of *E. coli*, *Microchim. Acta*, **168** 1–8 (2010).

Poetz, O., Schwenk, J. M., Kramer, S., Stoll, D., Templin, M. F., Joos, T. O., Protein microarrays: catching the proteome, *Mech. of Ageing Dev.* **126**, 161–170 (2005).

Prasad, P. N., *Introduction to Biophotonics*, Wiley, Hoboken, NJ (2003).

Schweitzer, B., Wiltshire, S., Lambert, J., O'Malley, S., Kukanskis, K., Zhu, Z. R., Kingsmore, S. F., Lizardi, P. M., Ward, D. C., Immunoassays with rolling circle DNA amplification: A versatile platform for ultrasensitive antigen detection, *Proc. Nat. Acad. Sci. USA* **97**, 10113–10119 (2000).

Shapiro, H. M., *Practical Flow Cytometry*, 3rd edition, Wiley-Liss, New York (1995).

Shoemaker, D. D., Linsley, P. S., Recent developments in DNA microarrays, *Curr. Opin. Microbiol.* **5**, 334–337 (2002).

Steinberg, J. L., Yeo, W., Zhong, S., Chang, J. Y. H., Tam, J. S., Chan, P. K. S., Leung, N. W. Y., Johnson, P. J., Hepatitis B virus reactivation in patients undergoing cytotoxic chemotherapy for solid tumours: Precore/core mutations may play an important role, *J. Med. Virol.* **60**, 249–255 (2000).

Stern, E., Vacic, A., Li, C., Ishikawa, F. N., Zhou, C., Reed, M. A., Fahmy, T. M., A nanoelectronic enzyme-linked immunosorbent assay for detection of proteins in physiological solutions, *Small* **6**, 232–238 (2010).

Stewart, C. C., Goolsby, C., Shackney, S. E., Emerging technology and future developments in flow cytometry, in R. Riley, ed., *Hematology–OncologyClinics of North America*, W.B. Saunders, Philadelphia, 477–495 (2002).

Sukhanova, A., Nabiev, I., Fluorescent nanocrystal-encoded microbeads for multiplexed cancer imaging and diagnosis, *Crit. Rev. Oncol. Hematol.* **68**, 39–59 (2008).

Toussirot, É., Saas, P., Deschamps, M., Pouthier, F., Perrot, L., Perruche, S., Chabod, J., Tiberghien, P., Wendling, D., Increased production of soluble CTLA-4 in patients with spondylarthropathies correlates with disease activity, *Arthritis Res. Ther.* **11**, R101 (2009).

Van Blaaderen, A., Vrij, A., Synthesis and characterization of colloidal dispersions of fluorescent, monodisperse silica spheres, *Langmuir* **8**, 2921–2931 (1992).

Welsh, J. B., Zarrinkar, P. P., Sapinoso, L. M., Kern, S. G., Behling, C. A., Monk, B. J., Lockhart, D. J., Burger, R. A., Hampton, G. M., Analysis of gene expression profiles in normal and neoplastic ovarian tissue samples identifies candidate molecular markers of epithelial ovarian cancer, *Proc. National Acad. Sci. USA* **98**, 1176–1181 (2001).

CHAPTER 11

NANOPHARMACOTHERAPY

Pharmacotherapy is the treatment of diseases through the use of drugs. Pharmacotherapy is thus an interdisciplinary area of healthcare that is responsible for ensuring the safe, appropriate, and economical use of drugs in patients. Through the control of the delivery of a drug to its intended target, nanopharmacotherapy has the potential to optimize drug efficacy while minimizing side effects, thereby dramatically improving health outcomes for patients at a lower overall cost [Devalapally et al., 2007]. This chapter presents general aspects of nanopharmacotherapy and discusses how nanomedicine provides benefits in enhancing therapeutic efficiency, while minimizing adverse side effects of a therapy. In particular, important factors determining the quality of a therapy discussed here are pharmacokinetics and pharmacodynamics; in each case the impact of nanotechnology is presented. Section 11.1 presents an overview of nanopharmacotherapy. Section 11.2 provides examples of different modes of nanoformulation for pharmacotherapy. Section 11.3 introduces pharmacokinetics and the impact of nanoformulations on it. Section 11.4 covers in a greater depth the area of biodistribution and tissue targeting. Section 11.5 introduces pharmacodynamics and discusses how nanocarriers can influence pharmacodynamics. Section 11.6 provides a discussion of more specific methods to produce controlled drug release or drug activation using nanoformulation.

Introduction to Nanomedicine and Nanobioengineering, First Edition. Paras N. Prasad.

11.1. NANOPHARMACOTHERAPY: AN OVERVIEW

The benefits of nanotherapy have been described in various earlier chapters, starting from Chapter 1. This section presents an integrated overview of nanotherapy, focusing particularly on the nanopharmacotherapy aspect. In regard to any drug therapy, the key issues to be addressed are [Devalapally et al., 2007]:

- *Solubility of the Drug.* The dispersibility of a drug is an important concern because it determines what fraction of it is able to be absorbed into the vascular system with subsequent systemic delivery to target sites. Most drugs are hydrophobic and therefore pose a problem in being dispersed in an aqueous-based biological medium. A nanocarrier-based delivery can overcome this problem by tailoring of the nanoparticle surface to introduce hydrophilic groups for them to be dispersible in an aqueous medium.
- *The Chemical and Metabolic Stability of the Drug.* The stability of a drug in biological media and subsequent systemic availability to drug metabolizing enzymes are also of major concern. A nanocarrier formulation, whether in the form of nanoencapsulation or incorporated in the core of a dendrimer, can act as a drug reservoir that is shielded from various direct chemical degradation and metabolic enzymes in the liver and other organs.
- *Specificity of Drug Delivery.* As has been discussed in Chapter 7 on biotargeting, nanopharmacotherapy provides the ability for both passive and active targeting including access at the subcellular and organelle level. This feature provides two benefits: (i) increase of local concentration of a therapeutic agent at the intended site of action and (ii) minimization of any undesired damage to healthy tissue sites.
- *Toxicity of Drugs.* The toxicity of any drug formulation is also an important factor in determining its therapeutic value. Nanopharmacotherapy provides benefits by reducing the toxicity of a specific drug. A nanoformulation with biocompatible coatings on the surface can reduce the exposure of healthy tissues to the drug. Once the drug reaches the targeted the site of drug action, controlled release by various mechanisms can be attained and this is discussed later in this chapter.

The general benefits of a nanopharmacotherapy are summarized in Figure 11.1.

The biodispersion aspect has already been discussed above. Nanoformulation of hydrophobic drugs has played a significant role in enhancing the efficacy of a drug. An example is Taxol, where the delivery of this drug in nanoencapsulated form enhances its absorption into the systemic circulation, thus leading to improved bioavailability. Biocompatibility of the encapsulated drug or a therapeutic agent can be readily introduced by surface modification

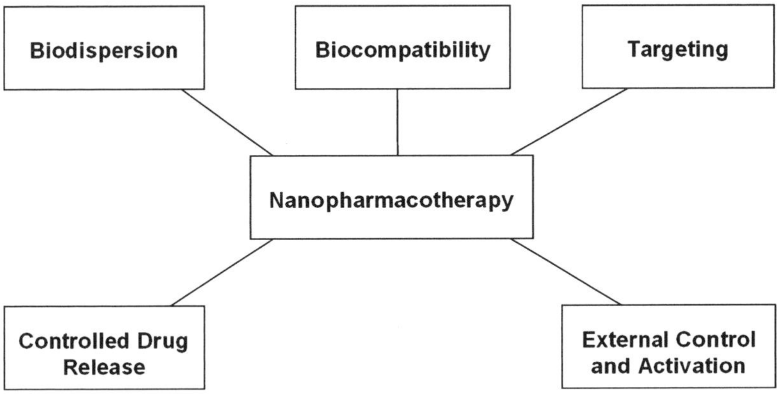

Figure 11.1. Benefits of a Nanopharmacotherapy.

of the nanoparticle or appropriate chemical manipulation of the end group of a dendritic structure. The tissue targeting aspect of a nanotherapy has already been discussed in detail in Chapter 7.

Controlled release of a drug is a very important benefit of a nanoformulation. This feature can be introduced in a number of ways. First, nanostructures or nanoparticles can be designed to effect size-based exclusion of the encapsulated drug. Here, controlled porosity of a nanoparticle or externally accessible free space of a nanostructure (such as a dendrimer) can be manipulated to allow for a sustained release only of the drug molecules. Another mechanism that provides the capability to deliver intracellular pharmacotherapy is the release of the drug, once the nanocarrier is localized inside the target cell by the mechanism of receptor-mediated endocytosis, discussed in Chapter 7. This release may be in response to a change in pH, which may enhance the porosity of the nanocarrier to release the drug. Another application of pH responsiveness is the release of genetic material for gene therapy, which is attached to the nanoparticle surface by electrostatic binding. This type of release is discussed in Chapter 14, entitled "Gene Therapy."

Another type of drug release may be activated by site-specific enzymatic cleavage of the chemical unit attaching the drug to the nanocarrier by using overexpressed enzymes at the targeted sites. In the case of gene therapy, it could also be triggered by endosomal capture of the nanocarriers.

External control and activation of a therapeutic action is also a major advantage of nanotherapy. A nanoformulation lends itself conveniently for external control by using a number of mechanisms, depending on the nature of therapy. In the case of pharmacotherapy introduced by a chemical, the drug can be released by using a thermoresponsive nanocarrier where heating induces swelling and opening of pores and releases the drug. Heating can be introduced by radio frequency (RF), magnetically, or by using light. These functionalities have been discussed in Chapter 5. Another type of externally controlled therapy is where the therapy is actually activated externally by using RF, magnetic, or optical field. The examples discussed in more detail are cancer therapy, discussed in Chapter 13, that involves magnetic therapy with magnetic nanoparticles or photodynamic therapy using light.

Finally, another major advantage of using nanotherapy is the ability to introduce multimodal therapy. A nanoformulation conveniently allows for simultaneous introduction of a number of functionalities as discussed in Chapter 5, each of which can be introduced to effect a specific mode of therapy in tandem. The combined therapeutic action can result in synergistic efficacy with lower incidence of adverse effects. Examples of this type of multimodal therapy are discussed in the chapter on nanotechnology for cancer therapy (Chapter 13) where chemo-, magnetic, photothermal, and photodynamic therapy can be combined to result in a significant advancement in the benefits of cancer therapy.

The two principal modes of nanoformulation for nanotherapy are (i) use of nanocarriers for drug delivery and (ii) nanoformulation itself acting as a therapeutic agent. These two modes are discussed in the next section, 11.2.

The two important factors that determine the overall therapeutic efficacies are:

- *Pharmacokinetics.* This describes the effect of the biological environment on a drug. This topic is discussed in detail in Section 11.3, and the biodistribution of a drug throughout the various organs, tissues and cells is covered in Section 11.4.
- *Pharmacodynamics.* This concept involves details of study of effect of a drug in the tissues and cells targeted. Section 11.5 provides a detailed discussion of pharmacodynamics.

11.2. MODES OF NANOFORMULATION FOR NANOPHARMACOTHERAPY

The two principal modes of nanoformulation are described in Figure 11.2. In one principal mode, a nanostructure is simply used as a nanocarrier to transport a drug effectively to the target biological site (e.g., cancer cells). Physical entrapment of a drug within a nanostructure has been extensively used for drug delivery. The main advantage of this approach is the simplicity of physical encapsulation, because no chemistry is needed for attachment to the nanocarrier. Another major advantage is that this mode of delivery does not involve any chemical modification of the drug. Thus regulatory approval (e.g., approval by the Food and Drug Administration in the United States) is simpler, if the approach of physical encapsulation involves an already approved drug. A major challenge in considering the physical encapsulation is the occurrence of uncontrolled release (leaching out) of the drug during systemic circulation, even before localization at the targeted site.

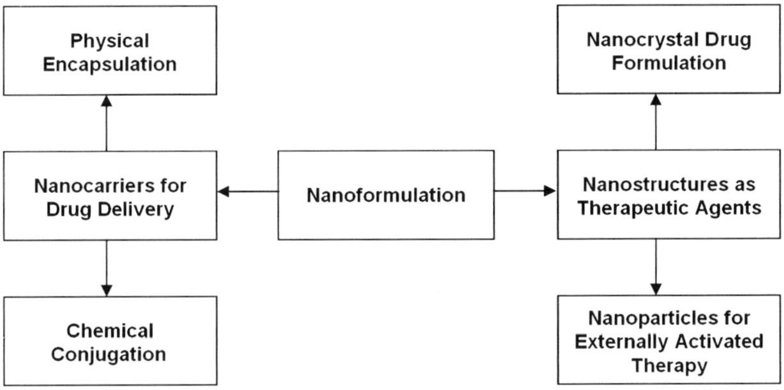

Figure 11.2. The principal modes of nanopharmacotherapy.

Chemical conjugation may require a chemical modification of the drug to provide a chemical binding site at which it can be attached to an appropriate chemically linkable group in the nanostructure. However, one has to ascertain that such a chemical modification does not alter the therapeutic property of the drug and also that the chemical conjugation can be cleared by some mechanism to release the drug, if it is necessary for therapeutic action. In some cases, such as photodynamic therapy, to be discussed in Chapter 13 for cancer treatment, it may be sufficient to have the drug chemically linked to the surface of a nanoparticle where it can have access to oxygen to generate the therapeutic action. The regulatory clearance for this type of nanoformulation can be more involved, as the chemically conjugated drug–nanoparticle system may be considered as a chemically different therapeutic agent than the drug alone.

In the other mode of nanoformulation, the nanostructures themselves perform an active therapeutic function and thus are not just a carrier. One example is a nanocrystal or a nanoparticle formulation of the drug itself. This method has wide utility in the usage of hydrophobic drugs that can be used in the form of a stable nanocrystal dispersion in an aqueous phase and be delivered to the targeted cells. In the intracellular environment, the hydrophobic regions provided by appropriate protein segments can then dissolve these nanocrystals to produce the free drug. As discussed in Chapters 4 and 13, we have demonstrated this approach for photodynamic therapy.

Another example of therapeutically active nanoparticles is where the functionality of the nanoparticle is used to externally activate a therapy. Magnetic therapy using magnetic nanoparticles in an applied magnetic field falls in this category. Photothermal therapy using gold nanoparticles in the presence of light of wavelength of their plasmonic resonance (see Chapter 5) is another example.

11.3. PHARMACOKINETICS

Pharmacokinetics, often abbreviated as PK, focuses on the processes that allow a drug to gain access to the inside of the body, and then the processes of biodistribution from the vascular space and clearance through metabolism, and/or excretion [Joseph et al., 2010; DiPiro et al., 2010]. It is often described by mathematical models that consider the time course of drug absorption into the plasma after a systematic delivery, drug distribution inside the body, and drug elimination from the body by metabolism and excretion, which is also summarized in Table 11.1. Thus PK describes the relationship between (a) the drug input defined by dosage form, dose, frequency, and administration route and (b) the drug concentration in the plasma and other body tissues achieved with time.

The two important parameters are the total clearance (Cl) and volume of distribution (V). These two independent parameters determine the maximum and minimum concentrations (in percentage of the administered dosage) of the drug reached in the plasma. The ratio of the Cl/V provides the elimination rate (k) and the half-life $t_{1/2}$ ($t_{1/2} = 0.693/k$) of drug elimination. For a one-compartment model, this term $t_{1/2}$ is the time at which the concentration in the plasma reduces to half the value of the maximum concentration. It thus provides a quantitative assessment of the time course of drug concentration decline in the plasma. For this reason, it is also called plasma half-life of the drug. For mathematical treatment of the time course in PK, very often a model involving a single or several interrrelated compartments is used. A simple one compartment model assumes that the drug distributes uniformly and instantaneously throughout all body areas, after a dose is administered, as is shown in Figure 11.3a.

In the one compartment model, the body is represented as a single compartment (one box) that consists of all body tissues and fluids. However, many drugs do not distribute instantaneously and uniformly everywhere in the body. For example, an intravenous injection will produce a pattern where the drug rapidly reaches highly profused organs such as liver and kidneys and then slowly distributes to other body tissues (peripheral), including the intended

TABLE 11.1. Description of Pharmacokinetics

Administration of drugs

↓ Absorption

Drug concentration in body fluid (plasma)

↓ Distribution

Drug distribution in body

↓ Metabolism and excretion

Drug elimination

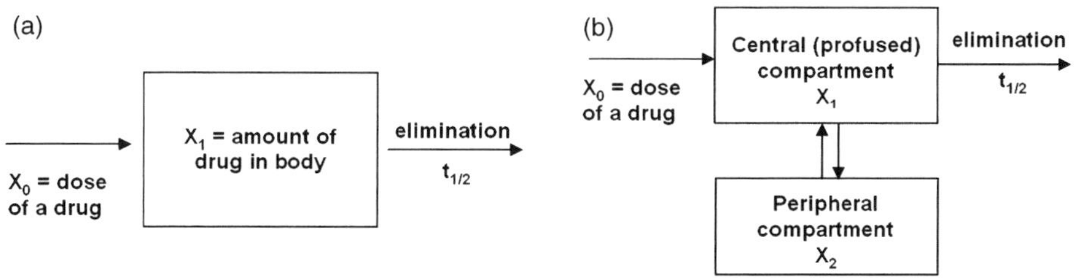

Figure 11.3. (a) One-compartment model. (b) Two-compartment model.

(targeted) biologic site. It may conform in this case to a two-compartment model as shown in Figure 11.3b. Here X_2 is the amount of drug in the peripheral compartment.

PK studies require blood sample collection over a series of timed intervals after the drug administration, along with measuring the amount of drug in the sample. For this purpose, radioactive labeling of the drugs has been frequently utilized. However, many drugs such as photosensitizers for photodynamic therapy (to be discussed in Chapter 13) are fluorescent. Thus a fluorescence technique can also be utilized, particular if fluorescence is in the biological optical transparency window of near-IR spectral range.

The use of nanoparticle delivery can provide significant advantage by producing highly favorable PK. As discussed in earlier chapters, the size and surface characteristics have a major influence on the performance of nanoparticles as drug carriers. First, encapsulation of hydrophobic drugs by nanoformulation leads to more extensive absorption and systemic availability. Also, liver metabolism is significantly reduced, thus leading to their prolonged circulation and hence increased $t_{1/2}$.

As has been discussed previously, the cut-off size for renal clearance is ~5 nm; thus the use of larger size nanoparticles can reduce the rate of renal clearance. Furthermore, surface modification with PEG (polyethylene glycol) prolongs circulation of nanoparticles. This feature has also been discussed earlier in relation to RES capture. The charge on the surface of the nanoparticles also plays an important role in avoiding RES capture, minimizing opsonization, and prolonging circulation time. The nanoparticle delivery can readily be tracked *in vivo* to study PK. We have used near-IR emitting quantum dots containing nanoparticles, systemically delivered through the tail vein of a mouse, for *in vivo* tracking of targeted nanoparticles [Yong et al., 2010].

11.4. BIODISTRIBUTION

Biodistribution describes the pattern of distribution of a drug through different tissues in the body. It thus provides very useful information on the extent of localization of a drug at the intended biological site in need of therapy.

Multiple factors contribute to the drug concentration in different tissues including plasma protein binding, hydrophobicity, and the affinity of a drug for membrane protein transporters that influence cellular uptake and efflux.

A study of biodistribution also can reveal the extent of success for a targeted scheme, discussed in Chapter 7, entitled "Biotargeting." Traditionally, the biodistribution profile is obtained by using a labeled (radioactively or fluorescently) drug that is administered in an animal. Then the animal is sacrificed, and various organs are harvested and analyzed for the drug by using the label signal quantification.

The nanoparticle technology provides the ability to produce the most favorable biodistribution for a therapy by means of the features of both active and passive (enhanced permeability and retention effect) targeting (Chapter 7). In addition, the capability of *in vivo* multimodal imaging, being readily incorporated in a nanoparticle platform, allows one to monitor in live animals the biodistribution pattern and assess the success of targeting.

11.5. PHARMACODYNAMICS

Pharmacodynamics (PD) is the next step after pharmacokinetics, because it deals with the effect of a drug on the body, whether therapeutic or an adverse reaction. Thus it provides information on the time course and the intensity of drug action in the body. PD relates the drug concentration in the body to both the intended effect (efficacy of a therapy) as well as unintended (side or adverse effects).

Drugs (D) produce the biological effects, in general, by binding to receptors (R), which are mostly proteins. These receptors may be enzymes, ion channels, cytoskeletal proteins, hormones, or neurotransmitters. In some cases, the receptors can also be DNA, RNA, or membranes. Thus the key steps in PD are as follows:

$$\text{D} + \text{R} \rightarrow \text{D–R complex} \rightarrow \text{Secondary intracellular signaling pathways} \\ \rightarrow \text{Biological effect}$$

The magnitude of a biologic response is, therefore, proportional to the number of D–R complexes. In this step of binding, the limiting factor is either (a) the number of the drug molecules (concentration of the drug) at the target site or (b) the number of receptor sites. In fact, PD may involve a number of intermediate steps, triggered by binding to a primary receptor, to provide pathways for a therapeutic action. Some of these steps may create a bottleneck toward producing the therapeutic action and may create an alternate pathway that produces unwanted effects. In addition, they may be the rate-determining step to yield the time constant for producing efficacy. Understanding the intermediate steps and their binding kinetics is an important aspect of PD.

A nanoparticle formulation is of significant value in favorably manipulating PD. Increasing the local concentration of the nanoparticles by targeting the desired site, grafting a biorecognition element on the nanoparticle surface to influence binding with the receptors, and affecting the therapeutic pathways as well as controlled and externally activated release of the drug, are some examples of the ways of controlling PD. Finally, surface modification of the nanoparticle can also be used to reduce the adverse effect of a drug and enhance its efficacy.

11.6. CONTROLLED RELEASE BY EXTERNAL ACTIVATION

An important feature of nanopharmacotherapy is controlled release of a drug which over the dosing period leads to sustainability of a therapeutic action. An appropriately designed nanoparticle also provides an opportunity for externally activated release. This release could be passive and dependent on tissue conditions, or it could be active where the release time and site are determined by a clinician. This section describes some of the methods that can be utilized to produce drug release.

pH-Controlled Release. A change of pH may lead to structural changes in the nanoparticle to induce a passive release of the carried drug. pH-responsive polymers serve as a protection of drug payload that enhances the drug release in an acidic environment (e.g., tumor site) [Sankaranarayanan et al., 2010; Lee et al., 2010]. Another example is the release of a genetic material, electrostatically bound to the nanoparticle surface, where the electrostatic binding is affected by a change of the pH. We have used this approach for intracellular delivery of DNA or siRNA, electrostatically bound to a nanoparticle. We introduced the use of gold nanorod–siRNA nanoplexes that silenced the dopaminergic signaling pathway in the brain for drug addiction therapy [Bonoiu et al., 2009]. The pH change is derived from the intracellular environment being different. This method is discussed in Chapter 14, entitled "Gene Therapy."

Release by Enzymatic Cleavage. A drug can be chemically conjugated to a nanoparticle by a linking group that can be cleaved by an appropriate enzyme overexpressed at the targeted biological site (e.g., tumor). Anticancer drug(s) can be loaded into stimulus–responsive polymers, either in the nanoparticle core or the shell domain [Singh et al., 2011]. An example of a passive release is the action of protease enzymes, overexpressed in many cancer cells, that can be used for cleaving an appropriate linker to release the drug [Jiang et al., 2004].

Thermally Activated Release. A thermoresponsive nanoparticle can be induced to actively release the drug by either (a) thermally swelling to widen pores for entrapped drug molecules to leach out or (b) by shrinking to increase extrusion of drug [Xu et al., 2008]. The heating can be affected by light using photothermal effect or by an AC magnetic field in the case where magnetic nanoparticles are co-encapsulated in the nanoformulation. For example, a particular wavelength of light is strongly absorbed at the plasmonic resonance of metallic nanoparticle so that release of the content can be controlled by thermal effect produced by an appropriate a laser source [Troutman et al., 2009; Hu et al., 2008].

Optically Triggered Release. In this case of active release, the drug molecule is chemically linked to a nanoparticle formulation by a photolabile group that can be photochemically cleaved [Hartner et al., 2007]. To reach a deep tissue site, it is desirable to use two-photon activation using a near-IR wavelength. We have utilized this method to release nitric oxide from a water-soluble nitrosyl complex [Zheng et al., 2008]. The nitric oxide (NO) release induced by one- or two-photon irradiation can be used as a photolabile drug for cancer therapy. This light-activated therapy is discussed in Chapter 13.

HIGHLIGHTS OF THE CHAPTER

- Pharmacotherapy is the treatment of diseases through the use of drugs and is thus responsible for ensuring a safe, appropriate, and economical use of drugs in patients.
- Nanopharmacotherapy deals with the utilization of nanoparticle technology for pharmacotherapy. Some of the key benefits of nanoopharmacotherapy are (i) increased dispersibility of drugs, particularly hydrophobic; (ii) enhanced drug stability against chemical and enzymatic degradation in body; (iii) specificity of drug delivery in targeted site; (iv) reduced toxicity of a drug; and (v) controlled and activated release of drugs.
- The two principal modes of nanoformulation for nanopharmacotherapy are: (i) use of nanocarriers for drug delivery and (ii) nanoformulation itself acting as a therapeutic agent.
- The main advantage of physical entrapment of a drug is the simplicity of physical encapsulation, because no chemistry is needed for attachment to the nanocarrier.
- Chemical conjugation of a drug requires a chemical modification of the drug to provide a chemical binding site at which it can be attached to an appropriate chemically linkable group in the nanostructure. One has to ascertain that such a chemical modification does not alter the therapeutic property of the drug.
- Pharmacokinetics (PK) describes the effect of the biological environment on a drug.
- PK focuses on the processes that allow a drug to gain access to the inside of the body, and then it describes the processes of biodistribution

from the vascular space and clearance through metabolism and/or excretion.

- Two important parameters, the total clearance (Cl) and volume of distribution (V), determine the maximum and minimum concentrations (in percentage of the administered dosage) of the drug reached in the plasma.
- The ratio of the Cl/V provides the elimination rate (k) and the half life $t_{1/2}$ ($t_{1/2} = 0.693/k$) of drug elimination.
- In the one compartment model of PK, the body is represented as a single compartment (one box) that consists of all body tissues and fluids. However, many drugs do not distribute instantaneously and uniformly everywhere in the body, in which case multicompartment models are used for PK.
- PK studies require blood sample collection over a series of timed intervals after the drug administration and measuring the amount of drug in the sample. For this purpose, radioactive labeling of the drugs has been frequently utilized.
- The use of nanoparticle delivery can provide significant advantage by producing highly favorable PK. The size and surface characteristics have a major influence on the performance of nanoparticles as drug carriers.
- The charge on the surface of the nanoparticles plays an important role in avoiding RES capture, minimizing opsonization, and prolonging circulation time.
- Biodistribution describes the pattern of distribution of a drug through different tissues in the body. It thus provides very useful information on the extent of localization of a drug at the intended biological site in need of therapy.
- The nanoparticle technology provides the ability to produce the most favorable biodistribution for a therapy by means of the features of both active and passive targeting.
- Pharmacodynamics (PD) deals with the effect of a drug on the body, whether therapeutic or an adverse reaction. It thus provides information on the time course and the intensity of drug action in the body.
- Drugs (D) produce the biological effects, in general, by binding to receptors (R) which are mostly proteins. The magnitude of a biologic response is proportional to the number of D–R complexes.
- A nanoparticle formulation is of significant value in favorably manipulating PD by: (i) increasing the local concentration of the nanoparticles by targeting the desired site, (ii) grafting a biorecognition element on the nanoparticle surface to influence binding with the receptors, and (iii) affecting the therapeutic pathways as well as controlled and externally activated release of the drug.

- An important feature of nanopharmacotherapy is controlled release of a drug, which over the dosing period leads to sustainability of a therapeutic action.
- A pH-controlled release is achieved by using a pH change at the biological site of action to induce a passive release of the carried drug from the nanoparticles.
- A release by enzymatic cleavage of a drug chemically conjugated to nanoparticles is produced by an appropriate enzyme, overexpressed at the targeted biological site.
- A thermally activated release can be achieved in a thermoresponsive nanoparticle by either (a) thermally swelling to widen pores for entrapped drug molecules to leach out or (b) shrinking to increase extrusion of the drug.
- The heating can be affected by light using photothermal effect or by an AC magnetic field producing magnetic heating.
- An optically triggered release is achieved by releasing the drug molecule chemically linked to a nanoparticle formulation by a photolabile group that can be photochemically cleaved.

EXERCISES

True or False

1. A hydrophobic drug can be administrated directly since it can be dissolved in the biological medium.

2. A tumor site is less acidic than the healthy tissue.

3. A nanoparticle's surface modified with polyethylene glycol can prolong the *in vivo* circulation time.

4. Gold nanoparticle can be used as a nanotherapy agent by conducting the external heat to the targeted site.

5. Heat of nanoparticle can be generated by the radio frequency, magnetic field, and light.

Multiple Choice

6. Which of the following CANNOT trigger the release of drug from the nanoparticle?

(a) Laser
(b) Strong magnetic bar
(c) Protease
(d) High temperature

7. Which of the following abbreviation describes a passive uptake of nanoparticle?
   (a) RES
   (b) EPR
   (c) PEG
   (d) PD

8. What does pharmacokinetics tell you?
   (a) The toxicity of drug
   (b) The amount of drug inside the body
   (c) The effect of drug to the body
   (d) The number of drug receptor in the body

9. Which of the following is NOT a benefit of nanopharmacotherapy?
   (a) Controlled drug release
   (b) Targeting
   (c) Biodispersion
   (d) Smaller size

Answers: 1. F; 2. F; 3. T; 4. T; 5. T; 6. b; 7. b; 8. b; 9. d

REFERENCES

Bonoiu, A. C., Mahajan, S. D., Ding, H., Roy, I., Yong, K.-T., Kumar, R., Hu, R., Bergey, E. J., Schwartz, S. A., Prasad, P. N., Nanotechnology approach for drug addiction therapy: Gene silencing using delivery of gold nanorod-siRNA nanoplex in dopaminergic neurons, *Proc. Natl. Acad. Sci. USA* **106**, 5546–5550 (2009).

Devalapally, H., Chakilam, A., Amiji, M. M., Role of nanotechnology in pharmaceutical product development, *J. Pharm. Sci.* **96**, 2547–2565 (2007).

DiPiro, J. T., Spruill, W. J., Wade, W. E., Blouin, R. A., Prueme, J. M., *Concepts in Clinical Pharmacokinetics*, American Society of Health-System Pharmacists, Bethesda, MD (2010).

Hartner, S., Kim, H. C., Hampp, N., Phototriggered release of photolabile drugs via two-photon absorption-induced cleavage of polymer-bound dicoumarin, *J. Polym. Sci. Part A: Polymer Chemi.* **45**, 2443–2452 (2007).

Hu, S. H., Tsai, C. H., Liao, C. F., Liu D. M., Chen, S. Y., Controlled rupture of magnetic polyelectrolyte microcapsules for drug delivery, *Langmuir* **24**, 11811–11818 (2008).

Jiang, T., Olson, E. S., Nguyen Q. T., Roy, M., Jennings, P. A., Tsien, R. Y., Tumor imaging by means of proteolytic activation of cell-penetrating peptides, *Proc. Natl. Acad. Sci. USA* **101**, 17867–17872 (2004).

Joseph, T. D., William, J. S., William, E. W., Robert, A. B., Jane, M. P., *Concepts in Clinical Pharmacokinetics*, American Society of Health-System Pharmacists, Bethesda, MD (2010).

Lee, S. M., Ahn, R. W., Chen, F., Fought, A. J., O'Halloran, T. V., Cryns, V. L., Nguyen, S. T., Biological evaluation of pH-responsive polymer-caged nanobins for breast cancer therapy, *ACS Nano* **4**, 4971–4978 (2010).

Sankaranarayanan, J., Mahmoud, E. A., Kim, G., Morachis, J. M., Almutairi, A., Multi-response strategies to modulate burst degradation and release from nanoparticles, *ACS Nano* **4**, 5930–5936 (2010).

Singh, N., Karambelkar, A., Gu, L., Lin, K., Miller, J. S., Chen, C. S., Sailor, M. J., Bhatia, S. N., Bioresponsive mesoporous silica nanoparticles for triggered drug release, *J. Am. Chem. Soc.* **133**, 19582–19585 (2011).

Troutman, T. S., Leung, S. J., Romanowski, M., Light-induced content release from plasmon-resonant liposomes, *Adv. Mater.* **21**, 2334–2338 (2009).

Xu, F. J., Li, J., Yuan, S. J., Zhang, Z. X., Kang, E. T., Neoh, K. G., Thermo-responsive porous membranes of controllable porous morphology from triblock copolymers of polycaprolactone and poly(*N*-isopropylacrylamide) prepared by atom transfer radical polymerization, *Biomacromolecules* **9**, 331–339 (2008).

Yong, K. T., Roy, I., Hu, R., Ding, H., Cai, H. X., Jing, Zhu., Zhang, X. H., Bergey, E. J., Prasad, P. N., Synthesis of ternary CuInS2/ZnS quantum dot bioconjugates and their applications for targeted cancer bioimaging, *Integrative Biol.* **2**, 121–129 (2010).

Zheng, Q., Bonoiu, A., Ohulchanskyy, T. Y., He, G. S., Prasad, P. N., Water-soluble two-photon absorbing nitrosyl complex for light activated therapy through nitric oxide release, *Mol. Pharm.* **5**, 389–398 (2008).

CHAPTER 12

THE HUMAN CIRCULATORY SYSTEM AND THERANOSTICS

The human circulatory system is composed of *the cardiovascular system*, including heart and interconnected blood vessels consisting of arteries, capillaries, and veins, and *the lymphatic system*, which is comprised of lymphatic capillaries, vessels, and lymph nodes. The cardiovascular system and the lymphatic system have already been briefly introduced in Chapter 2 (Section 2.9). Here these two systems are revisited and discussed in detail in relation to theranostics. The flow of blood in the human body involves a complex network of blood vessels and multilevel fluid dynamics to deliver oxygen and nutrients to various tissues and organs in the body, as well as to remove the waste from them. Thus, the study and manipulation of blood flow provide a valuable approach for assessment of the state of a specific tissue or organ and necessary intervention if needed. The lymphatic system regulates the flow of lymph, the interstitial fluid lying within the interstices of various tissues, unidirectionally toward the heart. The main functions of the lymphatic systems are (a) to aid the immune system in destroying pathogens and metastasized cancer cells and (b) as to filter wastes before returning the lymph to the heart. Thus the monitoring (mapping) of the lymphatic system and lymphatic targeting with nanoparticles provide an effective way to diagnose and treat cancer and various infections. This chapter introduces blood fluidics and describes therapeutics, based on monitoring and control of blood flow. Again, the emphasis

Introduction to Nanomedicine and Nanobioengineering, First Edition. Paras N. Prasad.

is on the application of nanotechnology to monitor and manipulate the blood flow. Section 12.1 provides an introduction to blood fluidics. Section 12.2 gives an account of the circulatory system-based disease profiling. Section 12.3 describes the various methods used to monitor blood flow and map out their dynamic distribution and passage through tissues of various organs. Section 12.4 discusses some potential therapeutic approaches using nanoparticles to manipulate blood flow. Section 12.5 describes applications of nanoparticles in lymph node mapping. Section 12.6 provides an account of lymphatic targeting of drugs using nanocarriers.

12.1. BLOOD FLUIDICS AND CARDIOVASCULAR SYSTEM

The cardiovascular system, shown in Figure 12.1, involves a complex closed-loop, three-dimensional network of vessels in the circulatory system to distrib-

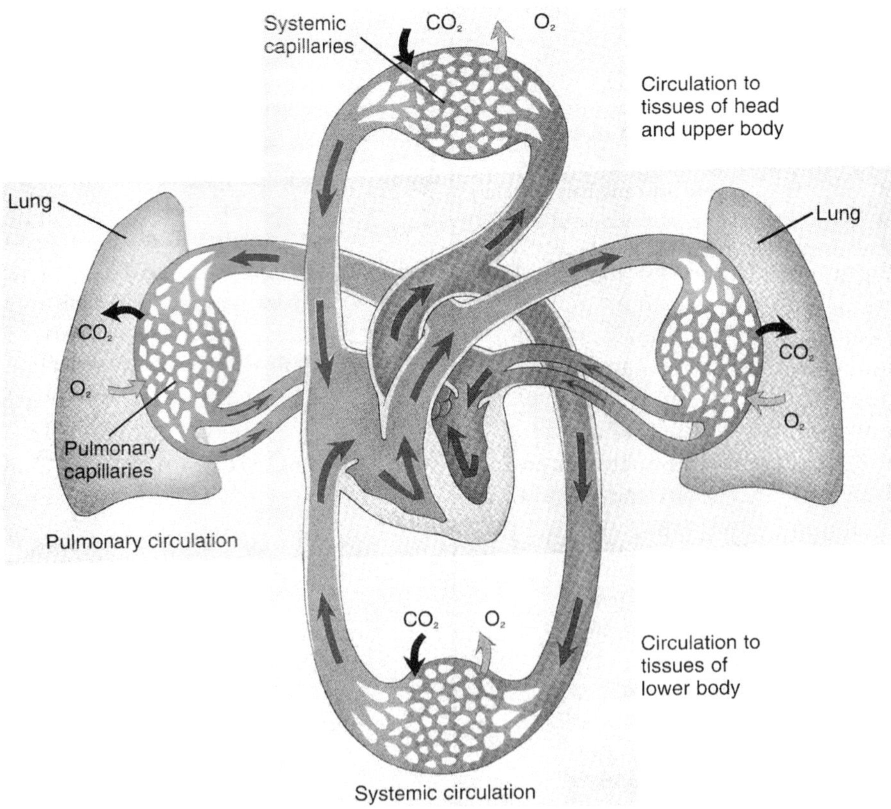

Figure 12.1. Schematics of the cardiovascular system. (Reproduced with permission from Thibodeau and Patton [2003]).

ute and recycle blood throughout the body. The dimensions of the vessels span multiple length scales, ranging from the largest artery (aorta), with a diameter of approximately 2 cm, to arteries and veins with diameters of several millimeters, to arterioles, venules and capillaries, which have diameters as small as tens of microns (see Table 12.1). In addition, blood consists of a heterogeneous suspension of red and white blood cells (erythrocytes and leukocytes, respectively) and platelets, which constitute a complex fluid known as blood plasma that also contains numerous organic (mostly proteins and lipoproteins) and inorganic solutes [Popel and Johnson, 2005].

Various types of blood cells have already been discussed in Chapter 2. These constituents of blood are of different shapes and sizes as described in Chapter 2, and they give rise to a complex flow dynamics. The high deformability of these cells and their ability to aggregate produce a non-Newtonian viscosity for blood. The hematocrit, which is the percentage by volume of packed red blood cells in a given sample of blood, is nominally 40–45%. The rheological properties of blood depend on many factors, including the diameter of the blood vessel, the flow velocity, the hematocrit, the finite size of the blood cells, their elastic properties, the aggregation and deformation of the blood cells and so on. Blood exhibits non-Newtonian fluidic behavior especially at very low shear rates, when the red blood cells clump together into larger particles. Blood also exhibits non-Newtonian behavior in the microvasculature, as the cells squeeze through small branches and capillaries and a cell-free skimming layer reduces the effective viscosity which is known as the Fahraeus–Lindquist effect. In smaller vessels with a diameter comparable to the size of the blood cells, blood is usually considered as a two-phase fluid. As the vessel diameter decreases further, the transport of RBCs and plasma are often divided into separate domains. For the smallest capillaries, flow becomes irregular and is determined by the transport of individual blood cells. The heart pumps blood to the circulatory system in a cyclic fashion, ejecting and filling itself with blood in alternating cycles called systole and diastole, respectively. This cyclic pumping creates a characteristic pulsatile pressure and flow in the aorta and arteries. Two dimensionless numbers are commonly used to characterize blood flow: the Reynolds number (*Re*)

$$\mathrm{Re} = \frac{d\bar{\mathbf{V}}\rho}{\eta} \tag{12.1}$$

and the Womersley number (α) [Womersley, 1955; McDonald, 1974],

$$\alpha = d\sqrt{\frac{2\pi f}{\eta}} \tag{12.2}$$

Here, $\bar{\mathbf{V}}$ is the average flow velocity, d is a characteristic length (e.g., diameter of the vessel), f is a characteristic frequency, and ρ and η are the density and viscosity of the fluid, respectively.

The Reynolds number is proportional to the ratio of the inertial force to the viscous force in a fluid and can be used to determine whether a flow is either laminar or turbulent. Viscous stress within a fluid acts to stabilize and organize the flow into a laminar pattern, while excessive fluid inertia acts to disrupt an organized flow pattern, transforming it into chaotic turbulent behavior. Fluid flows are laminar when Re < 2000 and are completely turbulent when Re > 4000. For Reynolds numbers between 2000 and 4000, the flow is transitory between laminar and turbulent. Typical Reynolds numbers for blood flow in humans range from ~1 in small arterioles to approximately 2500 in the aorta.

The Womersley number represents the ratio of the oscillatory inertial force to the shear force in a fluid. When $\alpha < 1$, the frequency of pulsation is low enough to allow a laminar parabolic velocity profile to develop during each cycle. In this case, the flow is nearly in phase with the instantaneous pressure gradient and can be predicted using Poiseuille's law. When $\alpha > 10$, the frequency of pulsation is large enough that the mean flow lags the pressure gradient by approximately 90 degrees and the corresponding velocity profile is relatively flat (plug flow).

The Reynolds and Womersley numbers can be used, in conjunction with clinical assessment of blood flow and pressure and cardiovascular imaging, to quantify hemodynamic behavior and performance of critical elements of the circulatory system. As noted above, these numbers are related to blood flow turbulence and velocity fluctuations, which can play a role in hemolysis, platelet activation, and thrombus formation. These numbers are used, for example, with measurements of flow and pressure across the aortic valve and through the aorta to quantify valve performance (e.g., onset and degree of stenosis) and ejection efficiency from the heart (Stalder et al., 2011). Table 12.1 lists nominal values of these parameters for different blood vessels.

TABLE 12.1. Reference Values for Mean Velocity $\bar{V}$, Diameter d, Shear Rate $\dot{\gamma}$, Reynolds Number Re, and Womersly Number α in the Human Vascular Network

Vessel	$\bar{V}$ (m/s)	d (mm)	$\dot{\gamma}$ (s^{-1})	Re (–)	α (–)
Aorta	0.4	25	130	2500	260
Arteries	0.45	4	900	450	40
Arterioles	0.05	0.05	8000	0.5	0.5
Capillaries	0.001	0.008	1000	0.002	0.1
Venules	0.002	0.02	800	0.01	0.2
Veins	0.1	5	160	125	50
Vena cava	0.38	30	100	2800	310

Source: Charm and Kurland, [1974].

12.2. CIRCULATORY-SYSTEM-BASED DISEASE PROFILING

Common impediments of blood circulation in several organs are discussed below.

Heart. The heart is a hollow muscular organ pumping blood into the vessels. The human heart has four chambers, two superior (atria) and two inferior (ventricles). The heart works in cycles, where the contraction (systole) is followed by relaxation (diastole). During the systole, blood is ejected from heart and pumped into both pulmonary and systemic circulation. During the diastole the ventricles are filled with blood in preparation for the next contraction phase. At the same time, the blood supply to the cardiac muscle occurs through the coronary arteries, which are compressed during the systole and therefore can be filled with blood only during the diastole. This mechanism of blood supply produces a lower pressure gradient in the coronary arteries and makes blood circulation vulnerable to interruptions.

A blockage of coronary blood flow causes coronary heart disease, which affects about 14 million men and women in the United States alone. The most common cause of disease is due to cholesterol plugs, which impede the blood flow in coronary arteries—a condition known as atherosclerosis. Significant disruption or blockage of the coronary blood flow results in heart attacks, frequently ending in death.

Pulmonary System. The pulmonary system is a complex and extensively branched network of blood vessels in lungs, in which blood flows from the heart to the lungs and then returns back to the heart. In the pulmonary circulation, the oxygen-depleted (venous) blood is supplied from the right ventricle of heart to the pulmonary trunk, which consequently branches out to pulmonary arteries, lobar arteries, and arterioles. Finally it passes through interweaving network of capillaries called pulmonary capillary beds, where an exchange of oxygen and carbon dioxide occurs, transforming the venous blood into arterial blood. In the lungs, the capillary beds are extremely developed and their total cross-sectional area is roughly equal to that in the rest of the circulatory system.

The pathologies in the pulmonary system include the following:

1. *Pulmonary Hypertension.* This results from constriction of the pulmonary arteries and is associated with an elevated pressure in the blood vessels of the lungs. This abnormally high blood pressure increases the load on the heart, which has to pump against a higher resistance to blood flow in pulmonary circulation.
2. *Pulmonary Embolism.* This is a condition that occurs when one or more pulmonary blood vessels are clogged. The usual cause for pulmonary embolism is a blood clog brought to the lungs by the bloodstream from

veins in legs. Occasionally, this disease can be caused by other substances, such as pieces of tumor, air bubbles, fat from the marrow of a broken bone, and others.

3. *Pulmonary Artery Aneurysm.* This relatively rare disease involves widening of the pulmonary blood vessels and is caused by structural cardiac and vascular abnormalities, pulmonary hypertension, infection, and blood vessel inflammation [Gleich and Weizenecker, 2005].

Brain. The brain in the central nervous system (CNS) is another organ whose functioning requires ample blood supply—about 15–20% of cardiac output. The blood enters the brain via two types of arteries: internal carotid arteries and vertebral arteries. The major arteries are connected in the brain with an arterial circle, known as the Circle of Willis. This interconnection enables a collateral blood flow in order to compensate for reduced blood supply from contributing arteries.

The disturbance of blood supply to the brain leads to damage of vulnerable brain tissues. In particular, an up-regulation of blood flow can raise intracranial pressure and mechanically damage the brain. On the other hand, a restricted blood flow to the brain, known as ischemia, could cause death of cells. Cerebral ischemia is a common cause of brain damage, partial loss of its functions, and ultimately death.

The Kidneys. The kidneys (renes) function as a filtering station by excreting metabolic waste from the blood. Despite their small size (about 4–5 inches long and 2–3 inches wide), they receive about 20% of the cardiac output [Gleich and Weizenecker, 2005]. The kidneys are connected to the blood supply through the renal arteries, which progressively branch into the arterioles and capillaries. Together, the two kidneys contain about 100 miles of blood vessels.

Similarly to processes that occur in blood vessels in the heart, renal circulation is vulnerable to atherosclerosis. Atherosclerosis leads to development of stenosis, a condition in which a blood vessel becomes abnormally narrowed. The renal circulation stenosis elevates blood pressure, which can damage blood capillaries and impair proper filtration, ultimately leading to a gradual loss of kidney function if not treated.

Besides the impact on specific organs outlined above, the blood circulation is routinely examined and monitored in relation to a diverse group of diseases. For instance, deep venous thrombosis is a condition when blood clots appear in deep veins of the legs, thighs, or pelvis. These clots interrupt the blood flow and can be a source of pain, swelling, and the inflammatory process, called thrombophlebitis. Furthermore, the blood clots can loosen up and travel through the bloodstream, causing embolism, such as pulmonary embolism, or cerebral ischemia outlined above.

The growth of a tumor is usually accompanied with the generation of a new blood vessel infrastructure to support needs of the tumor cells. The growth of the new blood vessels branching from already existing ones is called angiogenesis. This process is a prerequisite for tumor growth, because the delivery of essential nutrients and gases to the abnormally proliferating cancer cells is mediated by the newly formed blood vessel network. Moreover, the spread of cancer by metastasis occurs by the migration of metastatic cancer cells through the blood capillaries, which mandates significant efforts to study and monitor blood flow associated with tumors. Angiogenesis inhibitors have already been used in cancer therapies. Moreover, the endothelial cells of tumor-associated blood vessels are often more vulnerable to therapeutic treatment than cancer cells, which makes them attractive targets for novel cancer therapies.

The Lymphatic System. The lymphatic system, shown in Figure 12.2, is a part of the circulatory system, composed of lymphatic vessels and lymph nodes. The lymphatic system caries a fluid called lymph, which is derived from the blood. Lymphatic vessels eventually return the lymph back to the blood circulation. On the way to blood circulation, the lymph passes the lymph nodes, located in different areas throughout the body, including the neck, chest, armpit, abdomen, pelvis, and groin. The lymph nodes contain immune cells, which destroy various germs, pathogens, and tumor cells when these are present in the lymph. Assessment of lymph nodes is a routine step in the

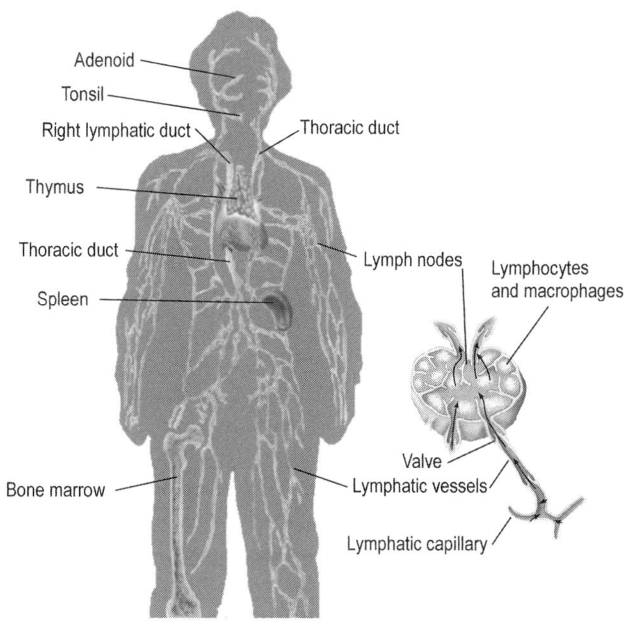

Figure 12.2. Diagram of lymphatic system, showing enlarged lymph nodes and capillaries.

treatment of both infectious diseases and cancer. Thus an active area of investigation is lymph node mapping using various imaging techniques. This topic is covered in Section 12.5. An emerging approach for evaluation of local immune response in distinct lymph nodes relies on uptake of nanoparticles by the macrophages, followed by MRI scans.

12.3. METHODS TO MONITOR BLOOD FLOW

From the discussions presented in Section 12.2, it is evident that characterization of blood flow is crucial for biomedical assessment of the condition of the cardiovascular system and the function of various organs and body parts. As described in Section 12.1, three different patterns of blood flow can be distinguished: (1) high Reynolds number flow, in large vessels such as the aorta and human larger arteries; (2) laminar flow, in arterioles, capillary, and venules; and (3) single-file flow in small capillaries, where the red blood cells (RBCs) are compelled to deform into a folded or parachute-like configuration.

The modern arsenal of the blood flow imaging techniques utilizes different imaging modalities. The traditional technologies for blood flow characterization and imaging include ultrasound and laser Doppler flowmetries [Aaslid et al., 1982; Dirnagl et al., 1989], Doppler OCT [Chen et al., 1997], magnetic resonance imaging [Calamante et al., 1999], positron emission tomography [Heiss et al., 1994], laser speckle imaging [Dunn et al., 2001], and photoacoustic imaging [Kim et al., 2010; Beard, 2011]. Many of these techniques can be significantly advanced with the use of nanotechnology, because systemically administered nanoparticles can be transported by the blood stream within the circulatory system to serve as exogenous contrast agents for different imaging techniques. In another approach, subpopulation of blood cells, such as macrophages, erythrocytes, or platelets, can be stably labeled by association with nanoparticles and imaged selectively in the blood vesicles. This approach is invaluable for identification, mapping, and characterization of inflammatory processes and dynamics of the blood circulation. Some principal methods for determination of blood flow are described here, along with the prospect of nanoparticle-based contrast enhancement.

Doppler Ultrasound Flowmetry. Doppler ultrasound flowmetry is a widely used technique for evaluation of blood flow through major arteries and veins in the arms, legs, and neck [Aaslid et al., 1982; Brown et al., 1989]. This method employs emitted and reflected sound waves to see how blood flows through a blood vessel. This method can determine blocked or reduced blood flow due to narrowing in the major arteries of the neck that could cause a stroke. It can also detect blood clots in leg veins (deep vein thrombosis), which may break loose and travel to a pulmonary artery in the lung, causing blockage of the blood flow to the lungs (pulmonary embolism). During the Doppler ultrasound test, a hand-held transmitter/transducer of sound waves is applied to the skin above a blood vessel, sending and receiving sound waves, which are bounced off flowing blood cells. The movement of blood cells causes a change in the pitch of the reflected sound waves (the Doppler effect). This change in the reflected sound waves can be analyzed using a computer, providing information about the blood flow through the blood vessels.

A new ultrasonic imaging modality, referred to as power doppler harmonic imaging (PDHI), color-encodes changes in the acoustic backscatter signal amplitude and the motion of ultrasonic scatterers between insonifying pulses to obtain clear and accurate anatomical images. PDHI is particularly adept at imaging contrast agents, because harmonic echoes produced by nonlinear contrast agents are much higher than that produced by blood and tissue [Ngo et al., 2000]. Ideally, a contrast agent should demonstrate long circulatory persistence and resistance to ultrasonic destruction while displaying significant echogenicity. The majority of systemic contrast agents are microbubble-based formulations designed to provide blood pool enhancement. They provide excellent echogenicity, but a short-lived effect as microbubbles are susceptible to destruction by ultrasound and are rapidly eliminated by the lungs. New-generation ultrasound contrast agents consist of gaseous perfluorocarbon (PFC) bubbles encapsulated within polymer shells. Perfluorocarbon (PFC) nanoparticles, which are a combination of a poorly water-soluble fluorinated gas encapsulated within a biodegradable and nontoxic polymer, have been shown to exhibit long circulation times and stability to insonification, albeit with poor echogenicity [Ngo et al., 2000]. Recently, polymeric contrast agents have been developed [Pisani et al., 2006, 2008] substituting the gaseous PFC by a liquid one. These new ultrasound contrast agents consist of nanocapsules of poly(lactide-co-glycolide) (PLGA) encapsulating a liquid core of perfluorooctyl bromide (PFOB). Liquid PFOB yields capsules that are more resistant to pressure changes and mechanical stresses. They have shown promises in enhancing the blood pool contrast [Diaz-Lopez et al., 2010]. In nanoemulsions, PFOB has been used as a blood pool ultrasound contrast agents for the ventricular function and wall motion abnormalities. PLGA was chosen, because it is approved by the FDA for use in sutures and drug delivery devices. Also, it is biodegradable and has good biocompatibility [Shive et al., 1997; Diaz-Lopez et al., 2010]. Additionally, the polymeric shell improves the stability of the capsules as compared to ultrasound contrast agents stabilized by a monomolecular layer of surfactant [Pisani et al., 2008; Diaz-Lopez et al., 2010]. Figure 12.3 presents ultrasound images of mouse tumors obtained in a nonlinear imaging mode (tissue harmonic imaging, THI) using PFOB/PLGA core–shell nanocapsules as imaging contrast agents.

Laser Doppler Flowmetry. Laser Doppler flowmetry (LDF) is a noninvasive method to measure the microcirculatory blood flow in neural, muscle, skin, bone, and intestine tissues [Obeid et al., 1990; Dirnagl et al., 1989]. Similarly to Doppler ultrasound flowmetry, the LDF utilizes Doppler shift (the frequency change that waves undergo when reflected by moving objects—red

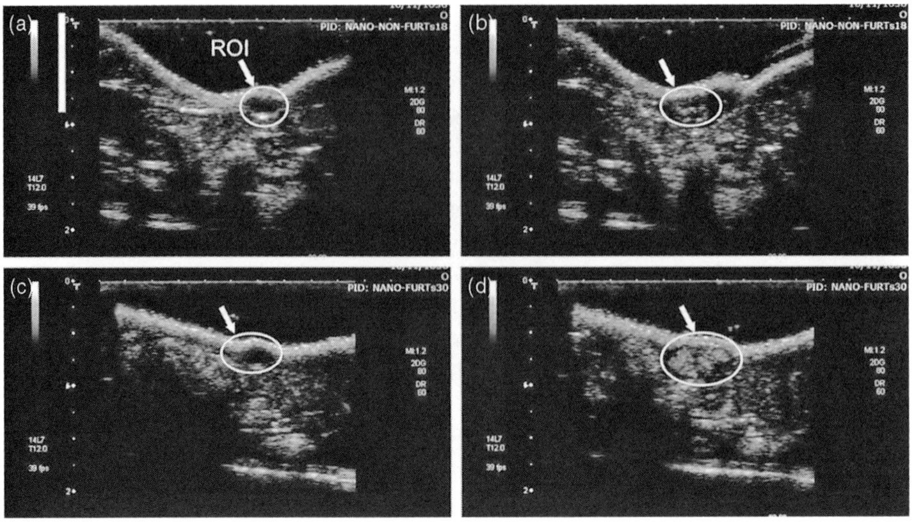

Figure 12.3. Ultrasound images of mouse tumors obtained in a nonlinear imaging mode (THI) before injection (a, c) and after intra-tumoral injection of 200 mL of a 50 mg mL/L nanocapsules. Suspension, plain nanocapsules (b) and PEGylated nanocapsules (d). The tumor is indicated as the region of interest (ROI) represented by a circle. Detection of nanocapsules in nonlinear imaging mode (tissue harmonic imaging THI) implies that a nonlinear response is present. (Reproduced with permission from Diaz-Lopez et al. [2010].)

blood cells in the case of blood flow), but light waves are employed instead of ultrasound ones. The emitted light from a low-power laser source is reflected by the target and then fed to a processing computer. Measurement of the RBC motion is performed continuously in the outer layer of the studied tissue, with little or no influence on physiologic blood flow. The result of measurement shows the flux of red cells, defined as the number of red cells multiplied by their velocity, and is reported as microcirculatory perfusion units. Systemically administrated nanoparticles, such as gold nanorods or one with high refractive index, can be used as contrast enhancers in LDF, exploiting reflection of light in the near infrared (NIR) spectral range, where RBC scattering is optimal and background tissue scattering is minimal.

Doppler OCT. Doppler OCT (optical Doppler tomography, ODT) provides flow velocity measurement in a sample using Doppler frequency shift as the contrast mechanism in the OCT imaging modality [Chen et al., 1997]. OCT was discussed in Chapter 8, "Multimodal Biomedical Imaging." The high spatial resolution of OCT and high velocity sensitivity of Doppler OCT provide a powerful tool for mapping fluid flow, allowing structural imaging along with flow velocity measurement. Typical applications of Doppler OCT include studying embryo cardiac dynamics in developmental biology, vascular pathology and treatment response, and flow velocimetry in micro-fluidic channels [Yang et al., 2003; Yazdanfar et al., 2003]. Nanoparticles that provide an enhanced contrast in OCT (i.e., gold nanoparticles) would be evidently capable of enhancing the contrast in ODT.

Speckle Method. When a diffuse object is illuminated with coherent laser light, the backscattered light will form an interference pattern on the detector. This pattern of bright and dark areas is called a speckle pattern. When there is movement in the object, the speckle pattern will change over time fluctuations in their intensity. These fluctuations can be used to provide information on the movement of the object [Draijer et al., 2008]. A simple way of accessing this information is to image the speckle pattern, where the fluctuations cause a blurring of the speckle, leading to a reduction in the local speckle contrast. Thus velocity distributions are coded as speckle contrast variations. The same information can be obtained by using the Doppler effect, but producing a two-dimensional Doppler map requires scanning; speckle imaging provides the same information without the need to scan [Briers, 2007]. Laser speckle contrast techniques use this change in speckle pattern to visualize blood flow and tissue perfusion.

Photoacoustic Imaging. Photoacoustic (PA) imaging, already discussed in Chapter 8, is a hybrid imaging modality, combining the high-contrast and spectroscopic-based specificity of optical imaging with the high spatial resolution of ultrasound imaging. The PA image can be regarded as an ultrasound image in which the contrast depends not on the mechanical and elastic properties of the tissue, but also on its optical properties, specifically optical absorption [Beard, 2011]. As a consequence, it offers greater specificity than conventional ultrasound imaging, with the ability to detect hemoglobin and other light-absorbing chromophores in a blood flow. Since PA signal does not depend on light scattering but only on absorption, it allows greater penetration depth than purely optical imaging modalities, which are strongly affected by tissue light scattering.

As discussed in Chapter 8, a variety of targeted contrast agents, which absorb at visible and NIR wavelengths, can provide additional sources of spectrally selective PA contrast and be used to image and characterize blood flow. Examples of nanoparticle contrast enhancers, also already discussed in Chapter 8, are metallic nanorods or shells, but are currently limited to preclinical use [Beard, 2011].

Magnetic Particle Resonance Tomographic Imaging. Magnetic particle imaging (MPI) is a tomographic imaging technique that enables real-time high-resolution 3D imaging of magnetic nanoparticles flowing in blood vessels (Figure 12.4). The operation principle of MPI is based on the nonlinear magnetization response that is typical for iron-oxide-based magnetic nanoparticles. Magnetic nanoparticles are already clinically approved as contrast agents for

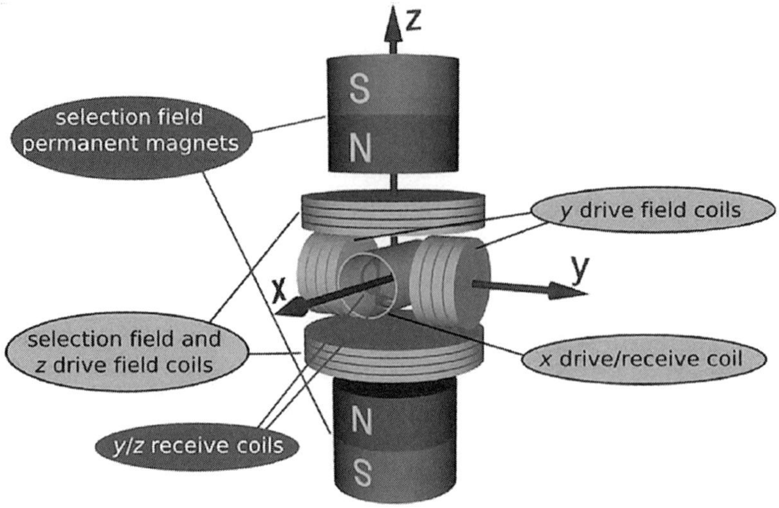

Figure 12.4. MPI imaging setup. Mouse specimen supported in x drive/receive coil cylinder (32-mm diameter). The selection field is generated by both permanent magnets and paired coils in the z direction. Drive field coils can move the single field-free point (FFP) in all three spatial directions. (Reproduced with permission from Weizenecker et al. [2009].)

magnetic resonance imaging (MRI) following intravenous injection, as already discussed in Chapter 8. They would also be suitable for MPI [D.E. Markov et al., 2010]. The MPI concept is different from MRI because it relies on the nonlinearity of the magnetization curves of ferromagnetic material and on the fact that the particle magnetization saturates at some magnetic field strength. If an oscillating magnetic field, called the "modulation field," is applied with frequency f_1 and sufficiently high amplitude A, the magnetic material will exhibit a magnetization $M(t)$, where t is time. $M(t)$ contains not only the drive frequency f_1, but also a series of harmonic frequencies. These higher frequencies can be easily separated from the received signal by means of appropriate filtering [Gleich and Weizenecker, 2005]. In other words, when subjected to an oscillating magnetic field with frequency f_1, the nanomaterial generates an output signal at odd multiple harmonics $n{\cdot}f_1$ of the base frequency. Analysis of the spectral intensity of these harmonics results in a highly sensitive method of detection of the magnetic material. Spatial encoding is realized by using a coil arrangement that allows for a field-freepoint motion across the otherwise magnetically saturated field of view. In a basic MPI system, as shown in Figure 12.4, a magnetic gradient field is being created in such a way that there is only one point in the 3D imaging volume where the magnetic field is zero. This spot is called the field-free point (FFP) and only at this point the magnetization of the nanoparticles, if present, can be changed by a superimposed oscillating magnetic field. Outside the field-free point, the nanoparticles are in magnetic saturation and no change in magnetization occurs upon the application of an oscillating magnetic field. The change in magnetization, induced in the field-free point, can be detected using receive coils. The detected MPI signal is proportional to the concentration of the nanoparticles in the field-free point. To turn this finding into an imaging method, the field-free point is moved around until the complete imaging volume is covered. Detection and reconstruction of the resulting signals lead to volumetric images of the concentration of the nanoparticles in the 3D imaging volume. Thus, this method results in images with high spatial and temporal resolution as well as high contrast, which scales linearly with the tracer concentration [Gleich and Weizenecker, 2005; Weizenecker et al., 2009; Markov et al., 2010]. MPI holds great promise for vascular imaging and could be the modality of choice for the diagnosis of coronary artery diseases.

Figure 12.5 presents an example of the MPI imaging of mouse cardiovascular system, acquired in combination with MRI. Such a combination is shown to provide complementary information on the circulatory system [Weizenecker et al., 2009].

It is important to note that the noticeable MPI signal is mainly generated by particles with minimum diameter of magnetic core of 30 nm [Gleich and Weizenecker, 2005]. When coated with dextran derivatives for stabilization, these iron oxide particles have an average hydrodynamic diameter of 50–60 nm. This results in a short blood half-life of several minutes due to rapid excretion via the reticuloendothelial system (RES), which in fact limits the applicability of such compounds in MPI. The use of longer blood half-time tracer materials would make MPI highly suitable for perfusion imaging and image-guided drug delivery (interventional) applications that require repeated

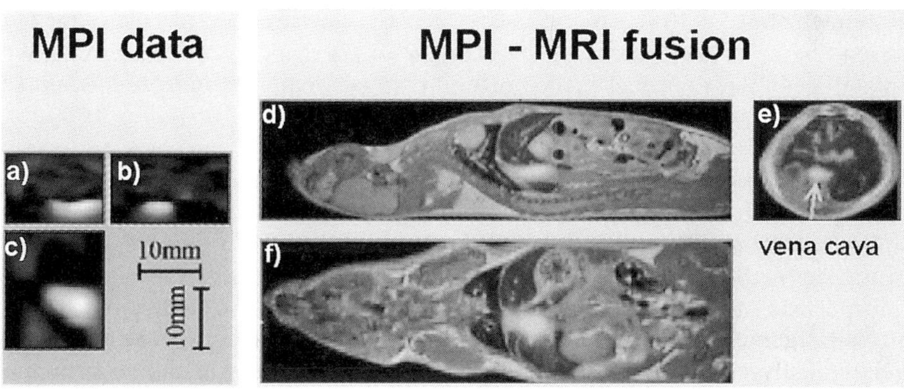

Figure 12.5. MPI/MRI imaging of the mouse cardiovascular system. The left-hand side shows dynamic MPI images obtained ~7 after bolus injection, while the right-hand side shows their fusion with static MRI images. Pairs (a, d), (b, e) and (c, f) show the images in the same planes with orthogonal orientations. The spatial and temporal resolution enable resolution and identification of all heart chambers as well as parts of the vessel tree. (Modified with permission from Weizenecker et al., 2009.)

imaging and/or continuous monitoring. To enable this type of *in vivo* application, the tracer needs to be modified while ensuring particle stability to RES in terms of biocompatibility and nontoxicity. It has been suggested that cell carriers such as erythrocytes are promising candidates for circumventing these limitations [Pierige et al., 2008, Rossi et al., 2005]. Recently, it has been demonstrated that autologous red blood cells (RBCs) can be successfully utilized to encapsulate magnetic nanoparticles to address biocompatibility and to avoid rapid clearance of the MRI contrast agent [Antonelli et al., 2008, Markov et al., 2010]. Such biomimetic constructs preserve the main properties of the native cells that escape the RES clearance, as well as the properties of the nanoparticles required to demonstrate great MRI performance.

12.4. THERAPEUTIC APPROACHES UTILIZING MANIPULATION OF BLOOD FLOW

Magnetic Control of Blood Flow. A potential application of magnetic particles involves their use in controlling blood flow. A sufficient concentration of biocompatible magnetic NPs can be injected into the bloodstream to enable the control of blood flow using an external field source as shown in Figure 12.6. The nanoparticles would aggregate in the presence of a sufficiently strong magnetic field (and field gradient), which could be used to slow down or completely block blood flow. This technique can be applied to control bleeding from sites located on the extremities and/or from lacerations involving the femoral or axillary vessels, and also from major vascular or visceral injuries in the body cavities. The main advantage of this approach is that it could selectively arrest bleeding locally, only at the site of an injury or lesion without affecting other healthy areas. As such, it could be an effective treatment for tumors by selectively blocking blood flow to the cancerous tissue. Another advantage of this approach is the portability of permanent magnet field sources and low cost.

Monocyte Delivery of Nanoformulations. Monocytes, a type of white blood cell, are a part of a human body's immune system which move quickly to the site of inflammation or infection to provide an immune response. They are produced by the bone marrow from monoblasts circulating in the bloodstream for approximately one to three days.

Nanoformulations of anti-inflammatory drugs (either pure form or loaded in a biodegradable nanoparticle) can be targeted to the antigens/receptors on the surface of monocytes to be attached, and then they can be transported by monocytes for action at the site of inflammation to increase the local concentration of the anti-inflammatory drug as well as to provide sustained release [Batrakova et al., 2011; McMillan et al., 2011].

Another application of the nanomedicine approach is to use nanoparticle-based gene modulation, discussed in Chapter 14 ("Gene Therapy"), to directly

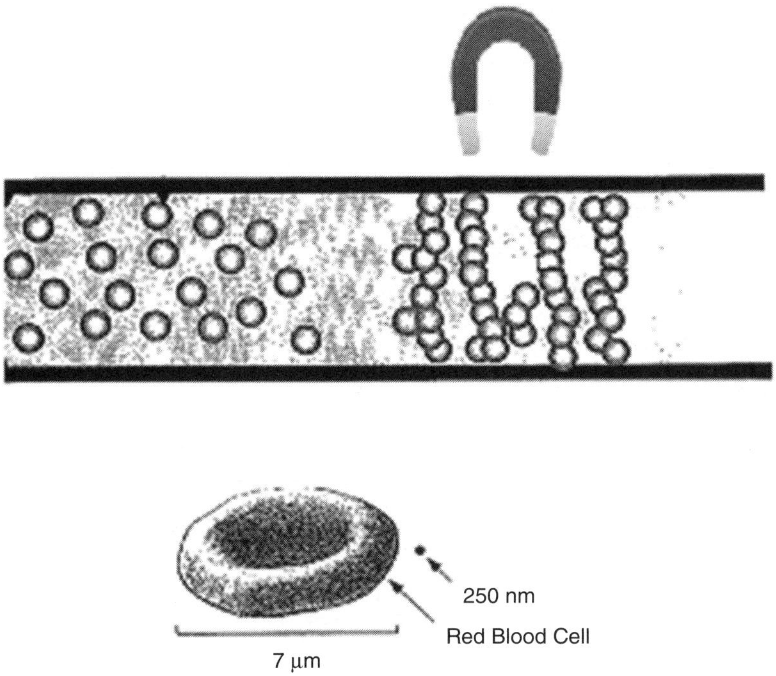

Figure 12.6. Magnetic tourniquet of nanoparticles: blood vessel carrying blood cells and magnetic nanoparticles. Nanoparticles aggregate to block blood flow in the presence of an applied field. (Reproduced with permission from Kotha and Sudarshan [2010].)

activate monocytes for increased production of anti-inflammatory cytokines such as TNFs and interleukin.

Macrophage Trafficking of Nanoformulations. Macrophages are produced by differentiation of monocytes in tissues to become tissue specific. Their primary function is to eliminate foreign substances, infectious microbes and cancer cells by attacking, destroying and ingesting them, a process called phagocytosis. These macrophages can be utilized for trafficking of nanoformulations of antimicrobial and antiviral drugs [Batrakova et al., 2011]. These macrophages can easily uptake nanoparticles with characteristics similar to a microbe or a virus, even without any specific protein (antibody) coating on the surface, and can internalize in a subcellular compartment such as endosomes, which is the site of localization of many microbes and viruses. An advantage of the macrophage delivery is that they are able to cross biological barriers in response to cytokine signaling, and deliver a nanoformulation directly to infected tissues. An example is the macrophage delivery of antiviral nanoformulation used by Gendelman and co-workers to cross the blood–brain barrier [Kadiu et al; 2011]. This macrophage delivery method of antimicrobial nanoformulation is detailed in Chapter 15, entitled "Nanotechnology for Infectious Diseases."

As Gendelman and co-workers point out, cells of mononuclear phagocyte lineage which include monocytes, macrophages, and dendritic cells can be used as Trojan horses to carry both anti-inflammatory and anti-infective medicines [McMillan et al., 2011].

12.5. LYMPH NODE MAPPING

A major pathway of spread of cancer is via the lymph nodes. Thus, lymph node mapping is valuable for diagnosis of the metastatic stage of cancer. The sentinel lymph node (SLN) is the first lymph node to drain the primary tumor, because the orderly progression of cancer is from the primary tumor to the SLN and then onto echelon lymph nodes. Hence a major focus has been placed on mapping of SLN [Ravizzini et al., 2009]. The status of SLN by mapping is useful as a part of surgical management, particularly for treatment of melanoma, breast, cervical, and head and neck cancers.

CT, MRI, and PET, all discussed in Chapter 8, have been used to identify abnormally enlarged lymph nodes caused by the presence of cancer [Luciani et al., 2006]. However, these imaging modalities are not suitable to determine which node is SLN.

Imaging of SLN using nanoparticles provide a number of benefits. The nanoparticles can be injected near the primary tumor, whereby they can be picked up by the adjacent lymphatics and transported to the SLN. The size, composition, surface charge, and surface coatings of nanoformulation can be manipulated to significantly influence the lymphatic uptake of subcutaneously injected nanoparticles [Nune et al., 2011]. A systematic study of the size dependence shows that particles greater than 100 nm remain largely at the injection site, while those between 10 and 80 nm are well taken up by the lymphatics. Particles smaller than 10 nm are absorbed by the lymphatic capillary network draining into the systemic circulation. The two approaches currently in use in the United States for the SLN imaging utilize isosulfan or "blue dye" for optical imaging and radionuclide-labeled sulfur colloid for radio imaging, with both of these imaging agents localizing in the SLN. However, both rely on presupposing the general region of the SLN. Other issues related to their usage are discussed in detail by Ravizzini et al. [2009]. In the case of the radionuclide method, limited resolution and exposure to radiation are other concerns.

Thus in search for newer imaging methods for lymph node mapping, nanoparticle-based imaging has received considerable attention. The two modalities of imaging where the nanoparticle technology for lymph node mapping has significantly impacted are MRI-imaging using magnetic nanoparticles and optical imaging using quantum dots. MRI-based imaging of lymph nodes, also called magnetic resonance lymphography (MRL), has utilized Gd-containing liposomes in animal models [Fujimoto et al., 2000]. The use of Gd-containing nanoparticles for MRI imaging has already been discussed in Chapter 8. Subcutaneous injection of these liposome nanoparticles results most likely in their trapping by macrophages, leading to their uptake by the lymph nodes. Ultrasmall supermagnetic ion oxide nanoparticles have also been utilized for lymph node mapping and detection of modal metastates [Hoffman et al., 2000]. A particular variant used for detection of SLN is dextran-coated iron oxide nanoparticles, Ferumoxtran-20 [Torchia et al., 2001]. Gd-containing dendrimers have also been used to image the lymphatic system and the SLNs of normal mice and mice with xenografted breast tumors [Kobayashi et al., 2004]. The smaller-sized dendrimers accumulate and are retained by lymph nodes.

Optical imaging using near-infrared (NIR) emitting nanoparticles have also been utilized for the SLN mapping. In this regard, quantum dots (QDs) have received considerable attention. The advantages of using QDs for optical imaging have already been detailed in Chapter 8. Kim et al. [2004] used NIR Cd-based QDs for *in vivo* imaging of SLN in breast tissue. Simultaneous multicolor imaging of five different lymphatic basins using five NIR QDs have also been reported by taking advantage of narrow emission profiles of QDs [Kobayashi et al., 2007].

Because of the toxicity concern with the Cd-based QDs, recent efforts have moved toward the application of heavy metal-free QDs. Bawendi and coworkers used NIR emitting core/shell/shell $InAs_xP_{1-x}/InP/ZnSe$ QDs for sentinel lymph node mapping [Zimmer et al., 2006]. Another Cd free QD reported for SLN mapping is $CuInS_2/ZnS$ [Pons et al., 2010]. In our group, we introduced the use of silicon QDs for SLN mapping [Erogbogbo et al., 2011]. Figure 12.7 demonstrates the application of nontargeted Si QDs with hydrodynamic radii as small as 20 nm that were tailored for SLN mapping. When

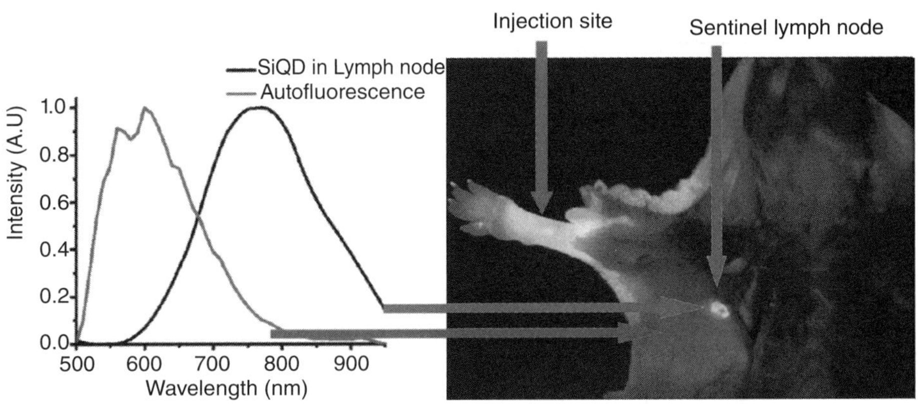

Figure 12.7. Sentinel lymph node imaging following localization of Si QDs in an axillary position. (Reproduced with permission from Erogbogbo et al. [2011].)

injected subcutaneously in the paw of a mouse, QDs traveled through the lymphatics and migrated to an axillary location that could be detected using the Maestro imaging system. The localization of Si QDs indicates the position of the SLN. Multiphoton excitation properties of the NIR Si QDs, along with reports of Si QD QY in the NIR above 60%, suggest that Si QDs are very suitable for SLN mapping.

Another direction in lymph node mapping is the simultaneous use of dual multimodalities such as radionuclide and optical imaging which can overcome the sensitivity limitation and depth limitations of the individual method [Kobayashi et al., 2007].

12.6. LYMPHATIC DRUG DELIVERY

Delivery of drugs to the lymphatic system can more effectively provide chemotherapeutic drugs to the regional lymph nodes, and by accumulation these can prevent the metastasis of tumor cells [Nishioka and Yoshimo, 2001; Nune et al., 2011]. The advantage of lymphatic delivery is that it avoids the RES capture of nanocarriers in the liver, which generally occurs in a systemic injection of nanoparticles into the bloodstream. The lymphatic delivery can be made by oral administration, as well as by intramuscular or subcutaneous injection.

The lymphatic system is, however, not easily accessible for delivery of drugs into the lymphatic tissues and the lymph nodes. As mentioned in Section 12.5, nanoparticle characteristics can be judicious optimized for enhanced lymphatic uptake. In addition, specific markers on tumor lymphatics can be utilized for targeting of nanoparticles carrying chemotherapeutic agents [Laakkonen et al., 2002]. Furthermore, by involving receptor–ligand or antigen–antibody interactions between the nanoparticle surface and the dendritic cells, which, when activated, can migrate to the lymph nodes, one can achieve lymphatic targeting. Drug-carrying nanoparticles, coupled with multimodal imaging capability, thus can be used for theranostics to diagnose and treat lymph node metastasis.

The review by Nune et al. [2011] provides examples of many types of nanoparticles used for lymphatic drug-delivery. Biodegradable nanoparticles of poly(lactide-co-glycolide) is an example used to deliver drugs and imaging agents to the lymphatic system. Multi-walled carbon nanotubes decorated with magnetic nanoparticles have also been utilized [Yang et al., 2009].

HIGHLIGHTS OF THE CHAPTER

- The human circulatory system is composed of the cardiovascular system including heart and interconnected blood vessels: arteries, capillaries and veins, and the lymphatic system.

- The lymphatic system regulates the flow of lymph, the interstitial fluid lying within the interstices of various tissues, unidirectionally toward the heart.
- The main functions of the lymphatic systems are to aid the immune system in destroying pathogens and metastasized cancer cells as well as to filter wastes before returning the lymph to the heart.
- Monitoring of the lymphatic system and lymphatic targeting with nanoparticles provide an effective way to diagnose and treat cancer and various infections.
- In the cardiovascular system the dimensions of the vessels span multiple length scales, ranging from the largest artery (aorta), with a diameter of approximately 2 cm, to arteries and veins (with diameters of several millimeters), to arterioles, venules, and capillaries, which have diameters as small as tens of microns.
- The high deformability of erythrocytes and leukocytes and their ability to aggregate give rise to a non-Newtonian viscosity of blood.
- Blood exhibits non-Newtonian fluidic behavior especially at very low shear rates, when the red blood cells clump together into larger particles.
- The Fahraeus–Lindquist effect describes the non-Newtonian behavior of blood in the microvasculature as the cells squeeze through small branches and capillaries and a cell-free skimming layer reduces the effective viscosity.
- As the vessel diameter decreases, the transport of red blood cells and plasma are often divided into separate domains. For the smallest capillaries, flow becomes irregular and is determined by the transport of individual blood cell.
- Reynolds number (Re), which is commonly used to characterize blood flow, is proportional to the ratio of the inertial force to the viscous force in a fluid and can be used to determine whether a flow is either laminar or turbulent.
- Womersley number (α), another characteristic of blood flow, represents the ratio of the oscillatory inertial force to the shear force in a fluid.
- The Reynolds and Womersley numbers can be used in conjunction with clinical assessment of blood flow, pressure, and cardiovascular imaging, to quantify hemodynamic behavior and performance of critical elements of the circulatory system.
- Heart, a hollow muscular organ pumping blood into the vessels, is supplied with blood through the coronary arteries and coronary veins.
- A blockage of coronary blood flow causes coronary heart diseases. The most common cause of disease is due to cholesterol plugs, which impede the blood flow in coronary arteries—a condition known as atherosclerosis.

- The pulmonary system involves a complex and extensively branched network of blood vessels in lungs, in which blood flows from the heart to the lungs and then returns back to heart.
- In the pulmonary circulation, the oxygen-depleted blood is supplied from the right ventricle of the heart to the pulmonary trunk, which consequently branches out to pulmonary arteries, lobar arteries, and arterioles.
- An exchange of oxygen and carbon dioxide occurs in an interweaving network of capillaries called pulmonary capillary beds.
- Pulmonary embolism is a condition that occurs when one or more pulmonary blood vessels are clogged, which is commonly caused by the blood clog brought to the lungs by the bloodstream from veins in legs.
- The widening of the pulmonary blood vessels, which can be caused by structural cardiac and vascular abnormalities, pulmonary hypertension, infection, and blood vessel inflammation, is known as pulmonary artery aneurysm.
- The blood enters the brain via two types of arteries: internal carotid arteries and vertebral arteries. The major arteries are connected in the brain with an arterial circle, known as the Circle of Willis.
- The disturbance of blood supply to the brain leads to damage of vulnerable brain tissues. In particular, an up-regulation of blood flow can raise intracranial pressure and mechanically damage the brain.
- Kidneys function as a filtering station by excreting metabolic waste from the blood and are connected to the blood supply through the renal arteries.
- Renal circulation is also vulnerable to atherosclerosis, which leads to the development of stenosis, a condition in which a blood vessel becomes abnormally narrowed.
- The renal circulation stenosis elevates blood pressure, which can damage blood capillaries and impair proper filtration, ultimately leading to a gradual loss of kidney function.
- The growth of a tumor is usually accompanied by the generation of a new blood vessel infrastructure to support needs of the tumor cells. This process is called angiogenesis.
- On the way to blood circulation, the lymph passes the lymph nodes, located in different areas throughout the body, including the neck, chest, armpit, abdomen, pelvis, and groin.
- Assessment of lymph nodes is a routine step in treatment of both infectious diseases and cancer.
- The traditional technologies for blood flow characterization and imaging include ultrasound and laser Doppler flowmetries, Doppler OCT, laser speckle imaging, photoacoustic imaging, and magnetic particle imaging.

- Doppler ultrasound flowmetry employs emitted and reflected sound waves to see how blood flows through a blood vessel. The movement of blood cells causes a change in pitch of the reflected sound waves (the Doppler effect). This change in a signal from the reflected sound waves can be analyzed using a computer, providing information about the blood flow through the blood vessels.
- A new ultrasonic imaging modality, Power Doppler Harmonic Imaging (PDHI), color-encodes changes in the acoustic backscatter signal amplitude and the motion of ultrasonic scatterers between insonifying pulses.
- New-generation ultrasound contrast agents consist of gaseous perfluorocarbon (PFC) bubbles encapsulated within polymer shells, PFC, and poly(lactide-co-glycolide) (PLGA) encapsulating a liquid core of perfluorooctyl bromide (PFOB).
- Laser Doppoler flowmetry (LDF) utilizes Doppler shift (the frequency change that waves undergo when reflected by moving objects) of light waves. The emitted light from a low-power laser source is reflected by the target and then fed to a processing computer.
- Systemically administrated nanoparticles, such as gold nanorods or one with high refractive index, can be used as contrast enhancers in LDF.
- Doppler OCT (optical Doppler tomography, ODT) provides flow velocity measurement in the sample using Doppler frequency shift as the contrast mechanism in the OCT imaging modality.
- The application of gold nanoparticle can enhance the contrast in Doppler OCT.
- When a diffuse object is illuminated with coherent laser light, the backscattered light will form an interference pattern on the detector called speckle pattern. Speckle method measures the change of intensity in the speckle patterns when an object moves.
- Photoacoustic (PA) imaging which combines the high-contrast and spectroscopic-based specificity of optical imaging with the high spatial resolution of ultrasound imaging has also been used to map blood flow.
- Magnetic particle imaging (MPI), which is based on the nonlinear magnetization response, enables real-time high-resolution 3D imaging of magnetic nanoparticles flowing in blood vessels.
- Magnetic control using magnetic particles can be employed for controlling blood flow. The magnetic nanoparticles would aggregate in the presence of a sufficiently strong magnetic field (and field gradient), which could be used to slow down or completely block blood flow through cancer tissues, effecting cancer killing.
- Nanoformulations of anti-inflammatory drugs can be targeted to the antigens/receptors on the surface of monocytes to be attached, and then they can be transported by monocytes for action at the site of inflammation to increase the local concentration of the anti-inflammatory drug as well as to provide sustained release.

- Macrophages can easily uptake nanoparticles with characteristics similar to a microbe or a virus which can be utilized for trafficking of nanoformulations of antimicrobial and antiviral drugs.
- An advantage of macrophage delivery is that they are able to cross biological barriers in response to cytokine signaling and deliver nanoformulation directly to infected tissues.
- Lymph node mapping is valuable for diagnosis of the metastatic stage of cancer, because the sentinel lymph node (SLN) is the first lymph node to drain the primary tumor.
- The nanoparticles can be injected near the primary tumor, whereby they can be picked up by the adjacent lymphatics and transported to the SLN.
- Delivery of drugs to the lymphatic system can more effectively provide chemotherapeutic drugs to the regional lymph nodes, and by accumulation they can prevent the metastasis of tumor cells.
- The advantage of lymphatic delivery is that it avoids the RES capture of nanocarriers in the liver which generally occurs in a systemic injection of nanoparticles into the blood stream.
- Drug-carrying nanoparticles, coupled with multimodal imaging capability, thus can be used for theranostics to diagnose and treat lymph node metastasis.

EXERCISES

True or False

1. Doppler ultrasound flowmetry can be used to reduce the chance of stroke by determining the blood blockage.
2. Photoacoustic imaging technique combines optical imaging with ultrasound imaging.
3. Laser Doppler flowmetry is an invasive approach to measure the microcirculatory blood flow in neural tissue as the tissue will be damaged by the laser.
4. Magnetic nanoparticles can be used to slow down or completely block blood flow under a sufficiently strong radiofrequency field.
5. Gadolinium derivative can serve as a contrast agent in MRI.

Multiple Choice

6. A system of arteries that regulates blood supply to the brain is known as
   (a) Central nervous system
   (b) Circle of Willis

   - **(c)** Lymphatic system
   - **(d)** Reticuloendothelial system

7. Which of the following is not true for a monocyte?
   - **(a)** A type of white blood cell
   - **(b)** Present in the site of inflammation
   - **(c)** Carry oxygen
   - **(d)** Produced by the bone marrow

8. The growth of a tumor is usually accompanied by the generation of a new blood vessels; this process is called
   - **(a)** Endocytosis
   - **(b)** Angiogenesis
   - **(c)** Atherosclerosis
   - **(d)** Stenosis

9. What is hematocrit representing?
   - **(a)** The ratio of the inertial force to the viscous force in the blood
   - **(b)** The ratio of the oscillatory inertial force to the shear force in the blood
   - **(c)** The percentage by volume of packed red blood cells
   - **(d)** The number of white blood cell in the blood

10. Which of the following is included in the pulmonary circulation system?
   - **(a)** Aorta
   - **(b)** Lung
   - **(c)** Kidney
   - **(d)** Tonsil

Answers: 1. T; 2. T; 3. F; 4. F; 5. T; 6. b; 7. c; 8. b; 9. c; 10. c

REFERENCES

Aaslid, R., Markwalder, T.-M., Nornes, H., Noninvasive transcranial Doppler ultrasound recording of flow velocity in basal cerebral arteries, *J. Neurosurg.* **57**, 769–774 (1982).

Antonelli, A., Sfara, C., Mosca, L., Manuali, E., Magnani, M., New biomimetic constructs for improved in vivo circulation of superparamagnetic nanoparticles, *J. Nanosci. Nanotech.* **8**, 2270–2278 (2008).

Batrakova, E. V., Gendelman, H. E., Kabanov, A. V., Cell-mediated drug delivery, *Expert Opin Drug Deliv. Epub* **8**, 415–433 (2011).

Beard, P., Biomedical photoacoustic imaging, *Interface Focus* **1**, 602–631 (2011).

Brown, H. S., Halliwell, M., Qamar, M., Read, A. E., Evans, J. M., Wells, P. N., Measurement of normal portal venous blood flow by Doppler ultrasound, *Gut* **30**, 503–509 (1989).

Calamante, F., Thomas, D. L., Pell, G. S., Wiersma, J., Turner, R., Measuring cerebral blood flow using magnetic resonance imaging techniques, *J. Cerebral Blood Flow Metab.* **19**, 701–735 (1999).

Charm, S. E., Kurland, G. S., *Blood flow, Microcirculation; Blood Flow Velocity; Measurement; Mathematical Models*, Wiley, New York (1974).

Chen, Z., Milner, T. E., Dave, D., Nelson, J. S., Optical Doppler tomographic imaging of fluid flow velocity in highly scattering media, *Opt. Lett.* **22**, 64–66 (1997).

Dıaz-Lopez, R., Nicolas Tsapis, N., Santin, M., Bridal, S. L., Nicolas, V., Jaillard, D., Libong, D., Chaminade, P., Marsaud, V., Vauthier, C., Fattal, E., The performance of PEGylated nanocapsules of perfluorooctyl bromide as an ultrasound contrast agent, *Biomaterials* **31**, 1723–1731 (2010).

Dirnagl, U., Kaplan, B., Jacewicz, M., Pulsinelli, W., Continuous measurement of cerebral cortical blood flow by laser-doppler flowmetry in a rat stroke model, *J. Cerebral Blood Flow Metab.* **9**, 589–596 (1989).

Draijer, M., Hondebrink, E., van Leeuwen, T., Steenbergen, W., Review of laser speckle contrast techniques for visualizing tissue perfusion, *Lasers Med. Sci.* **24**, 639–651 (2009).

Dunn, A. K., Bolay, H., Moskowitz, M. A., Boas, D. A., Dynamic imaging of cerebral blood flow using laser speckle, *J. Cerebral Blood Flow Metab.* **21**, 195–201 (2001).

Erogbogbo, F., Yong, K.-T., Roy, I., Law, W.-C., Zhao, W., Ding, H., Wu, F., Kumar, R., Swihart, M. T., Prasad, P. N., *In Vivo* targeted cancer imaging, sentinel lymph node mapping and multi-color imaging with biocompatible silicon nanocrystals, *ACS Nano* **5**, 413–423 (2011).

Fujimoto, Y., Okuhata, Y., Tyngi, S., Namba, Y., Oku, N., Magnetic resonance lymphography of profundus lymph nodes with liposomal gadolinium-diethylenetriamine pentaacetic acid, *Biol. Pharm. Bull.* **23**, 97–100 (2000).

Gleich, B., Weizenecker, J., Tomographic imaging using the nonlinear response of magnetic particles, *Nature* **435**, 1214–1217 (2005).

Heiss, W.-D., Graf, R., Wienhard, K., Löttgen, J., Saito, R., Fujita, T., Rosner, G., Wagner, R., Dynamic penumbra demonstrated by sequential multitracer PET after middle cerebral artery occlusion in cats, *J. Cerebral Blood Flow Metab.* **14**, 892–902 (1994).

Hoffman, H. T., Quets, J., Toshiaki, T., Funk, G. F., McCulloch, T. M., Graham, S. M., Robinson, R. A., Schuster, M. E., Yuh, W. T., Functional magnetic resonance imaging using iron oxide particles in characterizing head and neck adenopathy, *Laryngoscope* **110**, 1425–1430 (2000).

Kadiu, I., Nowacek, A., McMillan, J., Gendelman, H. E., Macrophage endocytic trafficking of antiretroviral nanoparticles, *Nanomedicine* **6**, 975–994 (2011).

Kim, S., Lim, Y. T., Soltesz, E. G., De Grand, A. M., Lee, J., Nakayama, A., Parker, J. A., Mihaljevic, T., Laurence, R. G., Dor, D. M., Cohn, L. H., Bawendi, M. G., Frangioni, J. V., Near-infrared fluorescent type II quantum dots for sentinel lymph node mapping, *Nat. Biotechnol.* **22**, 93–97 (2004).

Kim, C., Favazza, C., Wang, L. V., *In Vivo* photoacoustic tomography of chemicals: High-resolution functional and molecular optical imaging at new depths, *Chem Rev.* **110**, 2756–2782 (2010).

Kobayashi, H., Kawamoto, S., Sakai, Y., Choyke, P. L., Star, R. A., Brechbiel, M. W., Sato, N., Tagaya, Y., Morris, J. C., Waldmann, T. A., Lymphatic drainage imaging of breast cancer in mice by micro-magnetic resonance lymphangiography using a nano-size paramagnetic contrast agent, *J. Natl. Cancer Inst.* **96**, 703–708 (2004).

Kobayashi, H., Koyama, Y., Barrett, T., Hama, Y., Regino, C. A. S., Shin, I. S., Jang, B. S., Le, N., Paik, C. H., Choyke, P. L., Urano, Y., Multimodal nanoprobes for radionuclide and five-color near-infrared optical lymphatic imaging, *ACS Nano* **1**, 258–264 (2007).

Kotha, S., Sudarshan, T. S., Hemostatic composition, U. S. Patent 7,670,623 (2010).

Laakkonen, P., Porkka, K., Hoffman, J. A., Ruoslahti, E., A tumor-homing peptide with a targeting specificity related to lymphatic vessels, *Nat. Med.* **8**, 751–755 (2002).

Luciani, A., Itti, E., Rahmouni, A., Meignan, M., Clement, O., Lymph node imaging: Basic principles, *Eur. J. Radiol.* **58**, 338–344 (2006).

Markov, D. E., Boeve, H., Gleich, B., Borgert, J., Antonelli, A., Sfara, C., Magnani, M., Human erythrocytes as nanoparticle carriers for magnetic particle imaging, *Phys. Med. Biol.* **55**, 6461–6473 (2010).

McDonald, D. A., In: Nichols, W. W., D'Rourke, M. F. (Eds.), *Blood Flow in Arteries*, Edward Arnold, Southampton, pp. 102–108 (1974).

McMillan, J., Batrakova, E., Gendelman, H. E., Cell delivery of therapeutic nanoparticles, *Prog. Mol. Biol. Transl. Sci.* **104**, 563–601 (2011).

Ngo, F. C., Hall, C. S., Marsh, J. N., Fuhrhop, R. W., Allen, J. S., Brown, P., McLean, M. D., Scott, M. J., Wickline, S. A., Lanza, G. M., Evaluation of liquid perfluorocarbon nanoparticles as a blood pool contrast agent utilizing power Doppler harmonic imaging, *Ultrasonics Symposium, 2000 IEEE* **2**, 1931–1934 (2000).

Nishioka, Y., Yoshino, H., Lymphatic targeting with nanoparticulate system, *Adv. Drug Deliv. Rev.* **47**, 55–64 (2001).

Nune, S. K., Gunda, P., Majeti, B. K., Thallapally, P. K., Forrest, M. L., Advances in Lymphatic Imaging and Drug Delivery, *Adv. Drug Delivery Rev.* **63**, 876–885 (2011).

Obeid, A. N., Barnett, N. J., Dougherty, G., Ward, G., A critical review of laser Doppler flowmetry, *J. med. eng. technol.* **14**, 178–181 (1990).

Pierige, F., Serafini, S., Rossi, L., Magnani, M., Cell-based drug delivery, *Adv. Drug Deliv. Rev.* **60**, 286–295 (2008).

Pisani, E., Tsapis, N., Paris, J., Nicolas, V., Cattel, L., Fattal, E., Polymeric nano/microcapsules of liquid perfluorocarbons for ultrasonic imaging: physical characterization, *Langmuir* **22**, 4397–4402 (2006).

Pisani, E., Fattal, E., Paris J., Ringard, C., Rosilio, V., Tsapis, N., Surfactant dependent morphology of polymeric capsules of perfluorooctyl bromide: influence of polymer adsorption at the dichloromethane-water interface, *J. Colloid Interface Sci.* **326**, 66–71 (2008).

Pons, T., Pic, E., Lequeux, N., Cassette, E., Bezdetnaya, L., Guillemin, F., Marchal, F., Dubertret, B., Cadmium-free CuInS(2)/ZnS quantum dots for sentinel lymph node imaging with reduced toxicity, *ACS Nano* **4**, 2531–2538 (2010).

Popel, A. S., Johnson, P. C. Microcirculation and hemorheology, *Annu. Rev. Fluid Mech.* **37**, 43–69 (2005).

Ravizzini, G., Turkbey, B., Barrett, T., Kobayashi, H., Choyke, P. L., Nanoparticles in sentinel lymph node mapping, *Nanomed. Nanobiotechnol.* **1**, 610–623 (2009).

Rossi, L., Serafini, S., Pierige, F., Antonelli, A., Cerasi, A., Fraternale, A., Chiarantini, L., Magnani, M., Erythrocyte-based drug delivery, *Expert Opin. Drug Deliv.* **2**, 311–322 (2005).

Shive, M. S., Anderson, J. M., Biodegradation and biocompatibility of PLA and PLGA microspheres, *Adv. Drug Deliv. Rev.* **28**, 5–24 (1997).

Stalder, A. F., Frydrychowicz, A., Russe, M. F., Korvink, J. G., Hennig, J., Li, K., Markl, M., Assessment of flow instabilities in the healthy aorta using flow-sensitive MRI, *J. Magn. Res. Imag.* **33**, 839–846 (2011).

Thibodeau, G. A., Patton, K. T., *Anatomy and Physiology*, 5th edition, Mosby, St. Louis (2003).

Torchia, M. G., Nason, R., Danzinger, R., Lewis, J. M., Thliveris, J. A., Interstitial, M. R., lymphangiography for the detection of sentinel lymph nodes, *J. Surg. Oncol.* **78**, 151–156, discussion 157 (2001).

Weizenecker, J., Gleich, B., Rahmer, J., Dahnke, H., Borgert, J., Three-dimensional real-time *in vivo* magnetic particle imaging, *Phys. Med. Biol.* **54**, L1–L10 (2009).

Womersley, J. R., Method for the calculation of velocity, rate of flow and viscous drag in arteries when the pressure gradient is known, *J. Physiol.* **127**, 553–563 (1955).

Yang, D., Yang, F., Hu, J. H., Long, J., Wang, C. C., Fu, D. L., Ni, Q. X., Hydrophilic multi-walled carbon nanotubes decorated with magnetite nanoparticles as lymphatic targeted drug delivery vehicles, *Chem. Commun.* 4447–4449 (2009).

Yang, V. X. D., Gordon, M. L., Seng-Yue, E., Lo, S., Qi, B., Pekar, J., Mok, A., Wilson, B. C., Vitkin, I. A., High speed, wide velocity dynamic range Doppler optical coherence tomography (Part II): Imaging *in vivo* cardiac dynamics of *Xenopus laevis*, *Opt. Express* **11**, 1650–1658 (2003).

Yazdanfar, S., Rollins, A. M., Izatt, J. A., *In vivo* imaging of human retinal flow dynamics by color Doppler optical coherence tomography, *Arch. Ophthalmol.* **121**, 235–239 (2003).

Zimmer, J. P., Kim, S. W., Ohnishi, S., Tanaka, E., Frangioni, J. V., Bawendi, M. G., Size series of small indium arsenide–zinc selenide core–shell nanocrystals and their application to *in vivo* imaging, *J. Am. Chem. Soc.* **128**, 2526–2527 (2006).

CHAPTER 13

NANOTECHNOLOGY FOR CANCER

The most widely investigated therapeutic applications of nanotechnology are in cancer therapy. It continues to be a major thrust area of nanomedicine, receiving worldwide attention. In the United States, the National Cancer Institute (NCI) of the National Institutes of Health (NIH) has been a major sponsor of cancer nanotechnology which has produced significant advances in this field. Some nanoformulations have already made their presence felt in cancer treatment and many others are at various stages of clinical trials. A vast number of nanocarriers for cancer therapy have been reported, ranging from dendrimers to liposomes, to polymeric nanoparticles, and to various types of inorganic nanoparticles. Thus within the scope of this introductory book, it is not possible to provide a comprehensive review of the literature on this topic. Here, only selective examples will be provided to demonstrate the wide scope of cancer nanotherapy and its effectiveness.

This chapter starts with Section 13.1 providing an overview of the benefits of nanotechnology for cancer diagnostics and therapy. The subsequent sections cover the various types of cancer nanotherapies. Specifically, Section 13.2 describes chemotherapy and advancement in this modality using a nanoformulation. Section 13.3 introduces the concept of cancer gene therapy using nanoparticles. Section 13.4 discusses photodynamic therapy, another powerful method of cancer treatment. Section 13.5 presents magnetic therapy using magnetic nanoparticles. Section 13.6 covers photothermal therapy using primarily

Introduction to Nanomedicine and Nanobioengineering, First Edition. Paras N. Prasad.

metallic (plasmonic) nanostructures to locally generate heat by absorbing light. The magnetic and photothermal therapies are made possible solely by the application of nanotechnology. Advances in radiation therapy, made by the application of nanotechnology, are presented in Section 13.7. Following that, Section 13.8 presents a discussion on circulating tumor cells, and strategies for early diagnosis of cancer based on their detection. Finally, in Section 13.9 the concerted efforts of the interdisciplinary NCI Alliance for Cancer Nanotechnology to combat cancer have been briefly discussed.

13.1. BENEFITS OF CANCER NANOTECHNOLOGY

The following summarizes the principal benefits provided by the application of nanotechnology to the field of cancer.

Early Detection. Early detection of cancer has become a major goal of nanomedicine because survival rate for a number of cancers such as pancreatic is crucially dependent on it. As it has been described at several places in previous chapters, the use of a nanostructure platform allows one to incorporate multiple probes for multiplexed and multispectral imaging and sensing, whereby one can utilize detection based on multiple biomarkers. Multispectral imaging permits one to detect certain molecular changes even at a precancerous stage. Additional input can be provided by using a combination of imaging techniques discussed in Chapter 8 which yield structural, physiological, and morphological information on a particular disease manifestation.

An area that has generated considerable interest is optical biopsy (see Prasad [2003]; and Chapter 7). Traditional biopsy is invasive and requires taking a small portion of the tissue specimen to examine for signs of cancer. Optical biopsy utilizes interaction of light with the intact tissue (without removing it) in the body to detect cancer and is thus noninvasive. A nanoformulation with targeting ability (both active and the EPR-effect-derived passive targeting; see Chapter 7) and utilizing unique signatures of tumor-manifested changes in optical interactions (absorption, fluorescence, scattering) that can be collected through the body provides a major advancement for optical biopsy.

Figure 13.1 provides an example of cancer detection using our quantum dot formulation with targeted multiplex detection of cancer cells by *in vitro* fluorescence imaging. Figure 13.2 is an example of *in vivo* imaging of cancer tissue in a nude mouse, where the nanoparticles were injected into the tail vein.

Multimodal Therapy. A nanostructure platform allows one to use a number of different types of cancer therapies in tandem. Some of the principal modes of cancer therapy are listed in Figure 13.3. These modes are discussed in separate subsequent sections below. Each may have its own merit and thus a

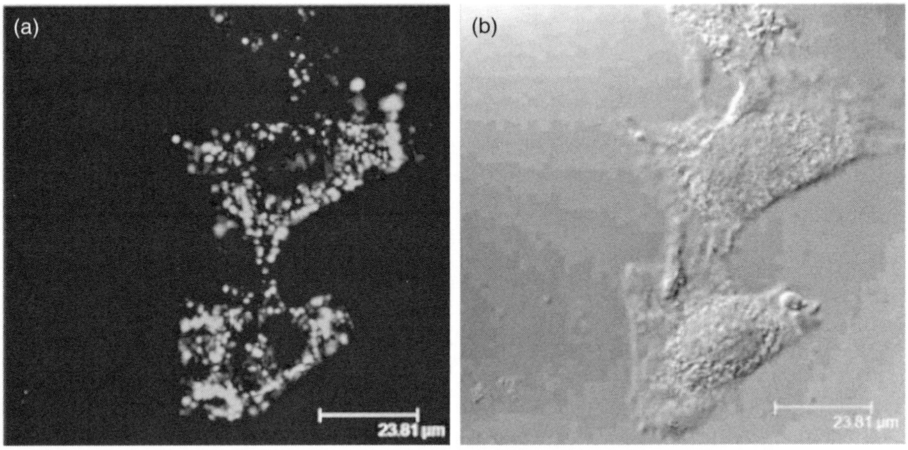

Figure 13.1. Confocal microscopic images of HeLa cells, treated with Tf-conjugated CdSe/CdS/ZnS QRs. (a) Fluorescence channel and (b) overlays of fluorescence and transmission channels. (Reproduced with permission from Yong et al. [2007].)

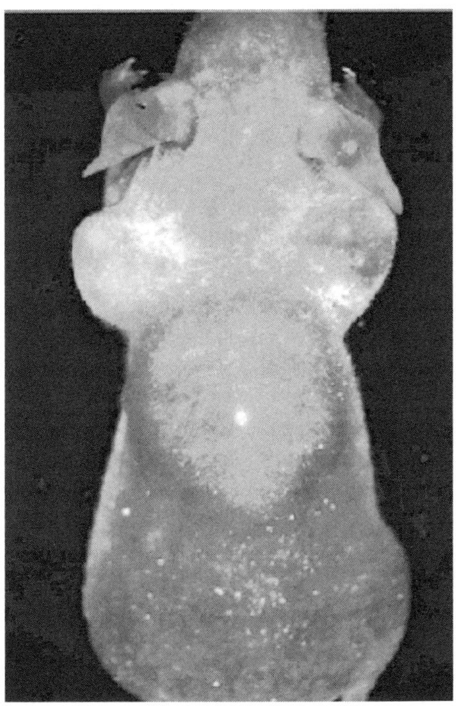

Figure 13.2. *In vivo* imaging of cancer, Reproduced with permission from Kumar et al. [2010].)

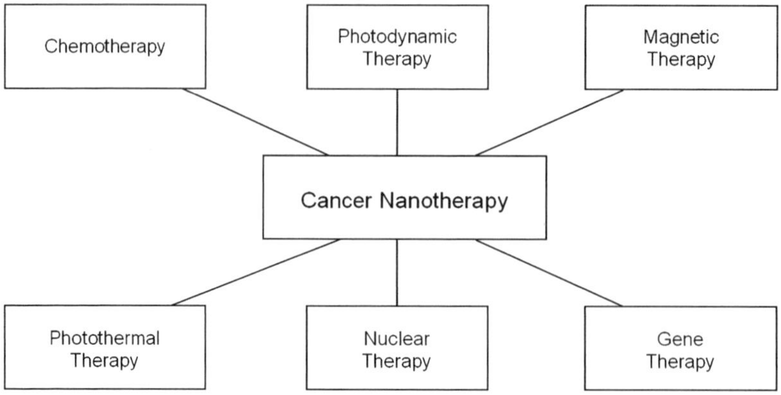

Figure 13.3. The various modes of cancer nanotherapy.

combination therapy provides a major gain in the treatment of cancer. It also minimizes the risk of individual therapy alone, because an appropriate combination may provide a balanced approach.

Image Guided Therapy/Surgery. The ability to guide a therapy/surgery and monitor its course is another major utility of a nanoplatform. The visualization of the nanocarriers at the tumor site and real-time monitoring of the tumor condition in the course of therapy allows for more effective treatment, as compared to conventional strategies, when the evaluation of therapy efficiency is performed post-treatment with a time delay.

Furthermore, one can use externally activated therapy to maximize its effectiveness and reduce any damage to healthy tissues, by activating it after the localization of nanocarriers at the targeted tumor site is confirmed.

The nanoplatform imaging of tumor margins following a surgical removal of the tumor is of considerable value for post-surgical assessment. This ability can eliminate any need for invasive post biopsy to ascertain that all cancer tissues are removed. Furthermore, it also provides an opportunity to treat the residual tumor tissues using other gentler methods listed in Figure 13.3.

Nanoparticle Sensitized Radiation Therapy. Radiation therapy uses high-energy radiation to kill cancer cells. At the same time, the use of a higher dosage of radiation can produce damage to healthy tissues, leading to impairment of some functions with traumatic consequences. A nanocarrier delivery of appropriate drug or genetic material can sensitize the tumor tissue so that it can be destroyed at a considerably lower dosage of radiation, thus considerably reducing the risk factor. An example is our work on the use of gene-silencing by nanoparticle delivery of si-RNA to the tumor to sensitize the cells for effecting radiation therapy at a lower dosage. This approach is described in Chapter 14 ("Gene Therapy").

13.2. CHEMOTHERAPY

Chemotherapy is the treatment of cancer with antineoplastic drugs that can be applied to a broad range of malignancies. However, it is also a harsh treatment that has pronounced side effects. Thus minimizing the side effects by reducing the dosage level is of significant value for chemotherapy. Controlling the pharmacokinetics, and pharmacodynamics (discussed in Chapter 11) of a chemotherapeutic agent by using a nanocarrier delivery method, is a major advancement. Nanocarrier delivery of a chemotherapeutic agent is the most advanced area of cancer nanotechnology. Already a number of nanoformulations are in clinical applications. To reiterate again from previous chapters, the benefits of nanocarrier delivery of a chemotherapeutic agent are improved bioavailability of a hydrophobic drug (as many chemotherapeutic drugs are hydrophobic), a favorable pharmacokinetics and biodistribution by the design and surface tailoring of the nanocarriers, and an enhanced pharmacodynamics produced by more effective drug–receptor interactions as well as by controlled drug release. The coupled features of targeting and controlled drug release also ensure minimal collateral damage and thus a lower risk of side effects.

A significant advantage is the gain in the therapeutic value of chemotherapy in the case of multidrug resistance (MDR) due to glycoprotein pumps that lead to expulsion of a drug from tumor cells [Galmarini and Galmarini, 2003]. A nanocarrier provides sustained, controlled, and targeted delivery of a chemotherapeutic agent by localizing in the cytoplasm or lysosomes without triggering the *p*-glycoprotein pump. Another mechanism of overcoming drug resistance stems from the direct effect of increased drug delivery into the cell. An example is the use of wax nanoparticles to deliver the entrapped antineoplastic drug paclitaxel, which were shown to overcome drug resistance in a human colon adenocarcinoma cell line [Suri et al., 2007].

A number of nanocarriers have been developed for the delivery of chemotherapeutic agents [Wang et al., 2009]. They are made of polymers, dendrimers, liposomes, micelles, organically modified silica (ORMOSILS), carbon nanotubes, and gold nanoparticles; however, the nanoparticle delivery systems so far approved by the FDA are based on liposomes, polymers, or micelles. Two prominent clinically approved nanoparticulate drug delivery systems are (i) *Doxil*® (liposomal-PEG doxorubicin; Ortho Biotech/ Schering-Plough), with FDA approval in 1995 for HIV-related Kaposi's sarcoma, metastatic breast cancer, and metastatic ovarian cancer [Gabizon et al., 2003], and (ii) *Abraxane*® (albumin-bound paclitaxel, Abraxis BioSciences), with FDA approval in 2005 for metastatic breast cancer [Blum et al., 2007]. In addition, some other nanoparticulate drug delivery systems approved for cancer chemotherapy and some others for clinical trials are listed below.

Polymeric.

Genexol-PM® (Methoxy-PEG-poly(D,L-lactide) taxol; Samyang, Korea): Approved in S. Korea for metastatic breast cancer. Phase II for pancreatic cancer in the United States.

Oncaspar® (PEG–L-asparaginase; Enzon): FDA approved for acute lymphoblastic leukemia.

Zinostatin stimalmer® (styrene maleic anhydrideneocarzinostatin): Has been used for primary unresectable hepatocellular carcinoma as an approved agent in Japan since 1994 (Astellas Pharma Inc., Osaka).

Neulasta™ (PEG-hrGCSF): FDA approval in 2002 to reduce the risk of infection in patients with some tumors receiving strong chemotherapy that decreases the number of infection-fighting white blood cells. In 2005, the FDA approved an expanded indication for its use in the first cycle of a chemotherapy regimen associated with a 17% or higher risk of neutropenia and subsequent cycles thereafter.

Liposomal.

DaunoXome® (liposomal daunorubicin; Gilead Sciences/ Diatos): FDA approval in 1996 for HIV-related Kaposi's sarcoma.

Myocet® (liposomal doxorubicin; Zeneus): FDA fast track designation for the treatment of human epidermal growth factor receptor 2 positive metastatic breast cancer; approved in Europe and Canada for treatment of metastatic breast cancer in combination with cyclophosphamide.

Mepact® (muramyl tripeptide phosphatidyl-ethanolamine): Approved in Europe for treatment of patients with nonmetastatic, resectable osteosarcoma; Phase III trials in United States.

Depocyt® (liposomal cytarabine): Granted accelerated approval in 1999 and full approval in 2007 by for the intrathecal treatment of lymphomatous meningitis, a life-threatening complication of lymphoma.

Some noteworthy anticancer drugs in polymeric nanoparticle formulations that are in clinical trials include paclitaxel poliglumex (Xyotax), *N*-(2-hydroxypropylmethacrylamide (HPMA) copolymer–camptothecin (MAG-CPT), and HPMA-DOX. Recently, there has been considerable focus on using biodegradable polymeric nanoparticles as drug carriers. For this reason, poly(lactic-co-glycolic acid (PLGA) has received considerable attention [Acharya and Sahoo, 2011]. PLGA nanoparticles have also been shown to provide sustained drug release and thus many chemotherapeutic drugs such as paclitaxel, doxorubicin, 5-fluorouracil, and dexamethasone have been successfully used with PLGA nanoparticles [Acharya and Sahoo, 2011]. PLGA is a copolymer of two monomeric units: lactic acid and glycolic acid; by varying the ratios of these two units, the degradation rate of PLGA can be controlled. The surface of the PLGA nanoparticles can also be modified to incorporate PEG groups to prevent opsonization and enhance circulation.

Many types of inorganic nanoparticles have also been used for delivery of anticancer drugs. The quantum dots (QDs), discussed in a number of chapters (5, 7, and 8) are one example [Qi and Gao, 2008; Bagalkot et al., 2007]. Their utility for fluorescence bioimaging, sensing, and *in vitro* diagnosis has already been discussed in earlier chapters. The advantage offered by the QDs is their ability to provide traceable drug delivery utilizing their fluorescence, which can be used to elucidate the pharmacokinetics and pharmacodynamics of the drug payload they carry. Another advantage offered by them is their small size, thus enabling their clearance by renal filtration. Bagalkot et al. [2007] used RNA aptamer-conjugated QDs for delivery of doxorubicin; here they used the FRET process (discussed in Chapters 5, 8, and 9) to monitor the carrying and subsequent release of doxorubicin. Kikkeri et al. [2009] took advantage of certain sugar molecules to selectively target receptors in specific tissues and organs to develop sugar-coated quantum dots that would deliver anticancer drugs selectively to the liver. Recently, our group also reported the delivery of doxorubicin with QDs as nanocarriers to specific lung cells, alveolar microphages in the lung [Chakravarthy et al., 2011]. A major concern in using QDs for drug delivery is their toxicity, particularly in the case of CdSe. Nie and coworkers [Gao et al., 2004], have worked on coating the QDs with polymers, which would prevent the leaking out of the toxic Cd ions from the CdSe QD. Another QD system reported for *in vitro* delivery of cancer drug is nontoxic ZnO nanoparticles coated with biodegradable chitosan [Yuan et al., 2010].

Another class of nanoparticles used for cancer drug delivery is carbon nanostructures discussed in Chapter 3. An example is carbon nanotubes that were used to deliver paclitaxel [PTX], widely used for cancer chemotherapy [Liu et al., 2008b]. They conjugated this drug to branched polyethylene glycol (PEG) chains on single-walled carbon nanotubes (SWNT) via a cleavable ester bond that produces water-dispersible SWNT-PTX conjugates. Their results revealed that the SWNT-PTX conjugate afforded higher efficiency in suppressing tumor growth than did the clinical paclitaxel in a murine mouse 4T1 breast cancer model. They attributed this enhancement to the prolonged blood circulation (due to the PEG group) and 10-fold higher tumor PTX uptake by SWNT delivery, derived most likely from passive targeting due to the EPR effect. The drug molecules were released into the reticuloendothelial system and were excreted via the biliary pathway, without any obvious toxicity.

Iron oxide magnetic nanoparticles have also been used as nanocarriers for anticancer drugs [Yu et al., 2008]. The relative merit of this carrier is its multifunctionality. These nanoparticles act as contrast enhancers for MRI imaging, as discussed in Chapter 8. They can be used also for magnetic therapy, as discussed below in Section 13.4. Thus the use of these nanoparticles as carriers for chemotherapeutic agents can provide dual modality therapy: chemo- and magnetic. In addition, Yu et al. coupled Cy5.5, a near-IR-emitting fluorophore, to thermally cross-linked iron oxide nanoparticles that delivered doxorubicin, thus providing fluorescence imaging (hence dual imaging) capability as well. Another advantage of using the iron oxide nanoparticles as drug carriers is that several formulations of iron oxide nanoparticles are already in various clinical trials, and these nanoparticles are also commercially available from a number of sources.

More recent efforts in using nanoparticle carriers for chemotherapy have focused on introducing active targeting function, rather than relying on the EPR effect which yields passive targeting [Davis et al., 2008]. Some commonly used antibodies for targeting recognize the HER-2 receptor, the transferrin receptor (TfR), or the prostate-specific antigen receptor, which are overexpressed on certain types of cancer cells [Wang and Thanou, 2010]. These antibodies exhibit high affinity to their corresponding receptors. Recent trends have been to use small-sized ligands rather than the large antibodies for targeting. Examples are antibody fragments consisting of the binding sites, aptamers, sugar molecules, peptides, and hormone analogues. PK2 (FCE 28069), which was composed of an HPMA–polymer–Gly-Phe-Leu-Gly–doxorubicin-conjugate containing the sugar galactosamine, was the first ligand-targeted nanoparticle in clinical trials. Another example is long-circulating doxorubicin-encapsulated immunoliposomes conjugated to human HER-2 antibodies [Nellis et al., 2005]. Aptamer-conjugated nanoparticles have also been pursued for targeted delivery [Bagalkot et al., 2007].

The introduction of an active targeting function, in general, may not lead to increased accumulation of the nanoparticles in the tumor beyond what is produced by the EPR effect. However, it does produce higher efficacy of therapy due to enhanced internalization of the nanoparticles into the cancer cell derived from receptor-mediated endocytosis. A factor to control in active targeting is the concentration of the targeting ligand on the surface. One would think that to increase the probability of nanoparticle interactions with the cell receptors, higher loading densities of the ligands on the nanoparticle surface are desirable. In contrast, the presence of high surface coverage by the ligands in place of the circulating PEG groups may result in the accumulation of nanoparticles in the liver and spleen (increased risk of RES capture). Thus it is necessary to balance the loading of the targeting groups versus the PEG groups on the nanoparticle surface.

Dendrimers are another major class of nanocarriers for drug delivery [Fox et al., 2009; Van der Poll et al., 2010]. A dendrimer architecture, discussed in chapters 3 and 4, allows drugs to be attached to the dendrimer's multivalent surfaces through chemical conjugation or electrostatic binding. Drugs can also be loaded in the cavities around the dendrimer core through a variety of interactions such as hydrophobic and hydrogen bonding. Baker and co-workers developed a polyamidoamine-based G-5 dendrimer, 5 nm in diameter, which was attached to the folate group for targeting and to methotrexate as the therapeutic agent [Kukowska-Latallo et al., 2005]. This dendrimer nanocarrier delivery was shown to be 10 times more effective than methotrexate alone in preventing tumor growth, at the same time exhibiting less systemic toxicity. Frechet and co-workers have reported the design of a robust, biodegradable PEGylated dendrimer based on a polyester–polyamide core for delivery of doxorubicin [Van der Poll et al., 2010]. This system was found to be as effective as Doxil against the C26 tumor model, with potentially less toxicity than Doxil. However, still more improvement needs to be achieved for using dendrimers as nanocarriers in order to reduce the cytotoxicity concern as well as to improve the biocompatiblity and biodistribution.

13.3. CANCER GENE THERAPY

Advances in genomic and proteomic research, coupled with the discovery of various pathways of cancer development, have led to the view that cancer is in essence a genetic disease. Two distinct classes of cancer associated genes have been identified, namely oncogenes and tumor-suppressor genes, with agonistic and antagonistic effects on cancer development, respectively. The origin of cancer is often attributed to the mutation of a single or multiple genes. Therefore, treatment of cancer at the genetic level, namely cancer gene therapy, has emerged as a highly promising treatment modality for cancer [Straton, 2011]. This therapy can be administered either alone or in combination with a traditional cancer therapy (e.g., chemo- or radiation therapy).

Gene therapy can be broadly classified into gene augmentation and gene knockdown therapies, which has been detailed later in Chapter 14. In both these cases, the therapeutic gene requires a carrier, or vector, to protect it from the harsh physiological environment and deliver it to target cells and further into the desired intracellular compartments [Templeton, 2009]. A variety of engineered nanoparticles, based on their nonimmunogenicity, multifunctionality, and target-specificity, can often be selected as an ideal gene carrier (non-viral vector). In the realm of cancer, gene augmentation therapy involves the exogeneous supplementation of a missing/malfunctioning tumor-suppressor gene in the body. The gene TP53, encoding for the protein p53, is the most-known tumor suppressor gene in the body, whose inactivation has been linked to the majority of human cancers. Therefore, several nanoparticle and liposome-mediated gene therapies of cancer have aimed the delivery of the TP53 gene into tumor cells [Brannon-Peppas et al., 2007; Xu et al., 2001]. Supplementation of tumor-suppressor genes leads to secondary effects such as rendering tumor cells vulnerable to apoptosis, overcoming of their multi-drug resistance, and so on, which eventually contribute to the obliteration of cancer cells [Brannon-Peppas et al., 2007].

On the other hand, gene knockdown therapies aim at stifling the expression of dominating oncogenes in the tumors. Here, small RNA molecules, such as antisense RNA and short interfering RNA (siRNA), are delivered with the aid of engineered nanoparticles, to suppress particular oncogenes, thus repairing damaged cellular regulation. Typically, such a repair triggers cellular death by the apoptotic pathway. For example, one of the most-studied oncogenes in the human genome is encoding for the protein, survivin, which imparts anti-apoptotic potential of cancer cells. Therefore, several nanoparticulate strategies for delivering anti-survivin siRNA, aimed at silencing of the survivin gene, have been developed. Furthermore, other oncogenes including RAS, bcl-2, myc, and so on, can all be targeted for anticancer treatment. Suppression of these genes in tumor cells leads to enhanced apoptosis, as well as sensitizing these cells to chemo- and radiation therapies. Our own group has demonstrated that nanoparticle-mediated silencing of the pro-apoptotic sphingosine kinase-1 (SPK-1 gene) in head-and-neck cancer cells leads to their radiosensitization (detailed later in Chapter 14).

Indirect gene therapy via DNA immunotherapy (discussed in Chapters 14 and 15) is another approach, involving the use of gene transfer technology to express antigens or cytokines in order to elicit an immune response against autologous tumor cells. More recently, T-cell receptors engineered to recognize tumor antigens have been incorporated into vectors and used to re-program T cells to now recognize autologous tumor cells [Brannon-Peppas et al., 2007]. Nanoparticles such as that of poly-lactide-co-glycolide (PLGA), cross-linked polyacrylamide, branched polyethylenimine (PEI), Chitosan, and so on, have been used to target DNA vaccines to immune cells such as antigen-presenting cells (APCs) and dendritic cells, which have been reviewed elsewhere [Brannon-Peppas et al., 2007].

13.4. PHOTODYNAMIC THERAPY

Photodynamic therapy (PDT) has emerged as an important treatment modality for cancer [Prasad, 2003; Celli et al., 2010; Rai et al., 2010]. It utilizes activation of an external chemical agent, called a photosensitizer or often also labeled as PDT drug, by light. The PDT drug is administered either intravenously or selectively to the malignant site as in the case of certain skin cancers. Then irradiation by light of a specific wavelength absorbed by the PDT photosensitizer produces reactive oxygen species that can destroy the tumor.

PDT treatment relies on the greater affinity of the photosensitizer for malignant cells. Hence, when a PDT drug is administered, both normal and malignant cells absorb the drug, but after a certain waiting period ranging from hours to days, the concentration of the PDT drug in the normal cell is significantly reduced. With the tumor-targeting agents attached to the PDT drug, this waiting period can be readily reduced to a matter of a few hours, while still maintaining a high degree of accumulation in the malignant cells. Subsequent illumination by light of an appropriate wavelength activates the PDT drug, which then leads to selective destruction of the malignant tissue by a photochemical mechanism (nonthermal, thus no significant local heating). In the case of cancer in an internal organ, such as the lung, light can be administered

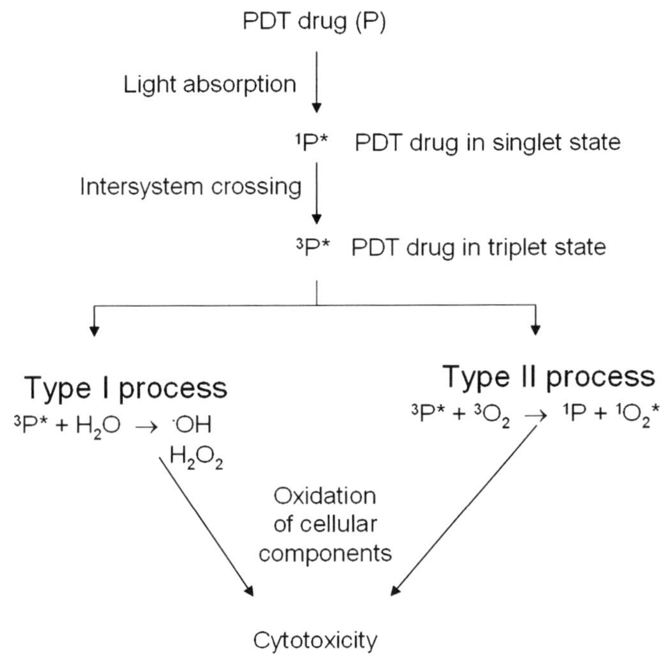

Figure 13.4. Mechanism of photodynamic therapy.

using a minimally invasive approach involving a flexible fiber-optic delivery endoscopic system, while a direct illumination method can be used for superficial malignancies such as skin cancer. Although any light source such as a lamp or a laser beam can be used to achieve the desired power density at the required wavelength, a laser beam is often used as a convenient source for this treatment, which also provides the ease of fiber-optic delivery. The flow sheet in Figure 13.4 illustrates the mechanism of photodynamic therapy [Prasad, 2003].

The light activation of a PDT drug is initiated by the absorption of light to produce an excited singlet state (S_1 or often written as $^1P^*$, where P* represents the excited photosensitizer), which then populates a long-lived triplet state T_1 (or $^3P^*$) by intersystem crossing. The singlet and triplet states have also been explained by the Jablonski diagram in Chapter 5. The photosensitizer now in a long-lived triplet state generates the reactive oxygen species by two types of processes. The type I process generates reactive species such as free radicals and peroxides, by electron or hydrogen transfer reaction with water or with a biomolecule, to produce cytotoxicity. In the type II process, the excited triplet state of the photosensitizer reacts with oxygen in the tissue and converts the oxygen molecule from its triplet ground state form to a highly reactive excited singlet-state form. The type II process is the major pathway for most photodynamic photosensitizers.

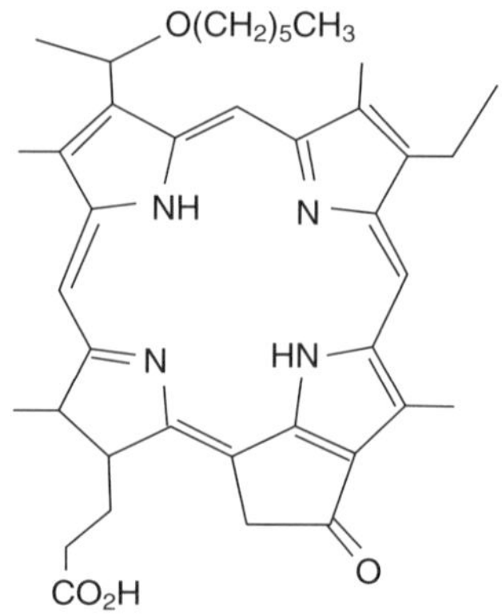

Figure 13.5. The structure of the PDT drug HPPH.

The two methods used for detecting the generation of singlet oxygen by an excited PDT photosensitizer are (i) spectroscopic and (ii) chemical. The spectroscopic method utilizes the phosphorescence emissions (explained in Chapter 5) from the excited singlet state of oxygen at ~1290 nm. The chemical method utilizes bleaching of absorption of a known singlet oxygen quencher such as 9,10-anthracenedipropionic acid (ADPA) [Bhawalkar et al., 1997]. The spectroscopic method has the advantage that it can also provide the information on the lifetime of the single oxygen state, which is an important factor for the photodynamic action.

The key factors determining photodynamic therapy and the impact of nanotechnology on them are described below.

Biodistribution and Bioavailability of the PDT Drug. Most PDT drugs are hydrophobic, where the delivery by using a nanocarrier can be of great value in improving their bioavailability. One example is the drug HPPH, shown in Figure 13.5, developed at the Roswell Park Cancer Institute in Buffalo. This drug is hydrophobic, and various formulations using surfactants have been used in the clinical trial. We have utilized both its nanocrystal formulation (described in Chapter 4) and its delivery by using an ORMOSIL carrier to improve its biodistribution. These formulations are discussed below.

Excited Triplet-State Population of the Photosensitizer. As presented in Figure 13.4, a key step in the photodynamic action is intersystem crossing to populate the triplet state, which subsequently provides a pathway to generate the cytotoxic singlet oxygen in the type II process. Thus, enhancing the intersystem crossing process to maximize the triplet population is important. A nanocarrier formulation can also provide enhancement of intersystem crossing by using external spin–orbit coupling induced by a heavy atom in the vicinity of the PDT drug. We demonstrated this concept by incorporating the

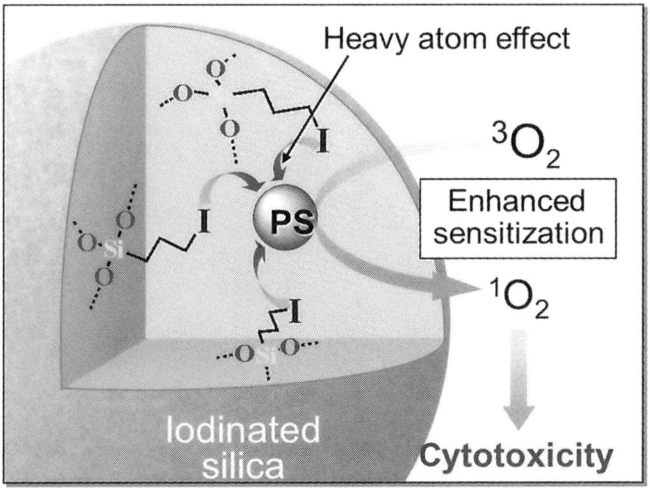

Figure 13.6. Illustration of an iodinated ORMOSIL nanoparticle encapsulating a photosensitizer, PS, which exhibits an intraparticle heavy-atom effect to enhance intersystem crossing to efficiently generate excited triplet state and subsequently singlet oxygen to produce cytotoxicity.

PDT drug, HPPH, in iodinated ORMOSIL nanoparticles, as shown in Figure 13.6 [Kim et al., 2009]. We used covalently iodine-concentrated organically modified silica (ORMOSIL) nanoparticles (diameter < 30 nm) and the hydrophobic photosensitizer, HPPH, embedded therein. Comparative studies with iodinated and noniodinated nanoparticles demonstrated that the intraparticle external heavy-atom effect due to iodine on the encapsulated photosensitizer molecules significantly enhanced the efficiency of $^1O_2$ generation and, thereby, the *in vitro* PDT efficacy.

Singlet Oxygen Generation. An efficient intersystem crossing, while leading to increased triplet-state population, does not in itself ensure efficient singlet oxygen generation. The triplet state may be deactivated by radiative (phosphorescence) or other competing nonradiative processes. To maximize the single oxygen generation pathway, the photosensitizer has to be in the oxygen environment. Also, the nanoparticle design should minimize any chemical groups that may react with singlet oxygen and consequently deplete its population. In our ORMOSIL approach, we have utilized three different ways of intracellular delivery of the hydrophobic PDT drug, HPPH, which are schematically shown below in Figure 13.7. In each of these cases, the porous structure of the ORMOSIL nanoparticles allows oxygen to be in contact with the photosensitizer and the singlet oxygen product to diffuse out to the cellular environment.

In one nanoformulation, ultrafine organically modified silica-based nanoparticles (diameter ~30 nm), physically entrapping water-insoluble HPPH, was

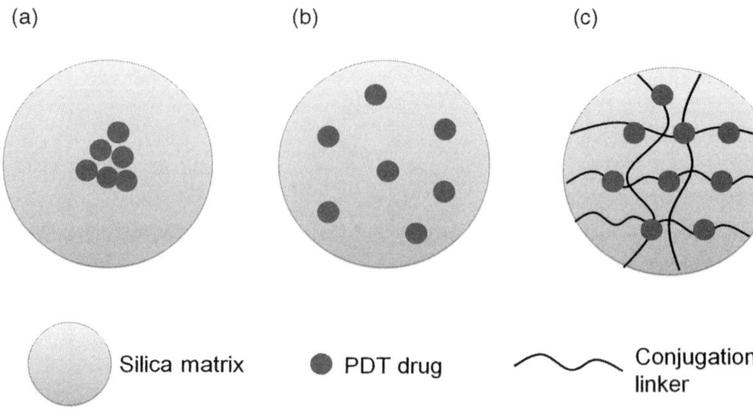

Figure 13.7. Schematic representation of three different ORMOSIL nanoformulations for delivery of a PDT drug: (a) physical entrapment of the PDT drug during the nanoparticle formation, (b) post-doping of the PDT drug through the pores of the ORMOSIL nanoparticles, and (c) chemical conjugation of the PDT drug to the ORMOSIL nanoparticles.

synthesized in the nonpolar core of micelles by hydrolysis of triethoxyvinylsilane [Roy et al., 2003]. The resulting drug-doped nanoparticles were spherical, highly monodispersed, and stable in an aqueous system. The entrapped drug was more fluorescent in the aqueous medium than the free drug, permitting the use of fluorescence bioimaging studies. Irradiation of the photosensitizing drug entrapped in the nanoparticles with light of suitable wavelength resulted in efficient generation of singlet oxygen, which was made possible by the inherent porosity of the nanoparticles. *In vitro* studies demonstrated an active uptake of the drug-doped nanoparticles into the cytosol of tumor cells. Significant damage to such impregnated tumor cells was observed upon irradiation with light of wavelength 650 nm.

Because of the porosity of the ORMOSIL nanoparticles, we also successfully post-doped them with the PDT drug, HPPH, by incubating the already prepared nanoparticles with the HPPH solution [Gupta et al., 2011]. The PDT efficacy of these post-doped nanoparticles in the *in vitro* study was found to be comparable to the one in which HPPH was entrapped during the nanoparticle preparation. A concern in physical encapsulation of the PDT drug, whether by entrapment or post doping described above, is the possibility of release of the drug during systemic circulation. To circumvent this problem, we also successfully chemically conjugated the PDT drug to the ORMOSIL nanoparticles [Ohulchanskyy et al., 2007]. This was accomplished by using an HPPH-conjugated ORMOSIL precursor, which was then mixed with another ORMOSIL precursor, co-hydrolyzed, and condensed using the sol–gel chemistry, as described in Chapter 4. In each of these three formulations represented in Figure 13.7, the inherent porosity of the ORMOSIL nanostructure allows oxygen to be in the vicinity of the photosensitizer. Furthermore, the ORMOSIL environment does not cause any depletion of the singlet oxygen formed and provides sufficient diffusion of the reactive specie to the cellular environment to produce cell lysing.

Light Penetration to the Tumor Site. As PDT is a light-activated process, reaching of light to the tumor site is absolutely essential. In the case of a thick tumor, light penetration through the tumor tissue becomes a major issue. For treatment of a thick tumor, it is thus highly desirable that the photosensitizer is activatable in the optical transparency window (~800–1000 nm) of tissues. Thus to meet this challenge, the three principal directions have been (a) development of new photosensitizers that absorb in this wavelength range of maximum tissue penetration [Prasad, 2003; Rai et al., 2010], (b) application of two-photon photodynamic therapy whereby even photosensitizers with shorter-wavelength absorption can be excited by two-photon absorption (TPA) using near-IR photons [Prasad, 2003] (the process of two-photon absorption has been discussed in Chapter 5), and (c) X-ray photodynamic therapy where the deep penetrating X-ray radiation acts to produce reactive species or energetic electrons that destroy the tumor.

An example of a near-IR absorbing photosensitizer is naphthalocyanine (N_c), which absorbs at wavelengths >800 nm and possesses an extremely high an extinction coefficient. Song et al. reported a novel nanoformulation by using tetra-*t*-butyl silicon naphthalocyanine bisoleate (SiN_cBOA) into a low-density lipoprotein (LDL) lipid core [Song et al., 2007]. Their *in vitro* studies on Hep G2 cells exhibited a preferential uptake of these N_c-LDL nanoparticles via LDL-receptor-mediated pathways. A preferential uptake by tumor tissue in nude mice was also confirmed *in vivo* by optical imaging. The tumor-to-normal muscle ratio reached a maximum of 8:1 at 2 h post-injection.

There has been a great deal of activities in the area of two-photon photodynamic therapy. The earlier efforts on this topic are described in the monograph on *Introduction to Biophotonics* [Prasad, 2003]. A great deal of work has been done in designing dendritic photosensitizers with strong two-photon absorption. Our collaborative work with Frechet's group utilized the concept of using dendritic antennas that are strong two-photon absorbers and transfer the two-photon excitation to the photosensitizer in the core, as shown in Figure 13.8 [Dichtel et al., 2004]. This scheme is similar to the one used by nature in photosynthesis using the chlorophyll antenna system.

This dendritic water-soluble photosensitizer was prepared by grafting two-photon-absorbing chromophores and water-solubilizing moieties to a known multivalent porphyrin photosensitizer. Efficient FRET energy transfer between the peripheral donor two-photon-absorbing chromophores and the central acceptor photosensitizer at the core of the dendrimer with >99% quenching of donor emission was obtained under two-photon excitation at 780 nm in an aqueous medium. Two-photon photosensitized production of

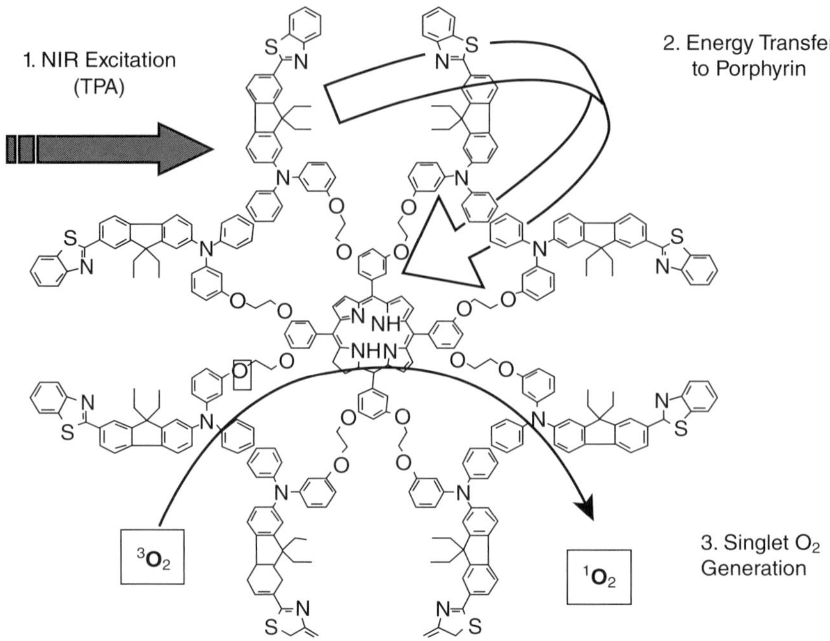

Figure 13.8. Structure and function of a two–photon dendrimer–photosensitizer for photodynamic therapy. (Reproduced with permission from Dichtel et al. [2004].)

singlet oxygen was detected by both chemical and singlet oxygen luminescence methods described above.

For two-photon photodynamic therapy, we have also utilized a nanoparticle co-encapsulation approach that does not involve chemical conjugation of the two-photon absorber and the photosensitizer and is thus simpler to implement [Kim et al., 2007]. The ORMOSIL nanoparticles co-encapsulated two-photon fluorescent dye nanoaggregates as an energy up-converting donor and a photosensitizing PDT drug as an acceptor. They combine two features: (i) aggregation-enhanced two-photon absorption and emission properties of a novel two-photon dye and (ii) nanoscopic fluorescence resonance energy transfer between this nanoaggregate and the HPPH photosensitizer. Stable aqueous dispersions of the co-encapsulating nanoparticles (diameter $\leq$ 30 nm) were prepared in the nonpolar interior of micelles by co-precipitating an organically modified silica sol with the PDT drug and an excess amount of the two-photon dye that forms fluorescent aggregates by phase separation from the particle matrix. Our results showed (i) indirect excitation of the photosensitizer through efficient two-photon excited intraparticle energy transfer from the dye aggregates in the intracellular environment of tumor cells and (ii) generation of singlet oxygen and *in vitro* cytotoxic effect in tumor cells by photosensitization under two-photon irradiation.

Another approach for two-photon PDT is to use upconverting rare-earth-containing nanoparticles that utilize sequential two-photon absorption as described in Chapters 5 and 8. Chatterjee and Yong [2008] used $NaYF_4$ nanocrystals co-doped with the rare-earth ions Yb^{3+} and Er^{3+} and coated with a polymeric polyethylene imine to convert 980-nm light. The upconverted green and red light excited the zinc phthalocyanine photosensitizer, noncovalently adsorbed on the nanoparticles, to produce photodynamic action.

In a similar approach, we developed nanoparticles that contain covalently conjugated photosensitizer(PS) HPPH in the ORMOSIL shell on surface of the upconverting nanophosphors ($NaYF_4$:Er^{3+},Yb^{3+}) [unpublished results]. These nanocrystals are able to upconvert light energy, emitting at ~550 and ~650 nm under excitation at ~980 nm. The 650-nm emission from $NaYF_4$:Er^{3+},Yb^{3+} nanocrystals spectrally overlaps with the absorption peak of the HPPH covalently conjugated into ORMOSIL shell. As a result, an energy transfer from the nanocrystals to PS occurs in these core–shell nanoparticles, sensitizing PS fluorescence and producing singlet oxygen. The latter produced toxicity to cancer cells *in vitro*.

Another direction to enhance tissue penetration for treatment of thick tumor is X-ray photodynamic therapy, where a deep-tissue-penetrating X ray is utilized to generate reactive oxygen species. In one such approach, Liu et al. [2008a] used LaF_3:Tb^{3+} nanoparticles conjugated to a meso-tetra(4-carboxyphenyl)porphine photosensitizer and carrying folic acid as the targeting group. The LaF_3:Tb^{3+} nanoparticles, acting as an X-ray scintillator, absorbs X rays, producing visible emission from the Tb^{3+} ions which overlaps with the absorption peaks in the PS to excited, the latter by energy transfer, whereby the PS then generates singlet oxygen. Nanobiotix, a Paris-based company (www.nanobiotix.com), uses another approach in which its X-ray absorbing nanoparticles produce energetic electrons to destroy the tumor.

Image-Guided Therapy. As discussed above, the initiation of light activation for therapy requires a waiting period during which the normal (healthy) tissues are clear of the PDT drug, while the PDT drug has remained accumulated in the tumor. It is thus highly desirable, both for enhancing the efficacy of PDT and for minimizing damage to the healthy tissues, to monitor the pharmacokinetics and biodistribution using imaging modalities. The two excellent reviews from Hasan's group [Celli et al., 2010; Rai et al., 2010] provide detailed accounts of combining imaging with PDT. This combination of various imaging modalities with PDT provides a number of advantages:

- Diagnosis of cancer and accumulation of the PDT drug in tumor.
- Monitoring of PDT.
- Post-therapy assessment, whether to reinitiate or try another type of therapy.

Most PDT photosensitizers are fluorescent, which can be utilized for optical imaging function. Thus, an approach can involve a compromise so that majority of photoexcitation of the photosensitizer (e.g., 75%) goes to produce singlet oxygen, but a small portion yields fluorescence for optical imaging [Rai et al., 2010]. This can readily be manipulated using a nanoparticle encapsulation strategy, whereby isolation of the photosensitizer from the external aqueous environment leads to enhancing its fluorescence quantum yield. However, it is more desirable to separate the two functions of imaging and PDT action, whereby one can use imaging first to ensure that the PDT drug is cleared of normal tissue and only present in the tumor, before initiating the PDT action by light irradiation. Chen et al. (2005) designed a bifunctional chromophore in which one group (indocyanine type IR emitting dye), excited at a longer wavelength, produced fluorescence for imaging, but no PDT action. The photosensitizer unit, excited at a shorter wavelength, produced the PDT action. The units were judiciously selected to minimize any FRET energy transfer from the short wavelength absorbing PDT photosensitizer to the fluorescent dye unit used for imaging.

Kopelman's group extensively used their PEBBLE concept, introduced in Chapter 3, for combined imaging and therapy [Koo et al., 2007; Tang et al., 2005]. The majority of their PEBBLE nanocarriers have been based on polymeric nanoparticles of polyacryl amide (PAA) which are slowly biodegradable. These nanoparticles have ranged in size from 30 to 70 nm; they exhibit low toxicity and hydrophilicity to allow for long plasma circulation time. In one formulation for *in vivo* studies, Kopelman and co-workers co-encapsulated an FDA-approved PDT drug, Photofrin, together with smaller-sized iron oxide nanoparticles which were used as MRI contrast agents [Koo et al., 2007]. In addition, they incorporated F3-peptide as the tumor targeting ligand and PEG for enhanced circulation. Rats bearing intracerebral 9 L gliosarcoma tumors were intraveneously injected with these nanoparticles, and the tumor was irradiated with a 630-nm laser radiation. The MR images revealed massive necrosis of the tumor.

Hasan's group has used liposome nanoparticles encapsulating benzoporphyrin-based photosensitizers to image, treat, and monitor PDT *in vivo* in a mouse model of disseminated ovarian cancer [Zhong et al., 2009]. This nanoformulation provided for fluorescence imaging, with a sensitivity of 86% for *in vivo* tumor detection using a microendoscope. At the same time, it provided effective PDT action, with an average decrease of 59% in the tumor volume.

Another nanocarrier used for PDT delivery is gold nanoparticles [Camerin et al., 2010]. In this work, the Zn(II)-phthalocyanine disulfide PDT drug was bound to gold nanoparticles and was intravenously injected in C57BL/6 mice, bearing a subcutaneously transplanted amelanotic melanoma. This nanoparticle–drug conjugate exhibited a greater selectivity of tumor targeting and a significantly enhanced PDT response. Another advantage of using gold nanoparticles as carriers is that they also provide opportunity for photothermal therapy with PDT. The photothermal therapy using gold nanoparticles is discussed in Section 13.6. In addition, gold nanoparticles also provide opportunities for a number of imaging modalities as discussed in Chapter 8.

A new approach using a carrier-free method for delivery of a hydrophobic drug is using its pure nanocrystals (nanosized crystals) form [Baba et al., 2007]. We produced HPPH nanocrystals using the reprecipitation method described in Chapter 4. The resulting drug nanocrystals were monodispersed and stable in an aqueous dispersion, without the necessity of any additional stabilizer (surfactant). These pure drug nanocrystals were taken up by the cancer cells with high avidity. Though the fluorescence and photodynamic activity of the drug were substantially quenched in the form of nanocrystals in the aqueous suspension, both these characteristics were recovered under *in vitro* and *in vivo* conditions. This recovery of drug activity and fluorescence is possibly due to the interaction of the nanocrystals with serum albumin, resulting in conversion of the drug nanocrystals into the molecular form. This was confirmed by demonstrating similar recovery in the presence of fetal bovine serum (FBS) or bovine serum albumin (BSA). Under similar treatment conditions, the HPPH in the nanocrystal form or in 1% Tween-80/water formulation showed comparable *in vitro* and *in vivo* efficacies.

13.5. MAGNETIC THERAPY

Magnetic therapy using magnetic nanoparticles has been widely investigated for cancer treatment. An advantage offered by this modality of therapy is that magnetic nanoparticles also can serve as a contrast enhancer for MRI, thus providing both therapy and MR imaging. Most of the work has focused on the use of the magnetic hyperthermia mechanism for thermal therapy, which requires an alternating (AC) magnetic field [Gazeau et al., 2008]. However, our group introduced the concept of magnetocytolysis, which involves cellular disruption by magnetic nanoparticles in a DC magnetic field such as the ones used in MRI [Bergey et al., 2002]. Thus the magnetic therapy can be classified in two categories, as shown in Table 13.1.

TABLE 13.1. Magnetic Therapy Classification

Magnetic Therapy	
AC magnetic field induced hyperthermia: Heating produced by magnetic nanoparticles leads to thermally induced tumor reduction	DC magnetic field induced magnetocytolysis: Intracellular localization and alignment of magnetic nanoparticles causes cellular disruption

Magnetic Hyperthermia. A large body of literature exists on the application of magnetic nanoparticles, particularly iron oxide, to produce hyperthermia and use it for thermal therapy [Berry, 2009; Gazeau et al., 2008; Laurent et al., 2008]. The physical mechanism of heat generation by a magnetic nanoparticle (typically <100 nm) in an AC magnetic field involves two sources: (a) Brownian relaxation, in which the energy coupled into the magnetic nanoparticle from the AC magnetic field dissipates due to rotation of the entire magnetic nanoparticle within the surrounding liquid medium, and (b) Neel relaxation, where the dissipation mechanism is the rotation of the magnetic moment within the magnetic core [Fortin et al., 2008].

Hyperthermia by local temperature rise between 42°C and 46°C leads to structural changes and changes in the enzymatic activities to produce necrosis and thus cell death. In addition, hyperthermia can also decrease cell growth and differentiation and induce apoptosis, which is initiated by a physical (heat) treatment as opposed to a chemical treatment in chemotherapy and thus have fewer side effects.

The quantifying parameter used to describe the efficacy of hyperthermia is specific absorption rate (SAR), also known as specific loss power (SLP), which is defined as the rate at which the electromagnetic energy is absorbed by a unit mass of the magnetic material and is expressed in the unit of calories per kilogram or watts per gram. It is proportional to the rate of the temperature increase. The main factor of the magnetic nanoparticles that determines SAR is their saturation magnetization (the maximum magnetization induced in them by increasing the magnetic field). In the case of nanoparticles, the saturation magnetization, apart from being determined by their chemical composition and crystal structure, is also dependent on the size, shape, surface modification induced by surface coating ligands, and polydispersity. SAR also depends on the external factors, such as the frequency f and the strength H, of the magnetic field. Thus in principle the SAR and the resulting heating efficiency can be increased by simply increasing the field strength H and the frequency f, but this would lead to inductive heating of the healthy tissue as well. Therefore, for effective hyperthermia, one keeps an optimized value of the field strength and the frequency [Hergt et al., 2004].

The size dependence of the nanoparticles has been extensively investigated. For smaller-sized nanoparticles, Neel relaxation, which is an intrinsic property, dominates. The variation of SAR with nanoparticle size is nonmonotonic, with the optimum value of size for iron oxide nanoparticles being around 15 nm [Fortin et al., 2008]. Monodispersed nanoparticles provide a significant increase of the heating efficiency compared to polydispersed systems. Hence a great deal of effort now is being devoted by various groups to develop and/or refine methods for narrowing down the size distribution of nanoparticles to effect thermal therapy with a lower dose of nanoparticles.

A number of mechanisms have been used for thermal therapy of cancer using magnetic nanoparticles, depending on the conditions of hyperthermia [Gazeau et al., 2008]. Under high hyperthermia producing temperature above

46°C, when nanoparticles are directly injected into the tumor tissue, thermoablation of the tissue causes tissue necrosis. Here the nanoparticles can either (a) act intracellularly by means of endocytotic uptake or (b) accumulate in the extracellular matrix of the tumor by the EPR effect. Under moderate hyperthermia condition, the temperature rise is in the range of 41–46°C. Under this condition the effects at cellular level, such as apoptosis and release of heat-shock proteins, can activate antitumoral immunity; at the tissue level, changes in pH and in the perfusion and oxygenation of the tumor microenvironment can take place.

With the help of active targeting of cancer cells with various targeting groups, there has been a considerable degree of interest recently in intracellular thermal therapy effected by receptor-mediated endocytosis of nanoparticles [Yoo et al., 2011]. The thermal therapeutic action is produced by internalization of the magnetic nanoparticles inside the intracellular endosomal compartments. Under these conditions, with the increased viscosity for the nanoparticles in the vesicles, it is expected that only Neel relaxation of magnetic moment will contribute to heating, with a vanishing magnitude of the Brownian contribution which is derived from the particle reorientation in the medium.

Another area of active investigation is the use of moderate hyperthermia as an adjunct to radiation and/or chemotherapy [Johannsen et al., 2006]. Hyperthermia allows an increase of perfusion in the tumor tissue and a higher oxygen content which makes the γ-ray radiation more powerful in destroying the tumor tissue. Thus a lower dosage of the γ-ray radiation can be used.

Another direction is to use the concept of self-regulation to control the temperature rise. This can be achieved by using magnetic nanoparticles that exhibit a Curie temperature of approximately 45–50°C [Gazeau et al., 2008]. The Curie temperature defines the temperature limit above which the net magnetic moment of the nanoparticles drastically drops, thus significantly reducing the saturation magnetization and SAR. The net result is a drop in heating, hence producing self-regulation action.

Magnetic Nanoparticles. Since the seminal work of Jordan et al. [1993] on hyperthermia treatment, most magnetic thermal therapy investigations have involved iron oxide nanoparticles in the superparamagnetic form, as discussed in Chapter 5. There are various types of iron oxide nanoparticles [Laurent et al., 2008]. The two forms used for hyperthermia are magnetite (Fe_3O_4) and maghemite (γ-Fe_2O_3). The magnetite form involves mixed oxidation states Fe^{3+} (ferric) and Fe^{2+} (ferrous) of iron, whereas maghemite consists of iron in only the Fe^{3+} oxidation state. The magnetic properties of both these forms of nanoparticles are quite comparable [Sun et al., 2004]. However, Fe_3O_4, due to the presence of the Fe^{2+} ion, is unstable in oxygen-rich environments because of the tendency of Fe^{2+} to oxidize to the Fe^{3+} state. Thus the shelf-life of an Fe_3O_4 colloidal dispersion is limited. In contrast, γ-Fe_2O_3 is quite stable. However, a great deal of work has been done using Fe_3O_4, and a number of commercial sources provide Fe_3O_4 nanoparticles. The reason is the simplicity of preparation and control of growth condition (to produce monodispersity) for Fe_3O_4.

These nanoparticle formulations are coated to improve biocompatibility as well as to provide linkage sites for attaching targeting and PEG groups. These coatings can be of natural materials such as dextran or proteins or phospholipids. Another advantage produced by the coating is to minimize the release of iron ions, whose accumulation can cause toxicity. For targeting, dextran-coated nanoparticles have been conjugated with targeting ligands. An example is the conjugation of breast cancer targeting chimeric L6 monoclonal antibody [DeNardo et al., 2005].

Materials with higher magnetization have also been investigated, because they will yield higher SAR [Gazeau et al., 2008]. An example is metallic iron nanoparticles that have higher magnetization compared with iron oxide. However, their biocompatibility is a major issue. Thus the use of these nanoparticles requires a biocompatible coating on them.

For utilizing the function of self-regulation of hyperthermia, nanoparticles of materials with Curie temperature around 50°C are being investigated [Gazeau et al., 2008]. An example is the substituted ferrites such as $Mn_{1-x}Zn_x Fe_2O_4$.

In Vivo Studies and Clinical Trials. In addition to extensive *in vitro* studies, many *in vivo* experiments have also successfully been conducted. Johannsen et al. [2006] have used magnetic-nanoparticle-induced hyperthermia in conjunction with external radiation in an orthotopic rat model of prostate cancer. They showed a significant and additive reduction of local tumor growth, and they have initiated clinical trials evaluating toxicity and tolerance ability of thermal therapy in patients with locally recurrent prostate carcinoma, after radiation therapy.

DC Magnetic Field-Induced Magnetocytolysis. This approach introduced by our group utilized γ-Fe_2O_3 nanoparticles encapsulated in an ORMOSIL nanoclinic (see Chapter 3) of size ~30 nm [Bergey et al., 2002]. The LHRH hormone analogue was attached to the nanoclinic for targeting MCF-7 and UCI-107 cells. The nanoclinic-cell suspension was exposed to a DC magnetic field, and magnetocytolysis was determined. LHRH-positive nanoclinics showed a greater magnetocytolytic effect on both the MCF-7 and the KB cells (both of which overexpress LHRH receptors) than on the UCI cells (which does not overexpress LHRH receptors). LHRH-negative nanoclinics had almost no lytic activity on any of the above cells.

To understand the mechanism of cellular disruption by magnetocytolysis, the magnetic behaviors of particles were investigated (in the presence and absence of KB cells) in the superparamagnetic range using a magnetometer (SQUID). A comparison of magnetization curves of LHRH-positive particles in the presence and absence of suspended KB cells confirmed a disruption of the cellular structure. The nanoclinics are bound to the cell membrane or inside the cell, and physical orientation in a magnetic field is then harder than for free nanoclinics. When the two curves reached the same plateau and the field is decreased, the resultant curves are equivalent. At this stage, all the nanoclinics can be considered as free particles. This data suggested that a probable mechanism for magnetocytolysis could be a physical orientation of the magnetic particles in the field, which induces a physical disruption of the cell. The data demonstrated the magnetocytolytic ability (cell lysing) of these nanoclinics and that this effect requires a direct contact with and/or internalization into the target cell. An AFM study also revealed disruption of the cellular structure.

Tumor Targeting of Magnetic Nanoparticles Across the Blood–Brain Barrier. Apart from magnetic hyperthermia, magnetic nanoparticles, owing to their excellent contrast-enhancing properties in magnetic resonance imaging (MRI), can serve as theranostic agents for combined diagnosis and treatment of cancer. Their versatility can be further emphasized by the fact that with proper size and surface functionalization, they can be made to carry anticancer drugs and ferry them across biological barriers. This is exemplified by the work of Miqin Zhang and colleagues, who have demonstrated the use of superparamagnetic iron oxide nanoparticles to deliver drugs across the blood–brain barrier (BBB) for the diagnosis and therapy of glioma, which is the most common type of brain tumor [Fang and Zhang, 2010]. Specifically, this iron oxide nanoparticle core was coated with the biocompatible polymer PEG, a fluorescence dye Cy5.5, and a targeting agent chlorotoxin. Cy5.5 emits light at near-infrared wavelengths, which has better penetration through biological specimens. For example, it can penetrate several centimeters through brain tissue. Inclusion of the optical probe within the magnetic nanoparticles will facilitate diagnosis of the target glioma using both MRI and fluorescence microscopy, thus allowing co-registration of two different imaging modalities for accurate diagnosis. This will also allow a surgeon to use the MRI scan as a roadmap to the fluorescently labeled glioma in the brain. Besides dual-modality imaging, the group is also actively investigating the concurrent delivery of a variety of cancer-therapeutic agents, such as siRNA, chemotherapy drugs, and so on.

13.6. PHOTOTHERMAL THERAPY

Photothermal therapy is now being recognized as a prospective cancer therapy modality, employing various nanoheaters that accumulate at tumor sites. This functionality and some of the nanoparticles used for producing photothermal effect have already been discussed in Chapter 5. The cell injury induced by local heating with these photothermal nanoheaters is most likely caused by both necrosis and cell membrane rupture, because the local temperatures could be raised above 70°C, high enough to denature proteins as well as disrupt protein, DNA, and RNA [El-Sayed, 2010]. These nanoparticles heat their local environment, when irradiated with laser light at wavelength of their absorption band. The adjacent healthy tissue without embedded nanoparticles is unaffected by the laser light alone, but cancer cells in the direct vicinity of the nanoparticles undergo hyperthermia and death, resulting in drug-free tumor remission.

A number of nanoparticle formulations have been used for photothermal therapy. The earlier work utilized organic dyes with strong light absorption, but having nonradiative decay of excitation as the main channel for returning to the ground state. However, the photostability of an organic dye is a major concern that has led to utilization of inorganic nanoparticles for photothermal therapy. In the case of these inorganic nanoheaters, the nanoparticles themselves play an active role in therapy, as they themselves absorb light and product heat. The majority of work on photothermal therapy has been done utilizing metallic nanostructures, specifically gold nanoshells and nanorods.

Pioneering work on these plasmonic nanoheaters has been done by Halas, West, and co-workers at Rice University, as well as by El-Sayed at Georgia Tech. They are described below in detail. More recently our group has utilized rare-earth-containing nanoparticles, where the optical excitation at a shorter wavelength produces emission in the 2.7- to 2.9-μm range which is strongly absorbed by the aqueous cellular environment to produce local heating. These types of nanoparticles for photothermal therapy are discussed here.

Plasmonic Photothermal Therapy. Plasmonic photothermal therapy (PPTT) takes advantage of the extremely high absorption coefficient of the plasmonic bands in the metallic nanostructures. For example, gold nanoparticles can absorb light four orders of magnitude higher than the best-absorbing organic dyes. Furthermore, these plasmonic excitations are efficiently converted by nonradiative processes into heat. Because the treatment of thick tumors requires larger optical penetration depth provided by light in the near-IR wavelength range, the choice of metallic nanostructure size and shape can readily lead to manipulation of the plasmonic absorption band to be in this region. Examples already discussed in Chapter 5 are gold nanoshells and gold nanorods. In the case of gold nanoshells, this manipulation can be made by a judicious choice of the shell-to-core ratio. Gold nanoshells were the first plasmonic structures introduced by Halas, West, and co-workers for photothermal therapy [Hirsch et al., 2003]. They were successfully tested in both *in vitro* and *in vivo* studies and are already in human trials following FDA approval for near-IR photothermal therapy in refractory head and neck cancers [http://clinicaltrials.gov/ct2/show/NCT00848042]. These nanoshells typically have a core of 60 nm and the outer shell of 70 nm with absorption coefficient per micron of ~36, a peak at 892 nm. When injected systematically, these particles accumulated in the tumor due to the EPR effect, producing passive targeting (see Chapter 7). Active tumor targeting can be added to the surface of the nanoshells by their appropriate functionalization and subsequent chemical conjugation with targeting moieties. The dimensions of these nanoshells can also be optimized to produce absorption and scattering, whereby the scattering property can be used for contrast enhancement in imaging. Halas, West, and co-workers demonstrated this by combining optical imaging using optical coherence tomography (OCT) with effective photothermal ablation of tumors [Gobin et al., 2007].

El-Sayed and co-workers have demonstrated a very effective photothermal therapy, together with imaging, using gold nanorods [Huang et al., 2008; Jain et al., 2007]. They claim that gold nanorods produce a tumoricidal effect with less light energy than the nanoshells. The increased effectiveness of the photothermal therapy produced by the gold nanorods was attributed to nearly 30 times higher absorption coefficient of the longitudinal plasmon resonance compared to that of the plasmon resonance in nanoshells [Jain et al., 2006]. For example, they report achieving cell death with immunotargeted gold nanorods in a 3-min exposure at 10 W/cm^2 *in vitro* [Dickerson et al., 2008], whereas gold nanoshells required a 7-min exposure at 35 W/cm^2 [Hirsch et al., 2003]. Again as described in Chapter 5, the longitudinal plasmon resonance in gold nanorods can be tuned to be in the near-IR region by appropriately selecting the aspect ratio of the gold nanorod. Gold nanorod photothermal therapy has also been demonstrated *in vivo* [El-Sayed, 2010]. In an oral squamous cell carcinoma implanted murine tumor, treatment with an 808-nm LED laser with a 6-mm beam achieved a 22°C temperature rise and a 25% reduction of the tumor in an intravenous injection and 57% reduction in direct injection in tumor. These authors also successfully used gold nanorods simultaneously for imaging. Kuo et al. [2010] have also used gold nanorods both as hyperthermia agents and as imaging probes.

Plasmonic Aggregates. Plasmonic photothermal therapy (PPTT) using either nanoshells or nanorods still requires excitation with high optical energy to produce significant damage of the diseased tissues; the photothermal transduction efficiency is dependent on the size of nanoparticles, with larger particles capable of a higher photothermal effect [Huang et al., 2008; Ekici et al., 2008]. This may limit clinical application of PPTT, since it requires high-power laser irradiation and larger particles with higher concentration in the targeted site to be efficient. However, the size of the nanoparticles cannot be significantly increased without affecting their stability in systemic circulation and their biodistribution pattern. As a result, the only way to achieve a large heating volume and, correspondingly, enhanced therapeutic effect in PPTT is to have clusters, aggregates, and arrays of nanoparticles of high concentration [Zeng et al., 2009]. Aside from the primary effect of heating the particles themselves, secondary effects include diffusion of heat in the surrounding media and vapor bubble nucleation, if the temperature at the surface of the particle exceeds the vaporization threshold of the media. The generation of vapor bubbles (plasmonic nanobubbles, PNB) is now considered to be a more efficient cytotoxic event in PPTT [Wagner et al., 2010; Lapotko, 2009]; efficiency of PNB generation is shown to be dependent on the size of nanoparticle clusters [Lukianova-Hleb et al., 2010; Kitz et al., 2011, Hleb et al., 2008]. The phenomenon of vapor bubble formation from gold nanoparticles upon exposure to pulsed irradiation is a very important aspect of the interaction of nanoparticles with light in PPTT. A development of plasmonic nanostructures that will provide the enhanced, controlled clusterization on the tumor site would result in significant enhancement of the efficacy of PPTT and is of undoubted importance to the field of cancer therapy.

Aggregation of the gold nanoparticles has been increasingly investigated as a factor enhancing the photothermal effect [Zeng and Murphy, 2009]. Nam et al. [2009] reported the design and synthesis of "smart" gold nanoparticles that can respond to pH change and form aggregates. When exposed to an acidic environment or internalized inside the cells, "smart" gold nanoparticles began to aggregate. This resulted in rapid formation of mixed charged particle surfaces and, correspondingly, the formation and growth of "smart" gold nanoparticle aggregates. The pH-induced aggregates shifted their absorption to far-red and NIR, which was exploited for photothermal cancer therapy. Enhanced accumulations of "smart" gold nanoparticles in cells by blocked exocytosis were observed and resulted in a lower power threshold for the efficient photothermal therapy *in vitro*.

Rare-Earth-Doped Nanoheaters. We recently introduced a novel optical nanoheater concept in which the heat in water-based biological systems is generated as a result of absorption of the nanoheater emission at ~2.8 μm. At this wavelength, water molecules exhibit strong vibrational absorption. As a result, 90% of emission with wavelength of 2.8 μm is absorbed in a 1-μm-thick water layer; most of the absorbed energy is converted to heat. Therefore, a single optical nanoheater particle emitting at 2.8 μm will produce much larger heating volume than a plasmonic nanoheater of the comparable size. Strong water absorption at ~3 μm is known to be used in biomedical applications. In particular, lasers operating at about 3 μm have been developed for applications in ablation surgery and dentistry [Kaufmann and Hibst, 1996]. Lasing at about 2.7–2.9 μm can be obtained utilizing the $^4I_{11/2} \rightarrow {}^4I_{13/2}$ transition of the Er^{3+} ions. We synthesized nanocrystals of $LiYF_4{:}Er^{3+}$ ($LiYF_4{:}Er^{3+},Yb^{3+}$) demonstrating the $0.98 \rightarrow 2.8$ μm conversion with the highest possible efficiency. This energy conversion in the synthesized nanocrystals was followed by the absorption of the generated ~2.8-μm emission and corresponding heating of the water-based surroundings. As a result of absorption of the 2.8-μm emission, the absorbed energy will be spread over a larger volume than in the case of PPTT. Because of this, a lower peak temperature value is expected which may not be necessarily favorable for PTT with nanophosphors, but can enhance therapeutic efficacy of PDT or chemotherapy.

13.7. NEUTRON CAPTURE THERAPY

Neutron capture therapy (NCT) is a radiation therapy; but as opposed to a conventional radiation therapy using X rays or γ rays where an external radiation source is used, NCT utilizes a neutron source to generate the tumor destroying radiation internally at the tumor site [Ichikawa et al., 2006]. The concept of NCT has already been introduced in Chapter 5. A major advantage is that the surrounding normal tissue does not get exposed to the radiation, which becomes a major issue in radiation treatment of tumors in deeper parts of the body. NCT involves the use of an isotope of a nucleus with a large neutron capture cross section which then emits high-energy radiation such as γ rays, α particles, or electrons that can interact with the local tumor tissue to destroy it. The elements prevalent in a biological medium, namely O, C, H, and N, have very low neutron capture cross section (<2 barn). In contrast, the neutron capture cross sections for ^{10}B and ^{157}Gd isotopes, so far used for NCT, are 3840 and 25,400 for thermal (slow) neutrons in the low energy range <0.4 eV. Thus at this low energy level, the neutron flux for NCT using these nuclei does not affect the intrinsic biological molecular contents that are made of C, H, O, and N. The natural abundance of ^{10}B isotope in the boron element is 20%, the remaining being ^{11}B. Thus, it has a high natural abundance. The ^{157}Gd isotope has a natural abundance of 15.7%, which again is high enough that naturally occurring Gd can be used effectively to capture relatively low energy (slow) thermal neutrons.

The concept of NCT was introduced in 1936, and the first clinical treatment started as early as 1951 (for historic accounts, please see Ichikawa et al. [2006]). The most extensive studies have been utilizing boron, the reason being the toxicity concern of gadolinium. The two widely used boron species are mercaptoundecahydro-closo-dodecaborate dianion (BSH) and boronphenylalanine (BPA), which have been used in clinical trials. When these boron species are delivered to tumor, the ^{10}B isotope captures the neutrons and emits ^{7}Li and α particles as the primary fission product of the neutron capture reaction with ^{10}B. It was reported that these α particles are relatively slow and give rise to closely spaced ionizing events that consist of tracks of sharply defined columns [Barth et al., 1990]. They have a path length of approximately 10 µrn, exhibit a high linear energy transfer, LET, and destroy a wide variety of biologically active molecules including DNA, RNA, and proteins. For these reasons, there is little, if any, cellular repair from α-particles-induced radiation injury. Since the ^{10}B(n, α)^{7}Li reaction produces a significant biological effect only when there is a sufficient fluence of thermal neutrons and a critical amount of ^{10}B localized around, on, or within the cell, the radiation produced can be extremely localized, thereby not affecting normal tissue components [Palmer et al., 2002].

The use of different molecular compounds of boron for early clinical trials at the Massachusetts General Hospital and the Brookhaven National Laboratory raised two major concerns for their success. Firstly, the penetration depth of the thermal neutrons is limited to 3–4 cm, and rapid attenuation in tissue due to absorption and scattering led to the conclusion that BNCT will only be applicable to superficial tumors. Secondly, low molecular weight boron compounds were freely diffusible and thus the desired concentration was not achieved in the tumor [Barth et al., 1990]. Thus it became apparent that the success of boron NCT naturally depends on selective accumulation of the boron species at the tumor site or inside the tumor cells for intracellular action. It is for this reason that various nanoparticle and dendritic structures have been produced and investigated for NCT, since a nanoparticle formulation can provide for both active and passive targeting of tumor, as discussed at various places earlier in this book.

A great deal of work has been focused on liposome nanoparticle delivery of boron [Martini et al., 2004; Koning et al., 2004]. Thirumamagal et al. [2006] used receptor-targeted liposomal delivery of boron-containing cholesterol mimics for boron NCT. Mortensen et al. [2006] utilized boron carbide nanoparticles <100 nm for T-cell-guided boron NCT. The *in vitro* thermal neutron irradiation of B16 melanoma cells incubated with these nanoparticles induced complete cell deaths. They found no evidence of nanotoxicity produced by the nanoparticles alone.

^{157}Gd captures thermal neutrons to emit long-range prompt γ rays, internal conversion electrons, X rays, and Auger electrons with a large total kinetic energy. The tumor killing effect in the case of ^{157}Gd NCT is produced by a combined action of γ rays and the electrons. In principle, ^{157}Gd NCT offers a number of advantages, some of which are:

- ^{157}Gd has the highest thermal neutron capture cross section, which is 66 times higher than that of ^{10}B.
- γ rays produced by ^{157}Gd after neutron capture has a quite long range (more than 100 μm) in contrast to the 9-μm range of the α particles produced by ^{10}B.
- The Auger electrons with short range and high energy can attack DNA in tumor cells.
- As discussed in Chapter 8, Gd^{3+} being paramagnetic can serve as a contrast enhancer for MRI, thus integrating NCT with diagnostics.

However, there has been a major concern about the toxicity of the Gd^{3+} ions. But chelation with ligands such as diethylenetriaminepentaacetic acid (DTPA) has been shown to yield a Gd-DTPA complex that is stable in the bloodstream and is nontoxic.

Various nanoformulations incorporating Gd have been used for NCT. Kobayashi et al. [2001] used an avidin-dendrimer Av-G6Gd for tumor targeting NCT of intraperitoneal disseminated tumor which could also be monitored by MRI. Av-G6Gd showed *in vitro* internalization into SH1N3 cells, 50 times greater than Gd-DTPA. Av-G6Gd also showed sufficient accumulation into SH1N3 cells *in vivo*. Gd has also been delivered using chitosan nanoparticles [Ichikawa et al., 2006].

13.8. CIRCULATING TUMOR CELLS

One of the major hallmarks of cancer is the ability of cancer cells to migrate from their origin (primary tumor) to distant sites via the circulatory system, a process known as metastasis. Herein, breakaway cancer cells enter the blood or lymph fluids by invasion/penetration through their vessel walls. The circulatory system is discussed in Chapter 12. Some of these traveling cells may possess the ability to again penetrate through their vessel walls and enter the nearest organ, where they nestle and proliferate, thus forming micrometastases. Since they circulate in the blood or lymph fluids prior to invading a tissue, they are known as circulating tumor cells (CTCs). These cells were first observed in 1869 by Australian physician Thomas Ashworth, in the blood of a man with metastatic cancer. CTCs have the unique capability to survive in circulation, just like blood cells, without being anchored to a host matrix [Hanahan and Weinberg, 2000].

Since the CTCs usually reflect the aggressiveness of cancer, their detection in the blood or lymph has enormous potential in the diagnosis, therapy, and real-time therapy monitoring for cancer. However, the number of CTCs that may be present in these fluids of patients suffering from metastatic cancer is negligibly small, of the order of 1–10 per milliliter of blood. In comparison, one milliliter of blood contains a few million white blood cells and a billion red blood cells. Therefore, detection of CTCs in the blood is an arduous task, which mandates the development of ultrasensitive *in vitro* diagnostic methods for their detection. Fortunately, recently few assays have been described, related to the detection and quantification of CTCs. These assays rely on a combination of enrichment and detection steps [Mostert et al., 2009].

The enrichment step usually involves separating and pooling tumor cells from a background of normal blood cells, either using morphologic criteria, such as size or density gradient, or via immunomagnetic separation using tumor-specific markers attached to magnetic nanoparticles. Some of the commercial enrichment techniques are RARE (RosetteSep-Applied imaging Rare Event), Histopaque, Percoll, ISET (Isolation by Size of Epithelial Tumor cells), Nucleopore, or MACS (Magnetic Activated Cell Sorting), among others. After enrichment, nucleic-acid-based techniques like reverse transcriptase (RT)-PCR or cytometric methods are applied to detect CTCs through putative tumor-specific markers. The usefulness of such CTC detection assays has been demonstrated in several metastatic cancers, such as breast colorectal, prostate, pancreatic, gastric, bladder, and lung cancers. More information about these assays can be found in Mostert et al. [2009].

13.9. NCI ALLIANCE FOR CANCER NANOTECHNOLOGY

In 2004, the U.S. National Cancer Institute (NCI) initiated the Alliance for Nanotechnology in Cancer program, in order to harness the power of nanotechnology to radically change the way cancer is diagnosed, treated, and prevented. Under this program, several Universities, Cancer Hospitals, and Research Institutes were funded to conduct interdisciplinary research aimed at translational medicine, merging the expertise of physicists, chemists, chemical engineers, and clinicians. In addition to evaluating the diagnostic and therapeutic potential of novel nanomaterials, new systematic protocols for product safety regulation also has been devised under this initiative. To implement this, NCI's Nanotechnology Characterization Laboratory (NCL) has joined forces with the National Institute of Standards and Technology (NIST) and the U.S. Food and Drug Administration (FDA) to develop standard preclinical toxicology, pharmacology, and efficacy assay cascades for nanomaterials.

Under this initiative, so far a number of promising nano-based diagnostic and therapeutic agents have been developed. Some examples include (a) a PAMAM dendrimer system for the delivery of methotrexate using folic acid targeting to tumor cells/tissues, developed in the lab of Dr. James Baker, (b) an aptamer-targeted polymeric particles that bind exclusively to the extracellular region of antigens expressed on prostate cancer (PCa) cells, developed in the lab of Dr. Robert Langer, (c) the Cyclosert™ delivery system based on cyclodextrin-containing polymers, developed in the lab of Dr. Mark Davis for the tumor-targeted delivery of camptothecin, (d) a targeted nanoparticles system to deliver doxorubicin to the $\alpha v\beta 3$-expressing tumor vasculature, developed in the lab of Dr. David Cheresh. Under this program, our own group has developed several quantum dot and gold-nanoparticle-based diagnostic/theranostic nanoplatforms for the early diagnosis and targeted therapy of pancreatic cancer. More information about this alliance can be found elsewhere [Farrell et al., 2011; Hinkal et al., 2011].

HIGHLIGHTS OF THE CHAPTER

- The use of a nanostructure platform allows one to incorporate multiple probes for multimodal and multispectral imaging and sensing, whereby one can utilize detection based on multiple biomarkers. The combined imaging probes provide molecular, structural, physiological, and morphological information on early stages of cancer development.
- The ability to guide a therapy/surgery and monitor its course is a major advantage of using a nanoplatform for cancer therapy. Furthermore, one can use an externally activated localized therapy to maximize its effectiveness and reduce any damage to healthy tissues.
- The benefits of nanocarrier delivery of a chemotherapeutic agent are improved bioavailability of a hydrophobic drug, a favorable pharmacoki-

netics and biodistribution by the design and surface tailoring of the nanocarriers, and enhanced pharmacodynamics produced by more effective drug–receptor interactions, as well as by controlled drug release.

- Nanoparticle delivery systems so far approved by the FDA for chemotherapy are based on liposomes, polymers, or micelles; the two prominent, approved nanoparticulate drug formulations are (i) *Doxil*® (liposomal-PEG doxorubicin) and (ii) *Abraxane*® (albumin-bound paclitaxel). Many other nanoparticulate drug delivery systems have been approved for chemotherapy and for clinical trials.
- A nanocarrier provides sustained, controlled, and targeted delivery of a chemotherapeutic agent by localizing in the cytoplasm or lysosomes of cancer cells, without triggering the *p*-glycoprotein pump, thus avoiding the multi-drug resistance (MDR) effect of such cells.
- Nanoparticles of poly(lactic-co-glycolic acid (PLGA) have received considerable attention owing to its biodegradability, biocompatibility, and ability to provide sustained drug release. Thus many chemotherapeutic drugs such as paclitaxel, doxorubicin, 5-fluorouracil, and dexamethasone have been successfully used with PLGA nanoparticles.
- Passive tumor targeting using nanoparticles rely on the enhanced permeability and retentivity (EPR) effect, which is characterized by leaky vasculature and poor lymphatic drainage associated with tumor tissues.
- Active tumor targeting using nanoparticles relies on the incorporation of targeting ligands, such as monoclonal antibodies, transferrin, folic acid, RGD peptide, and aptamers, on the nanoparticle surface, whose corresponding receptors are overexpressed in tumor cells/tissues.
- Photodynamic therapy (PDT) is based on light activation of photoactive drugs, known as photosensitizers, which lead to energy transfer to nearby molecular oxygen, in turn resulting in the formation of reactive oxygen species (ROS). ROS, such as singlet oxygen, are eventually responsible for the therapeutic action of PDT.
- The photosensitizer HPPH can be incorporated within the ORMOSIL nanoparticles, by either encapsulation (pre-loading), post-loading, or chemical conjugation. In each case, the PDT efficacy of the incorporated drug via the generation of singlet oxygen is maintained upon energy transfer to the surrounding molecular oxygen.
- Monodispersed nanocrystals of pure HPPH, prepared by reprecipitation, are "carrier-free" and taken up by the cancer cells with high avidity. Under similar treatment conditions, the HPPH in the nanocrystal form or in a 1% Tween-80/water formulation showed comparable *in vitro* and *in vivo* PDT efficacy.
- Magnetic nanoparticles provide magnetic therapy as well as can serve a as contrast enhancer for MRI. Most magnetic therapy has utilized alternating (AC) magnetic-field-induced hyperthermia to thermally destroy

the tumor cells. However, magnetic nanoparticles can also introduce magnetocytolytic therapy, which involves cellular disruption by magnetic nanoparticles in a DC magnetic field.

- Plasmonic photothermal therapy (PPTT) employs various nanoparticle heaters (nanoheaters) accumulated at tumor sites, to produce local heat at the tumor site, when irradiated with laser light at wavelength of the nanoparticle absorption band. The cell injury induced by local heating is most likely caused by both necrosis and cell membrane rupture.
- The majority of work on PPTT has utilized metallic nanostructures, specifically gold nanoshells and nanorods. A method to achieve a large heating volume and, correspondingly, enhanced therapeutic effect in PPTT is to use clusters, aggregates, and arrays of metallic nanoparticles of high concentration.
- Recently, rare-earth-containing nanoparticles have been introduced for PPTT, where the optical excitation at a shorter wavelength produces emission in the 2.7- to 2.9-μm range, which is strongly absorbed by the aqueous cellular environment to produce local heating.
- Neutron capture therapy (NCT) involves the use of a nonradioactive isotope with a large neutron capture cross section, which, when bombarded with a neutron beam, emits high-energy radiation such as γ rays, α particles, or auger electrons that can efficiently kill cancer cells at the targeted tumor sites.
- Boron-10 and gadolinium-157 have been used extensively for NCT owing to their high-neutron-absorption cross section. Nanoformulations for these two isotopes have also been used in NCT.
- Circulating tumor cells (CTCs) are cancer cells that break away from the primary tumor and invade other organs via the circulatory system to form metastases. Detection and quantification of CTCs from the blood can allow an early diagnosis of cancer, as well as monitor cancer therapy in real time.
- The U.S. National Cancer Institute (NCI) initiated the Alliance for Nanotechnology in Cancer program, in order to harness the power of nanotechnology to radically change the way cancer is diagnosed, treated, and prevented through multidisciplinary research.

EXERCISES

True or False

1. Drugs can only be encapsulated within nanoparticles.
2. Enhanced permeability and retentivity (EPR) effect is only associated with cancer and inflammatory sites.
3. Nanoparticles of poly-lactic-co-glycolide (PLGA) are nonbiodegradable.

4. Detection of circulating cancer cells (CTCs) help in the early diagnosis of cancer.
5. Multi-drug resistance (MDR) in cancer cells has no association with the *p*-glycoprotein pump present in these cells.
6. Presence of iodine enhances the singlet oxygen generation efficiency of photosensitizers in photodynamic therapy (PDT) of cancer.
7. Quantum dots have better photostability than organic fluorophores.
8. Disruption of cancer cells containing magnetic nanoparticles can be triggered by AC-magnetic field only.

Multiple Choice

9. Which of the following is/are property/ies of nanoparticles?
   (a) Multimodality
   (b) Controlled release of bioactive drugs
   (c) Targeted delivery
   (d) All of the above
10. Which of the following nanodrugs is not approved for treatment of cancer in humans?
   (a) Abraxane
   (b) Lunesta
   (c) Doxil
   (d) Genexol-PM
11. Which of the following ligand/s can be used to target nanoparticles to cancer cells?
   (a) Transferrin receptor
   (b) Folic acid receptor
   (c) Monoclonal antibodies
   (d) All of the above
12. The following component is not needed in photodynamic therapy (PDT) of cancer:
   (a) Carbon dioxide
   (b) Oxygen
   (c) Light
   (d) Photosensitizer
13. Quantum dots are not associated with which of the following?
   (a) Broad absorption spectra
   (b) Surface Plasmon resonance

(c) Narrow emission spectra
(d) Heavy-metal content

14. Light in the following electromagnetic region has maximum penetration through tissues:
(a) Ultraviolet
(b) Visible
(c) Near infrared
(d) Far red

15. Photothermal therapy is possible with the following nanostructure(s) of gold:
(a) Gold nanospheres
(b) Gold nanorods
(c) Gold nanocages
(d) All of the above

16. The following high-energy species is not generated in neutron capture therapy (NCT):
(a) Alpha particles
(b) Beta particles
(c) Gamma rays
(b) Auger electrons

Answers: 1. F; 2. T; 3. F; 4. T; 5. F; 6. T; 7. T; 8. F; 9. d; 10. b; 11. d; 12. a; 13. b; 14. c; 15. d; 16. b

REFERENCES

Acharya, S., Sahoo, S. K., PLGA nanoparticles containing various anticancer agents and tumour delivery by EPR effect, *Adv. Drug Deliv. Rev.* **63**, 170–183 (2011).

Baba, K., Pudavar, H. E., Roy, I., Ohulchanskyy, T. Y., Pandy, R., Prasad, P. N., A new method for delivering a hydrophobic drug for photodynamic therapy using pure nanocrystal form of the drug, *Mol. Pharma.* **4**, 289–297 (2007).

Bagalkot, V., Zhang, I., Levy-Nissenbaum, E., Quantum dot–aptamer conjugates for synchronous cancer imaging, therapy, and sensing of drug delivery based on bi-fluorescence resonance energy transfer, *Nano Lett.* **7**, 3065–3070 (2007).

Barth, R. F., Solloway, A. H., Fairchild, R. G., Boron neutron capture therapy of cancer, *Cancer Res.* **50**, 1061–1070 (1990).

Bergey, E. J., Levy, L., Wang, X., Krebs, L. J., Lal, M., Kim, K.-S., Pakatchi, S., Liebow, C., Prasad, P. N., DC magnetic field induced magnetocytolysis of cancer cells targeted by LH-RH magnetic nanoparticles in vitro, *Biomed. Microdevices* **4**, 293–299 (2002).

Berry, C. C., Progress in functionalization of magnetic nanoparticles for applications in biomedicine, *J. Phys. D: Appl. Phys.* **42**, 224003, 9 pages (2009).

Bhawalkar, J. D., Kumar, N. D., Zhao, C. F., Prasad, P. N., Two-photon photodynamic therapy, *J. Clin. Laser Med. Surg.* **15**, 201–204 (1997).

Blum, J. L., Savin, M. A., Edelman, G., Pippen, J. E., Robert, N. J., Geister, B. V., Kirby, R. L., Clawson, A., O'Shaughnessy, J. A., Phase II study of weekly albumin-bound paclitaxel for patients with metastatic breast cancer heavily pretreated with taxanes, *Clin. Breast Cancer* **7**, 850–856 (2007).

Brannon-Peppas, L., Ghosn, B., Roy, K., Cornetta, K., Encapsulation of nucleic acids and opportunities for cancer treatment, *Pharma. Res.* **24**, 618–627 (2007).

Camerin, M., Magaraggia, M., Soncin, M., Jori, G., Moreno, M., Chambrier, I., Cook, M. J., Russell, D. A., The *in vivo* efficacy of phthalocyanine–nanoparticle conjugates for the photodynamic therapy of amelanotic melanoma, *Eur. J. Cancer* **46**, 1910–1918 (2010).

Celli, J. P., Spring, B. Q., Rizvi, I., Evans, C. L., Samkoe, K. S., Verma, S., Pogue, B. W., Hasan, T., Imaging and photodynamic therapy: mechanisms, monitoring and optimization, *Chem. Rev.* **110**, 2795–2838 (2010).

Chakravarthy, K. V., Davidson, B. A., Helinski, J. D., Ding, H., Law, W. C., Yong, K. T., Prasad, P. N., Knight, P. R., Doxorubicin-conjugated quantum dots to target alveolar macrophages and inflammation, *Nanomedicine* **7**, 88–96 (2011).

Chatterjee, D. K., Yong, Z., Upconverting Nanoparticles as nanotransducers for photodynamic therapy in cancer cells, *Nanomedicine (Lond.)* **3**, 73–82 (2008).

Chen, Y., Gryshuk, A., Achilefu, S., Ohulchansky, T., Potter, W., Zhong, T., Morgan, J., Chance, B., Prasad, P. N., Henderson, B. W., Oseroff, A., Pandey, R. K., A novel approach to a bifunctional photosensitizer for tumor imaging and phototherapy, *Bioconjugate Chem.* **16**, 1264–1274 (2005).

Davis, M. E., Chen, Z., Shin, D. M., Nanoparticle therapeutics: An emerging treatment modality for cancer, *Nat. Rev./Drug Discovery* **7**, 771–782 (2008).

DeNardo, S. J., DeNardo, G. L., Miers, L. A., Natarajan, A., Foreman, A. R., Gruettner, C., Adamson, G. N., Ivkov, R., Development of tumor targeting bioprobes (^{111}In-chimeric L6 monoclonal antibody nanoparticles) for alternating magnetic field cancer therapy, *Clin. Cancer Res.* **11**, 7087–7092 (2005).

Dickerson, E. B., Dreaden, E. C., Huang, X., El-Sayed, I. H., Chu, H., Pushpanketh, S., McDonald, J. F., El-Sayed, M. A., Gold nanorod assisted near-infrared plasmonic photothermal therapy (PPTT) of squamous cell carcinoma in mice, *Cancer Lett.* **269**, 57–66 (2008).

Dichtel, W. R., Serin, J. M., Edder, C., Fréchet, J. M. J., Matuszewski, M., Tan, L.-S., Ohulchanskyy, T. Y., Prasad, P. N., Singlet oxygen generation via two-photon excited FRET, *J. Am. Chem. Soc.* **126**, 5380–5381 (2004).

Ekici, O., Harrison, R. K., Durr, N. J., Eversole, D. S., Lee, M., Ben-Yakar, A., Thermal analysis of gold nanorods heated with femtosecond laser pulses, *J. Phys. D: Appl. Phys.* **41**, 185501, 1–10 (2008).

El-Sayed, I. H., Nanotechnology in head and neck cancer: The race is on, *Curr. Oncol. Rep.* **12**, 121–128 (2010).

Fang, C., Zhang, M., Nanoparticle-based theragnostics: Integrating diagnostic and therapeutic potentials in nanomedicine, *J Control Release* **146**, 2–5 (2010).

Farrell, D., Ptak, K., Panaro, N. J., Grodzinski, P., Nanotechnology-based cancer therapeutics—promise and challenge—lessons learned through the NCI alliance for nanotechnology in cancer, *Pharm Res.* **28**, 273–278 (2011).

Fortin, J.-P., Gazeau, F., Wilhelm, C., Intracellular heating of living cells through Neel relaxation of magnetic nanoparticles, *Eur. Biophys. J.* **37**, 223–228 (2008).

Fox, M. E., Szoka, F. C., Fréchet, J. M. J., Soluble polymer carriers for the treatment of cancer: the importance of molecular architecture, *Acct. Chem. Res.* **42**, 1141–1151 (2009).

Gabizon, A., Shmeeda, H., Bareholz, Y., Pharmacokinetics of pegylated liposomal doxorubicin: Review of animal and human studies, *Clin. Pharmacokinet.* **42**, 419–436 (2003).

Galmarini, C. M., Galmarini, F. C., Multidrug resistance in cancer therapy: Role of the microenvironment, *Curr. Opin. Investig. Drugs* **4**, 1416–1421 (2003).

Gao, X., Cui, Y., Levenson, R. M., Chung, L. W. K., Nie, S., *In vivo* cancer targeting and imaging with semiconductor quantum dots, *Nat. Biotechnol.* **22**, 969–976 (2004).

Gazeau, F., Levy, M., Wilhelm, C., Optimizing magnetic nanoparticle design for nanothermotherapy, *Nanomedicine* **3**, 831–844 (2008).

Gobin, A. M., Lee, M. H., Halas, N. J., James, W. D., Drezek, R. A., West, J. L., Near-infrared resonant nanoshells for combined optical imaging and photothermal cancer therapy, *Nano Lett.* **7**, 1929–1934 (2007).

Gupta, A., Goswami, L. N., Ethirajan, M., Missert, J., Rao, K. V. R., Ohulchanskyy, T. O., Roy, I., Morgan, J., Prasad, P. N., Pandey, R. K., Organically modified silica nanoparticles as drug delivery vehicles in photodynamic therapy, *J. Porphyrins Phthalocyanines* **15**, 401–411 (2011).

Hanahan, D., Weinberg, R. A., The hallmarks of cancer, *Cell* **100**, 57–70 (2000).

Hergt, R., Hiergeist, R., Hilger, I., Kaiser, W. A., Lapatnikov, Y., Margel, S., Richter, U., Maghemite nanoparticles with very high AC-losses for application in RF-magnetic hyperthermia, *J. Magn. Magn. Mater.* **270**, 345–357 (2004).

Hinkal, G. W., Farrell, D., Hook, S. S., Panaro, N. J., Ptak, K., Grodzinski, P., Cancer therapy through nanomedicine, *Nanotechnol. Mag. IEEE* **5**, 6–12 (2011).

Hirsch, L. R., Stafford, R. J., Bankson, J. A., Sershen, S. R., Rivera, B., Price, R. E., Hazle, J. D., Halas, N. J., West, J. L., Nanoshell-mediated near-infrared thermal therapy of tumors under magnetic resonance guidance, *Proc. Natl. Acad. Sci. USA* **100**, 13549–13554 (2003).

Hleb, E. Y., Hafner, J. H., Myers, J. N., Hanna, E. Y., Rostro, B. C., Zhdanok, S. A., Lapotko, D. O., LANTCET: Elimination of solid tumor cells with photothermal bubbles generated around clusters of gold nanoparticles, *Nanomedicine* **3**, 647–667 (2008).

Huang, X., Jain, P. K., El-Sayed, I. H., El-Sayed, M. A., Plasmonic photothermal therapy using gold nanoparticles, *Lasers Med. Sci.* **23**, 217–228 (2008).

Ichikawa, H., Tokumitsu, H., Miyamoto, M., Fukumori, Y., Nanoparticles for neutron capture therapy of cancer, in *Nanomaterials for Cancer Therapy*, C. S. S. R. Kumar, ed., Wiley-VCH Verlag GmbH & Co., Weinheim (2006).

Jain, P. K., Lee, K. S., El-Sayed, I. H., El-Sayed, M. A., Calculated absorption and scattering properties of gold nanoparticles of different size, shape, and composition: Applications in biological imaging and biomedicine, *J. Phys. Chem. B* **110**, 7238–7248 (2006).

Jain, P. K., El-Sayed, I. H., El-Sayed, M. A., Au nanoparticles target cancer, *Nano Today* **2**, 18–29 (2007).

Johannsen, M., Thiesen, B., Gneveckow, U., Taymoorian, K., Waldofner, N., Scholz, R., Deger, S., Jung, K., Loening, S. A., Jordan, A., Thermotherapy using magnetic nanoparticles combined with external radiation in an orthotopic rat model of prostate cancer, *The Prostate* **66**, 97–104 (2006).

Jordan, A., Wust, P., Fähling, H., John, W., Hinz, A., Felix, R., Inductive heating of ferrimagnetic particles and magnetic fluids: Physical evaluation of their potential for hyperthermia, *Int. J. Hyperthermia* **9**, 51–68 (1993).

Kaufmann, R., Hibst, R., Pulsed Erbium:YAG laser ablation in cutaneous surgery, *Lasers Surg. Med.* **19**, 324–330 (1996).

Kikkeri, R., Lepenies, B., Adibekian, A., Laurino, P., Seeberger, P. H., In vitro imaging and in vivo liver targeting with carbohydrate capped quantum dots, *J. Am. Chem. Soc.* **131**, 2110–2112 (2009).

Kim, S., Ohulchanskyy, T. Y., Pudavar, H. E., Pandey, R. K., Prasad, P. N., Organically modified silica nanoparticles co-encapsulating photosensitizing drug and aggregation-enhanced two-photon absorbing fluorescent dye aggregates for two-photon photodynamic therapy, *J. Am. Chem. Soc.* **129**, 2669–2675 (2007).

Kitz, M., Preisser, S., Wetterwald, A., Jaeger, M., Thalmann, G. N., Frenz, M., Vapor bubble generation around gold nano-particles and its application to damaging of cells, *Biomed. Opt. Express* **2**, 291–304 (2011).

Kobayashi, H., Kawamoto, S., Saga, T., Sato, N., Ishimori, K., Konishi, J., Ono, K., Togashi, K., Brechbiel, M. W., Avidin-dendrimer-1B4M-Gd: A tumor-targeting therapeutic agent for gadolinium neutron capture therapy of intraperitoneal disseminated tumor which can be monitored by MRI, *Bioconjugate Chem.* **12**, 587–593 (2001).

Koning, G. A., Fretz, M. M., Woroniecka, U., Storm, G., Krijger, G. C., Targeting liposomes to tumor endothelial cells for neutron capture therapy, *Appl. Radiat. Isot.* **61**, 963–967 (2004).

Koo, Y.-E. L., Fan, W., Hah, H., Xu, H., Orringer, D., Ross, B., Rehemtulla, A., Philbert, M. A., Kopelman, R., Photonic explorers based on multifunctional nanoplatforms for biosensing and photodynamic therapy, *Appl. Opt.* **46**, 1924–1930 (2007).

Kukowska-Latallo, J. F., et al., Nanoparticle targeting of anticancer drug improves therapeutic response in animal model of human epithelial cancer, *Cancer Res.* **65**, 5317–5324 (2005).

Kumar, R., Ding, H., Hu, R., Yong, K.-T., Roy, I., Bergey, E. J., Prasad, P. N., *In vitro* and *in vivo* optical imaging using water-dispersible, noncytotoxic, luminescent, silica-coated quantum rods, *Chem. Mater.* **22**, 2261–2267 (2010).

Kuo, W.-S., Chang, C.-N., Chang, Y.-T., Yang, M.-H., Chien, Y.-H., Chen, S.-J., Yeh, C.-S., Gold nanorods in photodynamic therapy, as hyperthermia agents and in near-infrared optical imaging, *Angew Chem. Int. Ed.* **49**, 2711–2715 (2010).

Lapotko, D., Plasmonic nanoparticle-generated photothermal bubbles and their biomedical applications, *Nanomedicine* **7**, 813–845 (2009).

Laurent, S., Forge, D., Port, M., Roch, A., Robic, C., Elst, L. V., Muller, R. N., Magnetic iron oxide nanoparticles: Synthesis, stabilization, vectorization, physicochemical characterizations, and biological applications, *Chem. Rev.* **108**, 2064–2110 (2008).

Liu, Y., Chen, W., Wang, S., Joly, A. G., Investigation of water-soluble X-ray luminescence nanoparticles for photodynamic activation, *Appl. Phys. Lett.* **92**, 043901 (2008a).

Liu, Z., Chen, K., Davis, C., Sherlock, S., Gao, Q., Chen, X., Dai, H., Drug delivery with carbon-nanotubes for *in-vivo* cancer treatment, *Cancer Res.* **68**, 6652–6660 (2008b).

Lukianova-Hleb, E. Y., Hu, Y., Latterini, L., Tarpani, L., Lee, S., Drezek, R. A., Hafner, J. H., Lapotko, D. O., Plasmonic nanobubbles as transient vapor nanobubbles generated around plasmonic nanopartices, *ACS Nano* **4**, 2109–2123 (2010).

Martini, S., Ristori, S., Pucci, A., Bonechi, C., Becciolini, A., Martini, G., Rossi, C., Boronphenylalanine insertion in cationic liposomes for boron neutron capture therapy, *Biophys. Chem.* **111**, 27–34 (2004).

Mortensen, M. W., Sorensen, P. G., Bjorkdahl, O., Jensen, M. R., Gundersen, H. J., Bjornholm, T., Preparation and characterization of boron carbide nanoparticles for use as a novel agent in T cell-guided boron neutron capture therapy, *Appl. Radiat. Isotopes* **64**, 315–324 (2006).

Mostert, B., Sleijfer, S., Foekens, J. A., Gratama, J. W., Circulating tumor cells (CTCs): Detection methods and their clinical relevance in breast cancer, *Cancer Treatment Rev.* **35**, 463–474 (2009).

Nam, J., Won, N., Jin, H., Chung, H., Kim, S., pH-Induced aggregation of gold nanoparticles for photothermal cancer therapy, *J. Am. Chem. Soc.* **131**, 13639–13645 (2009).

Nellis, D. F., Ekstrom, D. L., Kirpotin, D. B., et al., Preclinical manufacture of an anti-HER2 scFv-PEG-DSPE, liposome-inserting conjugate. 1. Gram-scale production and purification, *Biotechnol. Prog.* **21**, 205–220 (2005).

Ohulchanskyy, T. Y., Roy, I., Goswami, L. N., Chen, Y., Bergey, E. J., Pandey, R. K., Oseroff, A. R., Prasad, P. N., Organically modified silica nanoparticles with covalently incorporated photosensitizer for photodynamic therapy of cancer, *Nano Lett.* **7**, 2835–2842 (2007).

Palmer, M. R., Goorley, G. T., Kiger, W. T., Busse, P. M., Riley, K. J., Harling, O. K., Zamenhof, R. G., Treatment planning and dosimetry for the Harvard–MIT Phase I clinical trial of cranial neutron capture therapy, *Int. J. Radiat. Oncol. Biol. Phys.* **53**, 1361–1379 (2002).

Prasad, P. N., *Introduction to Biophotonics*, Wiley, Hoboken, NJ (2003).

Qi, L., Gao, X., Emerging application of quantum dots for drug delivery and therapy, *Expert Opin. Drug. Deliv.* **5**, 263–266 (2008).

Rai, P., Mallidi, S., Zheng, X., Rahmanzadeh, R., Mir, Y., Elrington, S., Khurshid, A., Hasan, T., Development and applications of photo-triggered theranostic agents, *Adv. Drug Delivery Rev.* **62**, 1094–1124 (2010).

Roy, I., Ohulchanskyy, T. Y., Pudavar, H. E., Bergey, E. J., Oseroff, A. R., Morgan, J., Dougherty, T. J., Prasad, P. N., Ceramic-based nanoparticles entrapping water-insoluble photosensitizing anticancer drugs: A novel drug-carrier system for photodynamic therapy, *J. Am. Chem. Soc.* **125**, 7860–7865 (2003).

Song, L., Li, H., Sunar, U., Chen, J., Corbin, I., Yodh, A. G., Zheng, G., Naphthalocyanine-reconstituted LDL nanoparticles for *In-vivo* cancer imaging and treatment, *Int. J. Nanomed.* **2**, 767–774 (2007).

Straton, M. R., Exploring the genomes of cancer cells: Progress and promise, *Science* **331**, 1553–1558 (2011).

Sun, Y.-K., Ma, M., Zhang, Y., Gu, N., Synthesis of nanometer-size maghemite particles from magnetite, *Colloids Surf. A: Physicochem. Eng. Aspects* **245**, 15–19 (2004).

Suri, S. S., Fenniri, H., Singh, B., Nanotechnology-based drug delivery systems, *J. Occup. Med. Toxicol.* **2**, 16 (2007).

Tang, W., Xu, H., Kopelman, R., Philbert, M. A., Photodynamic characterization and *in vitro* application of methylene blue-containing nanoparticle platform, *Photochem. Photobiol.* **81**, 242–249 (2005).

Templeton, N. S., Nonviral delivery for genomic therapy of cancer, *World J Surg* **33**, 685–697 (2009).

Thirumamagal, B. T. S., Zhao, X. B., Bandyopadhyaya, A. K., Narayanasamy, S., Johnsamuel, J., Tiwari, R., Golightly, D. W., Patel, V., Jehning, B. T., Backer, M. V., Barth, R. F., Lee, R. J., Backer, J. M., Tjarks, W., Receptor-targeted liposomal delivery of boron-containing cholesterol mimics for boron neutron capture therapy, *Bioconjugate Chem.* **17**, 1141–1150 (2006).

Van der Poll, D. G., Kieler-Ferguson, H. M., Floyd, W. C., Guillaudeu, Jerger K., Szoka, F. C., Fréchet, J. M., Design, synthesis, and biological evaluation of a robust, biodegradable dendrimer, *Bioconjugate Chem.* **21**, 764–773 (2010).

Wagner, D. S., Delk, N. A., Lukianova-Hleb, E. Y., Hafner, J. H., Farach-Carson, M. C., Lapotko, D. O., The *in vivo* performance of plasmonic nanobubbles as cell theranostic agents in zebrafish hosting prostate cancer xenografts, *Biomaterials* **31**, 7567–7574 (2010).

Wang, M., Thanou, M., Targeting nanoparticles to cancer, *Pharmacol. Res.* **62**, 90–99 (2010).

Wang, X., Wang, Y., Chen, Z. G., Shin, D. M., Advances of cancer therapy by nanotechnology, *Cancer Res. Treat.* **41**, 1–11 (2009).

Xu, L., Tang, W. H., Huang, C. C., Alexander, W., Xiang, L. M., Pirollo, K. F., Rait, A., Chang, E. H., Systemic p53 gene therapy of cancer with immunolipoplexes targeted by anti-transferrin receptor scFv, *Mol Med.* **7**, 723–734 (2001).

Yong, K.-T., Qian, J., Roy, I., Lee, H. I., Bergey, E. J., Tramposch, K. M., He, S., Swihart, M. T., Maitra, A., Prasad, P. N., Quantum rod bioconjugates as targeted probes for confocal and two-photon fluorescence imaging of cancer cells, *Nano Lett.* **7**, 761–765 (2007).

Yoo, D. W., Lee, J. H., Shin, T. H., Cheon, J. W., Theranostic magnetic nanoparticles, *Acc. Chem. Res.* **44**, 863–874 (2011).

Yu, M. K., Jeong, Y. Y., Park, J., et al., Drug-loaded superparamagnetic iron-oxide nanoparticles for combined cancer imaging and therapy *in-vivo*, *Angew. Chem. Int. Ed. Engl.* **47**, 5362–5365 (2008).

Yuan, Q., Hein, S., Misra, R. D., New generation of chitosan-encapsulated ZnO quantum dots with drug: Synthesis, characterization and *in vitro* drug delivery response, *Acta Biomater.* **6**, 2732–2739 (2010).

Zeng, N., Murphy, A. B., Heat generation by optically and thermally interacting aggregates of gold nanoparticles under illumination, *Nanotechnology* **20**, 375702, 1–8 (2009).

Zhong, W., Celli, J. P., Rizvi, I., Mai, Z., Spring, B. Q., Yun, S. H., Hasan, T., *In vivo* high-resolution fluorescence microendoscopy for ovarian cancer detection and treatment monitoring, *Br. J. Cancer* **101**, 2015–2022 (2009).

CHAPTER 14

GENE THERAPY

Gene therapy, which covers a broad range of healthcare issues ranging from genetically inherited or acquired diseases to rejuvenative medicine, is regarded by many as a potential revolution in medicine [Verma and Somia, 1997; Mintzer and Simanek, 2009]. The treatment involves transfer of a genetic material to augment a useful but missing/malfunctioning gene or to block the expression of a disease causing gene (so-called bad gene). This chapter provides an overview of gene therapy. Section 14.1 describes the basics of gene therapy. Section 14.2 introduces the methods of gene delivery, with a particular emphasis on the use of nanocarriers as gene delivery vectors. Section 14.3 provides some examples of gene augmentation therapy. Section 14.4 describes some selected examples of applications of gene silencing. Section 14.5 describes indirect gene therapy by modulating the innate immune response. Section 14.6 illustrates transmucosal gene therapy, mainly using chitosan nanoparticles. Because the field of gene therapy is experiencing an explosion of activities, both in design of nanocarriers for nonviral delivery of genes as well as in applications, it is not possible for this monograph to provide an account of all developments. Thus, examples provided are mainly from our work. There are a number of books and reviews that can provide more detailed accounts of this field [Taira et al., 2005; Mintzer and Simanek, 2009].

Introduction to Nanomedicine and Nanobioengineering, First Edition. Paras N. Prasad.

14.1. THE PRINCIPLES, STEPS, AND IMPACT OF GENE THERAPY

The recently concluded Human Genome Project has substantially increased our knowledge of the molecular mechanisms of cancer and other gene-based diseases. In addition, several genes have been identified whose function in a living organism currently remains unknown. Concurrently, changes in the number of genes were found to be associated with a variety of disorders, some of which have been identified (suspected) as responsible for the onset of several human diseases. The application of gene therapy in the suppression or replacement of malfunctioning genes promises progress in understanding physiological roles of genes and in treating diseases at the genetic level. Thus, gene therapy can be classified in two broad categories, as shown in Table 14.1. The first one is gene augmentation therapy, where a specific DNA is delivered to the nucleus that can produce gene expression to provide the protein function, which is lacking to produce the disease manifestation. In many cases, it may be the lack of many genes that causes a disease manifestation. So the challenge is to deliver multiple genes that can concurrently express in the nucleus.

Table 14.1. Types of Nucleic Acid Therapy

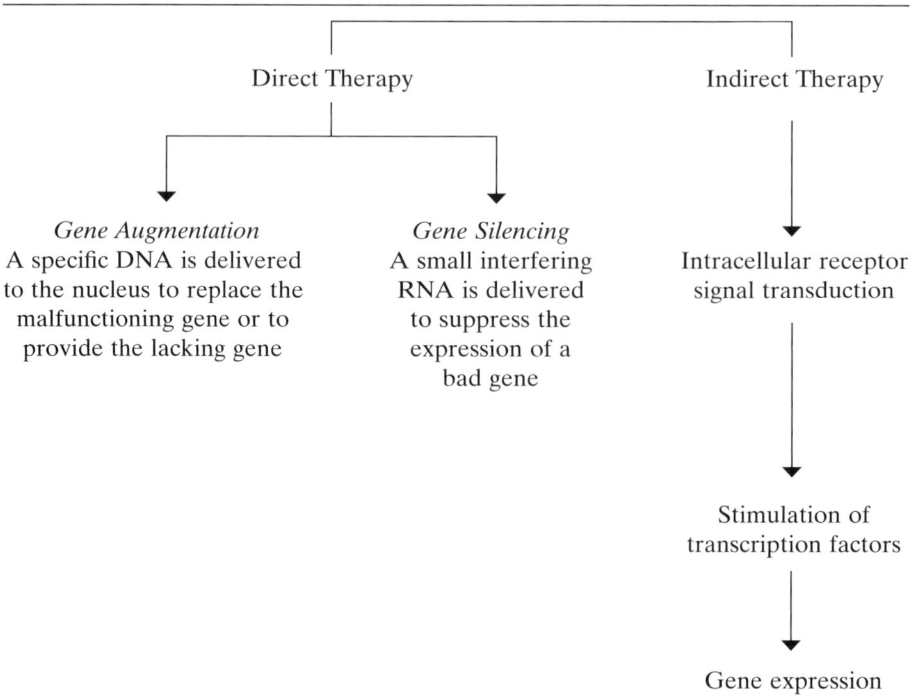

Table 14.2. Steps Involved in the Two Classes of Gene Therapy

Gene Augmentation	*Gene Silencing*
↓	↓
Delivery of DNA (plasmid or genomic)	Delivery of si-RNA
↓	↓
Protection against enzymatic degradation by DNAase	Protection against enzymatic degradation
↓	↓
Intracellular entry	Prevention of renal excretion
↓	↓
Localization in endo/lysosomes	Intracellular entry
↓	↓
Release into cytoplasm	Localization in endo/lysosomes
↓	↓
Entry into the nucleus	Release into cytosol
↓	↓
Gene transfection and expression	Complexation with m-RNA
	↓
	Inhibit translation into undesirable proteins

The second type of gene therapy involves delivering a small interfering RNA (siRNA) to the cell cytoplasm where it can bind irreversibly with the m-RNA, produced by transcription, to inhibit the translation process producing an undesirable protein. This therapy is commonly known as gene silencing or gene knockdown. These two types of therapies have been the focus; the steps involved in these two types of gene therapy are outlined in Table 14.2. In the case of gene augmentation therapy, the challenges are (i) prevention of enzymatic degradation of the external DNA delivered either systemically or topically, (ii) efficient cellular uptake, most likely by endocytosis, (iii) release of DNA from endosomes/lysosomes into cytoplasm, (iv) crossing the cytoplasm and overcoming the nuclear membrane barrier to enter the nucleus, and (v) successful gene transfection to produce efficient gene expression. As shall be discussed in the next section, a nanocarrier can be appropriately selected and surface functionalized to meet these challenges. In the case of gene silencing, major issues have been the stability of siRNA against enzymatic degradation and rapid renal clearance as well as poor cellular uptake. Again, an appropriately tailored nanocarrier to deliver siRNA can overcome these issues.

In the indirect approach also listed in Table 14.1, selected nucleic acid constructs (i.e., 5′PPP scrambled RNA or CpgDNA) are delivered intracellularly. Receptors (i.e., RIG-I, TLR 3, or TLR 7) in the cell recognize these RNA constructs as foreign and stimulate distinct signaling pathways that lead to transient gene expression. Although several possible usages are possible,

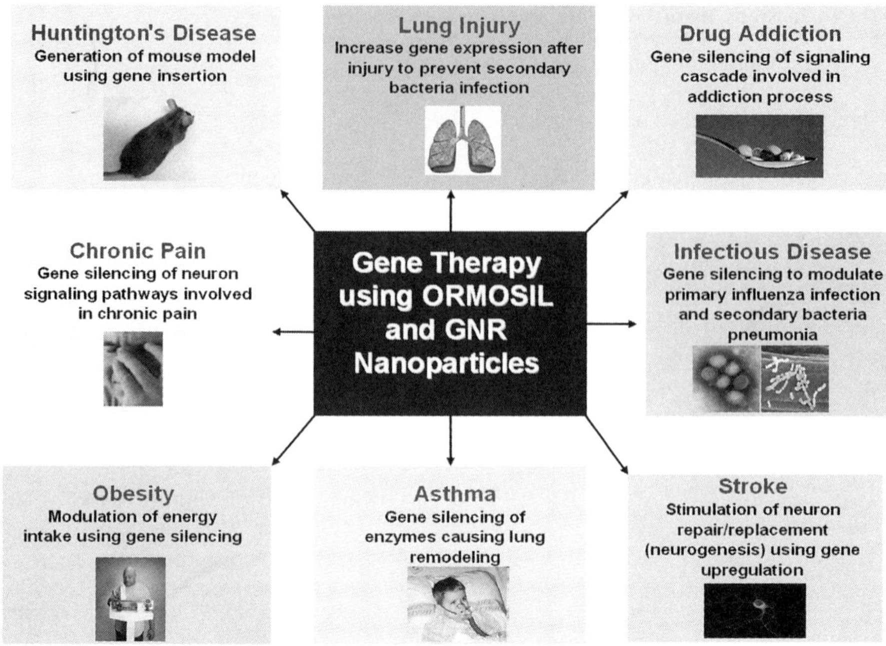

Figure 14.1. The wide scope of gene therapy.

enhancement of innate immune responses, specifically in instances where a pathogen inhibits host innate immunity, are currently the most studied. The term "indirect approach" is introduced here to differentiate it from the traditional gene therapy, labeled here as direct therapy, which directly affects the gene-directed process of transcription or translation. The delivered RNA does not bear any correlation with the genetic makeup of the organism, but rather is dictated by the external factors such as infections creating the disease manifestation. An example presented in Section 14.5 is that of an influenza virus infection where the delivered RNA mimics the nucleic acid of the virus and thus stimulates the antiviral innate immune response. The RNAs can act intracellularly, as in the case of an influenza virus, or act in an extracellular matrix to modulate intercellular signaling pathways.

Gene therapy promises to have a major impact on future healthcare by its potential application to a wide range of diseases. An illustration of this breadth of applications is provided by our gene therapy program shown in Figure 14.1. Section 14.4 discusses some of these applications.

14.2. METHODS OF GENE DELIVERY

The various methods utilized so far for gene delivery are summarized in Table 14.3. The methods of electroporation and gene gun have proven to be ineffi-

Table 14.3. Methods of Gene Delivery

Electroporation	*Gene Gun*	*Viral Vectors*	*Nonviral Nanocarriers*
Direct injection of DNA through the use of electrical impulse	Bombardment with gold nanoparticles to force DNA across cellular membrane	Engineered recombinant viruses as gene carriers to deliver DNA to the nucleus	Nanocarriers to deliver DNA to the cell

cient and regionally highly localized, which severely limits their applications. The viral delivery was extensively explored in earlier days of gene therapy. Both RNA and DNA viruses, specifically retroviruses and adenoviruses, were evaluated as possible gene carriers by modifying them genetically to eliminate their pathogenicity. However, a number of problems have been encountered in the usage of viral delivery. These include toxicity, immunogenicity, and limitation in scaling-up [Mintzer and Simanek, 2009]. In addition, there is a potential risk of excessive immunoresponse and insertional mutagenesis, raising the concern about activating oncogenes or silencing tumor-suppressor genes to initiate cancer. Furthermore, possibility also exists that the engineered virus may revert to the pathogenic wild type to infect cells. For these reasons, the emphasis now has shifted to nonviral nanocarriers.

Although a genetic material can be chemically conjugated to the nanoparticles through a linkage that can be enzymatically cleaved or cleaved by an external stimulus (such as light when the linker is photocleavable), a major effort has involved electrostatic binding. The DNA or siRNA are negatively charged because of the phosphate groups in their structure, which can electrostatically bind with a nanoparticle surface carrying a positive charge. This electrostatic complexation with the nanoparticle, yielding what is commonly known as a nanoplex (discussed in Chapter 3), also reduces the net negative charge of the genetic material which overcomes the barrier posed by the negatively charged cellular membranes, thus facilitating intracellular entry of these nanoplexes by endocytosis. Once in the lysosome, the more acidic pH environment of the lysosome causes this dissociation of the nanoplex, thus producing the release of the genetic materials for their subsequent entry into the nucleus (in case of DNA for gene augmentation therapy) or their action in cytosol (in case of siRNA for gene silencing). This scheme of electrostatic binding and subsequent action is shown in Figure 14.2.

A very widely used approach is to use nanoparticles with amino groups on the surface which can carry positive charge. The nonviral nanocarriers utilized have varied from organics, such as cationic lipids, polymers, and dendrimers, to inorganic nanoparticles, such as inorganic phosphates, carbon nanotubes,

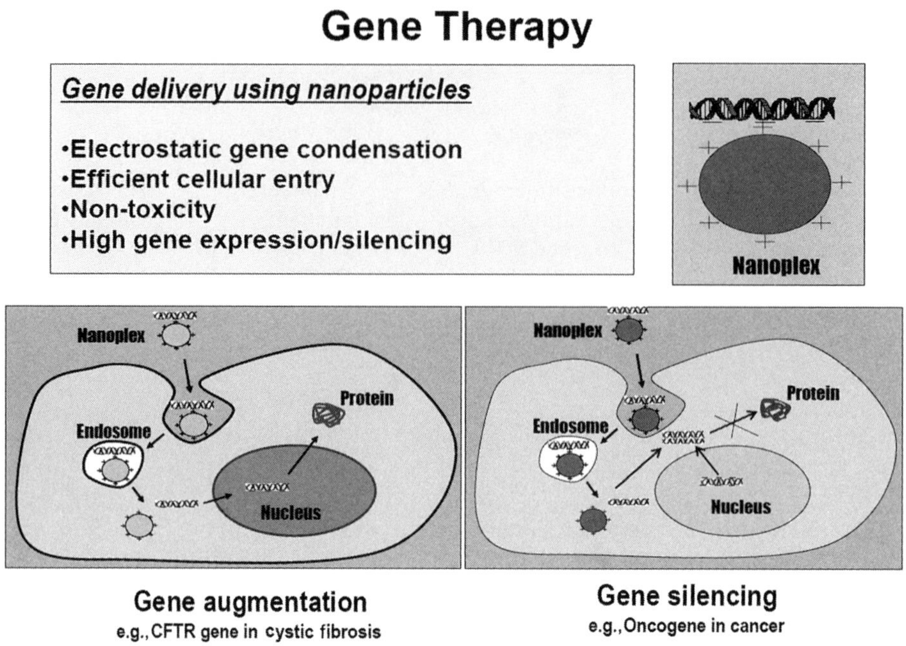

Figure 14.2. Nonviral gene delivery using the nanoplex technology.

silica-based nanoparticles, and metallic (gold) nanorods [Mintzer and Simanek, 2009]. In each of these cases, electrostatic complexation has been shown to reduce the chance of enzymatic degradation of the genetic materials, at the same time enhancing the cellular uptake. The efficiencies of cellular uptake and subsequent gene transfection can be further enhanced by incorporating targeting ligands on the nanoparticles.

Our own work has extensively used organically modified silica nanoparticles (ORMOSIL, see Chapters 3 and 4) and gold nanorods. These two nanocarriers (discussed in Chapter 3) are described here, and their applications in various gene therapies are discussed in Sections 14.3 and 14.4. We synthesized ORMOSIL nanoparticles with an average diameter of 20 nm by a synchronous hydrolysis of two precursors VTE and 3-aminopropyltriethoxysilane, which left amino groups on the surface of the nanoparticles. These nanoparticles exhibit the following features:

- They are not subject to microbial attack.
- They produce efficient binding with the DNA and protect double-stranded DNA against enzymatic digestion.
- These nanoparticles also show indications of being biodegradable through the biochemical decomposition of the Si–C bond.
- They have shown efficient *in vitro* and *in vivo* DNA transfection.

Our formulation of the ORMOSIL nanoparticles also encapsulated organic dyes as optical probes for optically monitoring each step in the DNA delivery in order to optimize the step. Specifically, the following steps in the listed sequence were optically monitored:

- Optical monitoring of binding, protection against enzymatic degradation and biodistribution of DNA using ORMOSIL nanocarriers
- Optical monitoring of release of DNA from the nanoparticles
- Optical monitoring of localization in the cytoplasm and subsequent entry into the nucleus
- Optical sensing of gene expression

Our ORMOSIL design shown in Figures 14.3 and 14.4 provided the ability to optically monitor these steps. The presence of both hydrophobic and hydrophilic groups on the precursor alkoxyorganosilane helps them to self-assemble both as normal micelles and reverse micelles, under appropriate conditions. The resulting micellar (and reverse micellar) cores can be loaded with biomolecules such as drugs, proteins, and so on. Such a system has a number of

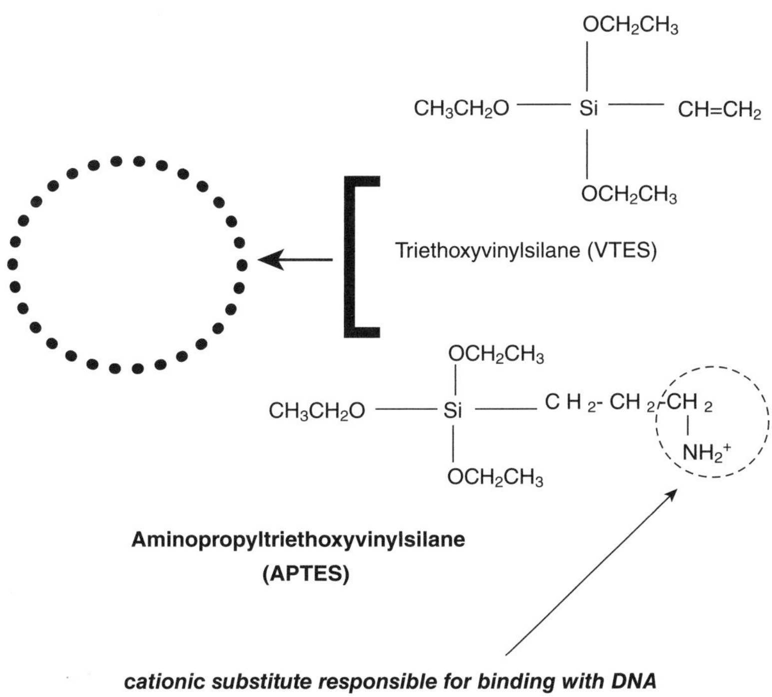

Figure 14.3. Organically modified silica (ORMOSIL) cationic nanoparticles as DNA carriers for gene therapy.

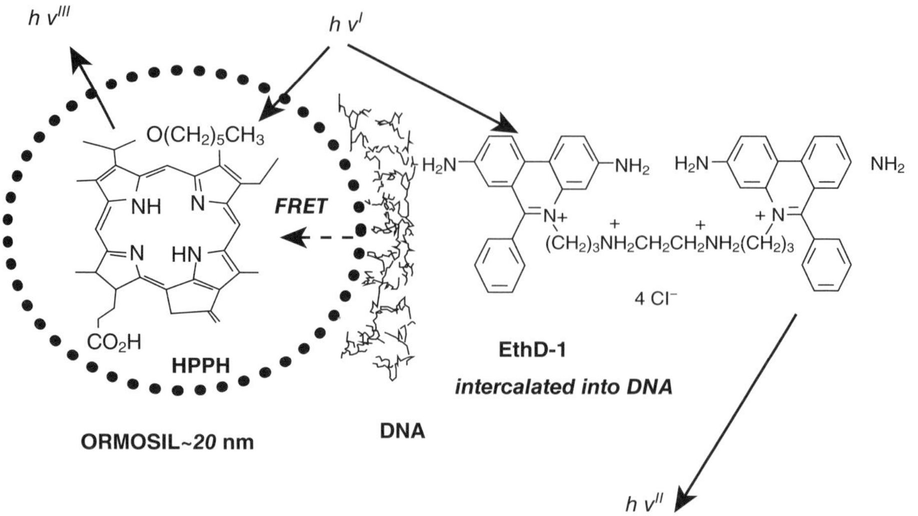

Figure 14.4. Optically trackable ORMOSIL nanoparticles for gene delivery. (Reproduced with permission from Roy et al. [2005].)

advantages: (i) They can be electrostatically complexed with the negatively charged genetic material; (ii) they can be prepared in oil-in-water microemulsions in which corrosive solvents such as cyclohexane and complex purification steps such as solvent evaporation, ultracentrifugation, and so on, can be avoided, and (iii) their organic groups can be modified further for attachment of targeting molecules. The presence of an organic group also imparts some degree of flexibility to the otherwise rigid silica matrix discussed for gene therapy, which is expected to enhance the stability of such particles in aqueous systems against precipitation.

The amino-functionalized nanoparticles were able to electrostatically condense DNA (both plasmid and genomic) and protect it from enzymatic degradation, as shown by gel-electrophoresis study. Gel-electrophoresis studies reveal that the particles efficiently complex with DNA and protect it from enzymatic digestion of DNase 1. The electrostatic binding of DNA onto the surface of the nanoparticles, due to positively charged amino groups, was also shown by intercalating an appropriate dye into the DNA and observing the Förster (fluorescence) resonance energy transfer (FRET) between the dye (energy donor) intercalated in DNA on the surface of nanoparticles, and a second dye (energy acceptor) inside the nanoparticles (Figure 14.3). Using confocal microscopy, we found that these fluorescently labeled nanoparticles extensively stain the cytoplasm of tumor cells in culture. Using calf-thymus (CT) DNA, separately irreversibly labeled with ethidium monoazide (EMA), we found that the nanoparticles release the genetic material inside the cytoplasm, which diffuses to the nucleus, a prerequisite for successful gene therapy.

In the case of delivery of a small size siRNA for gene silencing, a major challenge, compared to the larger size DNA delivery, is the stability of siRNA binding with the nanocarriers. The siRNA is also more prone to degradation by serum nucleases; also it exhibits poor cellular uptake and a more rapid renal clearance in the case of systemic delivery. However, we have successfully used a stable nanocomplex formed with a variety of nanocarriers. We have extensively used surface-coated gold nanorods as siRNA carriers, because of the biocompatibility of gold. The gold nanorods (GNR), produced by the colloidal method described in Chapter 4 contains a layer of surfactant, cetyltrimethylammonium bromide (CTAB), which makes it positively charged. However, to reduce the toxicity of CTAB, our approach has used a layer by layer coating of these produced gold nanorods with polyelectrolytes , as discussed in Chapters 3 and 4 and illustrated in Figure 3.10. In this layer by layer approach, the CTAB-coated gold nanorods (GNR) are coated with an anionic polyelectrolyte, poly(3,4-ethylenedioxythiophene/poly(styrenesulfate) (PEDT/PSS) which produces a net negative charge on the GNR. The excess PEDT/PSS are removed by centrifugation. Then the negatively charged GNR surface is reconverted to a positive charge by further coating with a layer of a cationic polyelectrolyte, poly(dicellyldimethylammoniumchloride (PDDAC). The resulting cationic GNR is then electrostatically complexed with siRNA to produce the nanoplexes.

Alternatively, we have also covalently conjugated GNR with a thiolated siRNA where the thiol group, for which the gold surface has a strong binding affinity, provides the linking. The use of gold nanorods also provides the ability to optically monitor binding with the siRNA, as well as the cellular uptake of the nanoplexes. Complexation between the GNR surface and the siRNA produces sufficient change in the local dielectric constant to produce a significant change in the localized longitudinal surface plasmon resonance (LSPR) peak of GNR [Bonoiu et al., 2009a]. These plasmon resonances have been discussed in Chapter 5. This shift was used to confirm binding with the siRNA and monitor the stability of the nanoplex. As discussed in Chapter 8 ("Multimodal Biomedical Imaging"), the plasmonically enhanced scattering at the plasmon resonance wavelength can be used for dark-field imaging. The dark-field imaging was used in our studies to visualize the cellular uptake of these nanoplexes [Bonoiu et al., 2009a].

A major concern still remains about the nondegradability of GNR. Gold is a very inert material and there is no evidence of GNR being biodegraded. Even though GNR is biocompatible and have shown no evidence of toxicity in both *in vitro* and *in vivo* studies, it still may remain a question as to what long-term effects GNRs may produce if they persist inside the body. For these reasons, there is an intense effort worldwide to use biodegradable nanocarriers which are eventually degraded and excreted from the body. In our own laboratory, the two nanocarriers we have more recently used are: (i) polymeric nanocarriers made of Chitosan, a biodegradable carbohydrate discussed in Chapter 3 and also in Section 14.6, and (ii) Si nanoparticle qauntum dots, described earlier in Chapters 3 and 4. Si is also enzymatically biodegradable because it can be converted into silicic acid.

Another class of widely investigated nanoparticulate for gene delivery involved are magnetic nanoparticles, with appropriate surface modification for complexation with genetic materials. Among the various types of magnetic nanoparticles existing, iron oxide (Fe_2O_3 and Fe_3O_4) nanoparticles have received the most attention owing to their biocompatibility. Magnetic nanocarriers of therapeutic genes have the added advantage that their dynamics *in vitro* and *in vivo* can be controlled by the use of external magnetic field (magnetophoretic control). The term "magnetofection" has been coined describing the enhanced *in vitro* transfection efficacy of plasmid DNA incorporated with magnetic nanoparticles, when they have been applied to cells under magnetophoretic control [Corchero and Villavarde, 2009]. This technique has also been commercialized under the trade name Magnetofection™. It has been found that the transfection efficacy of this system is about three times that of other available standard commercial gene transfection agents. In addition, this system has also successfully transfected few cell types, such as endothelial cells and embryonic stem cells, which are known to be difficult to transfect with other agents. Furthermore, the transfection efficacy of conventional nonviral gene delivery systems, such as lipofectamine and polyethyleneimine (PEI), has also been enhanced several fold *in vitro* following their co-incorporation with magnetic nanoparticles [Corchero and Villavarde, 2009]. Gene delivery using magnetic nanoparticles can also be extended to the *in vivo* realm, where magnetophoretic control can be exploited to ferry the transgenes across biological barriers and into the desired cells/tissue of choice. Examples are gene delivery across mucosal barriers, blood–brain barriers, and so on [Corchero and Villavarde, 2009].

Mirkin and co-workers introduced a new class of nanoparticles of genetic materials [Cutler et al., 2011] which are polyvalent nucleic acid nanostructures (PNANs). PNANs are comprised of only cross-linked and oriented nucleic acids and do not contain any core. The scheme for the preparation of PNAN is shown in Figure 14.5. These entirely nucleic acid nanostructures were reported to be capable of effecting high cellular uptake and gene regulation, without the need of any electrostatic complexation with another positively

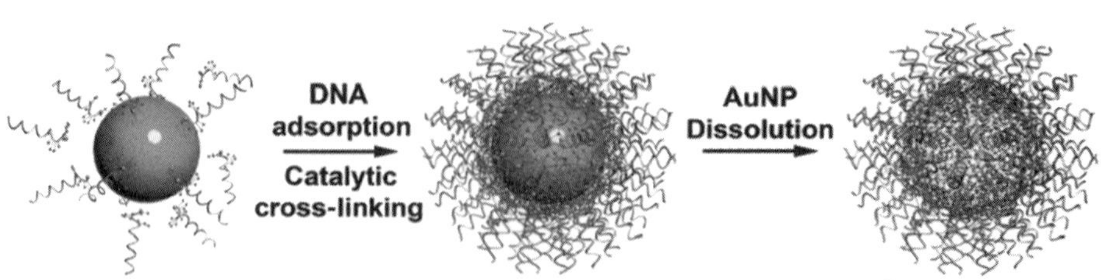

Figure 14.5. Schematics of preparation of PNANs from 3′-thiolated DNA. (Reproduced with permission from Cutler et al. [2011].)

charged nanocarrier. Moreover, they also exhibited nuclease resistance properties.

14.3. GENE AUGMENTATION THERAPY

An example of application of gene augmentation therapy is the treatment of genetic disorders. The potential for the treatment of genetic disorders has advanced tremendously with the ability to identify specific genes, whose defect or absence is responsible for a particular pathological condition. In genetically based diseases, specifically designed therapeutic genes, if successfully delivered into the appropriate cells, can provide a significant advancement in disease therapy.

Gene delivery using nanocarriers can be used to counteract brain pathological processes, augment compensatory mechanisms, and influence the brain regenerative capacities. Certain areas, in mature mammalian brain, such as subventricular zone (SVZ) of the lateral ventricle (LV), retain the capacity for neurogenesis, the process of generating neuronal cells of the brain. The SVZ contains a population of slowly dividing stem cells that generate faster-proliferating ("transit-amplifying") neural progenitor cells, the latter giving rise to brain neurons. The SVZ population of multipotent neural stem/progenitor cells (NSPCs) is increased when treated *in vitro* with exogeneous FGF2, which indicates that the intracerebroventricular administration of FGF2 peptide or FGF2-expressing adenovirus can stimulate FGF receptors by exogenous FGFs and influence the proliferation of the NSPC cells *in vivo*.

We successfully achieved *in vivo* transfection and modulation of the activity of neural cells in the brains of mice by stereotaxic injections of ORMOSIL nanoparticles complexed with plasmid DNA encoding for EGFP, into the mouse ventral midbrain and into lateral ventricle [Bharali et al., 2005]. Fluorescence targeting using EGFP vectors was used to visualize the extensive transfection of neuronal-like cells in substantia nigra and areas surrounding the lateral ventricle with the efficiency equal to or exceeding that obtained with a viral vector. An *in vivo* optical imaging technique using a fiber-based confocal fluorescent imaging system was used to show the retention of viability of the transfected cells. Transfection of a plasmid expressing the nucleus-targeting fibroblast growth factor receptor type 1 showed that this nuclear receptor can control the proliferation of the stem/progenitor cells in this region of the brain. Our results thus show that a nanocarrier gene delivery platform has promising future direction for effective therapeutic manipulation of the neural stem/progenitor cells, as well as for *in vivo* target brain therapy.

14.4. GENE SILENCING THERAPY

Several examples of the gene silencing therapy used in our research are provided below.

Gene Silencing for Brain Diseases. The blood–brain barrier (BBB), as discussed in Chapter 6, plays a crucial role in homeostasis of the central nervous system (CNS) by serving as the brain's protective shield against the entry of cells and soluble factors from blood to brain [Ellis et al., 2007]. This barrier is comprised of endothelial cells, astrocytes and pericytes as described in Chapter 6 [Teichberg, 2007]. The matrix-degrading metalloproteinases (MMP) are proteolytic enzymes that are instrumental in the turnover of the extracellular matrix (ECM) and are mediators of cell migration [Woessner, 1991]. These enzymes play an important role in disrupting BBB under neuro-inflammatory conditions, such as multiple sclerosis, human immunodeficiency virus type 1 (HIV-1), encephalitis (HIVE), meningitis, cerebral ischemia, and brain tumors [Ellis et al., 2007; Paul et al., 1998; Gidday et al., 2005]. Viral infections of the brain enhance MMP expression by several fold [Sundstrom et al., 2001; Wu et al., 2000].

MMP-9 is known to mediate the transmigration of inflammatory leukocytes across basement membranes, thus leading to proteolytic degradation of ECM, and is a predominant MMP in neuroinflammatory processes leading to BBB leakage [Persidsky et al., 1997].

The pathogenesis of HIVE is primarily attributed to infiltration of the CNS by HIV-infected mononuclear cells by migration through the brain microvascular endothelial cells (BMVECs). This leads to impairment of the BBB integrity [Persidsky and Gendelman, 1997; Annunziata, 2003; Langford et al., 2006].

The inhibition of MMP-9 activity via post-translational gene silencing using delivery of short interfering RNA (siRNA) molecules at BMVECs (see Chapter 6) can thus have a significant impact on reducing the BBB permeability. However, a major hurdle limiting the use of this technology is the lack of methods to safely and efficiently deliver siRNA molecules to their target [Whitehead et al., 2009] because in the free form, siRNA molecules have a very short half-life in physiological conditions, owing to their vulnerability for degradation by endogenous nucleases. A nanoparticulate carrier (vectors) will not only protect them from degradation in the biological milieu, but also steer them to desired cells/tissues as well as facilitate their cellular entry [Whitehead et al., 2009].

We evaluated the specificity and efficiency of quantum dot nanocarriers for MMP-9 gene silencing in a primary brain endothelial cell line, which is well known to be difficult to transfect [Bonoiu et al., 2009b]. Additionally, we measured functional/physiological changes as a consequence of MMP-9 gene silencing in these BMVECs. Our results showed that silencing of MMP-9 gene expression resulted in the up-regulation of extracellular matrix (ECM) proteins like collagen I, IV, and V and a decrease in endothelial permeability, as reflected by reduction of transendothelial resistance across the BBB, in a well-validated *in vitro* BBB model. MMP-9 gene silencing also resulted in an increase in expression of the gene tissue inhibitor of metalloproteinase-1 (TIMP-1). These studies demonstrate the application of a novel nanoparticle-based siRNA delivery system in modulating the MMP-9 activity in BMVECs and other MMP-9 producing cells. Thus, our results show the promise of using gene silencing to modulate MMP-9 for preventing neuroinflammation and maintain the integrity of the BBB.

Substance Abuse Therapy. Substance abuse is of major worldwide health concern. An example is opiate addiction, which involves activation of the dopaminergic signaling pathway in the brain, in which adenosine 3′,5′-monophosphate-regulated phosphoprotein (DARPP-32) plays a significant role [Murer and Moratalla, 2011]. Neurotransmitters such as dopamine activate protein kinase A (PKA)-mediated phosphorylation of DARPP-32. All observations indicate that the DARPP-32 is the central molecular "trigger" that underlies the neurobiological alterations related to drug abuse. Therefore, the suppression of DARPP-32 gene expression using silencing by its siRNA antagonist in dopaminergic neuronal (DAN) cells can lead to significant behavioral inhibition of the drug addiction process.

A major hurdle in this case is the brain, the larger organ for drug addiction therapy, which represents a particularly inaccessible organ for siRNA delivery owing to the presence of the blood–brain barrier (BBB). Our approach utilized gold nanorod–DARPP–32 siRNA complexes (nanoplexes) that target the dopaminergic signaling pathway in the brain [Bonoiu et al., 2009a]. The shift in the localized longitudinal plasmon resonance peak of gold nanorods (GNRs) was used to show their interaction with siRNA. Plasmonic enhanced dark-field imaging was used to visualize the uptake of these nanoplexes in dopaminergic neurons *in vitro*. Gene silencing of the nanoplexes in these cells was evidenced by a reduction in the expression of key proteins (DARPP-32, ERK, and PP-1) belonging to this pathway, with no observed cytotoxicity. Moreover, these nanoplexes were shown to transmigrate across an *in vitro* model of the BBB described in Chapter 6. Therefore, these nanoplexes appear to be suited for brain-specific delivery of appropriate siRNA for therapy of drug addition, as well as other substance abuse.

Gene Silencing for Enhancing Radiation Therapy. Radiation therapy (RT) is one of the most important treatment modalities of cancer. However, there are multiple limitations of RT. Therapeutic radiation doses cause severe collateral damage to the normal tissue to the point that once the patient receives radiation to one anatomical site, he or she cannot be treated by radiation if cancer recurs. In addition, acute toxicity interrupts RT, delaying cancer treatment. Chronic toxicity can lead to a variety of diseases related to the anatomical site that has been radiated. Narrowing of the coronary arteries requiring bypass surgery in young individuals, stroke secondary to carotid artery stenosis, dysphagia and aspiration pneumonia, sloughing of intestinal mucosa, radiation-induced cancer, immune depletion, and death are among the most serious complications of RT. Recent advances in technology have improved the tumor targeting of RT (e.g., intensity-modulated RT). However, the current approaches for RT are antiquated. There is a acute need for radiation dose reduction in cancer patients.

For recurrent cancer, repeated RT cannot be administered. These limitations of current radiation methods pose a desperate need for engineering approaches that will reduce total dose of ionizing radiation to eradicate cancer. One potential approach to reduce the dose of radiation is to silence a gene called sphingosine kinase-1 (SPK-1). High expression of SPK-1 and increased levels of its product, sphingosine-1-phosphate (S1P), are associated with aggressive tumor growth and radiation resistance. Elevated levels of SPK-1 and S1P are implicated in proliferation and anti-apoptotic activity in a variety of human cancers. Sinha and co-workers engineered a novel gene delivery system that has potential to revolutionize the way cancer is treated now by radiation [Sinha et al., 2011] . This unique theranostic approach of gene delivery is based on a synergistic combination of gold nanorods (GNR) and siRNA. We have shown that the gene-silencing approach reduces the expression of SPK-1 and allows potent RT of cancer at low, nontoxic radiation dosages. We demonstrated *in vitro* that a 21-nucleotide siRNA (HNB-001) engineered in our labs clearly inhibits cell growth and potentiates radiation sensitivity of cancer cell lines at an 8- 10-fold reduced radiation dosage [Masood et al., 2011]. Biodistribution and persistence of the GNR–siRNA complex within the tumor was monitored by SPECT imaging.

14.5. INDIRECT GENE THERAPY MODULATING INNATE IMMUNE RESPONSE

An example of application of indirect gene therapy is in treating infectious diseases. The infectious disease presented here is that of influenza, particularly the A/H1N1-virus-induced infections, commonly known as swine flu. The influenza A/H1N1 virus containing genome segments derived from avian, human, and porcine influenza viruses was first isolated in April 2009 and quickly spread globally, prompting the World Health Organization to declare a pandemic. Vaccination programs form the backbone of public health intervention strategies. However, limited manufacturing capacities can often delay the availability of pandemic influenza vaccine. Antiviral drugs, another tool against influenza, are of two classes: the M2 ion channel blockers (amantadine, rimantadine) and the neuraminidase inhibitors (oseltamivir, zanamavir). But, influenza viral strains resistant to both of these classes of antiviral drugs are becoming increasingly common, necessitating new preventive and therapeutic strategies. One new pharmacological approach involves augmentation, which is based on recognizing viral pathogens via the pathogen recognition receptors (PRRs). Recognition of pathogen-associated molecular patterns by PRRs results in rapid induction of antiviral cytokines, such as IFN-β, as well as cytokines responsible for the activation of adaptive immunity. Influenza viral RNA is detected by the cytosolic RNA sensor retinoic-acid-inducible gene I (RIG-

I). Following binding RNA (double-stranded (ds) or 5′PPP-single stranded (ss)), RIG-I undergoes a conformation change, allowing it to interact with IFN-β promoter stimulator 1 (IPS-1). The interaction of IPS-1 and RIG-I leads to the induction of type I IFN genes and innate immune response cytokines. Hence, activation of RIG-I by its 5′PPP-ssRNA ligand is an attractive alternative to existing prophylactic treatments. Also, because innate immunity is evolutionarily conserved and crucial for host survival independent of viral strain, viral resistance to this therapeutic approach is less likely to develop.

A major challenge in using 5′PPP-ssRNA to activate RIG-I is in the delivery of this ligand. We showed that gold nanorod (GNR) enhanced delivery of bioactive 5′PPP-ssRNA RIG-I ligand which resulted in up-regulation of type I IFN through stimulation of RIG-I [Chakravarthy et al., 2010]. Increased type I IFN production resulted in reduced concomitant viral replication. Our results demonstrated the successful internalization of GNR-5′PPP-ssRNA nanoplexes, up-regulation of antiviral responses, and reduction of replication of both a seasonal influenza A virus (A/Solomon Islands/03/06) and a 2009 H1N1 pandemic virus (A/California/08/09). These findings suggest that a nanotechnology-based approach to stimulate antiviral responses of the host innate immune system provides a unique approach to treat infectious diseases.

14.6. TRANSMUCOSAL GENE DELIVERY

The transmucosal administration of most genes/proteins is vastly hindered by their quick degradation by the mucosal enzymes and also due to the low permeability across the biological barriers arising from the polarity of the gene/proteins [Csaba et al., 2006]. The exquisite requirements for the delivery of biopharmaceuticals across the mucosal membrane often includes appropriate hydrophilic/lipophilic balance, protection of the gene by a carrier, and intimate interaction with the epithelium to facilitate transport, which is governed by mucoadhesive processes, association to epithelial membranes, and/or intraepithelial transport. Chitosan nanoparticles have been extensively investigated as an efficient delivery vehicle for various drugs as well as for genes owing to its excellent biopharmaceutical properties, bioadhesiveness, permeabilization capacity and acceptable toxicity profile [Fuente et al., 2008]. Chitosan (CS) also improves the colloidal stability of some hydrophobic nanocarriers, such as PLGA and lipid NPs. This greater stability of CS-coated NPs has resulted in a more important transport of associated active molecules, such as tetanus toxoid, across the nasal mucosa [Alpar et al., 2005]. Moreover, the stabilizing capacity of CS can be further improved upon PEGylation.

A great deal of attention is focused on new nanocarriers that can provide improved interaction with mucosal barriers. Among several drug-delivery technologies, CS-coated and CS-based systems have been studied with particular interest owing to the strong mucoadhesive effect of this polymer. As a coating material, CS has been used to modify PLA NPs, triglyceride NPs, liposomes, and nanocapsules [Alpar et al., 2005]. CS-coated carriers have been applied to nasal and oral delivery of polypeptides with promising results. In the field of vaccine delivery as discussed in Chapter 15, liposomes, nanocapsules, and NPs coated with CS have been loaded with antigens and significant improvements in the immunological responses on mucosal administration have been observed.

Keeping all the prerequisites for the delivery of genes across the mucosa, we have shown the pulmonary delivery of 5′PPP–ssRNA across the mucosal barrier, using the ultrasmall NIR fluorophore-conjugated chitosan nanoparticles [Unpublished data]. The overall aim here again is to demonstrate genetic stimulation of antiviral cytokines in the lungs for combating viral diseases such as influenza, with chitosan nanoparticles replacing gold nanoparticles as carriers. The conjugated fluorophore provides a mean of optically tracking the path of the RNA complexed nanoparticles across the mucosal barrier and into the lungs. Chitosan–5′PPP–ssRNA nanoplexes were administered in mice by tongue pull (tracheal delivery) with chest compression, just prior to receiving 40 viral particles of H1N1 influenza (PR/8 strain) delivered intranasally. Robust up-regulation of type I IFN through stimulation of RIG-I was observed in 60% of animals, suggesting that chitosan nanoparticles can act as highly efficient transmucosal carriers of therapeutic RNA for pulmonary gene therapy [unpublished data]. It may be noted that no toxic side effects on the treated animals were observed in these experiments.

HIGHLIGHTS OF THE CHAPTER

- Gene therapy promising to cure diseases at the fundamental (genetic) level is regarded by many as a potential revolution in medicine. It impacts a broad range of healthcare issues, ranging from genetically inherited or acquired diseases to age-related ailments.
- The primary focus of gene therapy has been on transfer of a genetic material to augment a useful but missing/malfunctioning gene (gene augmentation therapy) or to block the expression of a harmful, disease-causing gene (gene suppression or silencing therapy).
- The challenges for gene augmentation therapy are (i) prevention of enzymatic degradation of the externally delivered DNA, (ii) efficient cellular uptake, (iii) release of DNA from endosomes/lysosomes into cytoplasm, (iv) nuclear entry, and (v) effective gene expression.
- In the case of gene silencing, major issues have been the stability of siRNA against enzymatic degradation and rapid renal clearance, as well as poor cellular uptake.
- Genetic materials require carriers, or "vectors," such as engineered viruses or nanoparticles, to help them overcome the above challenges and deliver

them to target sites of interest. However, various biosafety issues associated with engineered viruses, such as immunogenicity and pathogenicity, have led to a higher emphasis on the development of nonviral nanocarriers of genes.

- The DNA or siRNA , being negatively charged because of the phosphate groups in their structure, can electrostatically bind with a nanoparticle surface carrying a positive charge. This nanoplex formulation reduces the net negative charge of the genetic material, which overcomes the barrier posed by the negatively charged cellular membranes, thus facilitating their intracellular entry by endocytosis.
- ORMOSIL nanoparticles with amino groups on the surface of the nanoparticles form stable nanoplexes with plasmid DNA via electrostatic interaction.
- Co-incorporation of fluorophores into the ORMOSIL nanoplexes allows for the use of fluorescence resonance energy transfer to ascertain electrostatic binding of DNA and provide for optical tracking of the path of the nanoplexes *in vitro* and *in vivo*.
- ORMOSIL nanoplexes, when stereotaxically delivered in the mouse brain, produced robust gene expression and showed significant promise toward the treatment of neurodegenerative diseases *in vivo* via the stimulation of neuronal stem/progenitor cells to differentiate into desired neuronal cell types.
- Gold nanorods (GNRs), surface-modified with cationic polyelectrolytes, can form stable electrostatic complexes with siRNA. The process of nanocomplexation can be monitored by observing a change in the peak wavelength of localized longitudinal surface plasmon resonance (LSPR) of the gold nanorods.
- Polyvalent nucleic acid nanostructures (PNANs), comprised of only cross-linked and oriented nucleic acids and not containing any core, have been shown to effect high cellular uptake and gene regulation as well as exhibit nuclease resistance properties.
- The GNR–siRNA nanoplexes could passively diffuse across the blood–brain barrier (BBB), as demonstrated using an *in vitro* transwell model of the BBB. Targeting the DARPP-32 signaling pathway in the brain, these GNR–siRNA nanoplexes have shown significant promise in the treatment of drug addiction, which is modulated by this pathway.
- The GNR–siRNA nanoplexes were also used to sensitize head-and-neck cancer cells and tissues to radiation therapy. Here, the nanoplexes targeted and silenced the anti-apoptotic sphingosine kinase (SPK) gene, known to infer chemo- and radioresistance to cancer cells.
- An alternate gene therapeutic approach involves the activation of innate immunity of the body following delivery of RNA molecules (e.g., 5′PPP–ssRNA), which produces antiviral cytokines such as type 1 interferons, in order to fight viral infections.

- Transmucosal delivery can be effected using mucoadhesive chitosan nanocarries. Pulmonary delivery of 5′PPP–ssRNA across the mucosal barrier was achieved using ultrasmall NIR fluorophore-conjugated chitosan nanoparticles.

EXERCISES

True or False

1. Gene augmentation therapy necessitates the delivery of DNA molecules to the nucleus of target cells.

2. Genetic materials in the body require protection from nucleases such as DNase.

3. Viral vectors are extremely safe for the treatment of human diseases.

4. Genetic materials very efficiently form complex (nanoplex) with negatively charged nanomaterials.

5. The pH within lysosomes is acidic.

6. ORMOSIL nanoparticles can be synthesized only in the reverse micellar medium.

7. Dark-field imaging can be used to visualize gold nanoparticles within cells.

8. Neurogenesis is the process of generating neuronal stem cells from mature neurons in the brain.

9. The blood–brain barrier (BBB) is a shield that protects the brain from the various molecules circulating in the blood.

10. Type 1 interferons fight against viral infections in the body.

Multiple Choice

11. A short interfering (si) RNA is a

(a) Double-stranded RNA
(b) Single-stranded DNA
(c) Single-stranded RNA
(d) None of the above

12. Successful gene therapy involves:

(a) Viral vectors
(b) Nonviral vectors
(c) Either of the above
(d) None of the above

13. Which of these steps is not necessary for successful gene silencing with siRNA?
    (a) Protection of the siRNA from enzymatic degradation
    (b) Release of the siRNA from the lysosomes
    (c) Entry of the siRNA in the nucleus of target cells
    (d) None of the above.

14. Complexation of siRNA on the surface of gold nanorods (GNRs) produce a significant change in the GNR-associated peak of
    (a) Fluorescence spectra
    (b) Localized longitudinal surface plasmon resonance (LSPR) spectra
    (c) Nuclear magnetic resonance spectra
    (d) Raman spectra

15. The "gene gun" technique of gene delivery requires the use of
    (a) Magnetic nanoparticles
    (b) Gold nanoparticles
    (c) Silica nanoparticles
    (d) Viral particles

16. The following biomolecule is involved in the formation of blood–brain barrrier (BBB)
    (a) FGF-2
    (b) IFN-α
    (c) MMP-9
    (d) PKA

17. Which of the following cells are associated with the DARPP-32 singalling pathway in the brain?
    (a) DAN cells
    (b) DMVECs
    (c) NHAs
    (d) None of the above

18. Which of the following genes confer radioresistance to cancer cells?
    (a) RIG-I
    (b) SPK-1
    (c) EGFP
    (d) None of the above

19. The requirements for the delivery of biopharmaceuticals across the mucosal membrane includes
    (a) Appropriate hydrophilic/lipophilic balance
    (b) Mucoadhesiveness

(c) Intraepithelial transport
(d) All of the above

20. Chitosan nanoparticles have the following feature(s):
- **(a)** Bioadhesiveness
- **(b)** Permeabilization capacity
- **(c)** Acceptable toxicity profile
- **(d)** All of the above

Answers: 1. T; 2. T; 3. F; 4. F; 5. T; 6. F; 7. T; 8. F; 9. T; 10. T; 11. a; 12. c; 13. c; 14. b; 15. b; 16. c; 17. a; 18. b; 19. d; 20. d

REFERENCES

Alpar, H. O., Somavarapu, S., Atuah, K. N., Bramwell, V. W., Biodegradable mucoadhesive particulates for nasal and pulmonary antigen and DNA delivery, *Adv. Drug Deliv. Rev.* **57**, 411–430 (2005).

Annunziata, P., Blood–brain barrier changes during invasion of the central nervous system by HIV-1, *J. Neurol.* **250**, 901–906 (2003).

Bonoiu, A. C., Mahajan, S. D., Ding, H., Roy, I., Yong, K. T., Kumar, R., Hu, R., Bergey, E. J., Schwartz, S. A., Prasad, P. N., Nanotechnology approach for drug addiction therapy: Gene silencing using delivery of gold nanorod–siRNA nanoplex in dopaminergic neurons, *Proc. Natl. Acad. Sci. USA* **106**, 5546–5550 (2009a).

Bonoiu, A. C., Mahajan, S. D., Ye, L., Kumar, R., Ding, H., Yong, K. T., Roy, I., Aalinkeel, R., Nair, B., Reynolds, J. L., Sykes, D. E., Imperiale, M. A., Bergey, E. J., Schwartz, S. A., Prasad, P. N. Effects of MMP-9 gene silencing by a quantum dot–siRNA nanoplex delivery system on the blood–brain barrier, *Brain Res.* **1282**, 142–155 (2009b).

Bharali, D. J., Klejbor, I., Stachowiak, E. K., Dutta, P., Roy, I., Kaur, N., Bergey, E. J., Prasad, P. N., Stachowiak, M. K., Organically modified silica nanoparticles: A nonviral vector for *in vivo* gene delivery and expression in the brain, *Proc. Natl. Acad. Sci. USA* **102**, 11539–11544 (2005).

Chakravarthy, K. V., Bonoiu, A. C., Davis, W. G., Ranjan, P., Ding, H., Hu, R., Bowzard, J. B., Bergey, E. J., Katz, J. M., Knight, P. R., Sambhara, S., Prasad, P. N., Gold nanorod delivery of an ssRNA immune activator inhibits pandemic H1N1 influenza viral replication, *Proc. Natl. Acad. Sci. USA* **107**, 10172–10177 (2010).

Corchero, J. L., Villavarde, A. Biomedical applications of distally controlled magnetic nanoparticles, *Trends Biotechnol.* **27**, 468–476 (2009).

Csaba, N., Garcia-Fuentes, M., Alonso, M. J., The performance of nanocarriers for transmucosal drug delivery, *Expert Opin. Drug Deliv.* **3**, 463–478 (2006).

Cutler, J. I., Zhang, K., Zheng, D., Auyeung, E., Prigodich, A. E., and Mirkin, C. A., Polyvalent nucleic acid nanostructures, *J. Am. Chem. Soc.* **133**, 9254–9257 (2011).

Ellis, R., Langford, D., Masliah, E., HIV and antiretroviral therapy in the brain: Neuronal injury and repair, *Nat. Rev. Neurosci.* **8**, 33–44 (2007).

Fuente, M., Csaba, N., Garcia-Fuentes, M., Alonso, M. J., Nanoparticles as protein and gene carriers to mucosal surfaces, *Nanomedicine* **3**, 845–857 (2008).

Gidday, J. M., Gasche, Y. G., Copin, J. C., Shah, A. R., Perez, R. S., Shapiro, S. D., Chan, P. H., Park, T. S., Leukocyte-derived matrix metalloproteinase-9 mediates blood–brain barrier breakdown and is proinflammatory after transient focal cerebral ischemia, *Am. J. Physiol. Heart Circ. Physiol.* **289**, H558–H568 (2005).

Langford, T. D., Letendre, S. L., Larrea, G. J., Masliah, E., Changing patterns in the neuropathogenesis of HIV during the HAART era, *Brain Pathol.* **13**, 195–210 (2006).

Masood, R., Roy, I., Zu, S., Hochstim, C., Yong, K. T., Law, W. C., Ding, H., Sinha, U. K., Prasad, P. N., Gold nanorod–sphingosine kinase siRNA nanocomplexes: A novel therapeutic tool for potent radiosensitization of head and neck cancer, Integr. Biol. (Camb). December 7 online publication (2011).

Mintzer, M. A., Simanek, E. E., Nonviral vectors for gene delivery, *Chem Rev.* **109**, 259–302 (2009).

Murer, M. G., Moratalla, R. Striatal signaling in L-DOPA-induced dyskinesia: Common mechanisms with drug abuse and long term memory involving D1 dopamine receptor stimulation. *Front Neuroanat.* **5**, 51–59 (2011).

Paul, R., Lorenzl, S., Koedel, U., Sporer, B., Vogel, U., Frosch, M., Pfister, H. W., Matrix metalloproteinases contribute to the blood–brain barrier disruption during bacterial meningitis, *Ann. Neurol.* **44**, 592–600 (1998).

Persidsky, Y., Gendelman, H. E., Development of laboratory and animal model systems for HIV-1 encephalitis and its associated dementia, *J. Leucocyte Biol.* **62**, 100–106 (1997).

Persidsky, Y., Stins, M., Way, D., Witte, M. H., Weinand, M., Kim, K. S., Bock, P., Gendelman, H. E., Fiala, M., Model for monocyte migration through the blood–brain barrier during HIV-1 encephalitis, *J. Immunol.* **158**, 3499–3510 (1997).

Roy, I., Ohulchanskyy, T. Y., Bharali, D. J., Pudavar, H. E., Mistretta, R. A., Kaur, N., Prasad P. N., Optical tracking of organically modified silica nanoparticles as DNA carriers: A nonviral, nanomedicine approach for gene delivery, *Proc. Natl. Acad. Sci. USA* **102**, 279–284 (2005).

Sinha, U. K., Schorn, V. J., Hochstim, C., Chinn, S. B., Zhu, S., Masood, R., Increased radiation sensitivity of head and neck squamous cell carcinoma with sphingosine kinase 1 inhibition, *Head Neck* **33**, 178–188 (2011).

Sundstrom, J. B., Mosunjac, M., Martinson, D. E., Bostik, P., Donahoe, R. M., Gavanis, A. A., Effects of norepinephrine, HIV type 1 infection, and leukocyte interactions with endothelial cells on the expression of matrix metalloproteinases, *Ansari AIDS Res. Hum. Retrovir.* **17**, 1605–1614 (2001).

Taira, K., Kataoka, K., Niidome, T., *Non-viral Gene Therapy Gene Design and Delivery*, Springer-Verlag, Tokyo (2005).

Teichberg, V. I., From the liver to the brain across the blood brain barrier, *Proc. Nat. Acad. Sci. USA* **104**, 7315–7316 (2007).

Verma, I. M., Somia, N., Gene therapy—promises, problems and prospects, *Nature* **389**, 239–242 (1997).

Whitehead, K. A., Langer, R., Anderson, D., Knocking down barriers: Advances in siRNA delivery, *Nat. Rev. Drug Discov*. **8**, 129–138 (2009).

Woessner, J. F. Jr., Matrix metalloproteinases and their inhibitors in connective tissue remodeling, *FASEB J.* **5**, 2145–2154 (1991).

Wu, D. T., Woodman, S. E., Weiss, J. M., Mc, M. C. M., D'Aversa, T. G., Hesselgesser, J., Major, E. O., Nath, A., Berman, J. W., Mechanisms of leukocyte trafficking into the CNS, *J. Neurovirol. Suppl.* **1**, S82 (2000).

CHAPTER 15

NANOTECHNOLOGY FOR INFECTIOUS DISEASES

As pointed out in Chapter 1, phenomenal progress in aviation and transportation has made the world a small community, and thus, significantly prone to rapid spread of infections across the globe. In addition, the emergence of multi-drug resistance and rapid mutation of many pathogens have made infectious diseases a major healthcare issue, requiring new strategies to combat them. Nanotechnology holds the promise of new approaches for early detection, prevention, and multi-drug therapy for infectious diseases, and this is the focus of this chapter. Section 15.1 provides a general discussion of nanoparticle-based theranostic applications for infectious diseases and the benefits of the nano approaches. Four specific pathogens have been chosen as representatives of viruses, bacteria, and parasites for discussion in the subsequent sections. Section 15.2 discusses HIV, as an example of a chronic systemic viral infection, while Section 15.3 covering influenza is an example of an acute respiratory tract viral infection. Section 15.4 examines tuberculosis, a bacterial infection on the rise worldwide. Finally, Section 15.5 covers malaria, as an example of a protozoan parasitic infection.

Some general references on this topic are Tallury et al. (2010), Peek et al. (2008), and Semete et al. (2010).

Introduction to Nanomedicine and Nanobioengineering, First Edition. Paras N. Prasad.

15.1. PATHOGEN INFECTIONS AND NANOPARTICLE-BASED APPROACHES

A major cause of death in the world is from infectious diseases, nearly surpassing those from heart diseases. A large portion of infection-related mortalities are in developing nations, as a result of a number of factors including poor hygiene conditions, lack of adequate healthcare, unaffordability of drug cost, and co-morbidities that can impair the immune system, either due to malnutrition or because of other immune-deficiency-causing infections such as HIV. This is why many pathogenic infections are also labeled as poverty-related diseases (PRD).

The list of pathogens and the diseases caused by them is vast, forming a major part of microbiology. It is also a growing list because of appearance of new strains of pathogens. It is not possible within the scope of this monograph to provide a comprehensive account of all types of microbial agents and their pathogenesis. Here the coverage will be limited to some representative microbial pathogens, which are major health threats worldwide in general and for developing countries in Africa in particular. These pathogens are listed in Figure 15.1. Fungi, a group that has several pathogenic representatives, and nonprotozoan parasites (i.e., helminthes) are not covered here, although several of the nanotherapeutic approaches discussed in this chapter may be applicable for these infections.

The cellular structures and sizes of viruses and bacteria have already been discussed in Chapter 2. Viruses do not have a cellular structure and are considered by many as nonliving; however, in a host they replicate and can also mutate. Bacteria are living species; however, they do not have mitochondria and can only survive in the appropriate environment to which they have adapted. Protozoa are single-cell organisms that are larger than bacteria and

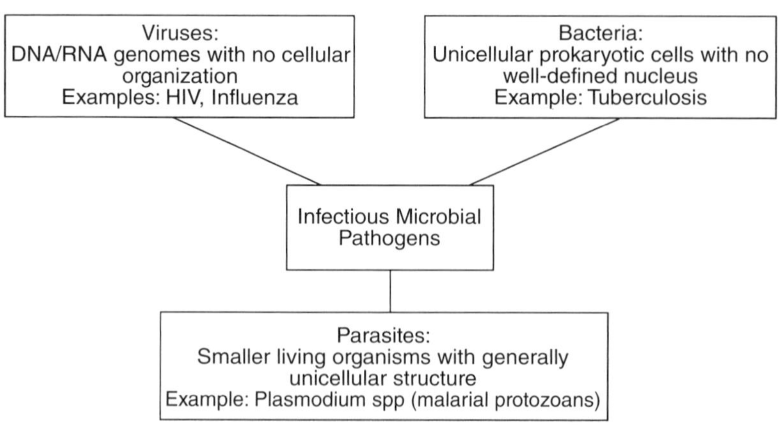

Figure 15.1. The various infection causing microbial pathogens discussed in this chapter.

have a eukaryotic cellular structure like humans. Some protozoa can live by themselves, whereas others, like malaria, are parasitic and require mosquitoes and humans hosts to complete their life cycle.

Although there are some common features in infections caused by these microbial agents, they differ in the pathways of infections (pathogenesis), thus needing special considerations. Some of the important differences are listed here.

Microstructure. As discussed above, the size increases in going from virus, to bacteria, to parasite. Also, the microstructures defined by the shape, surface charge, lipophilicity, and adhesion molecules vary broadly for different pathogens. This microstructure has important consequences. For example, a virus microstructure allows them to cross the various biological barriers, discussed in Chapter 6.

Mode of Transmission. Viruses can be airborne in a humidified environment and even be transmitted by inhalation, whereas a bacterial infection requires a closer contact, such as body fluid contamination and even through airborne sputum or nasal mucosal droplets. Parasites generally enter the bloodstream through the digestive track (intestine) or by another animal vector such as a female Anopheles mosquito whose bite intravenously inoculates malarial sporozoites.

Mode of Action. Viruses always replicate intracellularly. Bacteria, such as tuberculosis bacterium, are extracellular, intracellular, and within lesions that are formed by granulation tissue with a fibrous capsule known as a "tubercle." However, most bacteria are extracellular. Parasites may also have an intracellular phase, such as *Plasmodium* spp that cause malaria, though most of them are extracellular.

Microbial Pathogenesis. Microbial pathogenesis involves mechanisms based on host–pathogen interactions that manifest as clinical disease. Viral pathogenesis, in many instances, results from "collateral damage" secondary to the host antiviral response. Bacterial infections, in many instances, involve the secretion of toxins. Parasites, such as in the case of malaria, act by utilizing the host organ (liver) to initially replicate, then they reproduce within the red blood cells, finally rupturing them. There are numerous challenges in meeting the growing threat posed by multi-drug-resistant pathogens and continuously evolving mutants, as well as the emergence of new types of pathogens. Major issues with current approaches to microbial pathogens involve selectivity and specificity—for example, targeting the subcellular sites where the pathogens may be residing.

Nanotechnology with tailored nanoparticle composition, size, and surface characteristics can easily take advantage of their small size to target a specific microbial agent, as well as to concentrate in the subcellular site where the

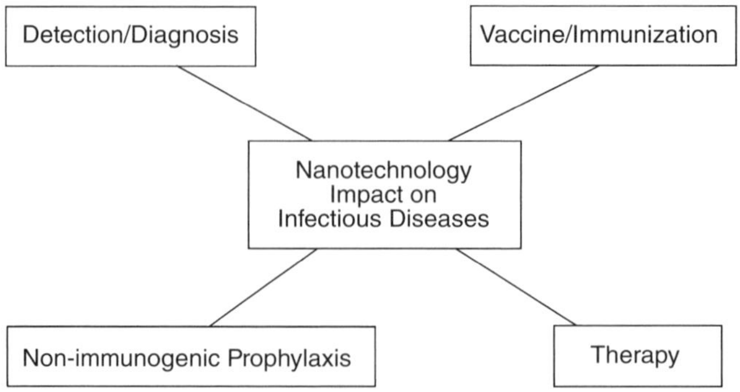

Figure 15.2. Impact of nanotechnology on diagnosis, prevention, and therapy of infectious diseases.

pathogen localizes. Figure 15.2 lists the areas of nanotechnology that can impact the current challenges posed by infectious diseases. In this section, we shall discuss the general aspects of benefits provided by nanotechnology. Specific applications in relation to HIV disease, influenza, tuberculosis, and malaria are covered in subsequent sections.

Intracellular/Subcellular Targeting. A major advantage of using nanoparticle delivery of diagnostics/prophylaxis/therapy is the ability to use it to target a pathogen at intracellular levels utilizing the microstructure and the receptor/antigen markers of the pathogen. In cases like HIV and tuberculosis, targeting at the organelle level is a promising direction to significantly enhance the therapeutic index. This aspect of targeted delivery will be further detailed in subsequent sections dealing with specific pathogens.

Rapid, Multiplexed Diagnosis with High Sensitivity and Specificity. Current diagnosis of infectious diseases involve the following methods either independently or in combination: (i) microbial culture growth pattern; (ii) microscopic examination of body fluid or excretion (sputum, blood, urine, stool), often in combination with biochemical and/or fluorescence staining using antibody labeling; (iii) biochemical tests using metabolic or enzymatic products characteristic of a pathogen; (iv) immunoassays such as enzyme-linked immunosorbent assay (ELISA), flow assay, and suspension bead technologies, discussed in Chapter 10; and (v) molecular diagnostics utilizing changes in genomic and proteomic expression patterns induced by pathogenesis, where PCR is used to amplify the target DNA or RNA.

Nanoparticles provide a multiple and multiplexed diagnostic platform utilizing a combination of detection methods, already discussed in Chapter 9 on biosensing. The large surface area offered by nanoparticles allows one to incorporate multiple pathogen-target-specific molecules/ligands to achieve ultrasensitivity, even at the single cell level. The increased sensitivity, together with enhanced intracellular uptake and organelle level targeting, coupled with their prolonged circulation time using PEGylation, provides the ability for rapid detection as opposed to the current diagnosis, which often takes days for the appropriate level of pathogen amplification to manifest. Increased sensitivity also enables the use of small volume samples. The prospect of using multiplexed detection, along with the targeting of many biomarkers, provides the ability to selectively identify specific strains of a pathogen, in order to be more specific therapeutically. The combined impact is reliable, rapid, specific, cost-effective, and easy-to-use diagnostics [Tallury et al., 2010].

In our own laboratory, the two approaches being pursued for rapid, ultra-sensitive, and multiplexed detection of pathogens are phase-sensitive surface plasmon resonance (SPR) utilizing gold nanorods to capture pathogen expressions, discussed in Chapter 9, and flow cytometry utilizing multiplexed microbead assay with quantum dot labeling, which has been discussed in Chapter 10.

Multimodal Therapy. As pointed above, the emergence of multidrug resistant pathogens, such as tuberculosis bacteria and malaria parasites, are serious health threats and can only be combated with a multipronged therapeutic strategy. Nanotechnology provides this promise. Figure 15.3 illustrates the various modes of therapy, made possible with the use of nanocarriers. A nanoparticle delivery allows one to achieve targeted delivery of multiple drugs with both hydrophobic and hydrophilic characteristics, as well as increase the circulation half-life and bioavailability of the drug. In addition, using stimuli-responsive nanoformulation, controlled and sustained—as well as sequential—release of multiple drugs can be effected. Furthermore, as discussed in Chapter 12, cells of monocyte/macrophage lineage can be utilized for trafficking of nanoformulations of antimicrobial and antiviral drugs for enhancing the tissue-targeted and therapy by providing subcellular (organelle specific) activity [Kadiu et al., 2011].

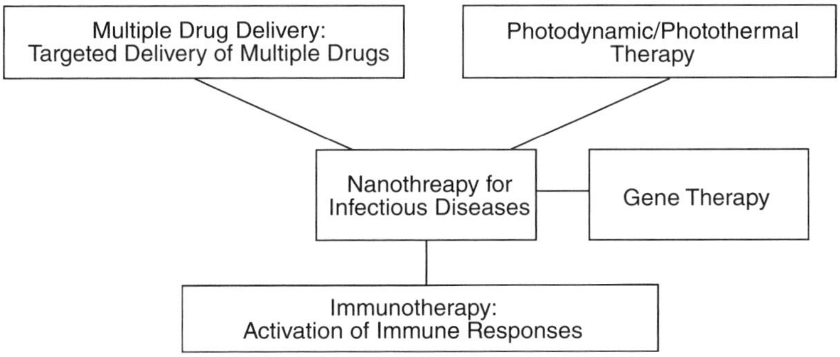

Figure 15.3. The various types of nanotherapies for infectious diseases.

Gene silencing therapy can be used to inhibit the replication of pathogens using siRNA, as discussed in Chapter 14 ("Gene Therapy"). For immunotherapy, nanoparticles can be used to deliver chemotherapeutic or immunomodulatory agents that can activate both innate and adaptive (antigen-specific memory) immune responses [Look et al., 2010]. For example, targeted nanoparticle delivery of RNA mimicking the foreign nucleic acid of a pathogen can stimulate the innate immune response. These RNAs can act intracellularly, as in the case of influenza antiviral response, thereby modulating intercellular signaling pathways. This approach and recent applications for effective inhibition *in vitro* of H1N1 influenza virus using gold nanorod delivery of 5′PPP–ssRNA are discussed in Chapter 14. Currently, we are utilizing the same strategy *in vivo* in a mouse model, but now using electrostatic complexation (nanoplex formation) of RNA with biodegradable chitosan nanoparticles.

Photodynamic therapy and photothermal therapy, described in Chapter 13 as major approaches for cancer therapy, have also been extensively explored for the treatment of pathogenic infections. Photodynamic therapy in the context of microbial infection is often called photodynamic antimicrobial therapy (PACT) [Jori et al., 2006; Nakonechny et al., 2010]. This approach is particularly promising in the case of antibiotic-resistant strains. The effect again is induced by photogeneration of reactive oxygen species (ROS), such as singlet oxygen, resulting from light absorption by a photosensitizer molecule. The ROS molecules act either intracellularly or extracellularly by attacking the membrane of the pathogen. Most research has been done *in vitro* because of the limitation by light penetration into tissues. Manipulation of size, shape, and charge on the nanoparticle delivery of photosensitizers can enhance PACT efficacy. Photosensitizers having absorption in the red region can be more effective and can directly be used to act on pathogens (e.g., malaria) in red blood cells, because blood is transparent in the red region. Examples are thiopyrylium photosensitizers [McKnight et al., 2007].

As discussed in Chapter 13, a nanoparticle approach to increase tissue penetration for PDT involves using upconverting nanoparticles with absorption in the near-IR region (975 nm), which has deep penetration in tissues [Lim and Soter, 1993]. These formulations generate emission in the visible range by a multiphoton upconversion mechanism, and the generated visible light excites the photosensitizer coupled to the upconverting nanoparticles, which subsequently generates ROS. Another promising approach that generates ROS in order to promote DNA damage in the pathogen using a photosensitizer dye with the ability to bind to DNA grooves is able to significantly inactivate pathogens [Ohulchanskyy et al., 2007]. Thus, a nanoparticle formulation of a photosensitizer, suitability surface functionalized to attach to the grooves of DNA, can be used to effect this mechanism of phototherapy.

Another exciting prospect is the use of nano X-ray therapy developed by Nanobiotix, Paris (www.nanobiotix.com), where the nanoparticles produce highly reactive electrons by X-ray absorption. X rays can penetrate any part of the body, thus this feature, coupled with active targeting of pathogens, can be highly effective in lysing microbial agents.

Another approach introduced by Nakonechny et al. [2010] makes use of luminol, which generates a chemiluminescence reaction that can be used to excite a photosensitizer. This process is termed chemiluminescent photodynamic antimicrobial therapy (CPAT). In their *in vitro* study, the investigators used liposome-encapsulated photosensitizers.

Photothermal therapy, also introduced in Chapter 13 in the context of cancer therapy, utilizes hyperthermia produced by light absorption to lyse cells. Pissuwan et al. [2010] have employed a photothermal effect produced by strong plasmonic absorption in functionalized gold nanoparticles (GNP) for controlling pathogenic bacteria. In their work, they use GNPs functionalized with a specific oligonucleotide complementary to a *Mycobacterium tuberculosis* sequence. With the use of appropriate gold nanorods, their longitudinal plasmon resonance can be tuned to be in the near-IR region to allow for deeper penetration into the tissues for photothermal effect to be achieved.

Finally, phototherapy (PACT) and photothermal therapy can also enhance the efficacy of chemotherapy in cases of tuberculosis. The mycobacteria have a thick lipid layer membrane that results in a large barrier for drugs to penetrate. PACT and photothermal effect produced at the membrane may make the lipid layer more permeable, thus allowing easy access by the drug.

Vaccine/Immunization. Immunization by vaccination is the mainstay of preventative therapy against many pathogens. Vaccination generally refers to introduction of an attenuated or inactive form of the pathogen to prevent infection caused by the more virulent microbial agent. Immunization refers to generation of both humoral (antibody) and cellular (cytolytic T lymphocyte) adaptive immune responses against a particular pathogen. The vaccine or immunization is generally introduced epicutaneously by injection, which may be painful, and requires both sterile materials and trained personnel. Other routes of administration of vaccines can be oral and intranasal. Appropriately tailored nanoparticles, loaded with the vaccine/immunization agent, can cross the skin and mucosal barriers to be delivered through the skin, orally or through the nasal delivery route. Thus, vaccines can be delivered by using penetration of its nanoformulation through tape-stripped skin [Nasir, 2009; Peek et al., 2008]. Nanoformulations also offer the prospect of simultaneously eliciting humoral, cellular, and mucosal immunity.

A number of studies have demonstrated the penetration of nanoparticles with size <100 nm through skin, typically through the follicular route. The nanoparticles can then be taken up by dendritic cells, which migrate to the regional lymph nodes. Mahe et al. [2009] have used a mouse model to test the nanomaterial penetration and immunogenicity. This mode of lymphatic delivery has been previously introduced in Chapter 12. A new direction has been the development of DNA vaccines, which contain a selective gene or genes that encode for antigens which result in immunity against the pathogen.

This approach shows promise for tuberculosis [Huygen et al., 1996] as well as for HIV-1 [Shinoda et al., 2004].

The advantages of a nanoparticle delivery platform for immunization are as follows [Peek et al., 2008; McMillan et al., 2011]:

1. Targeted delivery: As explained earlier, the vaccine can be targeted to dendritic cells for lymph node delivery; it can be targeted to the macrophage at subcellular sites, which hide intracellular pathogens (TB, HIV). In this regard, magnetic targeting using magnetic nanoparticles can also be applied.
2. Protection and sustained release: The vaccines can be protected against enzymatic degradation and be tailored for timed release to provide sustained activities. The release can also be activated by pH, temperature, or light.
3. Delivery of multiple immunization agents, hydrophobic and hydrophilic, as well as DNA vaccines: This can generate a multimodal response involving humoral, cellular and mucosal immunity.
4. Co-delivery of adjuvants such as aluminum salts, known to further enhance the immune response.

A number of different types of nanoparticles have been utilized for vaccine delivery, some of them utilizing virus nanoparticles or other biomaterials; others are synthetic nanoparticles. Virus-like particles (VLPs) are inactive viruses, with self-assembled viral envelope Protein but devoid of the accompanying genetic materials. Virosomes are another example. They also consist of a virus envelope, but to which additional components of the virus or another pathogen are attached or inserted. The VLPs and virosomes have been found to stimulate both cellular and humoral immunity [Peek et al., 2008; Grgacic and Anderson, 2006; Gluck et al., 2004]. For example, a malarial vaccine, made of VLP of the hepatitis B virus core antigen, containing proteins from the sporozite of the plasmodium parasite exhibits significant humoral and cellular immune responses, when formulated with Alhydrogel® [Oliveira et al., 2005]. Another example is an immune stimulating complex (ISCOM) that consists of ~40 nm cage-like particles made of a protein antigen, cholesterol, phospholipid, and the saponin adjuvant Quil A [Rimmelzwaan et al., 1994].

Examples of polymeric nanoparticles are those of PLGA and PLA, which are biocompatible and biodegradable and are FDA-approved for use in humans. These nanoparticles have already been described in Chapter 3, as well as at various places for drug and gene delivery. Their degradation rate can be tailored for sustained release, and they can offer protection to encapsulated antigens, delivered orally. They can also facilitate uptake by macrophages in the nasal-associated lymphoidal tissue, when administered intranasally to stimulate mucosal immunity [Illum, 2007]. Liposomes, composed of a phospholipid bilayer shell with an aqueous core, have also been used for vaccine delivery. Liposomes have also been discussed in Chapter 3. An antigen (or adjuvant) can be encapsulated in the core, buried within the lipid bilayer, or adsorbed on the surface for presentation to antigen presenting cells [Peek et al., 2008]. These formulations can simultaneously induce antibody (humoral) and cellular immune responses, and they have been used as adjuvants for malarial vaccines [Santos-Magalhaes and Mosqueira, 2010]. Some examples of inorganic nanoparticles used for vaccine delivery include calcium phosphate, silica, and gold [Peek et al., 2008]. Vaccine containing calcium phosphate nanoparticles have shown promising results and also exhibit adjuvant properties. Gold nanoparticles have been used in Phase I study of delivery of hepatitis B and malarial DNA vaccines [Fuller et al., 2006].

Nonimmunogenic Prophylaxis. Examples of these are antimicrobial creams or solution formulations, which are generally applied topically to provide protection. They are nanoemulsions that may bind electrostatically with the anionically charged pathogens. The energy released in this fusion process can destabilize the lipid membrane of the pathogen, thereby promoting microbial killing.

Metallic nanoparticles have also shown antimicrobial activities. Anionic gold nanoparticles were shown to effectively inhibit several influenza strains [Sametband et al., 2011]. It was suggested that the blocking of viral attachment to cell surface could be the primary inhibition mechanism, although viral fusion inhibition could not be ruled out. Silver nanoparticles also have been well recognized for their antimicrobial properties. They appear to be highly promising against new strains of bacteria, which are antibody-resistant. The small-sized silver nanoparticles ($\leq$10 nm) can act in three different ways: (i) They can attach to the surface of the bacterial cell membrane, disturbing its function; (ii) they can penetrate the bacteria to interact with its DNA; and/or (iii) silver ions released from the nanoparticles can produce bactericidal effects [Morones et al., 2005].

The microbial action of silver nanoparticles is also discussed in Chapter 20 ("Nanodentistry"). In that chapter its action against bacteria responsible for dental caries is presented.

15.2. HIV

HIV is a lentivirus which belongs to the families of Retroviridae. Its infection causes acquired immune deficiency syndrome (AIDS), a disease that impairs the body's immune system and establishes a chronic inflammatory state that is associated with cardiovascular and neurologic disease. Chronic HIV infection is associated with neurocognitive changes, even with antiretroviral treatment. HIV continues to be a growing pandemic, affecting more than 30 million people worldwide and with extensive morbidity and mortality. Recent advances in antiviral therapy of HIV have reduced the mortality rate by controlling the viral replication. However, still there is no vaccine protection against this viral infection, despite a large amount of resources being invested worldwide to develop protection and cure for this deadly infection.

The two different types of the HIV virus are HIV-1 and HIV-2, of which the HIV-1 is the more pathogenic and more efficiently transmitted [Levy, 2007] one. Both viruses are about 100- to 150-nm (nanoparticles) objects containing RNA genetic material with three structural genes, a nucleocapsid and a membrane that is, in part, derived from the host cell that the virion acquires when it "buds" out of the cell. There are two viral encoded glycoproteins on the outer membrane, gp120 and gp41, which recognize and adhere to the CD4 receptor and CCR5 or CXCR4 co-receptors, which are present on many cell membranes. Thus, the primary targets of this virus are macrophages, T cells and dendritic cells, which express CD4 cell surface receptor as well as CXCR4 and CCR5 co-receptors that bind to gp41. In addition, dendritic cells also bind to gp120 of HIV virus, without membrane fusion. These cells act as cellular reservoir sites for HIV. The macrophages are one of the most important carriers because these leukocytes transport HIV across the blood–brain barrier to infect the central nervous system.

Replication in the lymphoid tissue of the gastrointestinal tract involves local amplification of the virus at the mucosal site. The initial phase is usually very long because there are very few cells in the mucosa in which the virus can replicate. Subsequently migration of infected cells to the regional lymph nodes occurs, leading to a mild viral amplification in naive T cells from where they are disseminated to various lymphatic organs. The HIV virus genome exhibits a large number of polymorphisms, due to rapid mutation. This viral trait has been a major impediment in the development of an effective vaccine.

This section highlights the nanotechnology-based approaches to develop more effective diagnostics, vaccines, and retroviral therapy. Detailed accounts of nanotechnology approaches for HIV are presented in a number of reviews [das Neves et al., 2010; Mamo et al., 2010; Kim and Read, 2010].

15.2.1. Diagnosis

The current detection methods utilize ELISAs, described in Chapter 10, followed by Western blot [Hauck et al., 2010]. Rapid point-of-care testing (POCT) is available to perform HIV antibody detection with traditional enzyme immunoassays; the IgG or IgM antibodies to HIV in the samples reach the detection threshold 3–6 weeks following infection. Alternatively, for HIV diagnosis, detection of the viral capsid protein p24 by ELISA can be used as early as 16 days after infection. The p24 antigen assays can be significantly improved by real-time immunopolymerase chain reaction (IPCR) [Sutthent et al., 2003]. The direct viral RNA detection is even more sensitive and requires waiting for only ~9 days after infection for reliable detection. Thus, improved and more rapid HIV diagnosis requires:

- *Rapid Testing in Unprocessed Blood.* This will allow early detection in a resource-poor setting.
- *Rapid Quantification of Viral Load.* This will allow the physician to determine the severity of infection and to select appropriate treatment during the same office visit. The viral load is also a good indicator of the success of an antiretroviral therapy.

Nanotechnology-based diagnostic approaches can address both these needs. Some examples of them are provided below.

Nanoparticle Aggregation Assays. This method of detection, studied for decades, is simple and fast. It utilizes the principle of shift in the surface plasmon resonance of metallic nanoparticles (gold or silver) upon their aggregation, induced by the interaction between the antibodies conjugated to their surface and the target molecule. The target molecule plays the role of a bridge between several metallic nanoparticles. The aggregation can be detected by a change in the color produced by the shift of the plasmon resonance [Hirsch et al., 2003].

Halas, West, and co-workers used gold nanoshells, with their surface conjugated with antibodies, to detect the target antigen (rabbit-IgG) in saline, serum, and whole blood [Hirsch et al., 2003]. The sensitivity was found to be comparable to that of ELISA assay, but this method required minimal sample preparation and purification.

Genomic Detection. A simple method utilized by Mirkin and co-workers involves the use of metallic nanoparticles coated with specific nucleotide sequences, which are complementary to a target genome [Elghanian et al., 1997]. Again in the presence of the target genome complementary to that of the nucleotide on the metallic nanoparticles, hybridization occurs, leading to aggregation of nanoparticles and causing a shift in the plasmon resonance.

Multiplexed Detection. An advantage offered by nanotechnology-based diagnostics is the ability to simultaneously screen for multiple pathogens. As described above, HIV is often present with other comorbidities such as malaria or tuberculosis. This scenario is often found in some African nations. In such cases, multiplex detection is of great value. In addition, HIV has multiple strains. Thus, to develop an optimal therapeutic strategy, it is important to identify the particular strain causing the infection. An example is the SBA assay using flow cytometer, which has been described in Chapter 10. It utilizes labeling with quantum dots of different colors for multiplexed detection.

Biobarcode Amplifications (BCA). This method is based on detecting the p24 antigen of HIV-1. It utilizes anti-p24 antibody-coated microplates to capture the viral antigen, p24. The detection involves a monoclonal antibody with biotin label and streptavidin-coated nanoparticle-based biobarcode DNAs for signal amplification. For detection a chip-based scanometric method is used [Tang et al., 2007]. This method was found to be ~150-fold more sensitive than ELISA and detected HIV-1 infection three days earlier than ELISA.

15.2.2. Vaccines and Antimicrobial Drugs

Nanotechnology is playing an important role in the development of vaccines and antimicrobial drugs for preventing the spread of HIV. Since these viruses can avoid adaptive (cellular and humoral) host antiviral defenses by integrating into the DNA in macrophages, T cells, and dendritic cells, nanoparticles can promote cytosolic delivery of drugs or for intranuclear targeting. These nanoformulations can also provide controlled and sustained release. Furthermore, encapsulation can protect the tissue-targeted delivered of antimicrobial agents from degradation in the circulation, which is also associated with increasing their half-life. In addition, nanoparticle-based vaccines can be delivered through various routes such as intravenously, mucosally (oral and nasal), and subcutaneously. The nanoparticles can also serve as adjuvants. Finally, the formulations have also been applied into the vagina to prevent HIV transmission.

The various nanotechnology-based preventive strategies are listed in Figure 15.4. These strategies and their examples are discussed in a review by Mamo et al. [2010].

In protein/peptide vaccines, the immunogenic antigen can be either encapsulated in the core or absorbed (attached) on the surface of the nanoparticles. For dendritic cells, antigen encapsulation provides a more efficient delivery into the intracellular cytoplasm to derive a cellular immune response. On the

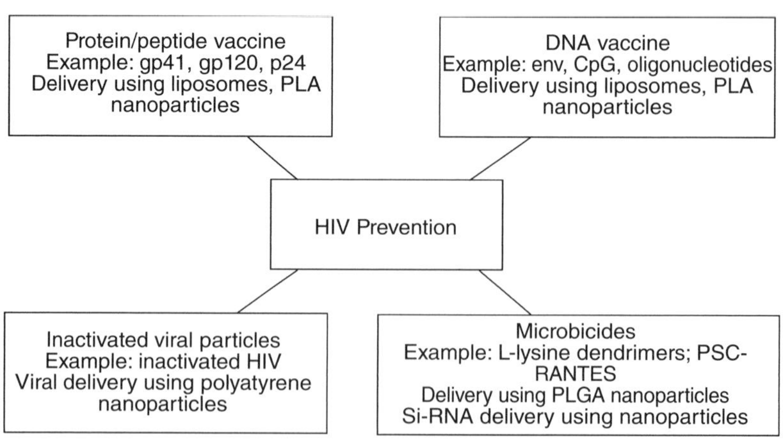

Figure 15.4. The various nanotechnology-based preventive strategies for HIV.

other hand, antigens on the surface of a nanoparticle delivered to the B cells generate more efficient antibody-specific immune response. The advantage of using a nanoparticle platform is their capacity to present antigens to dendritic cells where encapsulation works, as well as to B cells where surface absorption is desired.

As presented in Figure 15.4, liposomes have been utilized for delivery of HIV antigens, such as encapsulated gp160 and gp41 proteins [Mamo et al., 2010, and references therein]. Polymeric nanoparticles, PLA, and PLGA have shown promising potential for delivery of both protein and DNA-vaccines [Lu et al., 2009]. In the case of DNA vaccines, CpG oligonucleotides have been used for dendritic cell targeted vaccines. DermaVir, comprising HIV plasmid DNA material encapsulated in mannosylated PEI nanoparticles, has been used as a patch vaccine [Lisziewicz et al., 2006].

Inactivated HIV particles coated on polystyrene nanospheres of size ~350 nm have been used for nasal immunization [Miyake et al., 2004]. These engineered particles also promote an effective mucosal vaginal antibody response in mice. A promising direction is to use polymeric nanoparticles to combine the various preventive strategies such as utilizing inactivated viral particles, as well as the various vaccines in one nanoformulation, which may also serve as an adjuvant and exhibit microbicidal properties.

A large number of drugs, currently also in various stages of development, are to be used as antimicrobials [Rohan and Sassi, 2009]. PSC-RANTES, an amino terminus-modified analogue of the chemokine RANTES, which is a ligand for the HIV co-receptor CCR5, was encapsulated in PLGA nanoparticles of ~260-nm size and shown to exhibit significantly enhanced tissue uptake and localization at the basal layers of the epithelium, in an *ex vivo* cervical tissue model. It exhibits an antiviral activity by blocking the CCR-5 co-receptor, thus preventing entrance of HIV into cells. Gene silencing using siRNA has also been used as an intravaginal antiviral, using lipids and PLGA [Mamo et al., 2010, and references therein]. VivaGel is formed from the L-lysine dendrimer with a polyanionic outer surface [Rupp et al. 2007]. Cyclodextrin-drug complexes, nanogels, and SLNs (solid lipid nanoparticles) are also being considered in the development of novel antimicrobial formulations. Mannose (sugar) coated gold nanoparticles have also been explored as antivirals because of their capacity to interfere with the HIV-dendritic cell binding. A significant inhibition of HIV-1 binding to dendritic cells-SIGN expressing cells and trans-infection of T cell were observed with this treatment [Stefanidou et al., 2006].

15.2.3. Therapy

Nanotechnology provides the ability to use a multi-prong approach to treat the HIV infection. All the approaches listed in Figure 15.3 are being developed to combat HIV. Reiterating, the most promising is the ability of nanoparticles to provide intracellular targeting of macrophages, dendritic cells, and T cells to deliver multiple drugs to cytosolic nuclear compartments and organelles (e.g., endosomes of the macrophages), which act as the reservoir sites for the HIV virus replication. Also, the nanoparticles can be tailored to cross the blood–brain barrier, as we have previously demonstrated [Bonoiu et al., 2009], to act on the HIV virus in the central nervous system (CNS).

The current antiretroviral therapy utilizes combination antiretroviral therapy (cART), involving delivery of a combination of three or more drugs given in the same regimen to the patient. A major drawback to this therapeutic approach is the requirement for frequent (daily) drug administration, which, when compounded by poor patient compliance, results in recurrence of viral replication, increased viral load, and the development of resistance. The reviews by Mamo et al. [2010] and das Neves et al. [2010] provide excellent accounts of nanotechnology-based treatment.

Nanosuspensions of many HIV drugs have been extensively studied and have shown promise in the laboratory. 200-nm nanosuspensions of rilpivirine (TMC278), stabilized by polyethylene–polypropylene glycol (poloxamer 338) and PEGylated tocopheryl succinate ester (TPGS1000) studied in dogs, exhibited sustained release over 3 months, compared to a half-life of 38 h for the free drug [Baert et al., 2009]. In another study, nanosuspension of the HIV-1 protease inhibitor, Indinavir, stabilized by Lipoid E80 (surfactant system) were loaded in the macrophages and injected intravenously into mice, showing a high distribution into the lungs, liver, and spleen [Dou et al., 2009]. A single dose of the nanoparticle-loaded macrophages, when applied to a rodent mouse model of HIV brain infection, successfully crossed the blood–brain barrier and exhibited a significant antiviral activity.

Gendelman and co-workers have extensively pursued macrophage endocytic trafficking of crystalline antiretroviral nanoparticles (nanoART), produced by using high-pressure homogenization of ritonavir that was coated with a number of surfactants which included poloxamer [Kadiu et al., 2011]. The nano-ART exhibits significant uptake by the macrophages and the nano-ART-loaded macrophages successfully crossed biological boundaries (e.g., blood–brain barriers) in response to cytokine signaling to deliver the drug effectively to tissues and act subcellularly, dramatically reducing viral replication. The observed slow release can provide long-term antiretroviral efficacy.

Active targeting of macrophages by receptor-mediated internalization has also been demonstrated using various receptors such as formyl peptide, mannose, galactose, and Fe surface receptors. For example, the antiretroviral nucleotide analogue, stavudine, encapsulated in liposomes conjugated with mannose and galactose, exhibited significantly enhanced cellular uptake [Garg et al., 2006]. Another approach has utilized peptide nanocarriers conjugated to the drug for targeting macrophage HIV reservoirs [Wan et al., 2008].

Gene therapy approaches such as RNA interference (RNAi) have emerged as a therapeutic possibility for the elimination of HIV-1 [Ludwig LB. 2008; Surabhi and Gaynor, 2002; Ui-Tei et al., 2006]. Gene therapy for HIV has been used to insert a gene into the infected cell to interfere with viral infection or replication [Rossi et al., 2007; Haasnoot et al., 2007]. For example, siRNA targeted to viral mRNA can inhibit ("knock down") gene expression. A promising recent direction, as discussed in Chapter 14 ("Gene Therapy"), is nonviral nanoparticle-based delivery of siRNA. In our study, we used gold nanorods for siRNA delivery across the blood–brain barrier to treat HIV infection in the brain. We used a well-validated siRNA (si510) [Naito et al., 2006, 2007] that targets the poly A/TAR (transactivator of the HIV-1 LTR) site and suppresses viral replication. We evaluated the efficacy of the GNR–si510 HIV-1 siRNA nanoplex in suppressing viral replication in the HIV-1 infected monocytic cell line THP-1 by measuring p24 antigen levels and gene expression levels of HIV-1 LTR. Our results suggests that the GNR–si510 HIV-1 siRNA nanoplex is not only effective in delivering siRNA, but also in suppressing HIV-1 viral replication for a longer time period. Nanoparticles such as quantum rods (QRs) have also been easily been functionalized to incorporate siRNA forming stable nanoplexes that can be used for gene silencing [Mahajan et al., 2011a].

Nanoformulated antiretroviral drugs offer multifunctionality, i.e., the ability to package multiple diagnostic and therapeutic agents within the same nanocomposite, along with the added provisions of site-directed delivery, delivery across blood–brain barrier (BBB), and controlled release of therapeutics. We have stably incorporated two antiretroviral drugs, the protease inhibitors Saquinavir [Mahajan et al., 2009, 2010] and Amprenavir [Mahajan et al., 2011b], within a transferrin (TF)-conjugated quantum dot (QD), and evaluated the transversing ability of this nanoplex across an *in vitro* BBB model and analyzed its antiviral efficacy in HIV-1-infected monocytes [Mahajan et al., 2010, 2011b]. Such nanoplexes can play a pivotal role in HIV-1 therapeutics because they can increase drug solubility and enhancing systemic bioavailability. Moreover, following conjugation with biorecognition molecules such as transferrin (Tf), these drug-loaded nanoformulations can permeate across biological barriers such as the BBB. Thus, nanotechnology offers a new platform for therapeutic delivery of antiretrovirals to the CNS where HIV-1 is sequestered in patients with HIV-1-associated neurological disorders (HAND).

Another approach of RNA therapy we have pursued by our research team involves combined indirect and direct gene therapy using nanoparticle delivery of RNA to both stimulate the innate antiviral immune response and inhibit a viral "virulence factor" gene. This approach to treat the viral infection has already been discussed in Chapter 14, Section 14.5.

Immunotherapy involves various approaches used to modulate the immune response against the HIV. One example is the RNA-based approach as discussed above. Other approaches utilize nanoparticle delivery of cytokines such as IL-2, IL-7, and IL-15, or antigens [Pett, 2009]. The use of nanoparticle delivery for targeting antigens to dendritic cell surface receptors is a very promising direction. Thus, various polymeric nanoparticles have been studied *in vivo* for targeting dendritic cells with microbial antigens or DNA showing immune therapy promises. One example is the HIV p24 protein adsorbed on the surface of surfactant-free anionic PLA nanoparticles, which were efficiently processed by mouse dendritic cells to induce significant cellular and mucosal immune responses in mice [Aline et al., 2009].

15.3. INFLUENZA

Influenza, caused by a viral infection, has already been described in Section 14.5 in Chapter 14 ("Gene Therapy"). There is a constant threat posed by new strains of this virus, necessitating new diagnostic methods to rapidly identify them and develop strategies for their prevention and treatment. The focus here is on influenza A virus (IAV), which causes significant morbidity and mortality worldwide. Nanotechnology provides promising approach for accurate detection, prevention, and treatment of this infection.

The nanoparticle-based strategies are in general the same as described for the HIV-1 infection, except that the primary organ target is the lung, with the cellular target now being the bronchial epithelium and the alveolar macrophages (AMØ).

15.3.1. Diagnosis

Typical diagnosis of influenza involves virus isolation, antigen detection, serological responses, or PCR, which generally takes time and is expensive. Rapid influenza dipstick tests are available for point-of-care (POC) uses. These commercial immunochromatographic assays are useful for facilitating early treatment. Lateral flow assays are rapid and have the other advantages of being inexpensive and portable, with good shelf life. However, these diagnostic assays have low sensitivity due to the need to screen a large number of antigens. RT-PCR provides increased sensitivity but is time-consuming, expensive, and requires trained personnel, hence not suitable for POC.

A number of nanotechnology-based approaches show promise to overcome the current diagnostic limitations. Some examples are provided here.

Driskell et al. [2011] have proposed a one-step assay for detecting influenza virus using dynamic light scattering and gold nanoparticle (AuNPs) labels. They conjugated AuNPs to specific antibodies that exhibited aggregation induced by addition of the target virus. This aggregation approach has been also discussed earlier in relation to HIV detection. The aggregated AuNPs are detected by dynamic light scattering, where the observed mean hydrodynamic diameter can be correlated to virus concentration. The investigators have reported a one- to two-order-of-magnitude improvement of detection limit over the commercial diagnostic kits, without increasing the assay time.

Another approach utilized pentabody-activated nanoparticles [Mu et al., 2010] for detection of avian influenza virus (AIV). AIV-specific pentabody (pvHH3B) was conjugated to magnetic nanoparticles in order to capture AIV. Then gold nanopartices (GNPs), labeled with the anti-AIV mouse monoclonal antibody 3C8, were used as the detector. The mAB3C8-labeled GNPs bound to magnetic nanoparticles via AIV were separated by using a magnetic field, and the GNP-catalyzed oxidation of hydroquinone to quinone was studied by monitoring the optical density of quinone absorption to determine the AIV concentration. The sensitivity was claimed to be 10 times higher than that of the conventional double antibody sandwich ELISA.

We recently used gold nanorod-mediated capture of the expressed antigens and their coupling to the surface-plasmon resonance (SPR) sensor as providing a highly sensitive method to detect low levels of antigen expression for viral infection. This method has already been described in Chapter 9 ("Biosensing"). Another method utilizing flow cytometry and quantum dots with soluble bead assay for multiplexed detectors has also been discussed in Chapter 10, as well as in an earlier section of this chapter.

15.3.2. Vaccines

Vaccination is the primary strategy for reducing the morbidity and mortality associated with IAV. However, vaccines can be problematic because IAV displays "antigenic drift and shift." Current influenza vaccines utilize either split virus or subunit inactivated vaccines. Live attenuated influenza vaccines utilize cold-adapted strains for intranasal administration. These vaccines are safe and effective. Antigen-based vaccines have also been investigated; however, these antigens are often too small to be highly immunogenic. Virosome nanoparticles described earlier, which exhibits adjuvant characteristics allow antigen presentation in the context of major histocompatibility complex (MHC1) and MHC II and can induce both B- and T-cell adaptive immune responses. A virosome formulation in the form of Inflexal® V is currently in clinical study [Herzog et al., 2009].

15.3.3. Therapy

Currently , for influenza there are only four licensed antivirals. Amantadine and rimantadine targeted to M2 are delivered orally. The neuraminidase inhibitors that decrease viral spread are oseltamivir, delivered orally, and zanamivir, delivered by inhalation. However, as influenza regularly mutates, resistance to these antivirals is inevitable as has been observed. For example, the H1N1 was initially sensitive to oseltamivir, but over time became resistant to this drug. Thus, there is a clear need for new therapeutic approaches to combat the drug resistance and infection due to new strains.

Additional drugs at the investigative stage include the T-705 targeted to the RNA polymerase, peramivir targeted to the host sialic acid receptor, and CS-8958 targeted to the IAV neuraminidase. Other nanoparticles such as liposomes, PLA, and PLGA are being investigated as carriers of antiretroviral drugs for influenza using nasal inhalation.

A promising direction, which we are pursuing in collaboration with Dr. Knight's group, is nanoparticle-based nonviral nucleic acid and gene transfer strategies that target airways *in vivo* employing nanoplexes (an electrostatic complex of cationic nanoparticles (i.e., gold nanorods (GNR) or Chitosan) and anionic nucleic acids). The lung is especially well suited for these treatment strategies since direct contact with the environment provides a portal for inhalation of the nucleic acid conjugated nanoplatforms. The two 5′-triphosphate (PPP) RNA formulations (5′PPP-single-stranded (ss) RNA and 5′PPP–NS1 short hairpin (sh) RNA), which we have studied *in vitro*, stimulate IFN-I production through activation of RIG-I and indirectly inhibit NS1. 5′PPP–NS1shRNA also directly inhibits NS1, which results in a more intense antiviral effect. Additionally, a 5′PPP-NS1shRNA-generating cDNA nanoplex has been developed that continually produces NS1shRNA for a more sustained response [unpublished results from our laboratories].

15.4. TUBERCULOSIS

Pulmonary tuberculosis (TB) has been a major worldwide health threat for centuries and still kills nearly two million people annually. Each year, 40 million individuals will become infected by *Mycobacterium tuberculosis*, from which nearly 10 million will develop TB. Africa India, and China are currently facing an epidemic of TB. TB is on the rise in both developed and developing countries. Because of the length of the current therapy, the emergence of multiple drug resistant strains has become a major world health threat. This section highlights some of the recent advances in the areas of diagnostics, vaccine, and therapy for TB.

15.4.1. Diagnosis

Classically, TB is diagnosed by radiology, following culturing and identifying the bacterium from collected sputum samples. This diagnostic process is laborious and time-consuming. Microscopic analysis of the sputum prior to culturing is limited in its ability to detect bacteria in a large sample due to low numbers. Newer diagnostic tests include PCR analysis for specific mycobacterium DNA/RNA markers and assays for interferon-γ production from patient cells after exposure to specific *Mycobacterium* proteins. However, these tests require several days and specialized equipment to perform. Antibody-mediated tests for TB are currently becoming available in parts of the world and there are numerous research programs focused on the development of more specific biomarkers for this disease.

Given the rapidity of world travel, there is a growing need for a test capable of rapid, sensitive detection of active TB. Rapid diagnosis of pulmonary tuberculosis will involve analysis of multiple biomarkers in acquired body fluid samples such as induced sputum, urine, blood, and so on. Several potential biomarkers include: (i) interferon-γ and the proinflammatory/innate immune proximal cytokine, tumor necrosis factor-α , (ii) specific antibodies to bacterial antigens (active humoral immune response), and (iii) bacterial markers released into blood (ESAT-6, MTB-12, etc.). However, the detection and quantification of the levels of these biomarkers require multiple levels of instrumentation (cell response systems, immune-based detection, and flow cytometry) necessitating different processing of samples. However, the laboratory tests employing single analytes as "predictive" biomarkers for active tuberculosis has a poor sensitivity and specificity. Therefore, a system that has the capability to detect and quantify multiple markers would significantly advance the ability to rapidly diagnose active tuberculosis. In addition, these markers can also be utilized in monitoring the efficacy of treatment and resolution of the infection.

The ability to rapidly diagnose active TB would greatly be benefited by the availability of instrumentation that can be used under less than "ideal" conditions and that is capable of multiple analyses in a single "run." A colorimetric method utilizing functionalized gold nanoparticles has been proposed for rapid detection of TB using color change [Pissuwan et al., 2010]. This method utilized gold nanoparticles functionalized with a specific oligonucleotide complementary to a *Mycobacterium tuberculosis* sequence. The gold nanoparticles appear red in suspension due to the strong surface plasmon resonance absorption of green (~526 nm). This topic has been discussed in Chapter 5. When an aggregation of these nanoparticles occurs, such as induced by the addition of NaCl solution, its surface plasmon resonance shifts toward red and broadens, leading to a change in color to blue or purple. In the presence of target oligonucleotide of *M. tuberculosis*, hybridization prevents aggregation, leading to retention of red color following addition of NaCl solution. In negative or control samples, the addition of NaCl solution would lead to aggregation of gold nanoparticles, turning the suspension blue. Thus, it is a fast and relatively simple method, requiring only 2 h. However, a first round of PCR is required to increase the analyte concentration.

We are employing nanotechnology to develop a *M. tuberculosis* biomarker detection kit, together with a compact flow cytometer, where multiple biomarkers (immune response and bacterial markers) resulting from active tuberculosis can be simultaneously quantified on a single platform. The incorporation of the detection of these multiple TB biomarkers will afford a more "accurate" diagnosis, which will minimize false negative and positive diagnoses. In addition, by using multiple markers arising from different aspects of the infection, this system will be capable of differentiating a "tuberculin"-positive response due to exposure or vaccination from an "active" disease case.

Multicolored quantum dots (QDs), emitting at spectrally resolvable wavelengths, are bioconjugated to antibodies specific to the TB disease biomarkers. The ability to excite these multicolored nanoprobes with a single laser eliminates the need for multiple excitation lasers, which are used when exciting capture beads containing organic fluorophores. The foundation of this assay is the use of microbeads coated with the specific antibody or bacterial biomarker. This method is discussed in Chapter 10. The *M. tuberculosis*-specific markers in patient sample can then be captured on the microbeads and quantified using specific QD-labeled antibodies. Given the narrow emission band of the QDs and with the elimination of the fluorescent capture microbeads, six or more different biomarkers can be quantified simultaneously. The high photostability and narrow emission spectra overcomes the issue of spectral overlap between the multicolored QDs, allowing for the multiplexed detection of analytes, with minimum signal overlap.

15.4.2. TB Vaccine

The general description of nanoformulations of vaccine provided in Section 15.1 already has cited some important directions for vaccine development. Here a few specific examples for applications to TB are provided.

Cationic liposomes containing the *M. tuberculosis* antigen, EAST-6, the early secretory antigenic target protein, complexed with TLF agonists, have been tested as a prophylactic vaccine system, yielding considerable success [Zaks et al., 2006].

Another development is that of aerosol vaccines using the pulmonary route by inhalation. Aerosol delivery of the attenuated tuberculosis vaccine, Bacillus Calmette–Guértin (BCG), was tested in guinea pigs and found to enhance their resistance to *M. tuberculosis* infection [Garcia-Contreras et al., 2008]. The immune response generated was better than for a standard parental BCG formulation.

15.4.3. Therapy

Treatment of TB requires a multiple drug regimen consisting of two to four different drugs over a 6- to 12-month period. Primarily because of the long duration of treatment and inconsistent medication compliance, the appearance of drug-resistant strains of *Mycobacterium* is increasing. Thus, in addition to diagnosis, there is a need for real-time monitoring of the response of the disease to therapy. Although pulmonary tuberculosis is the primary form of the disease, the *Mycobacterium* can also disseminate throughout the body (extrapulmonary), particularly in patients with comorbidities. Therefore, the ability to monitor the treatment efficacy would be of great benefit for immune compromised individuals (i.e., AIDS).

Oral therapy using the currently employed antitubercular drugs (ATDs) is effective, but is still associated with a number of significant drawbacks. Since 80% of cases are isolated to just pulmonary TB, high drug dosages of ATDs are required because only a small fraction of the total dose reaches the lungs after oral administration. Additionally, the pulmonary drug concentration is cleared rapidly, requiring a continuous and frequent drug administration schedule. Many patients, particularly those in developing countries, find this regimen difficult to adhere to. ATD delivery schedules that can be administered via the pulmonary route and can avoid the daily dosing is predicted to have many advantages. Micron-sized particles for such inhalation delivery have been shown to be effective in animal models of TB. However, due to the size of these particles, "lung injury" can occur as a result of the long periods of therapy required for effective therapy. Nanotechnology, which offers the next generation in drug delivery vehicles, may be the solution to this therapeutic complication. A nanoplatform can be designed to (i) target specific cells and tissues, (ii) deliver multiple payloads for therapy, and (iii) produce little-to-no organ damage that is associated with micron-sized particle delivery systems.

This section outlines some of the approaches for TB treatment using nanotechnology. The directions covered are (i) multiple drug delivery, (ii) targeting organs and macrophage at subcellular level, (iii) immune therapy, and (iv) photodynamic therapy. They are described below in separate subsections.

Multiple Drug Delivery. The first line of drugs used for current treatment of TB is rifampicin (RIF), isoniazide (INH), pyrazinamide (PYR), and ethambutol (ETB). These agents have proven to be effective when taken according to prescription. However, in the case of noncompliance due to poor economic conditions, as happens in many developing countries in Africa, multiple drug-resistant (MDR) TB has emerged which are refractory to at least two of the first-line drugs, RIF and INH. Other issues are (a) metabolism of drugs before reaching the target site and (b) its bioavailability.

As discussed in Section 15.1, a nanoformulation provides a number of benefits for effective therapy using a combination of drugs. Appropriately tailored nanoparticles can deliver both hydrophobic and hydrophilic drugs. A large number of nanoformulations of TB drugs have been investigated [Sosnik et al., 2010; Ranjita et al., 2011]. They include a vast range of nanoparticles such as nanosuspensions, nanoemulsions, liposomes, polymeric nanoparticles, polymeric micelles, solid lipid nanoparticles (SLN), dendrimers and cyclodextrin inclusion complexes. For a detailed account of them, the readers are referred to the comprehensive reviews by Ranjita et al. (2011) and Sosnik et al. (2010). Here only selected examples are presented.

Nanosuspensions. These are nanoparticulates of the pure, solid form of the drug, which are prepared by grinding, milling, or homogenizing and then dispersed using surfactants to form stable colloidal suspensions. The methodology is simple and improves dispersibility. A major issue generally encountered is the large polydispersity (a broad distribution of sizes) of the particles, which significantly affect cellular uptake and bioavailability. Another approach used by Reverchon et al. applied supercritical carbon dioxide-assisted atomization to produce injectable RIF nanoparticles of controlled sizes [Reverchon et al., 2002]. A major concern in the process of production of nanosize particulates is that the drug physically or even chemically (by heat production) may be changed, affecting the dissolution rate and therapeutic efficacy.

Liposomes. Liposomes containing a phospholipid bilayer, which surrounds an aqueous core, have been utilized for anti-TB drug delivery. They are PEGylated to avoid RES capture and subsequent systemic clearance. They have shown reduced toxicity of drugs *in vitro* and *in vivo*, as well as enhanced anti-TB activities [Ranjita et al., 2011].

Niosomes are liposome-like vesicles made with cholesterol and nonionic surfactants. They form thermodynamically stable nanoparticles, which can incorporate hydrophilic drugs within the core and lipophilic drugs in the hydrophobic domain [Sosnik et al., 2010]. They have been investigated both *in vitro* and *in vivo* for TB drug delivery.

Polymer Nanoparticles. A specific suitable example is PLGA, which has been investigated as injectable implants for subcutaneous delivery in a mouse model as well as for oral delivery. PLGA nanoparticles have been simultaneously loaded with RIF, INH, PYZ, and ETB and administered orally in mice. The results showed that therapeutic levels were maintained for 5–8 and 9 days in the blood and plasma, respectively [Pandey and Khuller, 2006]. Swai and co-workers utilized PLGA nanoparticles of 200- to 300-nm size and encapsulate RIF, INH, PZA, and ETB with encapsulation efficiency of 50–65% [Swai et al., 2008]. In an *in vitro* assay, they observed the drugs to be released slowly, over days.

Targeting Organs and Macrophages at Subcellular Levels. The manifestations of TB infection are primarily in the respiratory system. Thus, local delivery of drugs to the lungs by inhalation is a major new direction of treatment of TB. Since alveolar macrophages act as the major cellular reservoir for TB, directly targeting these pulmonary cells by administration of neutralized nanoparticles into the airways is a very promising area for TB therapy. INH- and PYZ-loaded PLGA-neutralized nanoparticles of 186–290 nm have been studied. Therapeutic drug concentrations have been detected until day 6 for RIF and day 8 for both INH and PYZ [Pandey et al., 2003]. Additionally, nanoparticles in mannitol improved the *in vivo* uptake of the drug by alveolar macrophages in rat lungs [Ohashi et al., 2009]. Sugars such as mannose are recognized by lectin receptors on the surface of macrophages and, thus, can serve as a targeting ligand. Predictably, a pronounced increase in the macrophage uptake has been observed with the mannosylated nanocarriers. Ciprofloxacin-loaded mannose-modified liposomes have also been shown to exhibit a sharp increase in the uptake by the alveolar macrophages [Sosnik et al., 2010, and references therein]. Finally, an increased cellular uptake is observed with an increase in the mannose concentration.

Targeting the endosomes in the macrophage is another important therapeutic strategy as the pathogen resides in this subcellular (organelle) structure. Thus, using targeting by tailoring the characteristic of nanoparticles to mimic a pathogen (passive targeting) can result in their localization in the endosomes where these nanoformualtions can then effectively release the antibiotics, dramatically increasing local antibacterial concentrations without systemic effects. This approach has also been discussed in Section 15.1.

Immune Therapy. A general description of immune therapy has already been presented in Section 15.1.

In tuberculosis, the causative agent can evade the host immune system by its ability to avoid the primary defenses (lung based macrophages) response. Furthermore, these bacteria use the macrophage as an intracellular carrier site to avoid detection, while they remain quiescent (latent infection) until the conditions are better (i.e., stress-induced immune suppression) for an active infection. Even during an active infection, the bacterial infection can still be resistant to attack by macrophages that do not have the phenotype capable of killing the tuberculosis bacillus. Current technology in gene transfection offers a new direction to overcome the "unresponsive" action of these lung macrophages. Plasmids have been designed which can provide a strong "activation" signal that will enable the macrophage to become a more potent antibacterial cell, to more effectively eliminate the bacterial agent.

A nanoparticle platform based on a biodegradable material can encapsulate known anti-TB drugs, as well as carry a genetic surface payload, which will encode for the established macrophage activator, tumor necrosis factor alpha. In a study conducted by our laboratory, alveolar macrophages selectively "engulfed" nanoparticles (quantum dots) where a test drug (doxorubicin) was used [Chakravarthy et al., 2011]. In addition, we have also demonstrated the uptake of gold nanorods carrying a small RNA construct that inhibited the replication of the influenza virus in the target cells. Thus, a nanoplatform could potentially simultaneously activate and maintain that activation in alveolar macrophages and deliver antituberculosis drugs to intracellular regions of the cell in which the bacteria resides. The combined effect can effectively "kill and remove" the bacteria more efficiently than either the drugs or activation separately. The therapeutic strategy could effectively eradicate the bacterial infection in a shorter period of time and significantly reduce the ability of the bacteria to become latent in macrophages and antibacterial host responses.

Photodynamic Therapy. The concept of photodynamic therapy in relation to treatment of infectious diseases in general has already been introduced in Section 15.1. Phototherapy was actually applied to cutaneous tuberculosis in the early twentieth century. It can be very promising, if the issue of limited penetration depth of light in the tissue to reach the bacterial site can be resolved. As discussed in Section 15.1, to realize an access to deeper tissues, near-IR wavelength light is more suitable. For this reason, our laboratory is exploring the *in situ* generation of visible light by sequential multiphoton upconversion of the exciting near-IR radiation in nanoparticles containing rare-earth ions (Er, Yb). In one nanoparticle formulation, we have shown very efficient upconversion from 975-nm near IR (excitation wavelength) to 550 nm (green) and 640 nm (red) emission in Er-Yb co-doped NaF_2 nanoparticles of sizes ~30 nm. With appropriate surface functionalization, the macrophage will avidly take up these nanoparticle formulations, which will allow for targeting of the TB bacteria. Additionally, the nanoparticles can also be conjugated to a PDT photosensitizer so that shining light at 975 nm can generate visible light, which can be absorbed by the photosensitizer to generate ROS for inactivating (killing) the bacteria. Currently, we are pursuing the strategy of producing a nanoparticle for combined multi-drug delivery, immunotherapy, and PDT to develop a highly effective multipronged therapy.

15.5. MALARIA

Malaria is a very complex disease caused by one of the four species of plasmodium, a parasitic protozoan. These four species include *Plasmodium falciparum*, *Plasmodium vivax*, *Plasmodium malariae*, and *Plasmodium ovale*, which are all transmitted to humans by the bites of a female Anopheles mosquito vector. Of these, the most dangerous one is *P. falciparum*, which causes the most prevalent infection in Africa and is a major cause of mortality. Like tuberculosis, this disease is often seen in HIV patients as a result of the virus-induced immune suppression, which make the patient more prone to many other types of infections.

Like tuberculosis, the emergence of multi-drug-resistant (MDR) malaria worldwide in general, and developing countries in particular, has posed a serious health threat. This disease, mainly prevalent in Africa, India and China, has been labeled as poverty-related diseases (PRD) and, as a result, has been ignored by the Western countries until recently. However, the spread of this disease from the traditional regions of infection to the West is on the rise, because global travel has seen a phenomenal increase. For this reason, only recently has attention been paid to developing vaccines and new therapies to combat the malarial infections [Santos-Magalhaes and Mosqueira, 2010]. However, unfortunately there is still no vaccine in the market for malaria, although there are some recent promising developments [Crompton et al., 2010].

The complexity of the disease is derived from its complicated multistage pathogenesis [Miller et al., 2002]. In the initial stage of infection, the parasites are in the liver and infect the hepatic cells where they multiply. Following the intrahepatic replication, the parasites invade circulation, where they infect the red blood cells (erythrocytes). The parasite rapidly multiplies in these cells, which eventually burst allowing the malarial merozoite progeny to invade other cells. It is the release of these protozoan parasites from the erythrocyte and the resultant lysis of the red blood cells that is responsible for the manifestation of many of the classic symptoms of malaria (such as debilitating shaking chills).

Some of the challenges in treating malaria include:

- A complex multistage pathogenesis and long life cycle of the parasite
- Resistance to many of the drugs as well as injurious side effects of some new and more effective drugs
- Need for prolonged treatment

Nanotechnology provides promising approaches for this disease [Santos-Magalhaes and Mosqueira, 2010]. Many of these advantages are the same as discussed in Section 15.1. However, some benefits that are most significant for malaria are listed here:

- Targeting appropriate sites and cells at different stages of infections. As stated above, at the early stage of infection, the parasites replicate in hepatocytes, which fortunately are also in a primary organ (the liver) in which nanoparticles accumulate (passive targeting) following systemic delivery. Thus, localization of biodegradable nanoparticles in the liver and subsequent sustained release of antimalarial medications may be of tremendous value.
- Sustained release and improved pharmacokinetic profile of a multiple drug payload. This alleviates the need for frequent administration of drugs and makes effective treatment using multiple drugs.
- Enhanced circulation time of PEGylated nanoparticles. This feature allows nanocarriers of anti-malarial drugs to remain in the bloodstream for a long period, which enhances the action of the drug on the infected red blood cells and the membrane of the parasite residing intracellularly. The combined effect of this, together with sustained release, also saves on the cost of drugs, because smaller and less frequent dosages can be used.
- Reduced toxicity of antimalarial drugs. Some of the new drugs, proven to be more effective in treatment of MDR malaria, produce undesirable side effects. A nanoencapsulation of these drugs can dramatically modify their biodistribution and pharmacodynamics thereby reducing their toxicity.
- Minimizing the risk of transmitting infection from the mother to the child in the womb. Site-selective delivery of nanoencapsulated drugs provides the opportunity to treat infection in a pregnant woman at an early stage, thus minimizing the risk of spreading of the drug and the infection to the unborn child. This has been a major healthcare issue in the regions of high malaria incidence.
- Utilizing pH-sensitive nanocarriers to increase drug concentration at the intracellular site of the *Plasmodium* spp. To significantly enhance the therapy, it is necessary to produce high concentration of the drug in the intracellular parasitophorous vacuole. In order to access this target, a drug must traverse multiple membranes inside the red blood cells. A pH-sensitive nanocarrier can take advantage of the pH differences in the various intracellular compartments.

The following provides some account of current efforts in using nanotechnology to develop vaccines and more effective therapy for malaria.

15.5.1. Vaccines

As stated above, currently there is no vaccine for malaria. Intensive research in developing malarial vaccine has produced some progress because recent phase 3 trials have had some promising results [Agnandji et al., 2011]. There are numerous challenges in developing a vaccine for malaria; some principal ones include:

- The malaria parasite is intracellular. The infected cells do exhibit surface antigens indicative of malaria, which suggests the possibility of an immune-based therapeutic strategy. However, development of immune recognition of the malaria infected cells is hindered by a huge variability of these antigens. In this context, malarial parasites derive their variability both from genetic recombination in mosquitoes during the sexual development and from cell-to-cell variations in the expression of surface antigens [Chen et al., 2000]. Furthermore, the molecular makeup of a parasite can change at different stages of its life cycle in the human body. In general, the large reservoir of variant antigens gives the parasite a stealth ability to avoid immune recognition and elimination in the organism.
- In infected individuals, malaria proliferates extremely fast. The median sporozoite inoculum from a single mosquito bite is approximately 10 cells. The initial stage of parasite's multiplication occurs in liver cells, hepatocytes. Approximately one week following the infection, each infected hepatic cell produces approximately 30,000 merozoites. Released parasites then invade red blood cells and begin repetitive cycles of asexual reproduction at a rate that can produce up to 10^8 blood-stage parasites per milliliter of blood in a matter of a single week [Crompton et al., 2010; Simpson et al., 2002]. Such a rapid kinetics of proliferation produces certain challenges for capacity of immune system response.

However, there are clear indications that malaria *Plasmodium* is not evasive for the immune system. In endemic areas of intense malaria dissemination, in individuals who are continuously infected, a mild natural immunity to malaria can be generated [Marsh and Kinyanjui, 2006].

On the basis of their target, there are four types of malaria vaccines [Crompton et al., 2010]:

1. *Pre-erythrocytic Vaccines.* This vaccine targets the sporozoite at liver stages before the parasite reaches the bloodstream.
2. *Asexual Blood-Stage Vaccines.* This vaccine relies on recognition of parasites upon its entrance in a bloodstream.

3. *Combined Pre-erythrocytic and Blood-Stage Vaccines.* This vaccine is more general, and, as follows from its name, it combines targets of the earlier two types. Thus a part of the vaccine targets the infected hepatocytes, hence decreasing the number of merozoites released into the bloodstream. It allows the blood-stage vaccine more time to produce immunity before the fever threshold is reacted.
4. *Transmission-Blocking Vaccines.* This form of vaccination is used to target the antigens on parasite gametes, by a mechanism where antibodies are ingested by mosquitoes together with gametocytes, and disrupt certain stages of parasite's sexual development.

The most promising malarial vaccine approach is the development of a pre-erythrocytic vaccine that targets a protein expressed at this stage, the circumsporozoite protein (CSP), which covers the parasite surface. The vaccine involves the cloning of the *P. falciparum* CSP leading to the development of RTS,S vaccine which consists of hepatitis B surface antigen (HBsAg) particles. Twenty-five percent of the HBsAg peptide is fused to the central repeat and thromospondin domain of the CSP formulated in the adjuvant As01 [Crompton et al., 2010]. A recently reported Phase 3 trial of RTS,S/AS01 in African children has shown promising results [Agnandji et al., 2011]. This vaccine is believed to provide protection against malaria by temporarily reducing the number of merozoites emerging from the liver.

Another direction for pre-erythrocytic vaccines has been to induce T-cell immunity at the initial stage through either irradiated or genetically attenuated sporozoites, or through expression of *P. falciparum* hepatic-stage proteins in viral vectors [Crompton et al., 2010]. They induce cytotoxic CD8 T lymphocytes that kill parasite-infected hepatocytes.

In order to attack the parasite during erythrocyte infection, studies have focused on utilizing antibodies, IgG1 and IgG3 isotypes, which interact with activating Fc receptors. They may also block the binding of infected erythrocytes to the endothelial cells, thereby promoting opsonization and destruction of mesozoites and infected erythrocytes by phagocytic cells. There are, however, very few antigens in clinical development as vaccines to use during this phase of the infections. Crompton et al. [2010] provide an account of them.

Nanoparticles can provide a safer alternative to viral delivery of vaccine to the target site. In addition, these formulations can protect the vaccine from extracellular degradation. Active targeting to infected hepatocytes and parasite surface antigens will also considerably enhance the efficacy. In addition, nanocarriers such as liposomes, which are vesicles formed by an aqueous core surrounded by one or several phospholipid bilayers, have also been shown to serve as immunological adjuvants in malaria vaccine formulation [Alving and Richards, 1990; Alving, 2002].

Liposomes containing lipid A have exhibited significant humoral immune responses to the encapsulated malaria antigen (R32NS1) in mice [Verma et al., 1992]. The adjuvant effect of liposomal lipid A may be partly due to an increased antigen expression by macrophages and partly as a result of the recruitment of an increased number of macrophages that can then serve as antigen-presenting cells.

Another advantage provided by liposome-based malarial vaccine is that it can induce both humoral (antibody) and cell-mediated immune response by simultaneously activating the major histocompatibility complex (MHC) class 1 and MHC class II pathways [Santos-Magalhaes and Mosqueira, 2010]. However, a major concern in using liposomal nanocarriers is their *in vitro* and *in vivo* stability.

15.5.2. Therapy

The current therapy utilizes a number of antimalarial drugs. Some main drugs are:

1. Chloroquine (CQ). It has been widely used in many parts of the world, particularly where *Plasmodium* species have not developed resistance. It is relatively safe for adults, children, and even pregnant women. Primaquine (PQ) is added to kill persistent parasites in the liver, when the infection involves *Plasmodium vivax* or *Plasmodium ovale*. However, this requires a blood test to determine whether the patient has any glucose-6-phosphate dehydrogenase (G6PD) enzyme deficiency that may cause blood cells to break down, when treated with primaquine.
2. Doxycycline together with quinine (GN). This has been used to treat chloroquine-resistant malaria.
3. Atovaquone-proguanil. This is also a drug of choice in the case of chloroquine resistance. It has fewer side effects compared to quinine.
4. Mefloquine (MQ). MQ has also been employed as an antimalarial. Unfortunately, it has a number of side effects such as causing vivid dreams and producing psychological side effects.
5. Artemisinin, originated in Chinese herbal medicine, is named after the Artemisia annua plant from which it was extracted. This drug and its derivatives—for example, artemether (AM)—can produce a very rapid pharmaceutical response against *Plasmodium flaciparum* and *Plasmodium vivax*, which cover the wide geographic region from Asia to Africa and America. To prevent the buildup of drug resistance, a combination of several antimalarial drugs, one of which is a derivative of artemisinin, is administered, as per World Health Organization guidelines. Such artemisinin-combination therapies (ACTs) are currently recommended for the treatment of uncomplicated flaciparum malaria.

The advantages provided by using a nanocarrier to deliver drugs for malaria are already discussed in the earlier part of this section. Antimalarial drugs containing nanoparticles of dimensions <30 nm can access intracellular compartments of parasites through new permeability pathways in the infected red blood cells (RBCs), which are produced by increased permeability of the membranes in the infected RBCs [Charpian and Przyborski, 2008].

A wide variety of nanoparticles, both of naturally occurring materials and synthetic materials, have been used for delivery of antimalarial drugs. A detailed account can be found in the review by Santos-Magalhaes and Mosqueira [2010]. They range from liposomes, to solid lipid nanoparticles (SLN), to nanoemulsions, to chitosans, to cyclodextrin inclusion complexes, to nanosuspensions. The lipid-based systems, liposomes, and SLN are naturally attractive because of the involvement of liver. Here the specific nanoparticle discussed is liposomes.

Liposomal nanocarriers for antimalarial drug delivery have received considerable attention. Liposomes have already been discussed in a number of places in this chapter. An advantage of using liposomes is the ease of surface functionalization to place targeting and PEG groups, as well as to manipulate the surface charge which plays a significant role in intracellular uptake. A number of antimalarial drugs (CQ, PQ, QN, AM, etc.) have been encapsulated in neutral and long-circulating (PEGylated) liposomes, since the 1980s. These drugs are generally positively charged in physiological pH, and thus their encapsulation can be controlled by a pH gradient. The encapsulation provides protection against drug degradation in the blood, and it also reduces toxicity. These nanoencapsulations can be administered orally or through the intravenous route. Additionally, by increasing the length of the acyl chain of the phospholipid of liposome, the rate of release of the AM drug decreases, which allows for control of the drug release.

The various drugs have also been encapsulated in negatively charged liposomes. The negatively charged liposomes improve the encapsulation efficiency of the antimalarial drugs, which are positively charged under physiological pH. Another variation of liposome encapsulation is to coat them with chitosan, a shellfish-derived polysaccharide. These chitosan-coated liposomes are called chitosomes [Henriksen et al., 1997].

Targeted formulations of liposomes for antimalarial drug delivery have also been investigated. PQ has been linked to leucine and to peptides for the synthesis of covalent PQ-glycoprotein conjugates, which selectively target hepatocytes. These pro-drugs are then entrapped in liposomes. This formulation exhibits increased chemotherapeutic activity and reduced toxicity [Haynes et al., 2008]. Another study of peptide-targeted and long-circulating liposomes to target hepatocytes is by Longmuir et al. [2006]. The liver uptake of peptide-bearing and PEGylated liposomes was found to be more than 600-fold higher than that in the heart and more than 200-fold higher than that in the lungs or kidneys.

Finally, active targeting has also been achieved by using antibody-bearing liposomes. A high therapeutic efficacy was found using CQ-loaded $MA6F_{10}$ liposomes. This was attributed to specific interaction of the Mab F_{10}-fragment with the *Plasmodium*-infected erythrocytes that produced efficient internalization of the liposomes by the infected cells and subsequent specific binding of the MAb F_{10} liposomes to the parasites [Owais et al., 1995]. Antibody has also been covalently attached to the surface of liposomes.

HIGHLIGHTS OF THE CHAPTER

- The emergence of multi-drug resistance and rapid mutation of many pathogens, poor hygienic conditions, and nonavailability/unaffordability of effective drugs in developing countries, combined with the rapidity of world travel, have made infectious diseases a major healthcare issue across the globe, requiring newer strategies to combat them.
- The infections caused by three types of microbes: Viruses, bacteria, and protozoan parasites pose world threat.
- Viruses do not have a cellular organization and are considered by many as nonliving; however, in a host body they replicate and they can also mutate.
- Bacteria are living species; however, they do not have organelles, and they can only survive in the appropriate environment to which they have adapted.
- Protozoa are larger than bacteria and have a eukaryotic cellular structure like ours. Some larger protozoa can live by themselves, while others, like malaria, are parasitic and require host organisms like female Anopheles mosquitoes and humans in which to carry out their life cycle.
- Although there are some common symptoms in infections caused by microbial agents such as viruses, bacteria, and parasites, they differ in their microstructure (shape, surface charge, lipophilicity, and adhesion molecules), as well as in the pathways of infections (pathogenesis), thus needing special considerations and specific treatments.
- Nanotechnology holds the promise of new approaches for early detection, prevention, targeted delivery, multi-drug therapy, and theranostics for infectious diseases.
- A major advantage of nanoparticle delivery of diagnostics/prophylaxis/therapy is its ability to target a pathogen at the cellular and intracellular levels, utilizing their molecular structure and specific antigen markers. In diseases like HIV and tuberculosis, targeting at the organelle level is a promising direction to significantly enhance the therapeutic drug efficiency.
- Nanoparticles provide a rapid, ultrasensitive, and multiplexed diagnostic platform for infectious diseases, utilizing a combination of bioanalytical methods.
- Photodynamic therapy in the context of microbial infection is often called photodynamic antimicrobial therapy (PACT). This approach is particularly promising for the treatment of antibiotic-resistant pathogenic strains.

- Metallic nanoparticles such as that of gold and silver have demonstrated excellent antimicrobial properties via various mechanisms that may involve direct interaction of metallic nanoparticles with microbes, or light-induced photothermal therapy.
- A number of different types of nanoparticles have been used for vaccine delivery, some of them utilizing virus nanoparticles or other biomaterials, while others are synthetic nanoparticles.
- HIV is a lentivirus, which belongs to the family of Retroviridae. Its infection causes acquired immune deficiency syndrome (AIDS), a disease that impairs the body's immune system. Another severe symptom of AIDS is deterioration of the brain's function. This catastrophic latter stage is also known as NeuroAIDS.
- Liposomes and polymeric nanoparticles (e.g., PLA and PLGA) have shown enormous potential for delivery of both protein-based and DNA-based vaccines, along with antiretroviral drugs for the treatment of HIV infection.
- Gold nanorods have been used for delivery of siRNA across the blood–brain barrier (BBB) to treat NeuroAIDS produced by HIV infection in the brain, via the gene-silencing therapy approach.
- Nanoformulated antiretroviral drugs offer multifunctionality, that is, the ability to package multiple diagnostic and therapeutic agents within the same nanocomposite, along with the added provisions of site-directed delivery, delivery across the BBB, and controlled release of therapeutics.
- Typical diagnosis of influenza involves virus isolation, antigen detection, serological responses, or PCR, which generally takes time and is expensive. A number of nanotechnology-based approaches show promise to overcome the current diagnostic limitations.
- Nanoparticle-based nonviral nucleic acid and gene transfer strategies that target human airways *in vivo* employing nanoplexes (an electrostatic complex of cationic nanoparticles (i.e., gold nanorods (GNR) or chitosan) and anionic nucleic acids) are being actively explored for the treatment of influenza.
- Given the rapidity of world travel, there is a growing need for a test capable of rapid, sensitive detection of active TB. Rapid diagnosis of pulmonary tuberculosis will involve analysis of multiple biomarkers in body fluids samples such as induced sputum, urine, blood, and so on.
- A colorimetric method utilizing functionalized gold nanoparticles has been proposed for rapid and multiplexed detection of TB using a simple color change.
- Treatment of TB requires a multiple regimen consisting of two to four different drugs over a 6- to 12-month period. Primarily because of the long duration of treatment, the appearance of drug-resistant strains of *Mycobacterium* is increasing.

- Niosomes, investigated both *in vitro* and *in vivo* for TB drug delivery, are liposome-like vesicles made with cholesterol and nonionic surfactants. They form thermodynamically stable nanoparticles, which can incorporate hydrophilic drugs within the core and lipophilic drugs in the hydrophobic domain.
- A nanoparticle platform based on a biodegradable material can encapsulate known anti-TB drugs, as well as carry a genetic surface payload, which will encode for the established macrophage activator, tumor necrosis factor alpha, for the treatment of TB.
- Malaria is a very complex disease caused by any of the four species of *Plasmodium*, a parasitic protozoan, which include *Plasmodium falciparum*, *Plasmodium vivax*, *Plasmodium malariae*, and *Plasmodium ovale*. They are all transmitted to humans by the bites of a female Anopheles mosquito vector. Of these, the most dangerous one is *P. falciparum*, which causes the most prevalent infection in Africa and is a major cause of mortality.
- Some of the new antimalarial drugs, proven to be more effective in treatment of MDR malaria, produce undesirable side effects. A nanoencapsulation of these drugs can dramatically modify their biodistribution and pharmacodynamics, thereby reducing their toxicity.
- A promising strategy for malarial vaccine is to target the circumsporozoite protein (CSP), which covers the parasite surface and is expressed at a pre-erythrocytic stage.
- Nanoparticles can provide a safer alternative to viral delivery of vaccine to the target site and protect it from extracellular degradation, thereby enhancing the efficacy of vacination. Nanocarriers such as liposomes have also been shown to serve as immunological adjuvants in malaria vaccine formulation.
- A wide variety of nanoparticles, both of naturally occurring materials and synthetic materials, have been used for delivery of antimalarial drugs. Antimalarial drugs containing nanoparticles of dimensions <30 nm can access parasites intracellularly through increased permeability of the membranes in the infected red blood cells.

EXERCISES

True or False

1. Tuberculosis is caused by a parasite infection.
2. Viruses need to infect cells in order to proliferate.
3. RNA–nanoparticle complexes imitate the viral nucleic acid and can stimulate the innate immune response.

4. Nanoparticles can penetrate inside the bacteria and damage their interior.
5. Nanoparticle aggregation assays involve spectral shift in the surface plasmon resonance of metallic nanoparticles upon their aggregation on the targeted analyte.
6. Niosomes are the liposome-like vesicles formed mostly by cholesterol.
7. Malaria transmission blocking vaccines target sporozoite at liver stages before the parasite reaches the bloodstream.

Multiple Choice

8. Metallic nanoparticle can be used for HIV detection by the
   - **(a)** Nanoparticle decomposition
   - **(b)** Nanoparticle aggregation
   - **(c)** Nanoparticle ablation
   - **(d)** Electron transfer from the HIV to nanoparticle
9. In order to detect tuberculosis using gold nanoparticle, the surface of gold nanoparticles is functionalized with
   - **(a)** CD4 antibody
   - **(b)** Polyethylene glycol
   - **(c)** Complementary oligonucleotide
   - **(d)** siRNA
10. What is the major reason for implementation of the artemisinin combination therapies for treatment of malaria?
    - **(a)** Artemisinin, when used alone, has insufficient pharmaceutical efficiency.
    - **(b)** Artemisinin targets only extracellular parasites and cannot eliminate the entire pool of malaria parasite from the body.
    - **(c)** Malaria *Plasmodium* builds up a resistance against artemisinin when it used alone.
    - **(d)** Artemisinin is too toxic, and use of other drugs allows for diluting of its concentration.
11. Which of the following is a major concern of liposome-based malarial vaccine?
    - **(a)** Biodegradability
    - **(b)** *In vivo* stability
    - **(c)** Toxicity
    - **(d)** Expense

Answers: 1. F; 2. T; 3. T; 4. T; 5. T; 6. T; 7. F; 8. b; 9. c; 10. c; 11. b

REFERENCES

Agnandji, S. T., Lell, B., Soulanoudjingar, S. S., Fernandes, J. F., Abossolo, B. P., et al., First results of phase 3 trial of RTS,S/AS01 malaria vaccine in African children, *N. Engl. J. Med.* **365**, 1863–1875 (2011).

Aline, F., Brand, D., Pierre, J., Roingeard, P., Severine, M., Verrier, B., Dimier-Poisson, I., Dendritic cells loaded with HIV-1 p24 proteins adsorbed on surfactant-free anionic PLA nanoparticles induce enhanced cellular immune responses against HIV-1 after vaccination, *Vaccine* **27**, 5284–5291 (2009).

Alving, C. R., Design and selection of vaccine adjuvants: Animal models and human trials, *Vaccine* **20**, S56–S64 (2002).

Alving, C. R., Richards, R. L., Liposomes containing lipid A: A potent nontoxic adjuvant for a human malaria sporozoite vaccine, *Immunol. Lett.* **25**, 275–279 (1990).

Baert, L., van't Klooster, G., Dries, W., Francois, M., Wouters, A., Basstanie, E., Iterbeke, K., Stappers, F., Stevens, P., Schueller, L., Van Remoortere, P., Kraus, G., Wigerinck, P., Rosier, J., Development of a long-acting injectable formulation with nanoparticles of rilpivirine (TMC278) for HIV treatment, *Eur. J. Pharm. Biopharm.* **72**, 502–508 (2009).

Bonoiu, A. C., Mahajan, S. D., Ding, H., Roy, I., Yong, K. T., Kumar, R., Hu, R., Bergey, E. J., Schwartz, S. A., Prasad, P. N., Nanotechnology approach for drug addiction therapy: Gene silencing using delivery of gold nanorod-siRNA nanoplex in dopaminergic neurons, *Proc. Natl. Acad. Sci. U S A* **106**, 5546–5550 (2009).

Chakravarthy, K. V., Davidson, B. A., Helinski, J. D., Ding, H., Law, W. C., Yong, K. T., Prasad, P. N., Knight, P. R., Doxorubicin-conjugated quantum dots to target alveolar macrophages and inflammation, *Nanomedicine-UK* **7**, 88–96 (2011).

Charpian, S., Przyborski, J. M., Protein transport across the parasitophorous vacuole of *plasmodium falciparum*: Into the great wide open, *Traffic* **9**, 157–165 (2008).

Chen, Q., Schlichtherle, M., Wahlgren, M., Molecular aspects of severe malaria, *Clin. Microbiol. Rev.* **13**, 439–450 (2000).

Crompton, P. D., Pierce, S. K., Miller, L. H., Advances and Challenges in Malaria Vaccine Development, *J. Clin. Invest.* **120**, 4168–4178 (2010).

das Neves, J., Amiji, M. M., Bahia, M. F., Sarmento, B., Nanotechnology-based systems for the treatment and prevention of HIV/AIDS, *Adv. Drug Delivery Rev.* **62**, 458–477 (2010).

Dou, H., Grotepas, C. B., McMillan, J. M., Destache, C. J., Chaubal, M., Werling, J., Kipp, J., Rabinow, B., Gendelman, H. E., Macrophage delivery of nanoformulated antiretroviral drug to the brain in a murine model of NeuroAIDS, *J. Immunol.* **183**, 661–669 (2009).

Driskell, J. D., Jones, C. A., Tompkins, S. M., Tripp, R. A., One-step assay for detecting influenza virus using dynamic light scattering and gold nanoparticles, *Analyst* **136**, 3083–3090 (2011).

Elghanian, R., Storhoff, J. J., Mucic, R. C., Letsinger, R. L., Mirkin, C. A., Selective colorimetric detection of polynuclotides based on the distance-dependent optical properties of gold nanoparticles, *Science* **277**, 1078–1081 (1997).

Fuller, D. H., Loudon, P., Schmaljohn, C., Preclinical and clinical progress of particle-mediated DNA vaccines for infectious diseases, *Methods* **40**, 86–97 (2006).

Garcia-Contreras, L., Wong, Y. L., Muttil, P., Padilla, D., Sadoff, J., DeRousse, J., Germishuizen, W. A., Goonesekera, S., Elbert, K., Bloom, B. R., Miller, R., Fourie, P. B., Hickey, A., Edwards, D., Immunization by a bacterial aerosol, *Proc. Natl. Acad. Sci. USA* **105**, 4656–4660 (2008).

Garg, M., Asthana, A., Agashe, H. B., Agrawal, G. P., Jain, N. K., Stavudine-loaded mannosylated liposomes: *In-vitro* anti-HIV-1 activity, tissue distribution and pharmacokinetics, *J. Pharm. Pharmacol.* **58**, 605–616 (2006).

Gluck, R., Moser, C., Metcalfe, I. C., Influenza virosomes as an efficient system for adjuvanted vaccine delivery, *Expert. Opin. Biol. Ther.* **4**, 1139–1145 (2004).

Grgacic, E. V. L., Anderson, D. A., Virus-like particles: Passport to immune recognition, *Methods* **40**, 60–65 (2006).

Haasnoot, J., Westerhout, E. M., Berkhout, B., RNA inteference against viruses: Strike and counterstrike, *Nat. Biotechnol.* **25**, 1435–1443 (2007).

Hauck, T. S., Giri, S., Gao, Y., Chan, W. C. W., Nanotechnology diagnostics for infectious diseases prevalent in developing countries, *Adv. Drug Delivery Rev.* **62**, 438–448 (2010).

Haynes, S. M., Longmuir, K. J., Robertson, R. T., Baratta, J. L., Waring, A. J., Liposomal polyethyleneglycol and polyethyleneglycol-peptide combinations for active targeting to liver *in vivo*, *Drug Deliv.* **15**, 207–217 (2008).

Henriksen, I., Vaagen, S. R., Sande, S. A., Smistad, G., Karlsen, J., Interactions between liposomes and chitosan II: Effect of selected parameters on aggregation and leakage, *Int. J. Pharm.* **146**, 193–203 (1997).

Herzog, C., Hartmann, K., Künzi, V., Kürsteiner, O., Mischler, R., Lazar, H., Glück, R., Eleven years of Inflexal® V—A virosomal adjuvanted influenza vaccine, *Vaccine* **27**, 4381–4387 (2009).

Hirsch, L. R., Jackson, J. B., Lee, A., Halas, N. J., West, J. L., A whole blood immunoassay using gold nanoshells, *Anal. Chem.* **75**, 2377–2381 (2003).

Huygen, K., Content, J., Denis, O., Montgomery, D. L., Yawman, A. M., Deck, R. R., DeWitt, C. M., Orme, I. M, Baldwin, S., D'Souza, C., Drowart, A., Lozes, E., Vandensbussche, P., Van Vooren, J. P., Liu, M. A., Ulmer, J. B., Immunogeneicity and protective efficacy of a tuberculosis DNA vaccine, *Nat. Med.* **2**, 893–898 (1996).

Illum, L., Nanoparticulate systems for nasal delivery of drugs: A real improvement over simple systems? *J. Pharm. Sci.* **96**, 473–483 (2007).

Jori, G., Fabris, C., Soncin, M., Ferro, S., Coppellotti, O., Dei, D., Fantetti, L., Chiti, G., Roncucci, G., Photodynamic therapy in the treatment of microbial infections: Basic principles and perspective applications, *Lasers Surg. Med.* **38**, 468–481 (2006).

Kadiu, I., Nowacek, A., McMillan, J., Gendelman, H. E., Macrophage endocytic trafficking of antiretroviral nanoparticles, *Nanomedicine-UK* **6**, 975–994 (2011).

Kim, P. S., Read, S. W., Nanotechnology and HIV: Potential applications for treatment and prevention, *WIREs Nanomed. Nanobiotechnol.* **2**, 693–702 (2010).

Levy, J. A., *HIV and the Pathogenesis of AIDS*, 3rd ed., American Society of Microbiology Press, Washington, DC (2007).

Lim, H. W., Soter, N. A., *Clinical Photomedicine*, Marcel Dekker, New York (1993).

Lisziewicz, J., Kelly, L., Lori, F., Topical DermaVir vaccine targeting dendritic cells, *Curr. Drug Deliv.* **3**, 83–88 (2006).

Longmuir, K. J., Robertson, R. T., Haynes, S. M., Baratta, J. L., Waring, A. J., Effective targeting of liposomes to liver and hepatocytes *in vivo* by incorporation of a *Plasmodium* amino-acid sequence, *Pharm. Res.* **23**, 759–769 (2006).

Look, M., Bandyopadhyay, A., Blum, J. S., Fahmy, T. M., Application of nanotechnologies for improved immune response against infectious diseases in the developing world, *Adv. Drug Deliv. Rev.* **62**, 378–393 (2010).

Lu, J. M., Wang, X. W., Marin-Muller, C., Wang, H., Lin, P. H., Yao, Q. Z., Chen, C. Y., Current advances in research and clinical applications of PLGA-based nanotechnology, *Expert. Rev. Mol. Diagn.* **9**, 325–341 (2009).

Ludwig, L. B., RNA silencing and HIV: A hypothesis for the etiology of the severe combined immunodeficiency induced by the virus, *Retrovirology* **5**, 79 (2008).

Mahajan, S. D., Aalinkeel, R., Reynolds, J. L, Nair, B., Sykes, D. E,, Law, W. C., Ding, H., Bergey, E. J., Prasad, P. N., Schwartz, S. A., Nanotherapeutics using a HIV-1 poly A and transactivator of the HIV-1 LTR (TAR) specific siRNA, *Pathol. Res. Int.* **2011**, 1–9 (2011a).

Mahajan, S. D., Law, W. C., Aalinkeel, R., Reynolds, J. L., Nair, B., Yong, K. T., Roy, I., Prasad, P. N., Schwartz, S. A., Nanomedicine, in *Nanoparticle Mediated Targeted Delivery of Antiretrovirals to the Brain, Methods in Enzymology* Elsevier/Academic Press, New York (2011b).

Mahajan, S. D., Reynolds, J. L., Aalinkeel, R., Nair, B. B., Sykes, D. E., Roy, I., Ding, H., Prasad, P. N., Schwartz, S. A., Enhancing the transversing efficiency and efficacy of the anti retroviral drug "Saquinavir" using nanotechnology: Implications for nanotherapeutics in Neuro-AIDS, *Global Antiviral J.* **5**, 45–46 (2009).

Mahajan, S. D., Roy, I., Xu, G., Yong, K. T., Ding, H., Aalinkeel, R., Reynolds, J. L., Sykes, D. E., Nair, B. B., Lin, E. Y., Prasad P. N., Schwartz S. A., Enhancing the delivery of anti retroviral drug "Saquinavir" across the blood brain barrier using nanoparticles. *Curr. HIV Res.* **8**(5):396–404 (2010).

Mahe, B., Vogt, A., Liard, C., Duffy, D., Abadie, V., Bonduelle, O., Boissonnas, A., Sterry, W., Verrier, B., Blume-Peytavi, U., Combadiere, B., Nanoparticle-based targeting of vaccine compounds to skin antigen-presenting cells by hair follicles and their transport in mice, *J. Invest. Dermatol.* **129**, 1156–1164 (2009).

Mamo, T., Moseman, E. A., Kolishetti, N., Salvador-Morales, C., Shi, J., Kuritzkes, D. R., Langer, R., von Andrian, U., Farokhzad, O. C., Emerging nanotechnology approaches for HIV/AIDS treatment and prevent, *Nanomedicine* **5**, 269–285 (2010).

Marsh, K., Kinyanjui, S., Immune effector mechanisms in malaria, *Parasite Immunol.* **28**, 51–60 (2006).

McKnight, R. E., Ye, M., Ohulchanskyy, T. Y., Sahabi, S., Wetzel, B. R., Wagner, S. J., Shripchenko, A., Detty, M. R., Synthesis of analogues of a flexible thiopyrylium photosensitizer for purging blood-borne pathogens and binding mode and affinity studies of their complexes with DNA, *Bioorg. Med. Chem.* **15**, 4406–4418 (2007).

McMillan, J., Batrakova, E., Gendelman, H. E., Cell delivery of therapeutic nanoparti-ces, *Prog. Mol. Biol. Transl. Sci.* **104**, 563–601 (2011).

Miller, L. H., Baruch, D. I., Marsh, K., Doumbo, O. K., The pathogenic basis of malaria, *Nature* **415**, 673–679 (2002).

Miyake, A., Akagi, T., Enose, Y., Ueno, M., Kawamura, M., Horiuchi, R., Hiraishi, K., Adachi, M., Serizawa, T., Narayan, O., Akashi, M., Baba, M., Hayami, M., Induction of HIV-Specific Antibody response and protection against vaginal sHIV transmission by intranasal immunization with inactivated sHIV-capturing nanospheres in macaques, *J. Med. Virol.* **73**, 368–377 (2004).

Morones, J., Elechiguerra, J., Camacho, A., Holt, K., Kouri, J., Ramirez, J., Yacaman, M., The bactericidal effect of silver nanoparticles, *Nanotechnology* **16**, 2346–2353 (2005).

Mu, B., Huang, X., Bu, P., Zhuang, J., Cheng, Z., Feng, J., Yang, D., Dong, C., Zhang, J., Yan, X., Influenza virus detection with pentabody-activated nanoparticles, *J. Virol. Methods* **169**, 282–289 (2010).

Naito, Y., Nohtomi, K., Onogi, T., Uenishi, R., Ui-Tei, K., Saigo, K., Takebe, Y., Optimal design and validation of antiviral siRNA for targeting HIV-1, *Retrovirology* **8**, 80 (2007).

Naito, Y., Ui-Tei, K., Nishikawa, T., Takebe, Y., Saigo, K., siVirus: Web-based antiviral siRNA design software for highly divergent viral sequences, *Nucleic Acids Res.* **34**, W448–W450 (2006).

Nakonechny, F., Firer, M. A., Nitzan, Y., Nisnevitch, M., Intracellular antimicrobial photodynamic therapy: A novel technique for efficient eradication of pathogenic bacteria, *Photochem. Photobiol.* **86**, 1350–1355 (2010).

Nasir, A., Nanotechnology in vaccine development: A step forward, *J. Invest. Dermatol.* **129**, 1055–1059 (2009).

Ohashi, K., Kabasawa, T., Ozeki, T., Okada, H., One-step preparation of rifampicin/poly(lactic-co-glycolic acid) nanoparticle-containing mannitol microspheres using a four-fluid nozzle spray drier for inhalation therapy of tuberculosis, *J. Control Release* **135**, 19–24 (2009).

Ohulchanskyy, T. Y., Gannon, M. K., Ye, M., Skripchenko, A., Wagner, S. J., Prasad, P. N., Detty, M. R., Switched-on flexible chalcogenopyrylium photosensitizers. Changes in photophysical properties upon binding to DNA, *J. Phys. Chem. B* **111**, 9686–9692 (2007).

Oliveira, G. A., Wetzel, K., Calvo-Calle, J. M., Nussenzweig, R., Schmidt, A., Birkett, A., Dubovsky, F., Tierney, E., Gleiter, C. H., Boehmer, G., Luty, A. J. F., Ramharter, M., Thornton, G. B., Kremsner, P. G., Nardin, E. H., Safety and enhanced immunogenecity of a hepatitis B core particle *Plasmodium Falciparum* malaria vaccine formulated in adjuvant ISA 720 in a Phase I trial, *Infect. Immun.* **73**, 3587–3597 (2005).

Owais, M., Varshney, G. C., Choudhury, A., Chandra, S., Gupta, C. M., Chloroquine encapsulated in malaria-infected erythrocyte-specific antibody-bearing liposomes effectively controls chloroquine-resistant *Plasmodium* berghei infections in mice, *Antimicrob. Agents Chemother.* **39**, 180–184 (1995).

Pandey, R., Khuller, G. K., Oral nanoparticle-based antituberculosis drug delivery to the brain in an experimental model, *J. Antimicrob. Chemother.* **57**, 1146–1152 (2006).

Pandey, R., Sharma, A., Zahoor, A., Sharma, S., Khuller, G. K., Prasad, B., Poly(DL-lactide-co-glycolide) nanoparticle-based inhalable sustained drug delivery system for experimental tuberculosis, *J. Antimicrob. Chemother.* **52**, 981–986 (2003).

Peek, L. J., Middaugh, C. R., Berkland, C., Nanotechnology in vaccine delivery, *Adv. Drug Delivery Rev.* **60**, 915–928 (2008).

Pett, S. L., Immunotherapies in HIV-1 infection, *Curr. Opin. HIV AIDS* **4**, 188–193 (2009).

Pissuwan, P., Cortie, C. H., Valenzuela, S. M., Cortie, M. B., Functionalized gold nanoparticles for controlling pathogenic bacteria, *Trends Biotechnol.* **28**, 207–213 (2010).

Ranjita, S., Loaye, A. S., Khalil, M., Present status of nanoparticle research for treatment of tuberculosis, *J. Pharm. Pharmaceut.* **14**, 100–116 (2011).

Reverchon, E., De Marco, I., Della Porta, G., Rifampicin microparticles production by supercritical antisolvent precipitation, *Int. J. Pharm.* **243**, 83–91 (2002).

Rimmelzwaan, G. F., Osterhaus, A. D. M. E., A novel generation of viral vaccines based on the ISCOM matrix, in M. F. Powell, M. J. Newman, eds., *Vaccine Design: The Subunit and Adjuvant Approahces*, Plenum Press, New York (1994).

Rohan, L. C., Sassi, A. B., Vaginal drug delivery systems for HIV prevention, *AAPS J.* **11**, 78–87 (2009).

Rossi, J. J., June, C. H., Kohn, D. B., Genetic therapies against HIV, *Nat. Biotechnol.* **25**, 1444–1454 (2007).

Rupp, R., Rosenthal, S. L., Stanberry, L. R., VivaGel (SPL7013 Gel): A candidate dendrimer-microbicide for the prevention of HIV and HSV infection, *Int. J. Nanomed.* **2**, 561–566 (2007).

Sametband, M., Shukla, S., Meningher, T., Hirsh, S., Mendelson, E., Sarid, R., Gedanken, A., Mendelboim, M., Effective multi-strain inhibition of influenza virus by anionic gold nanoparticles, *Med. Chem. Commun.* **2**, 421–423 (2011).

Santos-Magalhaes, N. S., Mosqueira, V. C., Nanotechnology applied to the treatment of malaria, *Adv. Drug Delivery Rev.* **62**, 560–575 (2010).

Semete, B., Kalombo, L., Katata, L., Swai, H., Nano-drug delivery systems: Advances in TB, HIV and malaria treatment, in A. K. Mishra, A. Tiwari, S. B. Mishra eds. *Smart Biomolecular Medicine*, VBRI Press, Allahabad, India, pp. 15–52 (2010).

Shinoda, K., Xin, K. Q., Jounai, N., Kojima, Y., Tamura, Y., Okada, E., Kawamoto, S., Okuda, K., Klinman, D., Okuda, K., Polygene DNA vaccine induces a high level of protective effect against HIV-vaccinia virus challenge in mice, *Vaccine* **22**, 3676–3690 (2004).

Simpson, J. A., Aarons, L., Collins, W. E., Jeffery, G. M., White, N. J., Population dynamics of untreated *Plasmodium falciparum* malaria within the adult human host during the expansion phase of the infection, *Parasitology* **124**, 247–263 (2002).

Sosnik, A., Carcaboso, A. M., Glisoni, R. J., Moretton, M. A., Chiappetta, D. A., New old challenges in tuberculosis: Potentially effective nanotechnologies in drug delivery, *Adv. Drug Delivery Rev.* **62**, 547–559 (2010).

Stefanidou, M., Martinez, O., Hayes, M., Clavel, C., Champion, C., Penades, S., Shattock, R., Evaluation of gold manno-nanoparticles in a model for DC-SIGN binding and dissemination of HIV-1, Microbicides 2006 Conference, Cape Town, South Africa, 2006.

Surabhi, R. M., Gaynor RB., RNA interference directed against viral and cellular targets inhibits human immunodeficiency virus Type 1 replication, *J. Virol.* **76**, 12963–12973 (2002).

Sutthent, R., Gaudart, N., Chokpaibulkit, K., Tanliang, N., Kanoksinsombath, C., Chaisilwatana, P., p24 antigen detection assay modified with a booster step for diagnosis and monitoring of human immunodeficiency virus Type 1 infection, *J. Clin. Microbiol.* **41**, 1016–1022 (2003).

Swai, H., Semete, B., Kalombo, L., Chelule, P., Potential of treating tuberculosis with a polymeric nano-drug delivery system, *J. Control Release* **132**, e48 (2008).

Tallury, P., Malhotra, A., Byrne, L. M., Santra, S., Nanobioimaging and sensing of infectious diseases, *Adv. Drug Delivery Rev.* **62**, 424–437 (2010).

Tang, S., Zhao, J., Storhoff, J. J., Norris, P. J., Little, R. F., Yarchoan, R., Stramer, S. L., Patno, T., Domanus, M., Dhar, A., Mirkin, C. A., Hewlett, I. K., Nanoparticle-based biobarcode amplification assay (BCA) for sensitive and early detection of human immunodeficiency type 1 capsid (p24) antigen, *J. Acquir. Immune Defic. Syndr.* **46**, 231–237 (2007).

Ui-Tei, K., Naito, Y., Saigo, K., Essential notes regarding the design of functional siRNAs for efficient mammalian RNAi, *J. Biomed. Biotechnol.* **4**, 65052 (2006).

Verma, J. N., Rao, M., Amselem, S., Krzych, U., Alving, C. R., Green, S. J., Wassef, N. M., Adjuvant effects of liposomes containing lipid A: Enhancement of liposomal antigen presentation and recruitment of macrophages, *Infect. Immun.* **60**, 2438–2444 (1992).

Wan, L., Zhang, X., Pooyan, S., Palombo, M. S., Leibowitz, M. J., Stein, S., Sinko, P. J., Optimizing size and copy number for PEG-fMLF (*N*-formyl-methionyl-leucyl-phenylalanine) nanocarrier uptake by macrophages, *Bioconjug. Chem.* **19**, 28–38 (2008).

Zaks, K., Jordan, M., Guth, A., Sellins, K., Kedl, R., Izzo, A., Bosio, C., Dow, S., Efficient immunization and cross-priming by vaccine adjuvants containing TLR3 and TLR9 agonists complexed to cationic liposomes, *J. Immunol.* **176**, 7335–7345 (2006).

CHAPTER 16

REJUVENATION THERAPY

This chapter describes a future vision of nanomedicine to provide anti-aging solutions. As the world population is aging, we need revolutionary approaches to correct age-related degeneration of cells, tissues, and organs. The various directions that could evolve to an integrated approach to deal with age-related ailments are described here. This chapter also is a prelude to Chapters 17 on "Stem Cell Biotechnology" and 18 on "Tissue Engineering," the two major components of rejuvenation therapy. In this chapter these two topics are introduced, but details are left for subsequent chapters. Here, Section 16.1 provides an overview of the rejuvenation therapy. Section 16.2 presents "Free Radical Scavenging" to deal with excessive free radicals generated that, during aging, cannot be taken care of by the body's natural antioxidants. Section 16.3 describes "Chelation Therapy," which can be used to detoxify the body from excessive concentration of metals. Section 16.4 is on "Hormone Therapy" to restore important physiological functions or deal with age-related degeneration. In each case, the emphasis is on how a nanomedicine approach can enhance the therapy.

16.1. REJUVENATION THERAPY: FANTASY OR REALITY?

This section provides an overview of the exciting field, which promises longevity as well as improvement of the quality of life for the aging population of

Introduction to Nanomedicine and Nanobioengineering, First Edition. Paras N. Prasad.

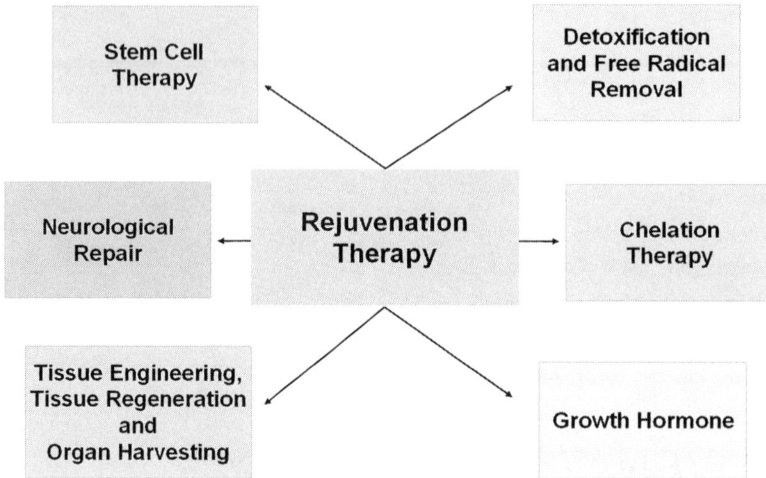

Figure 16.1. Integrated approaches for rejuvenation medicine.

the world. This therapy, in the current status, is often called regenerative therapy, whereby repair or regeneration of cells, tissues, and organ can restore their functions damaged or impaired due to aging. However, there can be a vision to look beyond this and dream of a day where the life span of humans can be significantly expanded. Some even dream of a human being living for 1000 years. It is a dream that may become a reality, because there are clear pathways to correct dysfunctions produced by the aging process. Even more important is to deal with the impairments like neurological diseases, failing of organs, slowing down of body functions, and increased susceptibility to infections. Thus, improving the quality of life should be a major goal of rejuvenation therapy. Although Indian Ayurvedic medicine and Chinese herbal medicine have long become an integral part of rejuvenation therapy, this chapter will limit itself only to recent nanotechnology-based biomedical developments.

Figure 16.1 presents integrated modern approaches for rejuvenation medicine. This section will briefly introduce them.

Stem Cell Therapy. This therapy utilizes pristine stem cells, which are blank cells without specialization, to make them differentiate and replace lacking or nonfunctional cells in the body. Until recent years, the stem cell research has been held back, because generation of stem cell lines from human embryos raised ethical concerns, but now this research area has been opened for full pursuit under new guidelines. The stem cell differentiation can be conducted *in vitro*, and then cells are implanted in body or nondifferentiated cells can be transplanted and then their differentiation occurs *in vivo*. Currently, stem-cell-based tissue regeneration strategies are being pursued for neurological repairs of brain function by replenishment of neurons in brain, or for cardiac repair by replenishment of cardiomyocytes. Other strategies utilize stem cells for growing of entire organs. Stem cell therapy, with this wide scope of potential, is discussed in Chapter 17, together with the impact of nanotechnology on it.

Neurological Repair. A major manifestation of the aging process is the impairment of various brain functions. Examples are Alzheimer's disease, Parkinson's disease, and amyotrophic lateral sclerosis (ALS), which are produced either by the lack of function of gene expression and/or by overexpression of certain proteins. Other age-related diseases can be derived from either (a) producing excessive reactive oxygen species that lead to increased oxidative stress or (b) accumulation of excess of metals such as Fe, Cu, or Zn. The treatment may require (i) stem cell therapy to replenish the neurons, the predominant cells of the central nervous system (CNS), (ii) gene therapy to augment or silence a gene as discussed in Chapter 14, or (iii) a combination of gene therapy and stem cell therapy, as well as free radical scavenging (discussed in Section 16.2) and chelation therapy (discussed in Section 16.3). The gene augmentation therapy to introduce lacking genes in the degenerative cells of the CNS in the brain has already been discussed briefly in the earlier Chapter 14 on "Gene Therapy."

Tissue Engineering. Tissue engineering is a highly active area that deals with tissue repair by tissue regeneration or tissue welding as well as replacement of the entire organ by new organs grown *in vitro* or *in situ* in the body. Thus, injured or damaged tissues can be repaired by regeneration of new tissues. An ill-functioning organ can be replaced by a new one. Nanotechnology is playing an important role in tissue engineering. Considering the tremendous amount of work going on in this area and its very broad impact, this topic is detailed in Chapter 18.

Free Radical Scavenging. As the body ages, it has to cope with increased oxidative stress. The oxidative stress is discussed in detail in Chapter 21 on "Nanotoxicity." The results are the production of highly reactive oxygen species, such as free radicals, peroxides, and superoxides, which can cause cell damage by oxidizing the cellular components. In a young body, the internal defense mechanism involves the naturally present anti-oxidants that undergo oxidation-reduction reaction (redox reaction) with these ROS to neutralize their effect. Aging brings about a significantly higher concentration of ROS and not enough antioxidants to overcome them, thus producing adverse effects. Section 16.2 discusses this topic, together with nanoparticles that can be used to either deliver supplemental antioxidants or themselves act as ROS scavengers (nano antioxidants).

Chelation Therapy. There is mounting evidence that metal imbalance (dishomeostasis) in the brain, produced by the aging process, plays an important role in development of neurodegenerative diseases. The metal ions suspected for this dysfunction are Cu, Fe, Zn, and Al. Chelation therapy utilizes chelating agents to scavenge these metals. However, a major challenge for the chelating agents to produce therapeutic action is first to cross the blood–brain barrier (BBB), bind with the metals, and then again cross the BBB in the way out to remove these metals from brain. Nanoparticle carriers can be designed to cross BBB as described in Chapter 6. In addition, nanoparticles may also suppress toxicity of the chelating agent itself. Thus, considerable efforts have been recently devoted to using carriers conjugated to metal chelators for chelation therapy. The chelation therapy is discussed in detail in Section 16.3.

Hormone Therapy. Hormones are lipid-phospholipid, peptide-based chemicals that are released by a cell or a gland in one part of the body and transported to interact with cells in other parts of the body to produce cell-type-specific responses for diverse body functions. Hormones are thus chemical messengers to carry a signal from one type of cell to another. Examples are insulin and growth hormone, which are protein macromolecules, and steroids, which constitute a major class of lipids [Ozsoy et al., 2009].

A healthy body produces hormones in adequate quantity to maintain vital channels of cell signaling. The naturally produced hormones are thus the endogenous hormones, intrinsically made inside the body. Because of inherent genetic defects and/or because of aging, a specific hormone may not be produced in sufficient quantity or not be fully functioning, producing various disease manifestations such as diabetes, bone loss, and so on. Hormone therapy introduces these molecules, produced synthetically or by using recombinant DNA technology; they are the exogenous hormones [Ozsoy et al., 2009]. Because a protein-based hormone involves a complex integration of a primary, secondary, tertiary, and quaternary structure for its conformation which determines its function, the stability of the conformation being crucially dependent on the environment, a major challenge in hormone therapy is to deliver exogenous hormones intact at the targeted sites. A nanocarrier delivery plays an important role in achieving the goal of providing stability of the protein structure and protecting it against enzymatic degradation, as well as yielding increased bioavailability at the target site. In addition, nanoparticle delivery is also showing promises for a noninvasive nasal (pulmonary) delivery of hormones as opposed to the traditional invasive and painful administration by parenteral injection. The topic of hormone therapy is discussed in more detail in Section 16.4.

16.2. FREE RADICAL SCAVENGING

Free radicals are molecular species that possess unpaired valence electrons and therefore are unstable and highly reactive. In biological systems, these radicals are derived from oxygen- and nitrogen-containing molecules and are collectively called reactive oxygen species (ROS) and reactive nitrogen species (RNS). A process called oxidative stress produces ROS and RNS, which include superoxide (O_2^-), the hydroxy radicals ($OH^•$), nitric oxide (NO), peroxynitrile ($ONOO^-$), and lipid hydroperoxides, which are powerful oxidants. Under normal conditions of their generation, the body's natural defense provides antioxidants that react with them to counter their oxidative-disruptive effect. In the course of aging, the oxidative stress increases, while the amount of antioxidants diminishes, creating an excess of ROS that can oxidize the cellular constituents and cause a disruption of cellular and tissue structures, thus impairing their functions. This ROS-initiated insult on the brain neurons and neuroglia can contribute to development of many neurological diseases.

To offset the adverse effect of ROS and RNS, supplemental antioxidants can be introduced which can act as free radical scavengers. They can be administered noninvasively by oral or nasal delivery. An example is vitamin E, for which oral formulations are available over-the-counter. Nanoparticle technology offers a considerable advantage in providing more effective free radical scavenging for both ROS and RNS. First, the large surface-to-volume ratio offered by nanoparticles is more effective in surface-facilitated reaction with free radicals. Next, the biotargeting can provide selective access to the sites of excessive oxidative stress which produce high levels of ROS and RNS and thus require correspondingly high concentrations or sustained activity of scavengers. The ability to design nanoparticles that can cross the blood–brain barrier is useful in effective free radical scavenging in the brain.

Two broad classes of nanoparticles can be used for free radical scavenging. They are summarized in Figure 16.2. A number of nanoformulations have been produced, which by themselves react with ROS and RNS to provide free radical scavenging. Cerium oxide, also known as ceria, has some unique features that make it very attractive for free radical scavenging [Singh et al., 2007; Rzigalinski et al., 2006]. In the solid phase, the cerium ions in the lattice can exist in either +3 reduced state or +4 fully oxidized states, providing an opportunity to interchange between these two states by redox (reduction–oxidation) processes. Thus, they can readily react with the powerful oxidants, ROS, by oxidizing from +3 to +4 state. This process in the solid state is also facilitated by the existence of lattice sites on the crystal structures that contain oxygen

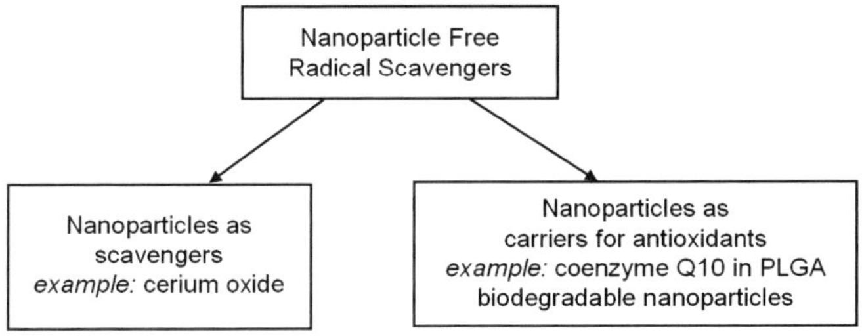

Figure 16.2. The two types of nanoparticle-based free radical scavengers.

vacancies (deficiencies or defects) caused by the loss of oxygen. Furthermore, the high hydrogen-absorbency capacity of the surface of ceria nanoparticles provides opportunities for reaction with H_2 and H_2O to regenerate Ce^{3+} for sustained ROS scavenging activities. It has also been known that the nanoparticle size of ceria makes more efficient redox reactions for many industrial processes [Rzigalinski et al., 2006]. The ideal size appears to be around 10 nm. It should be noted that a certain level of radical production is necessary for cell regulation; it is only the escalation of ROS which produces undesirable effects. Thus, an optimum stage of radical scavenging is desirable for producing the most benefit. The size and concentration dependence studies of ceria nanoparticles in one *in vitro* model confirm this hypothesis (Singh et al., 2007; Rzigalinski et al., 2006].

Both *in vitro* and *in vivo* studies have been conducted with ceria nanoparticles [Singh et al., 2007]. For the *in vitro* model of traumatic brain injury that produces pronounced oxidative stress, and thus significantly higher concentration of ROS, mixed cultures of cortical brain cells comprised of neurons and glia were used. The life span of these cells removed from the rat after traumatic brain injury exhibited a sixfold increase when treated with ceria. Using the free radical indicator dye dichlorofluorescein diacetate, it was demonstrated that nanoparticles could decrease free radical generation in the cell culture [Schubert et al., 2006]. The *in vivo* study was conducted with the fruit fly, *Drosophila melanogaster*, where adding 10-nm ceria directly to the food increased the median survival by 31% [Rzigalinski et al., 2006].

Carbon nanostructures, such as fullerenes (C_{60}) and carbon nanotubes described in Chapter 3, have also been reported to exhibit free radical scavenging ability [Dugan et al., 2001]. In these cases, the antioxidant property is suggested to be derived from the large electronegative centers in these nanostructures. Carboxy fullerenes, which are derivatives of C_{60} with a carboxylic group attached to the cyclopropane carbons of C_{60}, appear to provide improved scavenging.

One major issue to deal with nanoparticles as free radical scavengers is that they may themselves interact with the cell, generating oxidative stress and thus exhibit nanotoxicity. This subject is discussed in Chapter 21. Ceria nanoparticles of the size ~10 nm have not exhibited this behavior. In contrast, there have been several reports of nanotoxicity of carbon nanostructures [Fenoglio et al., 2006; Rzigalinski et al., 2006].

Nanoparticle formulations of conventional antioxidants, such as α-lipoic acid, have also been studied for their activities [Lee et al., 2009]. The α-lipoic acid is an important antioxidant because it is a strong scavenger for ROS, such as hydroxy radicals ($OH^{\bullet}$), hypochlorous acid (HClO), and trichloromethylperoxyl radicals ($CCl_3O_2^{\bullet}$), as well as for RNS, such as ONO. It is also a metal chelator by chelating Fe^{2+} and Cu^{2+}. Several studies point out to its neuroprotective effects against ROS induced damages. It also protects neurons against amyloid-induced cell death. The nanosphere formulation of six hydrophobic multiple α-lipoic acid containing compounds (mALAs) were prepared by the method of spontaneous emulsification [Bouchemal et al., 2004]. Their sizes were in the range of 200–600 nm. These nanospheres were found to be physically and chemically stable and maintained the antioxidant properties of α-lipoic acid.

Another major direction is incorporation of antioxidants in a nanocarrier. Here the emphasis has been the use of biodegradable polymeric nanoparticles in which the antioxidant molecules are incorporated either by physical entrapment or by chemical conjugation. An example provided here is the entrapment of coenzyme Q10 (CoQ10) in biodegradable nanoparticles of poly(lactide-co-glycolide (PLGA), a polyester [Ankola et al., 2007]. CoQ10 is a benzoquinone with 10 isoprenoid units. In addition to its vital function in the ATP synthesis process, it is also an antioxidant, scavenging free radicals and inhibiting lipid peroxidation. Studies point to its utility in prophylaxis and therapy of various disorders related to oxidative stress, such as hypertension.

Ankola et al. prepared the PLGA nanoparticle loaded with varying amounts of CoQ10 by emulsion method using quaternary ammonium salt, didodecyldimethylammonium bromide, as a stabilizer. They studied the effect of loading density on the entrapment efficiency and the particle size. They also tested the therapeutic efficacy in renal hypertensive animals, where this nanoformulation exhibited superior efficacy compared to the commercial formulation at an equal dose.

A major benefit offered by nanoparticle encapsulation is to protect the natural antioxidants from degradation during delivery. Nanoencapsulation has also been used for vitamin E (α-tocopherol) [Zigoneanu et al., 2008].

16.3. CHELATION THERAPY

Chelation therapy involves chelators that are defined by the American Society for Testing and Materials (ASTM) as "chemicals that form soluble, complex molecules with certain metal ions, inactivating the ions so that they cannot normally react with other elements or ions to produce precipitates or scale." The chelators are used to prevent excessive accumulation of metals that can produce toxic effects in different organs [Bolognin et al., 2009]. Thus, chelation therapy is the process of detoxification of the body from the metals where accumulation may even be poisonous. Some of these metals like mercury, lead, and arsenic may be introduced into the body by ingestion, inhalation, or skin contact. Chelation therapy is then essential to remove these metals. However, metals like iron, copper, and zinc, which are essential for life and play a central role in numerous biochemical pathways, when accumulated in excessive amounts, create an imbalance known as dyshomeostasis, which has been suggested to contribute to many types of neurodegeneration. This excessive accumulation may be a manifestation of the aging process.

Alzheimer's disease (AD), which is the most common cause of dementia in the elder population, has been suggested to be at least partially derived from the accumulation of Fe, Cu, and Zn. Although elevated levels of Fe in the brain of the patient with AD has been detected, the role of these metals in contributing to AD is still a controversial topic.

Parkinson's disease is another age-related illness producing tremor, rigidity, and posture instability, for which one of the main reasons has been suggested to be elevated levels of Cu, Fe, Zn and Al in the substantia nigra part of the brain. Even cardiovascular diseases have been attributed to the excessive metal accumulations. Chelation therapy has been extensively investigated to remove Fe, Cu, Zn, and Al, although no conclusive evidence exists for the efficacy of this therapy for AD, PD, or cardiovascular diseases. Chelators used for this purpose are molecules that can irreversibly bind with either a specific metal or with all of them. They are generally introduced either in the bloodstream by intravenous injection or by oral administration. However, for the neurodegenerative diseases, where the efforts have been to remove the excessive metals from the brain, these chelators have to cross the blood–brain barrier and exit after binding with the metal. Deferoxamine (DFO) is a chelator given intravenously to remove excessive iron. DFO is a hexadentate chelator preferentially binding with Fe, but also has an affinity for copper and zinc. It has been approved by the FDA for iron overload. Chelators like ethylene diaminetetraacetic acid (EDTA) have been commonly used.

Nanoparticle formulation for chelation therapy has been also extensively investigated. In this case the chelators can be chemically conjugated to the surface of the nanoparticles. An example is shown in Figure 16.3, where a

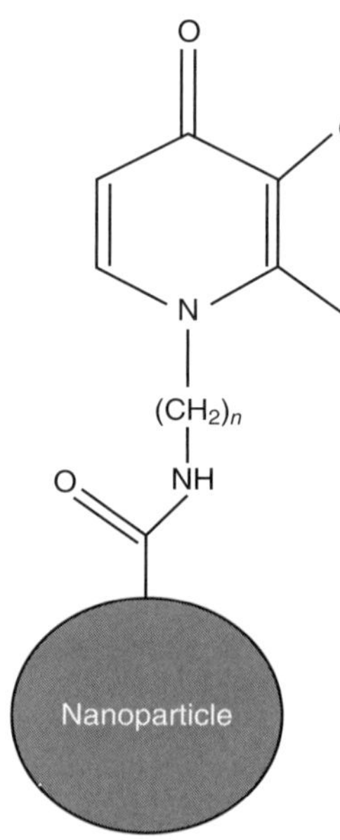

Figure 16.3. The iron chelator (MAPHP, 2-methyl-*N*-(3′-amino-propyl)-3-hydroxyl-4-pyridinone, $n = 3$) covalently linked to nanoparticles. (Reproduced with permission from Liu et al. [2006].)

chelator, MAEHP, is covalently linked to the nanoparticle through an amide linkage [Liu et al., 2006]. As discussed in Chapter 6, nanoparticles can be tailor made with the ability to cross the blood–brain barrier (BBB), where the surface-bound chelator can bind with the excessive metal.

Another advantage of the nanoparticle delivery across the BBB is that after chelator-metal binding, the nanoparticles can re-cross the BBB into the bloodstream. Thus, the nanoparticles can not only deliver a chelator to the brain but also remove the chelator–metal complex from the brain.

16.4. HORMONE THERAPY

As described above, hormones are the messengers in intercellular communications and play a vital role in functioning of many organs. An imbalance of hormones can lead to organ dysfunctions and cause many types of diseases. Diabetes, arthritis, and bone loss are some examples. Another is post-menopause symptoms created by the inactivity of a specific hormone, estrogen. Thus, hormone therapy is the introduction of exogenous hormones or hormone analogues to replenish the endogenous hormones whose activity is lacking. Because many of these hormones are relatively high-molecular-weight hydrophilic proteins, their delivery, stability, and bioavailability are quite challenging issues [Ozsoy et al., 2009]. These hormones may be introduced into the bloodstream by intravenous (i.v.) injections which are invasive and can be painful. A major issue for i.v. administration is the relatively short half-life of the proteins in the bloodstream as well as unwanted depositions that can require frequent administration and high dose that can produce unwanted side effects. The oral administration is also not effective because of the hepatogastrointestinal first-pass elimination.

Recent years have seen development of promising nanotechnology approaches for hormone therapies. There has been considerable interest in the nasal delivery of hormones using mucoadhesive nanoparticles, such as the ones made of chitosan, which is a positively charged polymer with a strong mucoadhesive property [Ozsoy et al., 2009]. The mucoadhesive property of the chitosan structures enables these nanoparticles to increase the retention time of the drug in the nasal cavity and open the tight junctions in the nasal mucosa.

Insulin is an important hormone, and insulin therapy is effectively used for treating diabetes by using parenteral injections. Several nasal formulations of insulin have been developed, but have not gained acceptance as yet [Ozsoy et al., 2009]. Other important hormones are steroids and estrogens, for which nanoformulations are also being pursued.

Human growth hormone (HGH) is a relatively high molecular weight (~22 kDa) protein comprised of a 191-amino-acid, single-chain peptide. It is secreted by the somatotroph cells of the pituitary gland, from where it is transported to other parts of the body to produce muscle growth and bone growth and stimulate the growth of all internal organs, excluding the brain. HGH also stimulates the immune system.

HGH deficiency in children produces retardation of growth, resulting in delay of sexual maturity and retardation of height. Synthetic HGH, administered through subcutaneous injection, has been the treatment for these growth disorders for children. There has been considerable interest to use the growth hormone as an anti-aging agent. The HGH therapy has been advocated by many for promoting cell growth, increasing vitality, and increasing bone densities, all of which are affected by the aging process. However, as yet, no definitive proof exists that HGH therapy produces anti-aging benefits. As a matter of fact, a great deal of concern has been raised about the dangers of HGH therapy, which can produce many adverse effects.

Despite the serious concern about the growth hormone therapy, research work using nanoparticle delivery of HGH has been carried out worldwide. The advantages are, again, prospect of noninvasive nasal delivery using mucoadhesive nanoparticles, instead of injection, protection of HGH against enzymatic degradation, and increase of its bioavailability. An approach used to produce nano-sized recombinant HGH particles utilizes supercritical fluid precipitation [Pyo et al., 2007]. In this approach, low-cost carbon dioxide is used as a supercritical fluid at 31.1°C temperature and 73.8 bar of pressure to precipitate out solute from a solution. For this purpose, HGH was dissolved in aqueous ethanol and the solution was sprayed into a vessel initially pressurized with CO_2. This method is commonly known as solution-enhanced dispersion by critical fluids (SEDS).

HIGHLIGHTS OF THE CHAPTER

- Rejuvenation therapy is a complex field that integrates detoxification of an organism, chelation therapy, hormone therapy, neurological repair, and emerging stem cell therapies, as well as tissue engineering and organs transplantation.
- The stem cell therapies utilize undifferentiated "blank" cells to produce fully specialized diverse cellular lineages, capable of replenishing damaged or dead cells in distinct organs and tissues.
- Neurological repair involves rehabilitation and regeneration of diseased or damaged nervous system. This broad field involves diverse approaches including stem cell therapy, gene therapy, and chemical detoxification.
- The free radicals in the organism are derived from either oxygen or nitrogen contents and are known as reactive oxygen species and reactive nitrogen species, respectively. In the course of aging, the oxidative stress increases, causing a substantial damage to cells and tissues in an organism.
- Two types of nanoparticles are utilized for free radical scavenging. One type (e.g., cerium oxide particles) acts as direct antioxidants, while another type represents inert, biodegradable antioxidant carriers.

- Excessive accumulation of metals can be toxic for organisms. In particular, some neurodegenerative diseases including Alzheimer's and Parkinson's have been suggested to correlate with elevated levels of Cu, Fe, Zn, and Al in the brain.
- Chelation therapy aims for detoxification of an organism exhibiting excessive concentrations of metals. Nanoparticle formulations have been developed for delivery of the chelators, including delivery across the blood–brain barrier.
- Hormonal therapy involves introduction of exogenous hormones or hormone analogues. Traditional administration of some hormones has necessitated frequent injections due to their short lifetime in a bloodstream.
- Nanocarriers enabling convenient and painless administration of hormones via nasal delivery as opposed to intravenous injections have been developed.

EXERCISES

True or False

1. Rejuvenation therapy requires treatment, integrating different biomedical levels including removal of toxic chemicals, replenishment of cells tissue engineering and organ transplantation.

2. Level of antioxidants in bloodstream is steady throughout the human lifespan.

3. Free radicals can damage the structure of cells in the tissue, but leave their function intact.

4. Oral administration is not feasible for many hormones.

5. Chelators can bind to mercury and then pull it out of the body.

6. Nanoparticles can be used as antioxidant carriers; however, they cannot directly neutralize oxidative damage.

7. Nanoparticles with coupled chelators can cross the blood–brain barrier in both directions.

Multiple Choice

8. Cerium oxide nanoparticle can be used as a ___ for free radical scavenging

(a) Biodegradable nanocarrier

(b) Imaging contrast agent

   - **(c)** Powerful oxidant
   - **(d)** Targeting agent

9. The aim of chelation therapy is to remove ___ from the body.
   - **(a)** Iron
   - **(b)** Copper
   - **(c)** Zinc
   - **(d)** Lead

10. Which of the following is *not* an advantage to use chitosan for hormone therapy?
   - **(a)** Strong mucoadhesive
   - **(b)** Nasal delivery
   - **(c)** Long retention time
   - **(d)** None of the above

Answers: 1. T; 2. F; 3. F; 4. T; 5. T; 6. F; 7. T; 8. c; 9. d; 10. d

REFERENCES

Ankola, D. D., Viswanad, B., Bhardwaj, V., Ramarao, P., Ravi Kumar, M. N. V., Development of potent oral nanoparticulate formulation of coenzyme Q10 for treatment of hypertension: Can the simple nutritional supplements be used as first line therapeutic agents for prophylaxis/therapy? *Eur. J. Pharm. Biopharm.* **67**, 361–369 (2007).

Bolognin, S., Drago, D., Messori, L., Zatta, P., Chelation therapy for neurodegenerative diseases, *Med. Res. Rev.* **29**, 547–570 (2009).

Bouchemal, K., Briancon, S., Perrier, E., Fessi, H., Nano-emulsion formulation using spontaneous emulsification: Solvent, oil and surfactant optimisation, *Int. J. Pharm.* **280**, 241–251 (2004).

Dugan, L. L., Lovett, E. G., Quick, K. L., Lotharius, J., Lin, T. T., O'Malley, K. L., Fullerene-based antioxidants and neurodegenerative disorders, *Parkinsonism Relat. Disord.* **7**, 243–246 (2001).

Fenoglio, I., Tomatis, M., Lison, D., Muller, J., Fonseca, A., Nagy, J. B., Fubini, B., Reactivity of carbon nanotubes: Free radical generation or scavenging activity? *Free Radical Biol. Med.* **40**, 1227–1233 (2006).

Lee, B. S., Yuan, X., Xu, Q., McLafferty, F. S., Petersen, B. A., Collette, J. C., Black, K. L., Yu, J. S., Preparation and characterization of antioxidant nanospheres from multiple α-lipoic acid-containing compounds, *Bioorg. Med. Chem. Lett.* **19**, 1678–1681 (2009).

Liu, G., Men, P., Harris, P. L. R., Rolston, R. K., Perry, G., Smith, M. A., Nanoparticle iron chelators: A new therapeutic Approach in Alzheimer Disease and other neurologic disorders associated with trace metal imbalance, *Neurosci. Lett.* **406**, 189–193 (2006).

Ozsoy, Y., Gungor, S., Cevher, E., Nasal delivery of high molecular weight drugs, *Molecules* **14**, 3754–3779 (2009).

Pyo, D., Lim, C., Cho, D., Oh, D., Size prediction of recombinant human growth hormone nanoparticles produced by supercritical fluid precipitation, *Anal. Bioanal. Chem.* **387**, 901–907 (2007).

Rzigalinski, B. A., Meehan, K., Davis, R. M., Xu, Y., Miles, W. C., Cohen, C. A., Radical nanomedicine, *Nanomedicine-Uk* **1**, 399–412 (2006).

Schubert, D., Dargusch, R., Raitano, J., Chan, S. W., Cerium and yttrium oxide nanoparticles are neuroprotective, *Biochem. Bioph. Res. Co.* **342**, 86–91(2006).

Singh, N., Cohen, C. A., Rzigalinski, B. A., Treatment of neurodegenerative disorders with radical nanomedicine, *Ann. N.Y. Acad. Sci.* **1122**, 219–230 (2007).

Zigoneanu, I., Astete, C. E., Monroe, W. J., Sabliov, C. M., Nanoparticles with entrapped α-tocopherol: synthesis, characterization, and controlled release, *Nanotechnology* **19**, 108 (2008).

CHAPTER 17

STEM CELL BIOTECHNOLOGY

This chapter covers stem cell biotechnology, which includes technology to isolate or generate stem cells and their application in a wide range of therapies and treatments for devastating injuries and damage caused by many degenerative diseases and age-related ailments. As already briefly introduced in Chapter 16, stem cell therapy is considered to make a major contribution to regenerative and rejuvenative therapies. Many have hailed it as a breakthrough biotechnology providing for repair of cells, tissues, and organs to restore an impaired function. At the same time, this biotechnology raised serious ethical questions about the most desirable sources of stem cells as discussed in this chapter, severely impeding its progress. Fortunately, recent breakthrough in reprogramming a differentiated cell to act like a stem cell has provided renewed impetus to this field. Section 17.1 provides an introduction to stem cells. Section 17.2 covers the recent breakthrough concept of cell reprogramming. Section 17.3 discusses gene transfection to produce desired cells of a specific type needed to be replaced or replenished. Section 17.4 introduces transdifferentiation of somatic cells, which involves direct conversion of differentiated cells into a different type of adult cells. Section 17.5 describes methods to isolate stem cells. Section 17.6 discusses the important aspect of tracking of stem cells to ensure a favorable biodistribution to the desired tissue site or organ. The topic of tissue engineering using stem cells, being hotly pursued with major efforts worldwide, is discussed separately in Chapter 18

Introduction to Nanomedicine and Nanobioengineering, First Edition. Paras N. Prasad.

("Tissue Engineering"). As in other chapters, a major focus here is the impact of nanotechnology on stem cell biotechnology.

17.1. STEM CELL BIOTECHNOLOGY: OVERVIEW

Stem cells are cells that have low or no differentiation, a process by which cells develop to become specialized-type cells like neuronal cells in the brain or epithelial cells of skin, and so on. In layman's term, these cells are like largely or completely blank CDs on which information can be written to utilize them for specific purposes. Similarly, these stem cells differentiate in a natural process to produce specialized cells of various organs to provide for body functions. They can also be utilized to replenish depleted cells of certain type due to a specific disease progression or to replace damaged cells; this is what constitutes stem cell therapy, which involves activation of lacking gene expressions in stem cells. Thus, a range of diseases such as various neurodegenerative, cardiovascular, diabetes, cancer, spinal cord injury, deafness, or vision impairment can be treated by stem cell therapy. In addition, tissues and organs can be repaired or regenerated. Three principal types of stem cells, receiving considerable attention, are listed in Figure 17.1.

Embryonic Stem Cells. These cells are obtained from the early state of embryo, 4–5 days post fertilization. At this stage, the embryo may consist of 50–150 cells. ESCs are like blank CDs or flash drive with no special files to perform a specific function. These ESCs are the best possible type of stem cells because they are pluripotent, which means that ESCs can be transformed to any type of cells by directed differentiation. However, ethical questions regard-

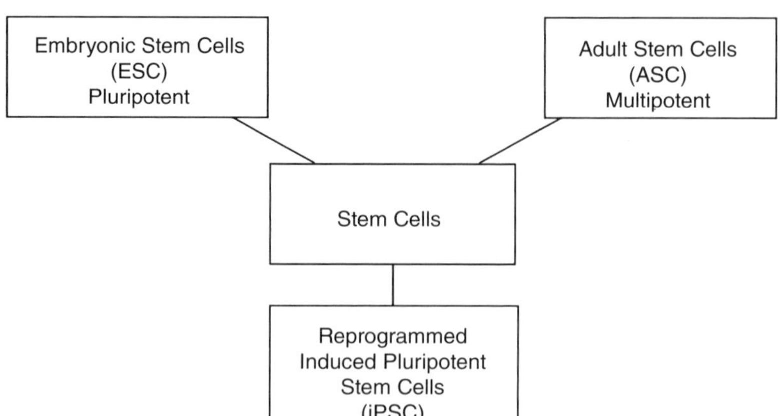

Figure 17.1. Principal types of stem cells discussed in this chapter.

ing the destruction of the embryo led to severe restriction of stem cell research using ESCs.

Adult Stem Cells (ASC). These cells are also known as somatic stem cells (SSCs), and they reside in specific organs and tissues such as skin, bone marrow, liver, neuronal tissues, and so on. Nature has stored them in various places in the body in order to replenish dying cells. Bone marrow stem cells already provide successful stem cell therapy: in fact, this is the only example of an FDA-approved stem cell therapy. The ASCs are multipotent in the sense that they can differentiate only in a limited type to produce only a few selective-type cells. For example, ASCs from bone marrow can differentiate into all the blood cell types but cannot produce neurons. It is this limitation that has led to the quest for how one can generate pluripotent stem cells, other than harvesting from the embryo.

Reprogrammed, Induced Pluripotent Stem Cells (iPSC). This is an exciting recent breakthrough discovered by Yamanaka and co-workers (Takahashi and Yamanaka, 2006; Takahashi et al., 2007), who found that differentiated cells from tissues can be reprogrammed into a new cell type, called induced pluripotent stem cells (iPSC), which appear to be similar to ESC. These iPSCs then can differentiate in selective ways to produce any type of cells. In the analogy of memory devices, such as a flash drive, reprogramming means erasing or inactivating the written files, so that new files can be written. This great discovery has provided a major impetus to stem cell research and raised great deal of optimism about stem cell therapy, because one does not have to resort to the use of ESCs, a highly controversial topic. The topic of iPSC is further discussed in Section 17.2.

In addition to these types of stem cells, there are also "intermediate"-type cells, such as fetal and infant stem cells. They have partially become differentiated, thus exhibiting characteristics in between those of ESCs and ASCs. These types of cells are not discussed here, because they are outside the scope of this monograph.

Nanotechnology can have a major impact on the development of stem cell biotechnology [Wang et al., 2009]. Some of the important areas advanced by utilization of nanoparticles are listed in Figure 17.2.

In this chapter we discuss all these areas in subsequent sections, except for tissue engineering, which deserves to be covered in more detail in a separate chapter, Chapter 18, because it is a more developed area.

17.2. CELL REPROGRAMMING

As described above, this is a breakthrough biotechnology that reprograms differentiated cells from a tissue to produce ESC-like pluripotent stem cells, called induced pluripotent stem cells (iPSC) [Takahashi and Yamanaka, 2006;

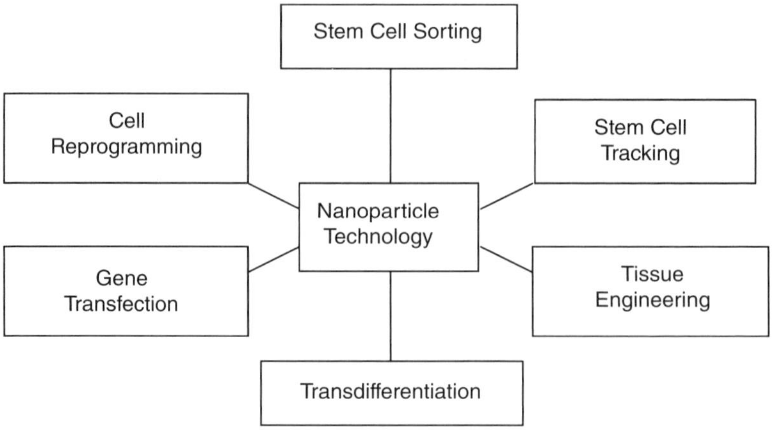

Figure 17.2. The impact of nanoparticle technology on stem cell biotechnology.

Takahashi et al., 2007]. The basic principle utilized in the original work was to introduce certain genes, the reprogramming genes, which are crucial for establishment or maintenance of the pluripotent state. A viral delivery of these genes was used.

The induced pluripotent stem cells exhibit properties very close to natural embryonic stem cells; these cells have numerous prospective biomedical applications. The advantage offered by iPSCs is that the cells of the same patient can be used for stem cell therapy, which strongly reduces or eliminates an immune response. Moreover, since the iPSCs technology does not involve embryo destruction, this system represents an attractive alternative to the embryonic stem cells from ethical perspective. Historically, iPSC production was originated from micromanipulation experiments. Micromanipulations with cells have shown that somatic cells can be reprogrammed into stem cells by transferring somatic cell nucleus to an enucleated unfertilized egg [Wilmut et al., 1997] or by fusion with embryonic stem cells [Cowan et al., 2005]. This indicated that unfertilized eggs and embryonic stem cells contain factors that can confer pluripotency onto the stem cells. Based on this, in a seminal study [Takahashi and Yamanaka, 2006], 24 transcription factors that are upregulated in embryonic stem cells and tumors were screened. In this study, it was found that there are a limited number of genes, specifically OCT3/4, Sox2, Klf4, and c-Myc whose simultaneous expression is sufficient for stem cell induction. All four of these genes code for transcription factors: (i) OCT4 stands for Octamer-Binding Transcription Factor 4, which is normally active in embryos. It is required for maintaining the pluripotency; (ii) Sox2 is a transcription factor required for proliferation of undifferentiated embryonic stem cells; (iii) c-Myc is a common transcription factor that regulates up to 15% of all genes and chromatin remodeling agent. c-Myc is also a proto-oncogene, and its overexpression can lead to cancerogenesis; (iv) Klf4 can serve as both transcription activator and repressor, depending on the type of promoter and interaction with other transcription factors. It contributes to rapid proliferation of the cells. Klf4 is often referred to as an oncogene.

Another study [Yu et al., 2007] used a different combination—OCT4, Sox2, Nanog, and Lin28—to produce iPSCs. The benefit of this combination is that the proto-oncogene c-Myc is replaced by "safer," from tumorigenic point of view, transcription factors Nanog and Lin28. Nevertheless, introduction of a foreign DNA to a cell can produce numerous side effects, including cancer.

In order to preserve the genomic integrity, a strategy alternative to DNA transfection to produce the iPSCs is to use synthetic RNA coding for these transcription factors [Rosa and Brivanlou, 2010] or protein-recombinant transcription factors that are chemically modified to facilitate cell entry [Zhou et al., 2009].

Recently a novel technique in the iPSCs production was utilized. This new approach relies on suppression of multiple genes by overexpression of specific micro RNAs. The micro RNAs are short noncoding RNA sequences that bind to the target mRNAs, inducing their degradation and preventing translation. In particular, the micro RNA 302–367 cluster is located in the human chromosome 4 and consists of five different miRNAs: miR-302a, miR-302b, miR-302c, miR-302d, and miR-367. These micro RNAs regulate a variety of processes important in stem cells biology, including cell cycle, apoptosis, differentiation, and maintenance of stem cell phenotype [Jesus et al., 2009]. It has been shown that the expression of micro RNA 302–367 cluster in combination with Sox2, Klf4, and OCT4 significantly improved the efficiency of iPSCs generation [Liao et al., 2011]. Furthermore, the groundbreaking study of Morrisey and co-workers has shown that the expression of micro RNA 302–367 cluster alone is sufficient for a rapid and efficient generation of iPSCs, even in the absence of exogenous transcription factors [Anokye-Danso et al., 2011].

As said earlier, these iPSCs are reported to be remarkably similar to the natural pluripotent embryonic stem cells in the following:

- iPSCs exhibit morphologic similarity to the ESC.
- Like ESCs, iPSC are capable of long-term self-renewal. They replicate themselves by dividing into the same nonspecialized cell type at a rate equal to that of ESCs.
- iPSCs expressed the same genes expressed in ESCs.
- Like human ESCs (hESCs), iPSCS demonstrate high telomerase activity, which is necessary to sustain cell division.
- iPSCs exhibit pluripotency similar to ESCs and have been differentiated into neurons and cardiomyocytes.

Nanotechnology application to production of iPSCs is still in the developmental stage. Our own program at Buffalo is engaged in utilizing our nanoparticle gene delivery approach, to manipulate adult cells. The nanoparticle approach is very exciting, because probes can be incorporated in the same nanoparticles to tract the transport, biodistribution, intracellular uptake, release, and expression of the plasmid.

17.3. GENE TRANSFECTION

Gene transfection forms the basis for stem cell therapy, whereby transfection of stem cells with specific genes produces the necessary cell differentiation to yield needed specialized cells. Nanoparticles can provide three distinct functions: (i) nonviral delivery of genes, (ii) tracking of stem cells, and (iii) targeted delivery of stem cells to specific organs or biological sites to minimize complications due to translocation.

The application of nanoparticles for gene delivery, in general, has already been discussed in Chapter 14. In this section we provide two examples of gene delivery to produce efficient differentiation in stem cells. These are (i) neurogenesis and (ii) angiogenesis.

Neurogenesis. Neurogenesis describes the process that generates fully developed brain cells, neurons, from the neural stem and progenitor cells. These new neurons are continuously produced predominantly in two regions of the brain: (i) the subventricular zone (SVZ) lining the lateral ventricles and (ii) subgranular zone (SGZ) in the hippocampus.

Many of the neurological degenerative diseases, which are particularly major healthcare issues in the aging population worldwide, result in loss of neurons, accompanied by deterioration of certain brain functions. Promotion of adult neural stem/progenitor cells to differentiate can repopulate degenerated brain areas and can treat these neurodegenerative diseases.

In our group, a new nanotechnology-based approach utilizing neuronal differentiation through an integrative nuclear FGF Receptor 1 signaling (INFS) pathway was developed . This pathway involves activation of neural genes by FGF Receptor 1 (FGFR1), which subsequently triggers differentiation of neural stem/progenitor cells (NS/PCs) into mature neurons. We used organically modified silica (ORMOSIL) nanoparticles carrying electrostatically bound plasmid FGFR1 DNA. Another necessary DNA component of the nanoplex was the FGF-2 sequence coding for FGFR1 ligand that promotes accumulation of FGFR1 in the cell nucleus. These nanoplexes, when injected bilaterally into 3-month-old adult mice using stereotoxic surgery, stimulated NS/PC differentiation *in vivo* [Stachowiak et al., 2009]. Furthermore, migration of differentiating cells from the SVZ to other areas of the brain such as the olfactory bulb, subcortical brain regions, and the brain cortex occurs. Thus, nanoparticle-mediated nonviral gene delivery can induce selective differentiation of NS/PCs, promising treatment of a broad range of neurological disorders.

Angiogenesis. Angiogenesis is the physiological process that describes the growth of new blood vessels from preexisting vessels. Ischemic diseases involve absolute or relative shortage of the blood supply to a specific organ or tissue site. The use of stem cells to differentiate to produce new blood vessels provides a valuable method for treating ischemic diseases.

Yang et al. [2010] used biodegradable polymeric nanoparticles to deliver a human vascular endothelial growth factor (hVEGF) gene to human mesenchymal (multipotent stem cells found in connective tissues) stem cells (hMSCs) and human embryonic stem cells. The treated stem cells exhibited significantly enhanced hVEGF production. They showed that scaffolds seeded with VEGF-expressing stem cells led to two- to four-fold higher vessel densities. In four weeks past intramuscular injection into mouse ischemic hindlimbs, hMSCs transfected with the VEGF genes exhibited significantly enhanced angiogenesis and limb salvage while reducing muscle degeneration and tissue fibrosis.

17.4. SOMATIC CELL TRANSDIFFERENTIATION

Transdifferentiation of somatic cells involves direct conversion of differentiated cells into a different type of adult cells, typically without undergoing dedifferentiation to iPSC (Figure 17.3).

The interconversion of cellular types requires a radical makeover of gene expression profiles, when numerous genes are silenced while others are activated [Blelloch, 2008]. Several cases of this conversion have been reported to occur in live organisms; moreover this process can be reproduced in cell culture by either expression or ablation of specific transcription factors [Graf and Enverm, 2009]. Examples include conversion of mouse fibroblasts into myoblasts upon transfection with skeletal muscle gene MyoD, conversion of

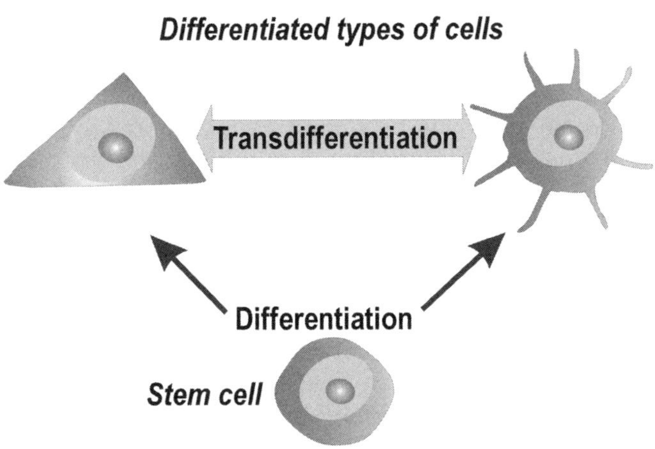

Figure 17.3. Schematics of cell differentiation processes.

glial cells to neurons by expressing Pax6, conversion of liver cells into islet b-cells by expression of Pdx1-VP16, change of pancreatic cells into a liver cell type upon activation of glucocorticoid receptors with dexamethasone, and many others [Graf and Enverm, 2009]. This ability for transdifferentiation is of significant value for regenerative medicine. Nanotechnology also enables delivery of genetic material to induce transdifferentiation as well as for labeling and tracking of transplanted cells.

17.5. STEM CELL SORTING

A major challenge for stem cell biotechnology is to isolate stem cells from a multi-type cell mixture using a low-cost, fast, and easily operated method [Wang et al., 2009]. The nanoparticle technology can provide useful approaches to separate the stem cells. A simple technique is to use magnetic nanoparticles, such as iron oxides discussed in various chapters of this monograph, to tag the stem cells and then use a magnetic field to collect them. This can be done in a flow system, such as the flow cytometer discussed in Chapter 10, where the magnetically labeled stem cells can be separated using a magnetic cell sorting scheme. Alternatively, the cell can be tagged with fluorescent nanoparticles and the trigger generated by the fluorescence signal can be used for electrostatic cell sorting.

Jing et al. [2007] used magnetic nanoparticles, combined with the CD34 antibody, to enrich peripheral blood progenitor cells from human blood circulation. For this, they used a continuous quadrupole magnetic flow sorter that consisted of a flow channel and a quadrupole magnet. There are a number of commercial progenetor cell labeling kits now available. Their results showed that the CD-34 expression cells, kG-1a, were strongly labeled and were isolated with a purity of 60–96% and a throughput of 61.7–9.30 × 10^4 cells/s.

In another study, the magnetic nanoparticles carrying a triple label (magnetic, fluorescent and isotope) were found to be internalized into hematopoietic stem and neural progenitor cells efficiently [Lewin et al., 2000]. They used magnetic labeling to recover progenitor cells. They also demonstrated that the labeled neural progenitor cells retained their ability for differentiating. The magnetic nanoparticles did not affect the viability and proliferation of the hematopoietic cells.

17.6. STEM CELL TRACKING

Tracking of stem cells transplanted into an organism is essential for successful development and optimization of stem cell therapies. Tracking of the injected stem cells is also important for reducing the side effects of the stem cell therapy. Embryonic stem cells have demonstrated tumorigenic potential; thus tracking in recipient body can avoid oncologic complications [Thomson et al.,

1998]. The stem cell monitoring is also important in damaged cardiac tissue to access their transmigration. Stem cells have the potential to differentiate into mature cardiac cells; however, if these cells migrate away from the heart, they produce pacemaker-like cells and generate focal arrhythmias with potentially fatal effects [Potapova et al., 2004]. Initially, the fate of transplanted stem cells and their involvement into the tissue regeneration were determined by using histological analysis [Pittenger et al., 1999]. Collection of tissues for histological studies, however, is often not compatible with live experimental models and with clinical applications. Histological analysis has an inherent limitation that it can be performed mainly at discrete time points, but not continuously. Several imaging modalities, including magnetic resonance imaging (MRI), multiphoton fluorescence microscopy, and fluorescence molecular tomography, have been used recently as an alternative and supplemental tool to traditional histological analysis [Schroeder, 2008]. Nanoparticle-based imaging can significantly contribute to tracking of stem cells noninvasively.

The two imaging modalities utilizing nanoparticles for tracking of stem cells that have been extensively investigated are (i) nanoparticle-enhanced magnetic resonance imaging (MRI) and (ii) quantum-dot-based optical imaging.

MRI is an efficient technique for stem cells tracking, which has the merit for monitoring of the transplanted cells in providing high-resolution 3D mapping and speed [Bulte et al., 2001; Chemaly et al., 2005]. To enable MRI detection of the stem cells, contrast agents are required which are efficient, nontoxic, nonimmunogenic, and stably associated with the transplanted cell. Various types of nanoparticles, including iron oxide particles [Lu et al., 2007], carbon nanotubes [Sitharaman et al., 2005], liposomes [Mulder et al., 2004], and micelles [Amirbekian et al., 2007], have been synthesized to obtain MRI contrast in the cells. Among these various types of nanoparticles, the superparamagnetic iron oxide is currently the most commonly used imaging material as also discussed in Chapter 8. The internalization of the iron oxide particles is facilitated by their coating with peptides or polymers; the labeled cells are then transplanted into the area of interest and their migration is tracked by MRI. Although the number of particles per cell can be rapidly diluted due to frequent divisions of the stem cells, the sensitivity of the state-of-the-art MRI technology is approaching single-particle detection per cell or per embryo [Shapiro et al., 2004, 2007]. The MRI technology has already been applied for medical imaging of the labeled stem cells in various tissues [McColgan et al., 2011; Zhu et al., 2006; Waters and Wickline, 2008; Kraitchman et al., 2003].

Quantum dot semiconductor nanoparticles (QDs), because of their unique photophysical properties as discussed in Chapter 5, which include high fluorescence intensity, resistance of fluorescence signal to photobleaching, and chemical stability, provide sensitive detection ability for long-term tracking. In combination with two-photon excitation [Larson et al., 2003] by near-IR irradiation that penetrates deep in the tissue and does not cause significant radiation damage, the QDs are very useful probes for tracking of stem cells transplanted in an organism. Furthermore, because of broad absorption and narrow emission spectra of QDs, different types of stem cells, transplanted into the same organism in course of a complex therapy, can be color-coded by QDs and tracked selectively by utilizing a single excitation source [Lin et al., 2007]. Stem cells have demonstrated a good tolerance for QDs; the QDs labeling seems to have little, if any, effect on the migration and differentiation of the stem cells transplanted into the body [Slotkin et al., 2007].

HIGHLIGHTS OF THE CHAPTER

- Stem cells are undifferentiated or precursor cells capable of indefinite self-renewal and differentiation into a wide variety of specialized cells.
- Stem cell therapies are used to replenish cells in injured or diseased organs as well as in tissue engineering.
- Embryonic stem cells, adult stem cells, and induced pluripotent stem cells are major types of the stem cells.
- The embryonic stem cells are pluripotent, which can be differentiated into any type of cells in an organism. The adult stem cells are multipotent as they can give a rise to a multiple, but limited number of cell lineages. Induced pluripotent stem cells exhibit all known properties of embryonic stem cells, except for source of their origin, which is reprogramming of formerly differentiated cells.
- A simultaneous expression of four genes—OCT3/4, Sox2, Klf4, and c-Myc—is sufficient for reprogramming of somatic cells (such as skin fibroblasts) into the induced pluripotent stem cells. Alternative approaches use expression of OCT4, Sox2, Nanog, and Lin28 genes or suppression of multiple genes by micro RNAs.
- Efficient and safe gene transfection of stem cells can be achieved by using nanoparticles for targeted gene delivery.
- Process of transdifferentiation involves direct interconversion of differentiated cells into a different type of adult cells, without undergoing dedifferentiation to a stem-cell-like phenotype.
- Nanotechnology also provides approaches for stem cell sorting and tracking in live organisms.

EXERCISES

True or False

1. Embryonic stem cells are usually collected on a third month of gestation.
2. The adult stem cells and induced pluripotent stem cells possess equal differentiation potential.

3. The induced pluripotent stem cells emulate properties of embryonic stem cells.
4. Transdifferentiation is the phenomenon when an adult specialized cell switches to another type of differentiated cell.
5. Post-transplant tracking of cells in live organism is required to exclude oncologic complications and ensure proper functioning of treated organs.

Multiple Choice

6. In which types of cells can adult stem cells be differentiated from bone marrow?
   (a) Diverse lineages of blood cells
   (b) Neurons
   (c) Epithelial cells
   (d) Fibroblasts
7. Select an advantage of the embryonic stem cells of adult stem cells.
   (a) Embryonic stem cells do not produce tumors.
   (b) Embryonic stem cells divide indefinitely.
   (c) Embryonic stem cells have higher differentiation plasticity.
   (d) Embryonic stem cells cause lower immune response.
8. Select a false statement.
   (a) Induced pluripotent stem cells can be obtained by transferring somatic cell nucleus to an enucleated unfertilized egg.
   (b) Induced pluripotent stem cells can be obtained by overexpression of few genes.
   (c) Induced pluripotent stem cells can be obtained by gene suppression.
   (d) Induced pluripotent stem cells can be obtained only from skin cells.

Answers: 1. F; 2. F; 3. T; 4. T; 5. T; 6. a; 7. c; 8. d

REFERENCES

Amirbekian, V., Lipinski, M. J., Briley-Saebo, K. C., Amirbekian, S., Aguinaldo, J. G. S., Weinreb, D. B., Vucic, E., Frias, J. C., Hyafil, F., Mani, V., Fisher, E. A., Fayad, Z. A., Detecting and Assessing macrophages *in vivo* to evaluate atherosclerosis noninvasively using molecular MRI, *Proc. Natl. Acad. Sci. USA* **104**, 961–966 (2007).

Anokye-Danso, F., Trivedi, C. M., Juhr, D., Gupta, M., Cui, Z., Tian, Y., Zhang, Y., Yang, W., Gruber, P. J., Epstein, J. A., Morrisey, E. E., Highly Efficient miRNA-Mediated Reprogramming of mouse and human somatic cells to pluripotency, *Cell Stem Cell* **8**, 376–388 (2011).

Blelloch, R., Regenerative medicine: short cut to cell replacement, *Nature* **455**:604–605 (2008).

Bulte, J. W. M., Douglas, T., Witwer, B., Zhang, S-C., Strable, E., Lewis, B. K., Zywicke, H., Miller, B., van Gelderen, P., Moskowitz, B. M., Duncan, I. D., Frank, J. A., Magnetodendrimers allow endosomal magnetic labeling and *in vivo* tracking of stem cells, *Nat. Biotechnol.* **19**, 1141–1147 (2001).

Chemaly, E. R., Yoneyama, R., Frangioni, J. V., Hajjar, R. J., Tracking stem cells in the cardiovascular system, *Trends Cardiovasc. Med.* **15**, 297–302 (2005).

Cowan, C. A., Atienza, J., Melton, D. A., Eggan, K., Nuclear reprogramming of somatic cells after fusion with human embryonic stem cells, *Science* **309**, 1369–1373 (2005).

Graf, T., Enverm, T., Forcing cells to change lineages, *Nature* **462**, 587–594 (2009).

Jesus, A. B. D., Lucena-Aguilar, G., Menendez, P., The miR-302-367 cluster as a potential stemness regulator in ESCs, *Cell Cycle* **8**, 394–398 (2009).

Jing, Y., Moore, L. R., Williams, P. S., Chalmers, J. J., Farag, S. S., Bolwell, B., Zborowski, M., Blood progenitor cell separation from clinical leukapheresis product by magnetic nanoparticle binding and magnetophoresis, *Biotechnol. Bioeng.* **96**, 1139–1154 (2007).

Kraitchman, D. L., Heldman, A. W., Atalar, E., Amada, L. C., Martin, B. J., Pittenger, M. F., Hare, J. M., Bulte, J. W. M., *In vivo* magnetic resonance imaging of mesenchymal stem cells in myocardial infarction, *Circulation* **107**, 2290–2293 (2003).

Larson, D. R., Zipfel, W. R., Williams, R. M., Clark, S. W., Bruche, M. P., Wise, F. W., Webb, W. W., Water-soluble quantum dots for multiphoton fluorescence imaging *in vivo*, *Science* **300**, 1434–1436 (2003).

Lewin, M., Carlesso, N., Tung, C.-H., Tang, X.-W., Cory, D., Scadden, D. T., Weissleder, R., Tat peptide-derivatized magnetic nanoparticles allow *in vivo* tracking and recovery of progenitor cells, *Nat. Biotechnol.* **18**, 410–414 (2000).

Liao, B., Bao, X., Liu, L., Feng, S., Zovoilis, A., Liu, W., Xue, Y., Cai, J., Guo, X., Qin, B., Zhang, R., Wu, J., Lai, L., Teng, M., Niu, L., Zhang, B., Esteban, M. A., Pei, D., MicroRNA Cluster 302–367 enhances somatic cell reprogramming by accelerating a mesenchymal-to-epithelial transition, *J. Biol. Chem.* **286**, 17359–17364 (2011).

Lin, S., Xie, X., Patel, M. R., Yang, Y. H., Li, Z., Cao, F., Gheysens, O., Zhang, Y., Gambhir, S. S., Rao, J. H., Wu, J. C., Quantum dot imaging for embryonic stem cells, *BMC Biotechnol.* **7**, 67 (2007).

Lu, C. W., Hung, Y., Hsiao, J.-K., Yao, M, Chung, T.-H., Lin, Y.-S., Wu, S.-H., Hsu, S.-C., Liu, H.-M., Mou, C.-Y., Yang, C.-S., Huang, D.-M., Chen, Y.-C., Bifunctional magnetic silica nanoparticles for highly efficient human stem cell labeling, *Nano Lett.* **7**, 149–154 (2007).

McColgan, P., Sharma, P., Bentley, P., Stem cell tracking in human trials: A meta-regression, *Stem Cell Rev.* **7**, 1031–1040 (2011).

Mulder, W. J. M., Strijkers, G. J., Griffioen, A. W., van Bloois, L., Molema, G., Storm, G., Koning, G. A., Nicolay, K., A liposomal system for contrast-enhanced magnetic resonance imaging of molecular targets, *Bioconjugate Chem.* **15**, 799–806 (2004).

Pittenger, M. F., Mackay, A. M., Beck, S. C., Jaiswal, R. K., Douglas, R., Mosca, J. D., Moorman, M. A., Simonetti, D. W., Craig, S., Marshak, D. R., Multilineage potential of adult human mesenchymal stem cells, *Science* **284**, 143–147 (1999).

Potapova, I., Plotnikov, A., Lu, Z., Danilo, P., Jr., Valiunas, V., Qu, J., Doronin, S., Zuckerman, J., Shlapakova, I. N., Gao, J., Pan, Z., Herron, A. J., Robinson, R. B., Brink, P. R., Rosen, M. R., Cohen, I. S., Human mesenchymal stem cells as a gene delivery system to create cardiac pacemakers, *Circ. Res.* **94**, 952–959 (2004).

Rosa, A., Brivanlou, A. H., Synthetic mRNAs: Powerful tools for reprogramming and differentiation of human cells, *Cell Stem Cell* **7**, 549–550 (2010).

Schroeder, T., Imaging stem-cell-driven regeneration in mammals, *Nature* **453**, 345–351 (2008).

Shapiro, E. M., Skrtic, S., Sharer, K., Hill, J. M., Dunbar, C. E., Koretsky, A. P., MRI detection of single particles for cellular imaging, *Proc. Natl. Acad. Sci. USA* **101**, 10901–10906 (2004).

Shapiro, E. M., Medford-Davis, L. N., Fahmy, T. M., Dunbar, C. E., Koretsky, A. P., Antibody-mediated cell labeling of peripheral T cells with micron-sized iron oxide particles (MPIOs) allows single cell detection by MRI, *Contrast Media Mol. Imaging* **2**, 147–153 (2007).

Sitharaman, B., Kissell, K. R., Hartman, K. B., Tran, L. A., Baikalov, A., Rusakova, I., Sun, Y., Khant, H. A., Ludtke, S. J., Chiu, W., Laus, S., Tóth, E., Helm, L., Merbach, A. E., Wilson, L. J., Superparamagnetic gadonanotubes are high-performance MRI contrast agents, *Chem. Commun. (Camb)*, 3915–3917 (2005).

Slotkin, J. R., Chakrabarti, L., Dai, H. N., Carney, R. S. E., Hirata, T., Bregman, B. S., Gallicano, G. I., Corbin, J. G., Haydar, T. F., *In vivo* quantum dot labeling of mammalian stem and progenitor cells, *Dev. Dyn.* **236**, 3393–3401 (2007).

Stachowiak, E. K., Roy, I., Lee, Y.-W., Capacchietti, M., Aletta, J. M., Prasad, P. N., Stachowiak, M. K., Targeting novel integrative nuclear FGFR1 signaling by nanoparticle-mediated gene transfer stimulates neurogenesis in the adult brain, *Integr. Biol.* **1**, 394–403 (2009).

Takahashi, K., Tanabe, K., Ohnuki, M., Narita, M., Ichisaka, T., Tomoda, K., Yamanaka, S., Induction of pluripotent stem cells from adult human fibroblasts by defined factors, *Cell* **131**, 861–872 (2007).

Takahashi, K., Yamanaka, S., Induction of pluripotent stem cells from mouse embryonic and adult fibroblast cultures by defined factors, *Cell* **126**, 663–676 (2006).

Thomson, J. A., Itskovitz-Eldor, J., Shapiro, S. S., Waknitz, M. A., Swiergiel, J. J., Marshall, V. S., Jones, J. M., Embryonic stem cell lines derived from human blastocysts, *Science* **282**, 1145–1147 (1998).

Wang, Z., Ruan, J., Cui, D., Advances and prospect of nanotechnology in stem cells, *Nanoscale Res. Lett.* **4**, 593-605 (2009).

Waters, E. A., Wickline, S. A., Contrast agents for MRI, *Basic Res. Cardiol.* **103**, 114–121 (2008).

Wilmut, I., Schnieke, A. E., McWhir, J., Kind, A. J., Campbell, K. H., Viable offspring derived from fetal and adult mammalian cells, *Nature* **385**, 810–813 (1997).

Yang, F., Cho, S.-W., Son, S. M., Bogatyrev, S. R., Singh, D., Green, J. J., Mei, Y., Park, S., Bhang, S. H., Kim, B.-S., Langer, R., Anderson, D. G., Genetic engineering of human stem cells for enhanced angiogenesis using biodegradable polymeric nanoparticles, *Proc. Natl. Acad. Sci. USA* **107**, 3317–3322 (2010).

Yu, J., Vodyanik, M. A., Smuga-Otto, K., Antosiewicz-Bourget, J., Frane, J. L., Tian, S., Nie, J., Jonsdottir, G. A., Ruotti, V., Stewart, R., Slukvin, I. I., Thomson, J. A., Induced pluripotent stem cell lines derived from human somatic cells, *Science* **318**, 1917–1920 (2007).

Zhou, H. Y., Wu, S. L., Joo, J. Y., Zhu, S. Y., Han, D. W., Lin, T. X., Trauger, S., Bien, G., Yao, S., Zhu, Y., Siuzdak, G., Scholer, H. R., Duan, L. X., Ding, S., Generation of induced pluripotent stem cells using recombinant proteins, *Cell Stem Cell* **4**, 381–384 (2009).

Zhu, J., Zhou, L., XingWu, F., Tracking neural stem cells in patients with brain trauma, *N. Engl. J. Med.* **355**, 2376–2378 (2006).

CHAPTER 18

TISSUE ENGINEERING

This chapter covers the topic of tissue engineering that has received a great deal of attention over past several decades. It utilizes assembly, proliferation, and differentiation of cells on a biocompatible scaffold mimicking the extracellular matrix, discussed in Chapter 2, to generate a functional tissue structure or even an organ. The cells used for this purpose can be stem cells or cells derived from a tissue structure needing to be repaired. Since the field of tissue engineering is large in scope, with a vast amount of literature and books already existing, the materials presented here are selective, focusing only on concepts and describing how nanotechnology is impacting this field. The outline of this chapter is as follows. Section 18.1 provides an overview of the scope of tissue engineering. Section 18.2 covers the area of tissue regeneration, both *in vitro*, followed by transplantation, and *in vivo*. Section 18.3 describes the various nanomaterials used as scaffolds for tissue regeneration. Section 18.4 provides examples of applications of nanofibers in tissue engineering. Section 18.5 describes nanoparticle delivery of biomolecules that provide molecular signals for cell growth, proliferation, and differentiation as well as morphogens that induce morphogenesis. It also covers their nanoparticle delivery. Section 18.6 presents magnetic field directed tissue engineering. Section 18.7 introduces the exciting new area of tissue/organ printing to artificially generate an entire tissue structure or a functioning organ by using printing technology for multicellular 3D assembly. Section 18.8 discusses the

Introduction to Nanomedicine and Nanobioengineering, First Edition. Paras N. Prasad.

topic of tissue bonding, which deals with tissue welding and tissue soldering. This topic has not normally been included in the traditional definition of the scope of tissue engineering. However, I feel that a broader scope of tissue engineering should include this topic, because it does provide promising approaches for tissue repair.

18.1. TISSUE ENGINEERING: OVERVIEW

Tissue engineering is an emerging multidisciplinary field with a continuously widening scope that deals with development of biological substitutes to regenerate, repair, and replace a damaged or nonfunctional tissue site or even an entire organ. It combines engineered biomaterial building blocks with microfluidic engineering to create artificial functional tissue structures. Often the term "Regenerative Medicine" is synonymously used with tissue engineering [Popat, 2011]. However, "Regenerative Medicine" generally deals with regeneration of tissues, while tissue engineering covers a wider scope, from tissue regeneration, to tissue/organ replacement, to tissue damage repair [Dvir et al., 2011]. An overview of the molecular mechanisms of tissue regeneration and tissue damage repair is provided by Andreadis [2006]. It highlights the development of high-throughput genomic and proteomic technologies for living skin substitutes and wound healing. This chapter also includes tissue bonding to repair/bond a torn tissue, which traditionally is not included in tissue engineering. However, artificial implants as mechanical replacements for a tissue (such as bones, hips, knees, etc.) are not included in this chapter, even though nanostructured materials have made a major impact on the development of these implant materials or as surface coating materials to improve on the function or biocompatibility of implants.

Figure 18.1 lists the areas of tissue engineering covered in this chapter.

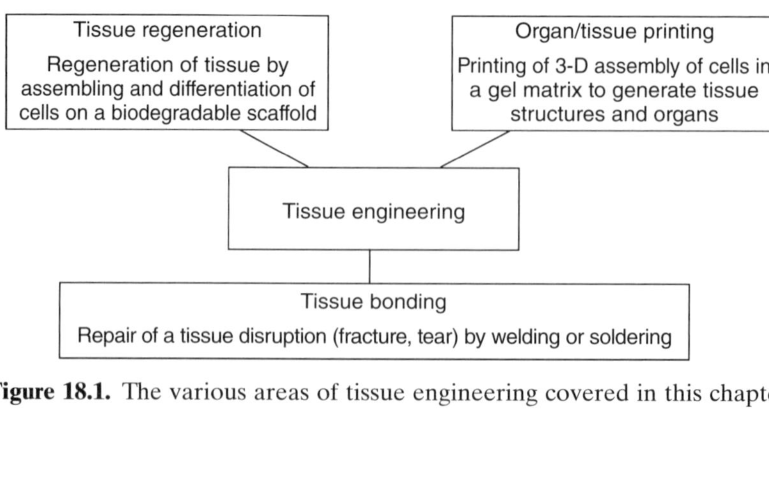

Figure 18.1. The various areas of tissue engineering covered in this chapter.

These areas are briefly defined and outlined in this section and then discussed in detail in separate sections.

Tissue Regeneration. This is the area of tissue engineering that has been investigated extensively over a number of decades. It focuses on regeneration of a specific tissue by assembling, and, if needed, further differentiation of cells on a biodegradable scaffold. As discussed in Chapter 2 ("The Human Body"), the cells in a tissue structure are assembled in an extracellular matrix (ECM) that serves a number of purposes, in addition to defining the connectivity of cells and the three-dimensional tissue structure. In tissue engineering, a biodegradable scaffold is thus used to mimic the role of the ECM. In the past, traditional approach, natural biomaterials or synthetic biodegradable polymers (such as polylactic acid, PLA) were used, which degrade within the human body after the formation of the desired and functional tissue structure. More recently, nanostructured materials have been the focus of next-generation scaffolds, providing a number of benefits. The nanostructured materials as suitable scaffolds are discussed in Section 18.3. The scaffolds can be used for *in vitro* (or *ex vivo*) regeneration of tissues, which are then implanted in the nonfunctional, damaged, or missing tissue areas *in vivo* for tissue regeneration.

An important class of nanostructures used for producing scaffolds are nanofibers. Section 18.4 discusses some specific examples of nanofiber preparation methods and their applications in tissue engineering.

Another important component is a group of molecules/macromolecules, which enables "biologically directed" growth by providing specific molecular signals, nutrients, and growth factors to lead to necessary cell growth, proliferation, and differentiation as well as promoting the proper three-dimensional (3D) tissue morphology. Important signaling biomolecules are hormones, growth factors, receptors, cytokines, histamines, and other messenger molecules. An important class of molecular signaling agents is morphogens, which are soluble molecules that can diffuse and carry molecular signals to induce morphogenesis, a biological process of developing 3D tissue shape. These cell signaling biomolecules are described in Section 18.5.

An important approach in tissue engineering is the use of an external field gradient, such as a magnetic field for field-directed self-assembly of an scaffold and magnetically labeled cells. An advantage of using magnetically directed and guided tissue engineering is that large scaffolds can be formed with a small amount of starting material, and the 3D tissue architecture (cell assembly) can be highly controlled. Magnetic labeling using superparamagnetic nanoparticles also provides opportunities for cell tracking and thus monitoring tissue regeneration. The magnetic-field-directed tissue engineering is discussed in Section 18.6.

An important emerging area of tissue engineering is tissue/organ printing. The prospect of growing on demand a functional complete organ in a laboratory has captured the imaginations of researchers for a long time. It is very close to becoming a reality. In addition, the prospect of using printing technology to use cells as "bioinks" for reconstructing a computer-generated 3D blueprint of an organ is very promising, even though skepticism exists about the real utility of organ printing with regards to their reliable functioning and long-term biocompatibility. This topic is detailed in Section 18.7.

Finally, Section 18.8 covers the area of tissue bonding, which can be utilized to join a ruptured or torn tissue. The two methods discussed are tissue welding and tissue soldering. Each case involves heating at the damaged tissue site.

18.2. TISSUE REGENERATION

The important components involved in the process of tissue regeneration are listed in Figure 18.2.

The components presented in Figure 18.2 are now discussed in detail.

Cells as Building Blocks. Only live cells can be used as building blocks for tissues. As described in Figure 18.3, these cells can be of two different types: (a) The ideal source is stem cells, which offer the versatility to be specifically differentiated, as discussed in Chapter 17, to generate any type of cells on demand. (b) Cells can also be extracted from a biopsied tissue specimen. Examples of biopsy-extracted cells are fibroblast cells from skin and chondrocyte cells from cartilage. In this case the tissue is minced, subsequently digested with enzymes (trypsin or collagenase) to remove the holding extracellular matrix, and then extracted by centrifugation or apheresis.

Depending on the source of cells, they are often classified in the following categories: (i) autologous cells, which we obtained from the same individual needing implantation of regenerated tissues, (ii) syngeneic or isogenic cells, which are from genetically identical persons, such as twins, (iii) allogeneic cells, which are from a donor of the same species, and (iv) xenogenic cells, which

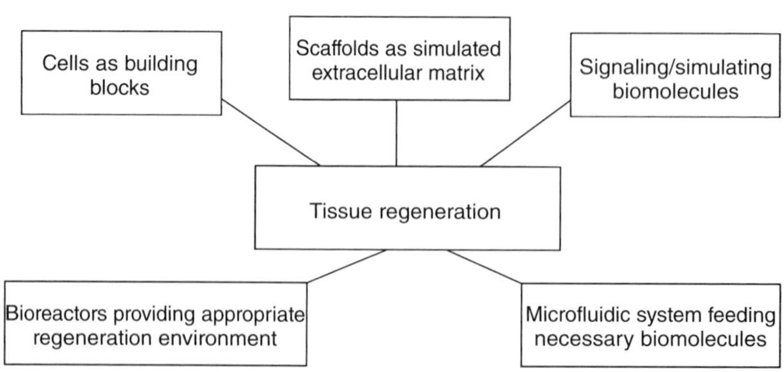

Figure 18.2. The main components involved in the process of tissue regeneration.

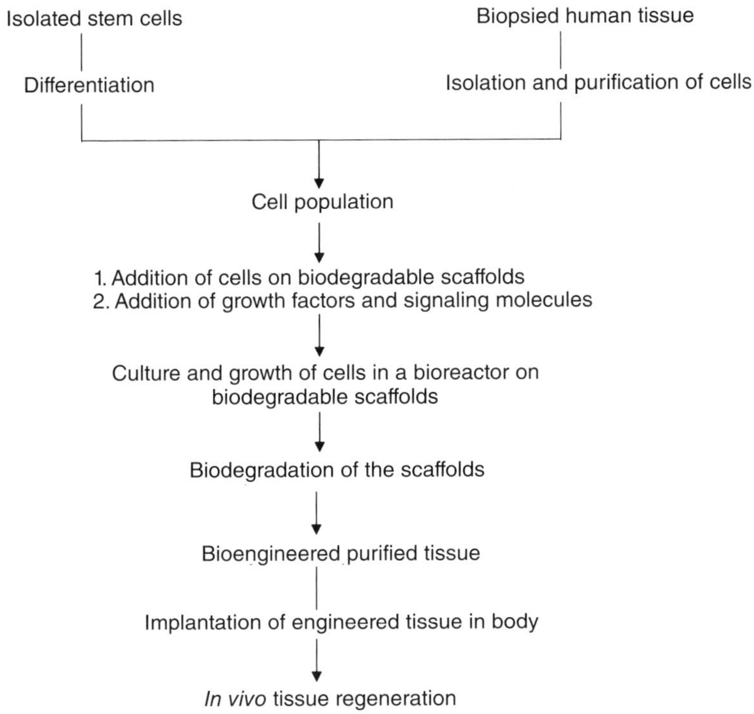

Figure 18.3. Schematics of tissue regeneration.

are from individuals of another species. Among these, tissues generated from autologous cells and xenogenic cells will have the minimum and maximum compatibility problems, respectively, while implanting in the body of the tissue acceptor. In addition, the cells can also be classified as primary, which directly come from a donor, or secondary, which are from a cell bank.

The majority of tissue regeneration work has used autologous cells from the tissue structure needing to be repaired by implant of regenerated tissues. However, recent development in stem cells, particularly the introduction of induced pluripotent stem cells described in Chapter 17, is opening up a new generation of tissue engineering using stem cells. It requires a proper understanding of how signal systems can be customized to control stem cell proliferation, differentiation, and organization to produce a specific tissue.

Scaffolds as Building Platforms. As described earlier, scaffolds serve as an artificial extracellular matrix medium for 3D assembling of cells. As described in Chapter 2, an extracellular matrix provides structures made up of adhesive and connective macromolecules. They allow transport of cell-signaling molecules. They contain structural proteins, collagens, which provide mechanical strength to tissues. Thus, some of the important functions performed by scaffolds which mimic the properties of ECM are:

- Providing for cell attachment and migration.
- Permitting diffusion of necessary biomolecules and vital nutrients.
- Providing mechanical strength (structural integrity) to allow for 3D assembling of cells in the desired tissue architecture.

In order to perform these functions, a scaffold thus must meet the following requirements:

- Good cell adhesion and binding to connective tissue.
- High porosity and adequate pore size for cell seeding and diffusion as well as diffusion of signaling biomolecules and nutrients.
- Biocompatibility and biodegradability. The scaffold should be absorbed in time by the surrounding tissues without needing any invasive removal procedure.
- Structural rigidity to support a given 3D tissue architecture of assembled cells.

A number of materials have been explored as scaffolds. The earlier work focused on the utilization of both natural and synthetic polymers. Examples of natural materials are collagens. Examples of synthetic polymers used as scaffolds are polylactic acid (PLA), polyglycolic acid (PGA), and polycaprolactone (PCL). They are biocompatible and biodegradable. For example, PLA, which is a polyester, degrades within the human body to convert to lactic acid, which is readily removed from the body. The degradation of PGA and PCL occurs by a similar mechanism, but the rates of degradation are quite different. These traditional approaches are not discussed here. Instead, Section 18.3, to follow, describes nanostructured materials, which have been the focus of recent activities.

Biomolecules. In order to produce a tissue structure from cells, it is not just sufficient to assemble them in certain 3D structures. In addition to assembling them, a number of bioactive molecules are needed which may promote cell differentiation, proliferation, and intra- as well as intercellular cell signaling. There are a variety of biomolecules, such as genetic materials (DNA, RNA), functional proteins, hormones, nutrients, and morphogens. These biomolecules have to be delivered to the cells in the scaffold matrix to promote tissue regeneration.

An example is a class of molecules called morphogens that promote morphogenesis, the process of development of tissue architecture from cell assembly and adhesion [Gilbert, 2000]. These morphogens are soluble molecules that carry molecular signals to control cell differentiation in a concentration-dependent manner to produce a 3D tissue structure. They act by binding to specific protein receptors on cells. An important class of morphogens are transcription factor proteins that determine the fate of specific proteins by their interaction with a specific DNA. Another important class of morphogens are biomolecules that control cell adhesion.

Bioreactors. A bioreactor provides a controlled physiological environment to promote cell assembly and tissue growth. In the *in vivo* tissue generation, it is the natural environment around the tissue that provides the appropriate physiological conditions for tissue growth. For *in vitro* tissue growth, a bioreactor is a chamber where many parameters, such as temperature, oxygen and CO_2 concentrations, and other biological stimuli, can be controlled. Several commercial sources provide applications-specific bioreactors.

Microfluidic Systems. These systems consist of networks of connected channels, pumps, mixes, and separations incorporated in a MEMS device. They are needed for two purposes: (i) microfluidic delivery of cells to control their positioning and (ii) controlled delivery of various biomolecules, such as cell-binding proteins. In addition, microfluidics can be used to develop precise arrangements of cells of multiple types in the scaffold by introducing agents that provide selective cell adhesion [Saltzman and Olbricht, 2002]. A third function of the microfluidics system is for local delivery to deliver a biomolecule to a precisely defined region of interiors of cells to induce a selected specific function.

18.3. NANOTECHNOLOGY IN TISSUE ENGINEERING

Nanotechnology is playing a vital role in the development of the field of tissue engineering. Most efforts have been focused on the development of nanostructured scaffolds that mimic the nanoscale interaction morphology and the mesoporosity of a real extracellular matrix (ECM). However, nanotechnology also provides opportunities to make a broader impact on tissue engineering. This broader scope of nanotechnology impact is illustrated in Figure 18.4. As

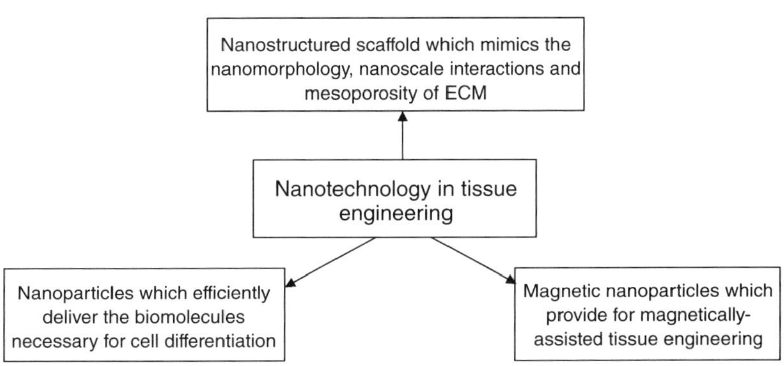

Figure 18.4. The various areas of tissue engineering impacted by nanotechnology.

stated earlier, a major focus in the application of nanotechnology to tissue engineering has been on the design of nanostructured scaffolds. A number of excellent reviews provide detailed descriptions of these efforts [Gelain, 2008; Goldberg et al., 2007; Borlongan et al., 2007; Zhang and Webster, 2009; Vasita and Katti, 2006]. Considering the introductory nature of this monograph, only an abbreviated account will be provided here.

As described in Section 18.1, a scaffold in tissue engineering provides an artificial substitute for an ECM, which is a multifunctional nanocomposite. The ECM is a nanostructured medium with a nanoscale morphology, which not only provides mechanical support for embedded cells, but is conducive to specific nanoscale interactions necessary for appropriate cell assembling, adhesion, migration, proliferation, differentiation, morphogenesis, and cell communication. An ECM medium has an intricate connecting nanostructure network of supramolecular structures containing connective tissues composed of structural protein fibers, such as collagens and elastins, ranging in dimensions from ten to several hundred nanometers. They are entangled to form a nonwoven mesh-type structure that imparts the necessary tensile strength and elasticity to the tissue. Fibronectin and laminin are examples of adhesive proteins that are also in the form of nanofibers. Even though the cells themselves are of dimensions in tens of microns, they respond to the ECM through nanoscale interactions with ECM proteins using the plasma membrane receptors, integrins. Nanoscale control of scaffold structure can allow the cells to receive both chemical and physical cues [Gelain, 2008]. Chemical cues are carried by the biomolecules that bind to cell membrane receptors. Physical cues can be nanoscale topography, because the cells have the ability to recognize topographical differences even at nanoscale, such as the presence of grooves and ridges. It has been shown that cell adhesion and proliferation are significantly improved with nanofibrous scaffolds [Xu et al., 2004].

Three principal types of nanostructured scaffolds are described here. They are (i) nanofibrous, (ii) nanocomposites, and (iii) nanopatterned.

Nanofibrous Scaffolds. These are comprised of nanofibers generally having diameters less than 100 nm. These fibers provide a behavior mimicking those of collagens, elastins, and so on. The nanofibrous scaffolds provide high surface to volume, high density of surface functionalization to manipulate nanoscale interactions, high diffusive capacities, and strong mechanical properties.

The three techniques available for producing nanofibers are: (i) electrospinning, (ii) self-assembly, and (iii) phase separation [Vasita and Katti, 2006]. These techniques are briefly described here.

(i) Electrospinning. This is a relatively simple method for producing nanofibers from polymeric biomaterials and offers control over thickness and composition of the nanofibers, as well as the porosity of the nanofiber meshes used as a scaffold. Figure 18.5 provides a representative schematic of the electrospinning setup, together with the SEM picture of the PLGA nanofibers pro-

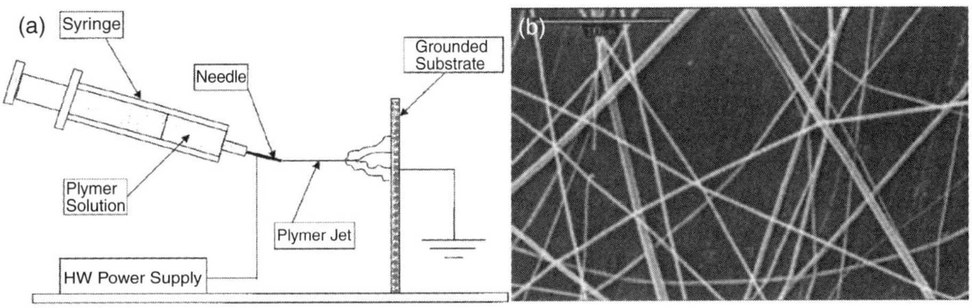

Figure 18.5. (a) A representative schematic of the electrospinning system. (b) PLGA nanofibers produced using this system. (Reproduced with permission from Katti et al., [2004].)

duced. In electrospinning, a polymer solution is held at the tip of a capillary tube by surface tension; then the application of an electrical potential imparts a charge to the polymer solution.

When the potential is increased to a certain critical value to overcome the opposing surface tension, a jet of polymer solution is ejected, which, by subsequent thinning in the air and solvent evaporation, eventually forms randomly-oriented nanofibers. These nanofibers can then be collected on a stationary or rotating grounded metallic collector.

The electrospinning technique has been applied to produce nanofibers of both natural and synthetic polymers. Examples of natural polymers used to produce nanofibers are collagen, gelatin, chitosan, hyaluronic, silk fibroin, and proteins [Vasita and Katti, 2006]. Examples of synthetic polymers processed by electrospinning are poly(lactic acid) (PLA), polyurethane (PU), poly(ε-caprolactone) (PCL), and poly(lactic-co-glycolic acid (PLGA).

Electrospinning can be combined with solvent selective phase separation described below to produce nanofibers with nanopores. For example, selection of a suitable single solvent can produce phase separation into a solvent-rich pore forming area and a solvent-poor area. Alternatively, solutions containing two different polymers can be used that phase separate within the fibers. Thus direct incorporation of active biomolecules as well as magnetic or optical nanoparticles can be made during the spinning process itself. Core–shell fibers can be produced for drug-delivery scaffolds in which the core provides storage of a functional biomaterial and the shell controls the release kinetics.

Seeding of a broad range of cell types on electrospun scaffolds has been achieved. Examples of these cell types are keratinocytes, chondrocytes, osteblasts, mesenchymal stem cells, fibroblasts, and endothelial cells. Examples of target organs or tissues generated are bone, skin, neural tissue, cartilage, blood vessels, heart, or lung [Gelain, 2008].

(ii) Self-Assembly. Self-assembly is a bottom-up approach to spontaneously assemble molecules, utilizing concerted action of weak and noncovalent

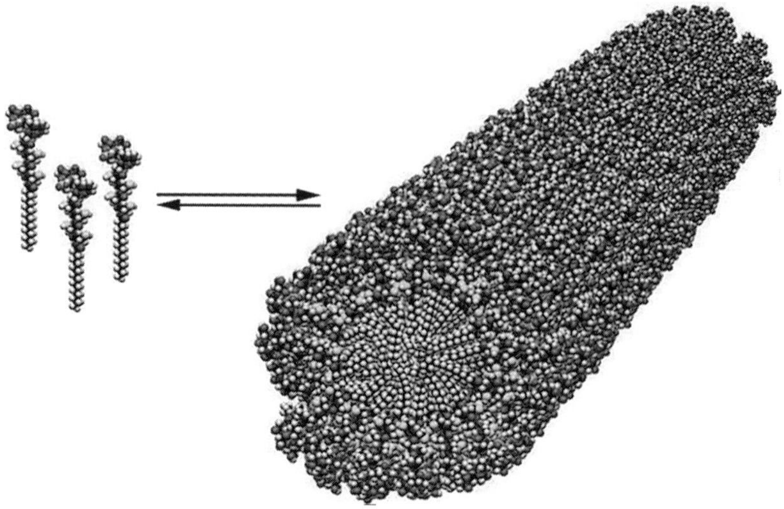

Figure 18.6. Schematics of self-assembly of RGD peptide-amphiphiles into a nanofiber of 7.6 ± 1 nm in diameter. (Reproduced with permission from Hartgerink et al. [2001].)

interactions, and produce nanofibrous hierarchical structures [Goldberg et al., 2007]. It has been used to produce peptide-based nanofibrous scaffolds. These self-assembling peptides are also called sapeptides. An example is the work of Stupp and co-workers [Hartgerink et al., 2001]. They used a peptide-amphiphile (PA) with the cell-adhesive ligand RGD (Arg-Gly-Asp) that self-assembled into supramolecular nanofibers as schematically shown in Figure 18.6.

Self-assembly of the long-chain nanopeptides containing both hydrophobic and hydrophilic units can be used to produce complex 3D nanorods, as illustrated in Figure 18.6, which can be further modified to support cellular adhesion and control cellular function. The nanofibers resulting from self-assembly consist of a hydrophobic core of alkyl residues and a hydrophilic exterior lined by peptide residues.

Nanofibrillar hydrogels can be formed with a slight modification of the peptide amphiphilic architecture [Silva et al., 2004]. Hydrogels are hydrophilic, 3D networks that swell in the presence of water. Thus, they can respond to many physiological stimuli such as pH, ionic strength, and temperature. Silva et al. successfully incorporated neural progenitor cells *in vitro* within the 3D network of nanofibrous hydrogels and showed them to rapidly differentiate into neurons, while the development of astrocytes was discouraged. Zhang and co-workers generated another class of nanofibrillar hydrogels with very high water content that supported a wide spectrum of mammalian cells, maintained the function of differentiated neural cells and chondrocytes, and promoted the differentiation of liver progenitor cells [Zhang, 2003; Goldberg et al., 2007].

An advantage of using peptide based self-assembly to produce nanostructured scaffold is that their building blocks are natural amino acids. Thus their degradation products can be reused by the body; hence they elicit negligible immune response. In addition, a self-assembling scaffold offers the advantage of spontaneously forming scaffolds embedding cells and thus placing functional motifs close to cell membrane receptors, at the same time selectively stimulating diverse cell signaling pathways. A shortcoming of the self-assembly method is that the scaffolds obtained are randomly oriented in the three dimensions. In some cases, the self-assembling process of peptides requires addition of salt solutions that could be harmful to the tissue to be regenerated, as found by Guo et al. in the case of reknitting of the injured spinal cord by self-assembling peptide nanofiber scaffolds [Guo et al., 2007]. Another concern is that because the sapeptides spontaneously form β-sheets with poor elastic properties, the resulting self-assembled scaffold can be fragile and fracture under deformation or stress. Thus future direction is producing nanostructured scaffolds by combining electrospinning and self-assembling.

(iii) Phase Separation. Another method produces nanofibrous foams as scaffolds [Ma and Zhang, 1999; Zhang and Ma, 2002], It utilizes thermally induced liquid–liquid phase separation. The steps involved in the phase separation technique are (i) dissolution of a polymeric material, (ii) liquid–liquid phase separation, (iii) polymer gelation control to produce nanoscale porosity, (iv) extraction of solvent from the gel with water, and (v) freezing and freeze-drying under vacuum. The resulting nanofiber properties were greatly influenced by the polymer concentration.

The nanostructured scaffold obtained by the phase separation technique has three levels of architecture: (i) macroporosity with controllable pore size and shape, (ii) variable interfiber distance obtained by changing polymer concentration, and (iii) tunable diameter of the nanofiber.

Zhang and Ma [2002] studied the interactions of the obtained nanofibrous scaffold with osteoblasts. They found increased cell adhesion and adsorption of fibronectin and collagen, which are necessary for cell and ECM interactions.

The advantage offered by this method is that it is simple and requires minimal equipment. The method is applicable to directly fabricate scaffolds for a desired anatomical-shape tissue structure.

Nanocomposites. Many tissue structures, particularly the hard tissues, are nanocomposites. For example, bone is a nanocomposite of a protein-based soft hydrogel template (such as collagens, laminin, fibronectin, vitronectin) and a hard inorganic component such as hydroxyapatite (HA), which is $Ca_{10}(PO_4)_6(OH)_2$. In fact, 70% of the bone matrix consists of HA in the nanocrystallite form which is 20–80 nm long and 2–5 nm thick. Thus using a nanocomposite formulation for the material composition of an artificial scaffold can mimic the structure, composition, and biological characteristics of natural bones. There have been a number of reports of using a 3D nanocomposite matrix of nano-HA/collagen for bone regeneration.

In one study, Hartgerink et al. [2002] used the self-assembly process to produce nanofibers from a peptide-amphiphile with the cell-adhesive ligand RGD, and then they utilized these nanofibers to directly nucleate and align HA on the long axis of the nanofibers. This nanofiber composite exhibited the same pattern as collagen and HA nanocrystals in the bone. Another example of the use of a nanocomposite scaffold is provided by the work of Osathanon et al. [2008], who developed a polymer/calcium phosphate composite for bone tissue engineering. These nanofibrous fibrin-based composites were found to promote osteoblast alkaline phosphatase activity as well as osteoblast marker gene expression to support bone maturation in a mouse calvarial defect model.

Cell Patterning. Nanostructured scaffolds have also been fabricated using patterning techniques such as lithography, which can provide surfaces with well-defined physiological and/or topological patterns [Borlongan et al., 2007]. Physiological patterns yield modulation of physical and chemical interactions between the cell and the surface, while the topological pattern are features such as grooves on the surface which can restrict the location and migration of cells. Thus patterning provides the ability to control the pattern of cell adhesion and modulate cell migration.

An example is the patterning of a polymer using lithography to produce patterns of hydrophilic and hydrophobic regions. Cells would then adhere only on the hydrophilic region pattern. Indeed fibroblast cells, when cultured on this type of patterned surface, showed prefactual adhesion to the hydrophilic area; they also elongated and aligned along the hydrophilic grooves that were rich in carboxylic acid groups [Umar et al., 2005].

18.4. NANOFIBERS FOR TISSUE ENGINEERING

This section provides examples of nanofibers utilized for engineering of various types of tissues. A variety of both natural polymers and synthetic materials in the nanofiber form have been utilized in tissue engineering. Examples of natural polymers used for this purpose are collagen, hyaluronic acid, gelatin, chitosan, elastin, silk, and wheat protein. Collagen is the most widely utilized biomaterial; its nanofibers have shown compatibility with a number of cells—for example, myoblasts and chondrocytes [Shields et al., 2004]. Furthermore, cross-linking in collagen type II scaffold can lead to improved mechanical properties. Chitosan nanofibers have also been produced in the nonwoven or aligned chitosan/poly(ethylene oxide) form by electrospinning. These nanofibers have exhibited enhanced attachment of human osteoblasts and chondrocytes, while the cells have maintained their characteristic morphology. Electrospun protein fibers have also been utilized for tissue engineering [Li et al., 2005]. Li et al. produced tropoelastic nanofibers that supported adhesion and proliferation of human embryonic palatal mesenchymal cells.

Examples of polymeric materials used in the form of nanofibrous scaffolds are PLA, poly(ethylene) terephthalate (PET) for blood vessel tissue engineering, PCL in neural and cartilage tissue engineering, and PLGA for bone and cartilage tissue engineering as well as for controlled biomolecule/drug delivery [Vasita and Katti, 2006].

Carbon nanofibers and carbon nanotubes are another class of synthetic materials that have been explored for tissue engineering [Zhang and Webster, 2009]. Sixty-nanometer-diameter carbon nanofibers exhibited significantly enhanced osteoblast adhesion. Single-walled carbon nanotubes have also been investigated for tissue engineering because of their excellent mechanical and conductive properties, making them appealing for neural tissue engineering.

18.5. NANOPARTICLE DELIVERY OF BIOMOLECULES

As described briefly in Section 18.1, delivery of a range of bioactive molecules plays an important role in tissue regeneration. These biomolecules can be of a wide range needed to provide a specific function. The use of stem cells as building blocks requires genetic materials for needed cell differentiation or transdifferentiation as discussed in Chapter 17. Cell proliferation needs growth factors and hormones. Introduction of molecular signaling agents, morphogens, promotes morphogenesis.

Controlled release of these biomolecules in time and space plays an important role [Saltzman and Olbricht, 2002]. As tissue regeneration involves a series of steps with molecular signals and cellular response unfolding in time through a sequence of spatial changes, controlled release is necessary. Controlled time release can provide the necessary sustained supply of biomolecules to be maintained during tissue regeneration. Delivery of the growth factors and hormones using nanoparticles with tailored porosity and biodegradability is a very desirable strategy. Spatial control of delivery can be introduced by using nanoparticles with specifically tailored surfaces to provide selective binding, thus allowing these biomolecules to be delivered in small volumes at specific locations. Nanoparticle control of delivery can also lead to controlled concentration profiles in order to stimulate directional or spatial changes. Targeted delivery of transcription factors that bind to specific DNA sequences can allow local control of gene expression in cells at the site of tissue regeneration.

Bone morphogenic proteins (BMPs) are an important class of growth factors used in bone formation and healing [Bessa et al., 2008]. Recombinant human BMPs (rhBMPs) have been widely used in regeneration of bone or cartilage. An example of nanoparticle delivery of BMPs is provided by the work of Wei et al. [2007], who used PLGA nanospheres for their delivery in PLA scaffolds with interconnected macroporous and nanofibrous architectures. They found that the nanocarrier delivered rhBMP-7 in a time-controlled manner and induced significant bone formation.

In yet another report, gold nanowries were functionalized with a cytokine and manipulated using an electric field to deliver and release their payload to a pre-specific cell with subcellular resolution in order to activate desirable signaling pathways [Fan et al., 2010; Dvir et al., 2011]. This control using an external electric or magnetic field, to be discussed in Section 18.6, can be used to localize biomolecules in a specific zone of a 3D construct, thus producing precise engineering of the cellular microenvironment by delivery factors to a specific cell type [Dvir et al., 2011].

In the case of stem cells used in tissue engineering, appropriate genetic materials can be delivered using nanoparticle carriers. Gene delivery using nanoparticle carriers has already been discussed in Chapter 14. Yang et al. [2010] developed poly(β-amino esters)–DNA nanoparticles and used them to genetically engineer stem cells for enhanced angiogenesis.

Finally, an exciting prospect is to use nanocarriers co-encapsulating multiple drugs with control release of each agent in a temporal fashion [Sengupta et al., 2005]. These nanoparticles can also be designed to be responsive to environmental changes such as pH, temperature, light, and mechanical stress to trigger drug release. Using nanoparticles surface functionalized with cell-specific ligands or signaling molecules, they can be immobilized on a scaffold's surface to enhance cell adhesion and guide cell migration [Shi et al., 2010].

18.6. MAGNETICALLY ASSISTED TISSUE ENGINEERING

Magnetic nanoparticles have been used, along with an applied magnetic field, for a variety of applications in tissue engineering [Cartmell and Dobson, 2009]. Figure 18.7 lists the important applications of nanomagnetism for tissue engineering.

These three areas are discussed here.

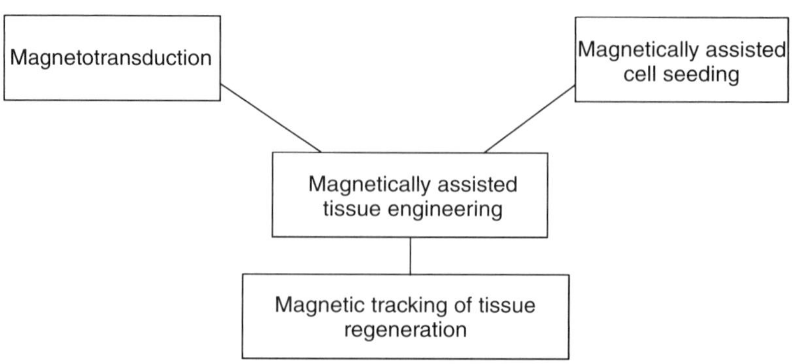

Figure 18.7. The various aspects of magnetically assisted tissue engineering.

Magnetotransduction. This describes the process in which a magnetic nanoparticle, attached either to the cell membrane or intracellularly to an internal structure, can act as a transducer of an applied magnetic field, enabling noninvasive control of cellular functions. Wang et al. [1993] introduced "magnetic twisting cytometry" (MTC), in which magnetic particles, coated with ligands that target different receptors on the cell surface, can produce selective accentuation in a magnetic field to control specific cellular processes.

Another development using magnetically induced mechanotransduction is a magnetic ion channel activation bioreactor, which permits application of magnetic fields to activate ion channels and thus control the production of tissue matrix components. Using this method, up-regulation of extracellular matrix-related gene expression has been demonstrated for both cartilage and bone [Dobson et al., 2006].

Cell Seeding. Magnetic nanoparticles can also be used for cell seeding onto a scaffold in a predetermined cell adherence pattern. Thus continuous cell sheets have been produced by holding protein-coated magnetic nanoparticle layers using a magnetic field [Ito et al., 2005]. Cell seeding has also been demonstrated on porous 3D scaffolds of collagen and poly-L-lactic acid [Shimizu et al., 2006].

Magnetic Tracking. Magnetic nanoparticle labeling of cells can be used to track cell proliferation, adhesion, and tissue regeneration by using magnetic resonance imaging.

Magnetic nanoparticle enhanced tissue engineering has been used for a variety of applications ranging from regeneration of skin to bones, and then to cartilage. Cartmell and Dobson [2009] describe some of these studies.

18.7. TISSUE/ORGAN PRINTING

Though elementary in theory, the bioengineering of a perfect human tissue/organ is fraught with numerous drawbacks. A critical requirement of engineering tissues on artificial scaffolds is that the rate of production of ECM components by the growing cells needs to match the rate of biodegradation of the scaffold material. It may not be strictly met always, leading to structural imbalance within the growing tissue. In addition, the scaffold material, while providing the backbone to the tissue, may not be able to structurally "direct" the tissue growth to reproduce all the structural complexities of the original tissue/organ. In addition, there is the issue of immunogenicity arising from any remnant scaffolding material or its biodegradation product in the body. To address these issues, a modified bioengineering approach, known as "tissue/organ printing," has been developed [Jakab et al., 2010]. This approach is fast gaining popularity because it promises to fabricate tissues/organs with minimal to no dependence on artificial scaffold materials. This approach is akin to regular printing using an inkjet printer, with the added provision for printing in three dimensions. The basics of this approach are as follows: A patient or normal human being will undergo whole-body medical imaging, from which a 3D image of the organ to be "printed" is reconstructed in the computer. This computer-generated organ "blueprint" will then be "fed" in the printer software and will serve as the graphic model that the printing process will try to reproduce. In addition, assemblies of live cells with various shapes (e.g., spheroids, sheets, cylinders, etc.) will serve as the "bioink" in the printer. Similar to tissue regeneration, these cells can be either isolated and differentiated stem cells, cells extracted from tissue biopsies, or cells taken from the cell bank. The printer will contain a variety of multicellular units that constitute the tissue/organ to be "printed," akin to different colors present in a printer. The printing process will lead to the fusion of the multicellular subunits on a gel matrix (the "bio-paper") in a 3D assembly, reconstructing along the provided organ "blueprint". The steps involved in tissue/organ printing are illustrated in Figure 18.8.

The promise of engineering fully functional human tissues/organs for on-demand repair/replacement of missing or damaged organs is evident from a large number of research articles and patents in this area, coupled with the commercialization of robotic bioprinters. For example, Organovo, Inc. (http://www.organovo.com/) has manufactured a 3D bioprinter based on printing of self-assembled multicellular subunits in a gel matrix. The company has so far

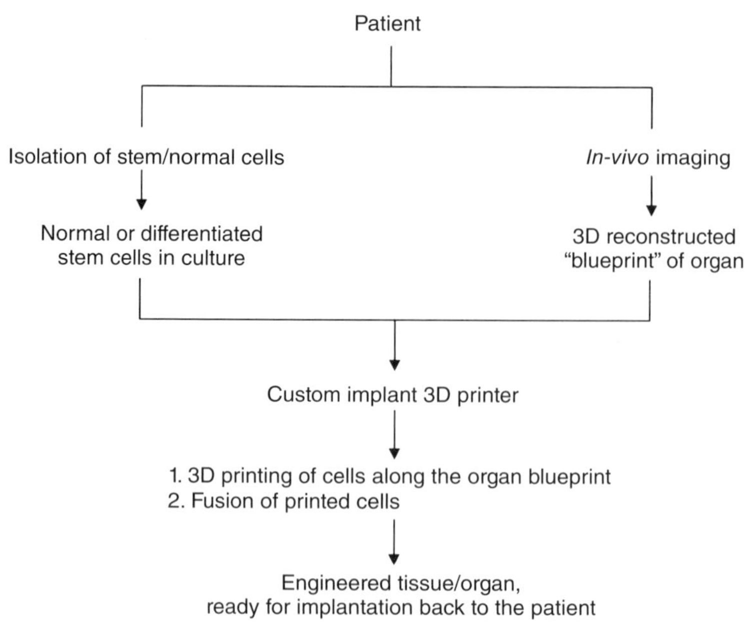

Figure 18.8. Schematics of tissue/organ printing.

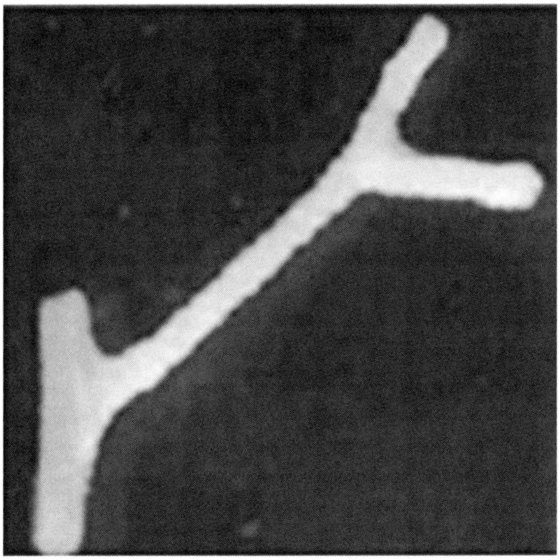

Figure 18.9. A branched vascular conduit, printed using spherical subunits. (Reproduced with permission from Jakab et al. [2010].)

successfully printed simple 3D tissue structures such as spheroids, cylinders, stents, and so on, which can be further assembled into more complex organ structures such as heart, kidney, and so on. Figure 18.9 shows the picture of a branched vascular conduit, printed using spherical tissue subunits.

However, for becoming a clinical routine in the future, this integrated bioengineering area will need compatible functioning of a number of automated components, such as 3D medical imaging and reconstruction, clinical cell sorter, microfluidic devices, robotic cell and cell-assembly dispensers, organ bioreactors, and so on. Noninvasive biomonitoring of the structural and functional maturation of the engineered organs prior to implantation is another important component to be delved into.

18.8. TISSUE BONDING

Tissue bonding employs the process of using laser energy to join or bond tissues. Tissue bonding has been mostly applied to soft tissues. The approaches to join or bond tissues are listed in Table 18.1.

Laser tissue welding was first demonstrated by Jain and Gorisch [1979], who used Nd:YAG laser light to seal rat arteries. Subsequent studies suggested that laser interaction could be used to heat a tissue sufficiently to denature proteins (collagens) in the tissue surfaces to form new connecting structures [Jain, 1984; Schober et al., 1986]. Most early studies of laser tissue welding employed CO_2

TABLE 18.1. The Approaches for Tissue Bonding

Tissue Bonding		
Direct Welding of Tissues:	Laser Soldering:	Dye/Nanoparticle-Enhanced Laser Soldering:
Local heating to ~60–80°C by laser energy absorption (photothermolysis) to denature collagen, uncoiling their native triple-helical structure and producing collagen bonding	Use of proteineous solder at the surfaces to be joined, followed by application of laser light to selectively heat the solder and seal it to the surrounding tissue	A dye/nanoparticle absorbing at the laser wavelength of soldering added to the solder to enhance selective absorption and subsequent heating of the solder and not of the nontarget tissue

lasers. The use of the CO_2 laser relied on water, the largest constituent of most tissues, absorbing strongly at its wavelength (10.6 μm). This strong absorption leads to a shorter optical penetration depth (~13 μm), limiting its use to extremely thin tissues. Also, under a CW laser exposure, a lateral spread of heating produces a large zone of injury.

Other lasers employed for laser tissue welding are argon-ion and Nd:YAG, which provide deeper penetration. Thus they produce deeper and more uniform tissue heating than that achieved by using a CO_2 laser. In the case of the Nd:YAG laser, the 1.320-μm laser output has been used, because at this wavelength both water and hemoglobin absorb. The use of pulsed lasers can minimize collateral thermal damage. However, the choice of laser wavelength and exposure parameters (energy, pulse duration, etc.) are clearly dependent on the tissue absorption, optical penetration depth, and the thermal relaxation time in the tissues to be welded [Prasad, 2003]. The optical penetration depth clearly has to be matched with the extent of the thickness to be welded to provide uniform heating.

The laser soldering technique utilizes laser light to fuse a proteineous solder to the tissue surface, thereby providing greater bond strength with less collateral damage compared to direct welding. Blood was the first material used as a solder. Subsequently, egg-white albumin followed by other proteins such as those derived from blood fibrinogen and other albumins were used as solder substitutes.

Dye-enhanced soldering was introduced to take advantage of the strong absorption of light by a selected dye and the efficient conversion of light into heat by the dye dispersed in the solder. This method also provided the benefit that an appropriate dye can be selected to match its absorption peak with the particular laser wavelength utilized. This method has allowed the ability to use the more common and relatively inexpensive 808-nm diode laser with the help of a biocompatible dye, indocyanine green (ICG) [Oz et al., 1990; Chivers, 2000]. In yet another approach, a polymer scaffold doped with serum albumin and ICG was used [McNally et al., 2000]. They found that the addition of the polymer membrane improved the welding strength and provided better flexibility compared to the use of albumin protein solder alone. The polymer scaffold makes the solder sufficiently flexible, allowing it to wrap around the tissue. Solders can be used for applications other than tissue bonding. Laser-assisted tissue sealing (LATS) can be used to seal bleeding surfaces of hemostasis (blood clotting). Anastomoses (sites where blood vessels have been rejoined surgically) that leak can be sealed and made impermeable.

Nanoparticles that absorb strongly and produce highly localized heating can facilitate laser induced welding and soldering. West and co-workers [Gobin et al., 2005] used gold nanoshells designed to strongly absorb light in the near IR to demonstrate tissue welding. The advantage offered by the metallic nanoshells and nanorods to produce photothermal effect for local heating to effect either welding or soldering is that they do not photobleach, a common problem with many organic dyes. In addition, as explained in Chapter 5, their plasmon resonance can be tuned to make these nanoparticles strongly absorbing at a selected wavelength that is out of the range of tissue absorption, thus minimizing any collateral damage. West and co-workers used gold nanoshells with an albumin solder in an *in vivo* rat skin wound-healing model. They were able to weld full thickness incisions in the dorsal skin of rats.

Laser welding or soldering can be used endoscopically and laparoscopically to extend the range of its applications to cases where sutures or staples cannot be used.

Applications of laser welding and soldering have been diverse [Bass and Treat, 1996]:

- *Cardiovascular Surgery:* Primary vascular anastomosis; sealing to reduce blood loss in vascular surgery.
- *Thoracic Surgery:* Sealing of air leaks after lung biopsy or wedge resection; sealing of the bronchial stump.
- *Dermatology:* Skin closure with improved cosmesis and faster healing.
- *Gynecology:* Repair of fallopian tubes.
- *Neurosurgery:* Welding and repair of peripheral nerves.
- *Ophthalmology:* Laser solder closure of incisions in the sclera and cornea.
- *Urology:* Closure of ureter, ureteroneocystostomy, urethra, and bladder.

HIGHLIGHTS OF THE CHAPTER

- Tissue engineering covers a wide range of scope which includes tissue regeneration, tissue/organ replacement, tissue damage repair, tissue/organ prinring, and tissue welding.

- For tissue regeneration, a biodegradable scaffold is used to mimic the role of the extracellular matrix (ECM).
- The important components involved in the process of tissue regeneration are: cells as building blocks, scaffolds as simulated extracellular matrix, signaling/stimulating biomolecules, bioreactors, and microfluidic system.
- Only live cells can be used as building blocks for tissues. These cells can be of two different types: (i) isolated stem cells and (ii) biopsied human tissue.
- Different source of cells can be classified into four categories: (i) autologous cells (obtained from the same individual needing implantation of regenerated tissues), (ii) syngeneic or isogenic cells (obtained from genetically identical persons) (iii) allogeneic cells (obtained from a donor of the same species), and (iv) xenogenic cells (obtained from individuals of another species).
- Scaffolds serve as an artificial ECM medium for 3D assembling of cells by (i) providing for cell attachment and migration, (ii) permitting diffusion of necessary biomolecules and vital nutrients, and (iii) providing mechanical strength (structural integrity) to allow for 3D assembling of cells in the desired tissue architecture.
- Both natural materials and synthetic polymers can be used as scaffolds. Examples of natural materials are collagens. Examples of synthetic polymers are polylactic acid (PLA), polyglycolic acid (PGA), and polycaprolactone (PCL).
- A number of bioactive molecules, such as DNA, RNA, functional proteins, hormones, nutrients, and morphogens, are needed to promote cell differentiation, proliferation, and intra- as well as intercellular cell signaling to assemble the cells into tissue structures.
- Morphogens promote morphogenesis, the process of development of tissue architecture from cell assembly and adhesion, which carry molecular signals to control cell differentiation in a concentration-dependent manner to produce a 3D tissue structure.
- A bioreactor provides a controlled physiological environment to promote cell assembly and tissue growth. In the *in vivo* tissue generation, it is the natural environment. For *in vitro* tissue growth, a bioreactor is a chamber where many parameters can be controlled.
- Microfluidic systems consist of networks of connected channels, pumps, mixes, and separations incorporated in a MEMS device. They are needed for two purposes: (i) microfluidic delivery of cells to control their positioning and (ii) controlled delivery of various biomolecules.
- Nanoscale control of the scaffold structure can allow the cells to receive both chemical and physical cues. Chemical cues are carried by the biomolecules that bind to cell membrane receptors. Physical cues can be nanoscale topography such as the presence of grooves and ridges. Three

principal types of nanostructured scaffolds are (i) nanofibrous, (ii) nanocomposites, and (iii) nanopatterned.

- Nanofibrous scaffolds are comprised of nanofibers, generally having diameters less than 100 nm. Three techniques available for producing nanofibers are: (i) electrospinning, (ii) self-assembly, and (iii) phase separation.
- Electrospinning is a simple method for producing nanofibers from polymeric biomaterials and offers control over thickness and composition of the nanofibers, as well as the porosity of the nanofiber meshes.
- Self-assembly is a bottom-up approach to (a) spontaneously assemble molecules, utilizing concerted action of weak and noncovalent interactions, and (b) produce nanofibrous hierarchical structures.
- The steps involved in the phase separation technique are: (i) dissolution of a polymeric material, (ii) liquid–liquid phase separation, (iii) polymer gelation control to produce nanoscale porosity, (iv) extraction of solvent from the gel with water, and (v) freezing and freeze-drying under vacuum.
- Bone is an example of nanocomposite of a protein-based soft hydrogel template. Seventy percent of the bone matrix consists of hydroxyapatite in nanocrystallite form which is 20–80 nm long and 2–5 nm thick. The material composition of an artificial scaffold (e.g., nano-HA/collagen matrix) can mimic the structure, composition, and biological characteristics of natural bones.
- Nanostructured scaffolds have also been fabricated using cell patterning techniques such as lithography, which can provide surfaces with well-defined physiological and/or topological patterns.
- Because tissue regeneration involves a sequence of steps with molecular signals and cellular response unfolding in time through a sequence of spatial changes, controlled release, provided by a nanoformulation, is necessary.
- Spatial control of delivery can be introduced by using nanoparticles with specifically tailored surfaces to provide selective binding, thus allowing these biomolecules to be delivered in small volumes at specific locations.
- PLGA nanospheres were used for their delivery of bone morphogenic proteins in PLA scaffolds with interconnected macroporous and nanofibrous architectures to induce significant bone formation.
- Gold nanowries were functionalized with a cytokine and manipulated using an electric field to deliver and release their payload to a pre-specific cell with subcellular resolution in order to activate desirable signaling pathways.
- Magnetic nanoparticles have been used, along with an applied magnetic field, for magetotransduction and magnetically assisted cell seeding in tissue engineering.
- Magetotransduction is a process in which a magnetic nanoparticle, attached either to the cell membrane or intracellularly to an internal

structure, can act as a transducer of an applied magnetic field, enabling noninvasive control of cellular functions.

- Another development using magnetically induced mechanotransduction is a magnetic ion channel activation bioreactor, which permits the application of magnetic fields to activate the ion channels and thus control the production of tissue matrix components.
- Magnetic nanoparticles can be used for cell seeding onto a scaffold in a predetermined cell adherence pattern by holding the protein-coated magnetic nanoparticle layers using a magnetic field.
- Magnetic nanoparticle labeling of cells can be used to track cell proliferation, adhesion and tissue regeneration by using magnetic resonance imaging.
- The process of tissue/organ printing involves (i) *in vivo* whole body imaging, (ii) 3D reconstructed "blueprint" of an organ, (iii) 3D printing of cells along the organ blueprint, (iv) fusion of printed cells, and (v) implantation of the engineered tissue/organ back in the patient.
- Three approaches of tissue bonding are developed: they are (i) direct welding of tissues, (ii) laser soldering, and (iii) dye/nanoparticle-enhanced laser soldering.
- Laser tissue welding involves the use of high-power laser (e.g., Nd:YAG, CO_2) for sealing the tissue.
- The laser soldering technique utilizes laser light to fuse a proteineous solder to the tissue surface, thereby providing greater bond strength with less collateral damage compared to direct welding.
- Dye-enhanced soldering employs the advantage of the strong absorption of light by the selected dye and the efficient conversion of light into heat by the dye dispersed in the solder.
- Nanoparticles such as gold nanoshell, which absorb strongly and produce highly localized heating, can facilitate laser induced welding and soldering.

EXERCISES

True or False

1. One of the disadvantages of using self-assembly technique to produce nanostructured scaffold is that the scaffolds obtained are randomly oriented.
2. Tissues generated from xenogenic cells will have maximum compatibility problems.
3. Bone is a nanocomposite of a protein-based soft hydrogel template. The use of poly(lactic acid) can mimic the structure.

4. Electrospinning is a simple method to produce nanofibers by using a very high frequency of electric field.

5. The use of the Nd:YAG laser in tissue welding relied on water, the largest constituent of most tissues, absorbing strongly at its wavelength (10.6 μm).

Multiple Choice

6. The regenerated tissues that we obtained from an identical twin are called
   - **(a)** Xenogenic cells
   - **(b)** Allogeneic cells
   - **(c)** Isogenic cells
   - **(d)** Autologous cells

7. Which of the following is not the function of scaffolds when mimicking the properties of ECM?
   - **(a)** Providing site for cell attachment
   - **(b)** Permitting diffusion of necessary biomolecules
   - **(c)** Promoting cell proliferation
   - **(d)** Providing mechanical strength

8. Which of the technique is not suitable for producing nanofiber for the nanofibrous scaffolds?
   - **(a)** Cell seeding
   - **(b)** Electrospinning
   - **(c)** Self-assemble
   - **(d)** Phase separation

9. The advantages of using nanoparticle for the delivery of biomolecules do not include
   - **(a)** Selective binding
   - **(b)** Concentration control of biomolecules
   - **(c)** Reduced amount of biomolecules
   - **(d)** Easy clearance

10. Which of the following cannot be used for tissue bonding?
    - **(a)** Nd:YAG laser
    - **(b)** Indocyanine green dye
    - **(c)** Poly(ethylene terephthalate)
    - **(d)** Gold nanoshell

Answers: 1. T; 2. T; 3. F; 4. F; 5. F; 6. c; 7. c; 8. a; 9. d; 10. c

REFERENCES

Andreadis, S. T., Experimental models and high-throughput diagnostics for tissue regeneration, *Expert Opin. Biol. Ther.* **6**, 1071–1086 (2006).

Bass, L. S., and Treat, M. R., Laser tissue welding: A comprehensive review of current and future clinical applications, in C. A. Puliafito, ed., *Laser Surgery and Medicine. Principles and Practice*, Wiley-Liss, New York, pp. 381–415 (1996).

Bessa, P. C., Casal, M., Reis, R. L., Bone morphogenetic proteins in tissue engineering: The road from laboratory to clinic, Part II (BMP delivery), *J. Tissue Eng. Regen. Med.* **2**, 81–96 (2008).

Borlongan, C. V., Masuda, T., Walker, T. A., Maki, M., Hara, K., Yasuhara, T., Matsukawa, N., Emerich, D. F., Nanotechnology as an adjunct tool for transplanting engineered cells and tissues, *Curr. Mol. Med.* **7**, 609–618 (2007).

Cartmell, S. H., and Dobson, J., The use of magnetic particles in tissue engineering in nanomaterials for the life sciences, in C.S.S.R. Kumar, ed., *Magnetic Nanomaterials* 4, Wiley-VCH, Weinheim, pp. 293–308 (2009).

Chivers, R., *In vitro* tissue welding using albumin solder: Bond strengths and bonding temperatures, *Int. J. Adhes. Adhes.* **20**, 179–187 (2000).

Dobson, J., Cartmell, S. H., Keramane, A., Haj, A. J. E., Principles and design of a novel magnetic force mechanical conditioning bioreactor for tissue engineering, stem cell conditioning, and dynamic *in vitro* screening, *IEEE Trans. Nanobiosci.* **5**, 173–177 (2006).

Dvir, T., Timko, B. P., Kohane, D. S., Langer, R., Nanotechnological strategies for engineering complex tissues, *Nat. Nanotechnol.* **6**, 13–22 (2011).

Fan, D., Yin, Z., Cheong, R., Zhu, F. Q., Cammarata, R. C., Chien, C. L., Levchenko, A., Subcellular-resolution delivery of a cytokine through precisely manipulated nanowires, *Nat. Nanotechnol.* **5**, 545–551 (2010).

Gelain, F., Novel opportunities and challenges offered by nanobiomaterials in tissue engineering, *Int. J. Nanomedi.* **3**, 415–424 (2008).

Gilbert, S. F., *Developmental Biology*, 6th edition, Sinauer Associations, Sunderland, MA (2000).

Gobin, A. M., O'Neal, D. P., Watkins, D. M., Halas, N. J., Drezek, R. A., West, J. L., Near infrared laser-tissue welding using nanoshells as an exogenous absorber, *Lasers Surg. Med.* **37**, 123–129 (2005).

Goldberg, M., Langer, R., Jia, X., Nanostructured materials for applications in drug delivery and tissue engineering, *J. Biomater. Sci. Polym. Ed.* **18**, 241–268 (2007).

Guo, J., Su, H., Zeng, Y., Liang, Y. X., Wong, W. M., Ellis-Behnke, R. G., So, K. F., Wu, W., Reknitting the injured spinal cord by self-assembling peptide nanofiber scaffold, *Nanomedicine* **3**, 311–321 (2007).

Hartgerink, J. D., Beniash, E., Stupp, S. I., Self-assembly and mineralization of peptide-amphiphile nanofibers, *Science* **294**, 1684–1688 (2001).

Hartgerink, J. D., Beniash, E., Stupp, S. I., Peptide-amphiphile nanofibers: A versatile scaffold for the preparation of self-assembling materials, *Proc. Natl. Acad. Sci. USA* **99**, 5133–5138 (2002).

Ito, A., Ino, K., Kubayashi, T., Honda, H., The effect of RGD Peptide-conjugated magnetite cationic liposomes on cell growth and cell Sheet harvesting, *Biomaterials* **26**, 6185–6193 (2005).

Jain, K. K., Gorisch, W., Repair of small blood vessels with the neodimium–YAG laser: A preliminary report, *Surgery* **85**, 684–688 (1979).

Jain, K. K., Sutureless extra-intracranial anastomoses by laser, *Lancet* **8046**, 816–817 (1984).

Jakab, K., Norotte, C., Marga, F., Murphy, K., Vunjak-Novakovic, G., Forgacs, G., Tissue engineering by self-assembly and bio-printing of living cells, *Biofabrication* **2**, 022001 (2010).

Katti, D. S., Robinson, K. W., Ko, F. K., Laurencin, C. T., Bioresorbable nanofiber-based systems for wound healing and drug delivery: Optimization of fabrication parameters, *J. Biomed. Mater. Res. B Appl. Biomater.* **70**, 286–296 (2004).

Li, M., Mondrinos, M. J., Gandhi, M. R., Ko, F. K., Weiss, A. S., Lelkes, P. I., Electrospun protein fibers as matrices for tissues engineering, *Biomaterials* **26**, 5999–6008 (2005).

Ma, P. X., Zhang, R., Synthetic nano-scale fibrous extracellular matrix, *J. Biomed. Mater. Res.* **46**, 60–72 (1999).

McNally, K. M., Song, B. S., Hammer, D. X., Heintzelman, D. L., Hodges, D. E., Welch, A. J., Improved laser-assisted vascular tissue fusion using solder-doped polymer membranes on a cyanine model, *Proc. SPIE* **3907**, 65–73 (2000).

Osathanon, T., Linnes, M. L., Rajachar, R. M., Ratner, B. D., Somerman, M. J., Giachelli, C. M., Microporous nanofibrous fibrin-based scaffolds for bone tissue engineering, *Biomaterials* **29**, 4091–4099 (2008).

Oz, M. C., Chuck, R. S., Johnson, J. P., Parangi, S., Bass, L. S., Nowygrod, R., and Treat, M. R., Indocyanine green dye enhanced vascular welding with the near infrared diode laser, *Vasc. Surg.* **24**, 564–570 (1990).

Popat, K., *Nanotechnology in Tissue Engineering and Regenerative Medicine*, CRC Press, Boca Raton, FL (2011).

Prasad, P. N., *Introduction to Biophotonics*, Wiley, New York (2003).

Saltzman, W. M., Olbricht, W. L., Building drug delivery into tissue engineering design, *Nat. Rev. Drug Discovery* **1**, 177–186 (2002).

Schober, R., Ulrich, F., Sander, T., Dürsclen, H., Hessel, S., Laser induced alteration of collagen substructure allows microsurgical tissue welding, *Science.* **232**, 1421–1422 (1986).

Sengupta, S., Eavarone, D., Capila, I., Zhao, G., Watson, N., Kiziltepe, T., Sasisekharan, R., Temporal targeting of tumour cells and neovasculature with a nanoscale delivery system, *Nature* **436**, 568–572 (2005).

Shi, J., Votruba, A. R., Farokhzad, O. C., Langer, R., Nanotechnology in drug delivery and tissue engineering: From discovery to applications, *Nano Lett.* **10**, 3223–3230 (2010).

Shields, K. J., Beckman, M. J., Bowlin, G. L., Wayne, J. S., Mechanical properties and cellular proliferation of electrospun collagen type II, *Tissue Eng.* **10**, 1510–1517 (2004).

Shimizu, K., Ito, A., Honda, H., Enhanced cell-seeding into 3D porous scaffolds by use of magnetite nanoparticles, *J. Biomed. Mater. Res. Part B: Appl. Biomater.* **77**, 265–272 (2006).

Silva, G. A., Czeisler, C., Niece, K. L., Beniash, E., Harrington, D. A., Kessler, J. A., Stupp, S. I., Selective differentiation of neural progenitor cells by high-epitope density nanofibers, *Science* **303**, 1352–1355 (2004).

Umar, Y., Thiyagarajan, M., Halberstadt, C., Gonsalves, K. E., Patterning of cells on bioresist for tissue engineering applications, in C. Laurencin, ed., *Nanoscale Materials Science in Biology and Medicine*, Materials Research Society, Warrendale, PA, pp. 303–309 (2005).

Vasita, R., Katti, D. S., Nanofibers and their applications in tissue engineering, *Int. J. Nanomedicine* **1**, 15–30 (2006).

Wang, N., Butler, J. P., Ingber, D. E., Mechanotransduction across the cell surface and through the cytoskeleton, *Science* **260**, 1124–1127 (1993).

Wei, G., Jin, Q., Giannobile, W. V., Ma, P. X., The enhancement of osteogenesis by nanofibrous scaffolds incorporating rhBMP-7 nanospheres, *Biomaterials* **28**, 2087–2096 (2007).

Xu, C., Yang, F., Wang, S., Ramakrishna, S., *In vitro* study of human vascular endothelial cell function on materials with various surface roughness, *J. Biomed. Mater. Res. A* **71**, 154–161 (2004).

Yang, F., Cho, S. W., Son, S. M., Bogatyrev, S. R., Singh, D., Green, J. J., Mei, Y., Park, S., Bhang, S. H., Kim, B. S., Langer, R., Anderson, D. G., Genetic engineering of human stem cells for enhanced angiogenesis using biodegradable polymeric nanoparticles, *Proc. Natl. Acad. Sci.* **107**, 3317–3322 (2010).

Zhang, L., Webster, T. J., Nanotechnology and nanomaterials: Promises for improved tissue regeneration, *Nano Today* **4**, 66–80 (2009).

Zhang, R., Ma, P. X., Processing of polymer scaffolds: Phase separation, in A. Atala, R. P. Lanza, eds., *Methods of Tissue Engineering*, Academic Press, San Diego, CA, pp. 715–724 (2002).

Zhang, S., Fabrication of novel biomaterials through molecular self-assembly, *Nat. Biotechnol.* **21**, 1171–1178 (2003).

CHAPTER 19

NANODERMATOLOGY AND NANOCOSMETICS

This chapter deals with the impact of nanotechnology on skin care. In a broad sense, dermatology deals with all aspects of skin care, including physical attributes that are covered by cosmetics. However, dermatology is also often used in a more restrictive sense of involving a therapeutic or surgical intervention to treat a skin condition, while cosmetology focuses on improvement of physical attributes through skin treatment. Section 19.1, presenting skin care delivery, revisits the skin barrier already discussed in Chapter 6 ("*Crossing the Biological Barriers*"), but now from the viewpoint of intradermal delivery of skin care chemicals. Section 19.2 presents an introduction to the impact of nanotechnology on skin care. Section 19.3 provides an account of the various nanoparticles utilized for nanodermatology and nanocosmetics. Section 19.4 gives an account of applications of nanodermatology. Section 19.5 describes applications of nanocosmetics. The chapter concludes with Section 19.6 discussing the nanotoxicity concern of using nanoformulations for skin care.

19.1. DELIVERY THROUGH SKIN

Even though skin has a large surface area, delivery of drug molecules through the skin can be extremely challenging because this barrier is formidable. Like all physiological barriers, the skin is extremely selective in allowing the permeation of molecules such as water and electrolytes while providing

Introduction to Nanomedicine and Nanobioengineering, First Edition. Paras N. Prasad.

protection against microorganisms, ultraviolet radiation, toxic agents, and mechanical insults. Therefore, this barrier acts as a significant impediment toward drug delivery via the skin, whether in the free form or attached to a nanoparticulate carrier [Kanikkannan et al., 2000]. The general anatomical organization of skin has already been presented in Chapter 6.

Various molecules and macromolecules can enter into the skin through three distinct pathways: (1) the intercellular pathway, which is through the lipid matrix occupying the intercellular spaces of the keratinocytes, (2) the transcellular pathway, which is through the keratinocytes, and (3) transappendageal pathway, which is across hair follicles, sebaceous glands, and sweat glands. Several attempts have been made, and are still under investigation, to develop efficient formulations to deliver molecules and macromolecules across the skin barrier. Substances smaller than 500 kDa, with sufficient oil solubility and high partition coefficient, can be absorbed into the skin. However, in contrast, larger molecules (molecular weight >500 kDa) cannot pass the cutaneous barrier [Bos and Meinardi, 2000]. To overcome these limitations, nanoparticles are being developed, which not only enhance the absorption of drugs into the skin but also release the drug in a controlled manner for a prolonged period of time.

19.2. SKIN CARE AND NANOTECHNOLOGY

Application of nanotechnology to skin care is an important area of nanomedicine since skin, as the outer protective covering, is the largest organ in the body. With a large surface area and being exposed to the environment, skin is the body's first defense line protecting the body from external invasion and environmental insult. In addition, the skin serves as a protective and impermeable layer to prevent an organism from losing water, minerals, and dissolved proteins. This barrier, however, can be accessed under appropriate conditions to provide for different forms of skin care as well as for transdermal delivery of drugs. Various skin patches are already in the market, allowing for absorption and transdermal delivery of drugs, which then enter the systemic circulatory system. The focus of this chapter is on skin care through dermatology and cosmetology, which involves only intradermal delivery; transdermal delivery will not be covered here. As a matter of fact, from the skin care perspectives, it is desirable that the topical formulation should not get into the circulation because it may lead to undesirable effects.

Dermatology is the area of medicine that deals with therapeutic or surgical intervention to treat a skin disease, disorder, or irritation. Cosmetics cover the use of formulations to improve aesthetic and physical attributes. However, the distinction between these two types of skin care is not always that sharp. For example, sunscreen lotions, which often utilize nanoparticle-based UV absorbers, provide protection against UV-induced skin cancer. However, the formulations also have to conform to aesthetic appeal—in other words, being transparent and not chalky. Another example is the skin disorder acne vulgaris. This medical condition requires treatment and thus is a focus of dermatology. However, it is not a disease that perturbs body functions, but only the physical appearance. A medicated topical cream formulation similar to many cosmetic products is prescribed for treatment. Despite the overlap between the domains of dermatology and cosmetics, government regulatory agencies, such as the Food and Drug Administration (FDA) in the United States, treat a product differently depending on whether it is dermatological or cosmetic, the latter being less rigorously regulated.

The application of nanotechnology produces many benefits for dermatology. For example, imaging using nanoparticles with targeting ability can be used to identify and map out a specific skin disease, such as cancer. Thus, real-time visualization of tumors and diagnosis of infections and malignancies can be achieved with minimal invasiveness. Tagged fluorescent quantum dots in animal models have been employed to visualize tumors in the skin and to harvest sentinel lymph nodes without using radioactive labeling [Kim et al., 2004].

Whether dermatological or cosmetic in nature, a nanoparticle formulation offers a number of benefits. Some of the primary ones are:

- *Enhancement of Chemical Stability of Active Components.* Encapsulation of an active component in a nanostructure prevents its degradation by various factors, such as external environments, oxygen and reactive oxygen species (ROS), or enzymes.
- *Enhancement of Absorption.* A nanoparticle formulation also provides a high level of concentration to lead to increased skin absorption of an active component.
- *Controlled Skin Penetration.* Different skin care applications may require different levels of skin penetration. A cosmetic formulation may need to be restricted to superficial layers, but a particular skin disease may require penetration to a specific depth in the skin structure. Nanoparticles can be tailored in composition, size, shape, and surface characteristics to provide this control of penetration depth.
- *Controlled Release of Active Components.* Nanoparticles with tailored porosity, core–shell structure and surface characteristics can provide controlled release of active components. This feature enables control of time-release and sustained action for longer period, maintaining the level of active component at the desired level. It also helps in minimizing any adverse effect produced by direct application of the active compound to the skin.
- *Adhesiveness, Occlusion, and Skin Hydration.* Nanoparticles provide considerable benefits derived from their good adhesive and occlusive effects. Because of their large surface-to-volume ratio, they have enhanced adhesive and occlusive properties. The smaller the size, the better are these

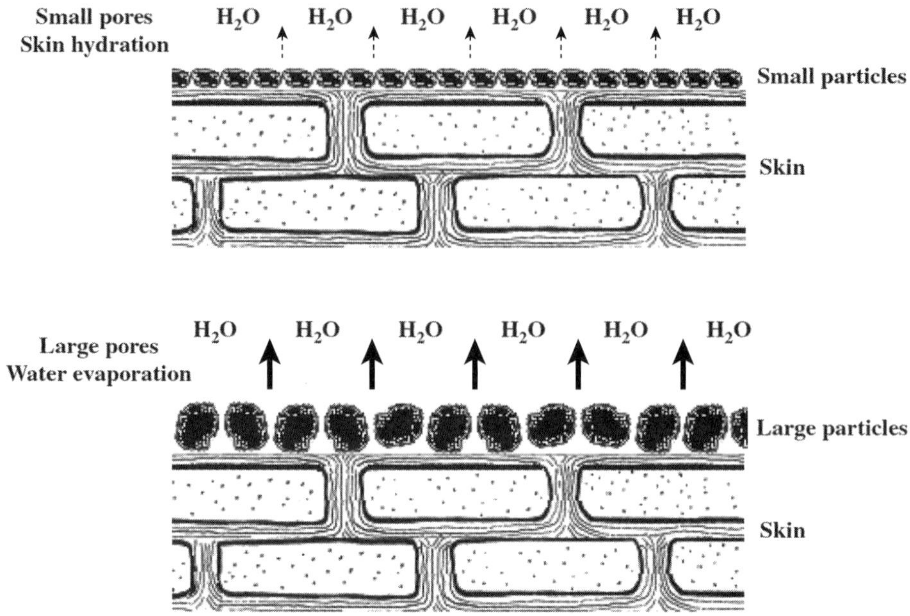

Figure 19.1. Dependence of occlusion effect on the particle size. (Reproduced with permission from Souto and Müller [2008].)

properties. Nanoparticles exhibit over 15-fold more occlusion than their microparticle counterpart [Schäfer-Korting et al., 2007]. The nanoparticles can form a monolayered film with an occlusive action to retard the loss of moisture from the skin, which can have many benefits to skin—including promoting of skin elasticity, the loss of which is a major effect of aging. This enhanced occlusion effect in preventing loss of moisture is illustrated in Figure 19.1.

- *Control of pH and Osmotic Effects.* Skin is generally acidic, but can exhibit a broad range of pH, from 4.0 to 7.0, and exhibits a marked resilience to pH variation. Any formulation that is strongly acidic or alkaline in nature can cause severe irritation and, perhaps, even skin damage. Appropriate choice of nanoparticle formulation can also control the pH. Also, nanoparticles, such as solid lipid nanoparticles, exhibit suitable osmotic behavior with body fluids.
- *Minimization of Skin Irritation and Allergic Reaction.* By design of biocompatible surface coatings, a nanoparticle formulation can be made to result in minimal skin irritation and allergic reaction.

To conclude this section, already a number of nanoformulations are available in the market for dermatological and cosmetic applications. Some specific applications are described in the next two sections. These applications are projected to grow many-fold in the near future.

19.3. VARIOUS NANOPARTICLES FOR DERMATOLOGY AND COSMETICS

This section provides an account of the various nanoparticles that have been extensively utilized for nanodermatology and nanocosmetic applications.

- *Polymeric Nanoparticles.* Polymeric nanoparticles have become the most popular drug delivery vehicles for most intended organs, with the skin being no exception. Chitosan, a natural polymer, is a popular component of nanoparticulate carriers owing to several beneficial features such as anticoagulant activity, wound-healing, and antimicrobial properties. In addition, biocompatible and biodegradable synthetic polymers such as poly-lactide-co-glycolide (PLGA) and poly-ε-caprolactone (PCL), which degrade through the natural pathways, have also found applications in dermal drug delivery, albeit with limited utility [Desai et al., 2010].
- *Solid Lipid Nanoparticles (SLN).* SLNs, discussed in Chapter 3, are composed of high-melting lipids such as triglycerides, and they are solids at room temperature. Owing to their biocompatibility, they have received considerable attention as carriers of active agents for dermatological and cosmetics usage. The surface of SLN is covered by a surfactant shell, which stabilizes the dispersion. The advantages of SLNs in drug delivery include (a) protection of encapsulated agents from enzymatic degradation, (b) prevention of transepidermal water loss, and (c) controlled and sustained release of the encapsulated active agent. SLNs for the skin topical application are made from biocompatible lipids such as glycerol behenate (Compritol® 888 ATO), glycerol palmitostearate (Precirol® ATO 5), or wax cetyl palmitate [Desai et al., 2010].
- *Titanium Dioxide and Zinc Oxide Nanoparticles.* Nanomaterials of titanium dioxide (TiO_2) (particle size up to several micrometers) and zinc oxide (ZnO) (particle size 30–200 nm) are key ingredients for many cosmetics and sunscreen nanoparticle formulations [Mu and Sprando, 2010], because these high bandgap materials efficiently reflect and scatter UV light. Moreover, both these nanoparticles, irrespective of their size, shape, and surface properties, are found mainly restricted in the outermost surface of the epidermis and cannot penetrate into deeper skin layers [Gamer et al., 2006].
- *Silver Nanoparticles.* Silver nanoparticles are attracting interest for a range of biomedical applications owing to their potent antibacterial activity. It has recently been demonstrated that they alter wound inflammatory events, facilitate early phase of wound healing, reduce wound infections,

decrease matrix metalloproteinase activity, and are cost-effective. Therefore, they could be exploited in developing better dressings for wounds and burns. The key to its broad-acting and potent antibacterial activity is the multifaceted mechanism by which silver nanoparticles act on microbes [Chaloupka et al., 2010].

- *Magnetic Nanoparticles.* Ultrasmall magnetic nanoparticles made up of iron derivatives (magnetic, paramagnetic or super-paramagnetic) have potential application in skin delivery. It has been shown that the rigid magnetic nanoparticles smaller than 10 nm can passively penetrate the skin through the SC lipid matrix and hair follicle orifices [Bartoli et al., 2007]. In addition to passive diffusion, one unique advantage of magnetic nanoparticles is that following topical application in the skin, their dynamics, skin-penetration, and localization can be controlled by the application of external magnetic field.

19.4. NANODERMATOLOGY

Nanoparticle-based delivery of active agents has been utilized for a variety of dermatological applications. Some of the principal applications are listed in Figure 19.2. Because nanodermatology is a rapidly growing area, the list of applications will continue to expand. This section briefly discusses the applications listed in Figure 19.2.

Antimicrobial Nanoparticles. Nanoparticles have been utilized for antimicrobial purposes as a substitute for antibiotics. The advantage is that a topical cream formulation can be applied to the site of infection or damage. It has proved to be particularly useful for wound management [Atiyeh et al., 2007]. Nanocrystalline silver has extensively been utilized in the formulation of wound dressing as an antimicrobial barrier layer. The examples are provided by Acticoat™ and Acticoat 7, which utilize a silver mesh containing silver

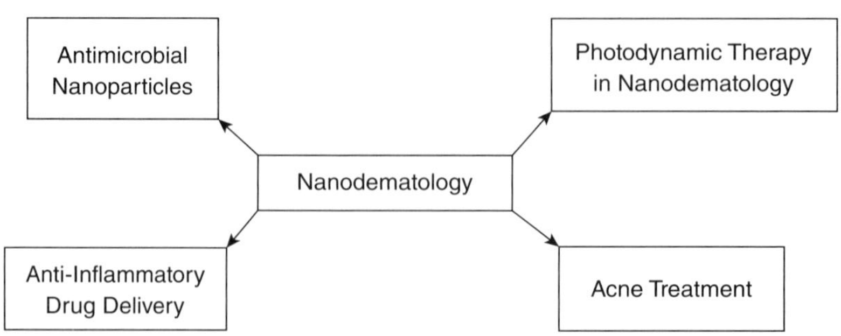

Figure 19.2. Some principal applications of nanodermatology.

nanocrystals [Fong and Wood, 2006]. Silver exhibits antiseptic, antimicrobial, and anti-inflammatory properties.

Controlling microorganisms is an essential step in wound healing. Especially in the case of chronic wounds, high levels of bacterial counts, the presence of multi-resistant organisms, and bacterial films seriously impact wound healing, because bacteria compete with host cells for nutrients and oxygen. Moreover, the presence of bacteria mediates an inflammatory response through diverse molecular mechanisms that in severe cases can lead to chronic nonhealing wounds. The wound microenvironment can support the formation of mixed bacterial flora embedded in a polysaccharide matrix, known as biofilms. The biofilms are extremely resistant to conventional antibiotics, which further slows the healing process.

The Acticoat nanoformulation produces antimicrobial activity of nanocrystalline silver, at the same time assisting to maintain a moist environment [Fong and Wood, 2006]. This combined feature promotes wound healing. The smaller size of silver in its nanocrystalline form leads to great wound surface contact with silver to increase bioactivity and silver solubility. The nanosilver containing metallic cluster form of silver produces silver cations, which exhibit strong antimicrobial effect by blocking the cellular respiration and disrupting the function of bacterial cell membranes. These ions also bind and denature bacterial DNA and RNA, leading to inhibition of cell replication.

Nitric oxide-releasing nanoparticles have also been utilized for antimicrobial effect [Abramovits et al., 2010]. These nanoparticles have demonstrated successful microbial activity against methicillin-resistant *staphylococcus aureus* in murine wound models. Nanoparticle-based nitric oxide therapy has shown potential for many wound-healing complications such as those derived from burns, diabetes, and obesity [Martinez et al., 2009].

Photodynamic Therapy in Nanodermatology. Photodynamic therapy (PDT) is a light-induced therapy to destroy abnormal cells, such as cancer cells. This therapy has already been discussed in detail in Chapter 13 ("Nanotechnology for Cancer"). Photodynamic therapy has been applied in dermatology for treating cancer, as well as psoriasis [Rao et al., 2010; Fritsch et al., 1998].

Skin, being an external readily accessible organ, is highly suitable for being treated by PDT, which requires the diseased tissue to be accessible to light. Thus, a primary cancer therapy application of PDT is in the treatment of skin cancer, where a photosensitizer can be used in the form of a topical formulation to be applied directly to the skin. 5-Aminolevulinic acid (ALA) has been extensively investigated for PDT for various nonmelanoma skin cancers, as well as for nonmalignant neoplastic diseases, which produce benign proliferative skin conditions [Rao et al., 2010; Ibbotson, 2002]. Examples of the latter class are viral warts and psoriasis [Tandon et al., 2008]. Even application to acne treatment has been performed. Psoriasis is one of the most common human, immune-mediated skin diseases, which is characterized by excessive growth and aberrant differentiation of keratinocytes [Lowes et al., 2007]. The most common form of psoriasis produces red, scaly, and raised plaques that can be treated by PDT.

ALA by itself is not the active photosensitizer, but rather is a pro-drug that is enzymatically transformed inside the cell to the active form, protoporphyrin IX. Another variant of ALA used for skin PDT is methylaminolevulinate (MAL). Both these compounds are approved for clinical use in several countries, including the United States. ALA is hydrophilic and thus its transport across the cell membrane for intracellular action by PDT is limited. However, the intracellular delivery of ALA can be facilitated by the use of appropriately designed nanocarriers. A suitable nanocarrier utilized widely for this purpose is a dendrimer structure [Abramovits et al., 2010; Battah et al., 2007]. Another photosensitizer, phthalocyanine conjugated to gold nanoparticles, has been studied *in vivo* for PDT of amelanotic melanoma [Camerin et al., 2010].

Anti-Inflammatory Drug Delivery. Anti-inflammatory drugs can be divided into two categories: (i) corticosteroids such as hydrocortisone and (ii) nonsteroidal anti-inflammatory drugs (NSAIDs) such as naproxen. Corticosteroids reduce inflammation by binding to glucocorticoid receptors. The NSAIDs reduce inflammation by inhibiting the enzyme cyclooxygenase needed for the production of proinflammatory prostaglandins.

Reports on nanoparticle-based topical delivery of anti-inflammatory drugs are extensive [Prow et al., 2011]. An example of nanoparticle delivery of a steroid is provided by the work of Jensen et al. [2010], who utilized solid lipid nanoparticles for lipid polarity controlled release of corticosteroid. Flufenamic acid (FFA) is an example of nonsteroidal anti-inflammatory drug, enhanced by nanoformulation. Luengo et al. [2006] reported topical delivery of FFA by encapsulation in PLGA nanoparticles.

Acne Treatment. Acne vulgaris is a skin disorder that affects a large population, particularly in the teenage. An effective treatment for mild to moderate acne is topical administration of all-*trans* retinoic acid (RA). However, this topical use of RA has a number of side effects, such as sensitivity to sunlight, eczematous irritation, and erythema, as well as dryness and exfoliation, thus limiting the efficacy of RA therapy.

Nanoparticle encapsulation of RA reduces the side effects by controlled release of RA and follicular penetration. A solid lipid nanoparticle formulation containing RA with high encapsulation efficiency has been developed [Castro et al., 2007].

19.5. NANOCOSMETICS

The area of nanotechnology applications for cosmetics is very large, with a huge commercial potential. Current estimates of nanocosmetic market in the United States alone are estimated to exceed over 60 billion dollars. The appli-

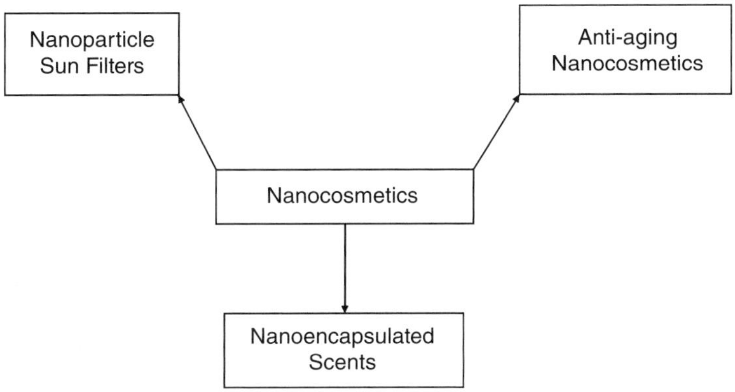

Figure 19.3. Some important applications of nanotechnology to cosmetics.

cations are of wide range, from sunscreen lotions to anti-aging nanocosmetics. Figure 19.3 lists some important areas of nanocosmetics. These are briefly described below.

Nanoparticle Sun Filters. Sunscreen lotions containing TiO_2 and ZnO nanoparticles are already in the marketplace under various trade names. These nanoparticles are in the range of 20–60 nm and have wide bandgaps to absorb in the UV region [Wiechers and Musee, 2010]. Because of their small size, these particles do not significantly scatter light and, thus, provide a transparent coating. TiO_2 and ZnO nanoparticles provide a broad coverage against ultraviolet light because it absorbs both UVA and UVB radiations in the sunlight. However, TiO_2 provides a more effective protection against UVB, while ZnO provides a better protection against UVA, which produces more UV-induced aging effect. Additionally, particle size is very important to control. There are also indications that TiO_2 nanoparticles smaller than 20 nm can penetrate deep inside the skin to interact with the immune system [Hagedorn-Leweke and Lippold, 1998].

Another formulation called OPTOSIL™ by Oxonica Ltd. contains TiO_2 nanoparticles with <1% manganese. It provides a balanced protection against both UVA and UVB and, at the same time, eliminates the generation of harmful free radicals [Wakefield et al., 2004].

An alternative approach utilizes molecular sunscreens, which can be either UVA or UVB filters depending on their chemical composition. However, the environmental stability of the organic materials always is an issue. Furthermore, they are more susceptible to be absorbed on skin and diffuse through the skin barrier to get into the bloodstream. Lipid nanoparticles loaded with these molecular absorbers can provide a more effective way of utilizing them as a vehicle for molecular sunscreens. In addition, the lipid nanoparticles also have shown by themselves UV-blocking ability due to efficient scattering of UV light to produce a synergistic effect [Müller et al., 2002].

Anti-aging Nanocosmetics. In the case of skin, the visual appearance of aging is clearly manifested. Important visual signs are wrinkles, dryness, and discoloration of skin. As discussed in Chapter 16 ("Rejuvenation Therapy"), the aging process involves lack of nutrients, antioxidants and accumulation of toxins. In the case of skin, it also involves loss of moisture to cause dryness of skin, which contributes to the loss of elasticity of skin, thus producing wrinkles.

A major focus of modern cosmetics is to produce anti-aging formulations, which nourish the skin, deliver antioxidants to scavenge reactive oxygen species (ROS), and prevent the loss of moisture. These nanoformulations have been in the cosmetic market since 1986. Two such products to cite as examples are a formulation by L'Oreal and a liposome formulation by Christian Dior. They contain a wide range of active agents, such as coenzyme Q10 (CoQ10), vitamin A, and the popular antioxidant, vitamin E. CoQ10 is a potent antioxidant that protects the cells against ROS. The nanoparticles increase the concentration of these active agents in the epidermis, with no signs of any acute or chronic toxicity [Touitou et al., 1994]. Fullerenes have also been considered for use in anti-aging cosmetic formulation because of their ROS scavenging ability [Lens, 2009]. However, potential toxicity concerns still remain an issue.

More recent formulations have utilized solid lipid nanoparticles. The lipid nanoparticles can contain a relatively high loading of CoQ10.

Retinoids are another group of anti-aging agents, which work through nuclear hormone receptors [Prow et al., 2011]. They are used to reduce wrinkles and treat photo-aging. A common retinoid is tretinoin (all-trans retinoic acid), which is unstable in the presence of light and oxygen. Unfortunately, retinoid therapy causes skin irritations such as erythema, peeling, dryness and pain. Encapsulation and controlled release of tretinoin in solid lipid nanoparticle (SLN) have shown to help minimize both these problems. For example, vitamin A palmitate, which eventually converts into tretinoin, has been formulated into SLN [Pople and Singh, 2006].

Solid lipid nanoparticles (SLN) containing creams have shown effectiveness in maintaining skin hydration and retaining viscoelasticity of skin, thus retarding wrinkle formation [Wissing and Müller, 2003].

Nanoencapsulated Scents. Nanoformulation encapsulation of perfume allows for longer lasting fragrance. This is derived from the feature of a nanoformulation that allows for controlled, prolonged release. Perfumes have been encapsulated in solid lipid nanoparticles [Wissing et al., 2000].

The merit of a prolonged release of perfumes is the ability to provide a once-a-day application, with the fragrance lasting for many hours. The release of a perfume can be slowed down by incorporating it in a SLN instead of an oil droplet (emulsion) as it has traditionally been formulated [Wissing et al., 2000; Hommoss et al., 2005]. Even after 6 h, only 75% of the perfume was released from SLN, whereas 100% was released from the emulsion.

Another advantage offered by nanoencapsulation is that it improves on the durability and environmental stability of fragrances that deteriorate under changing environment (light, air, humidity, etc.). Another direction broadening the application of perfume using the nanoparticle technology is to incorporate both the perfumes and chemicals that capture unpleasant odor in the air. This dual modality particle can be used to provide a pleasant environment.

19.6. NANOTOXICOLOGY OF THE SKIN

Although highly promising, the widespread dermatological applications of nanoparticles also necessitate a thorough evaluation of toxicological parameters related to their interaction with the skin. The potential toxic effects of nanoparticles on the skin may depend on a number of factors, such as the method of synthesis, size, shape (spheres/rods), surface properties, dosage, and so on. For example, silver nanoparticles, which are widely used for antimicrobial and wound healing properties on the skin, have been found to cause dose-dependent acute and subchronic dermal toxicity in guinea pigs [Korani et al., 2011]. Moreover, titania and zirconia nanoparticles, two widely used components in sunscreen lotions, have been observed to induce cytotoxicity and genotoxicity in skin keratinocytes, respectively [Smijs and Bouwstra, 2010]. Other potential toxic outcomes of such nanoparticles include triggering of immune response, oxidative stress, and carcinogenicity.

In general, nanoparticles are reported to be localized in the outer layers of the skin, such as the stratum corneum surface, and also in the hair follicle openings, whereas penetration to the deeper layers of the skin is not frequently observed. However, some studies have reported their localization in the deeper layers, such as the viable epidermis. For smaller nanoparticles (diameter 4–5 nm), particularly quantum dots, penetration into the dermis has also been reported [Smijs and Bouwstra, 2010]. Deeper skin penetration carries the risk of absorption of the nanopartcles into the systemic circulation, with eventual assimilation in vital organs such as the liver and brain. Further studies are required for a comprehensive evaluation of potential toxicological effects of various nanoparticles on the skin.

HIGHLIGHTS OF THE CHAPTER

- The skin barrier is extremely selective in allowing the permeation of useful molecules such as water and electrolytes, while providing protection against microorganisms, ultraviolet radiation, toxic agents, and mechanical insults. Therefore this barrier acts as a severe impediment toward drug delivery via the skin.
- Various molecules and macromolecules can enter into the skin through three distinct pathways: (1) the intercellular pathway, which is through

the lipid matrix occupying the intercellular spaces of the keratinocytes, (2) the transcellular pathway through the keratinocytes, and (3) transappendageal pathway, which is across hair follicles, sebaceous glands, and sweat glands.

- Dermatology is the field of medicine, which deals with therapeutic or surgical intervention to treat a skin disease, disorder, or irritation. Cosmetics cover the use of formulations to improve aesthetic and physical attributes.
- Whether dermatological or cosmetic in nature, a nanoparticle formulation offers a number of benefits, some which are: enhancement of chemical stability of active components; enhancement of absorption; controlled skin penetration; controlled release of active components; adhesiveness, occlusion, and skin hydration; control of pH and osmotic effects; minimization of skin irritation and allergic reaction.
- Nanoparticle-based (polymeric nanoparticles, solid lipid nanoparticles, titanium dioxide and zinc oxide nanoparticles, magnetic nanoparticles) delivery of active agents has been utilized for a variety of dermatological applications, such as antimicrobial drug delivery, photodynamic therapy of skin diseases, anti-inflammatory drug delivery, and acne treatment.
- Important areas of nanocosmetics are nanoparticle sun filters, anti-aging nanocosmetics, and nano-encapsulated scents.
- Although highly promising, the widespread dermatological applications of nanoparticles also necessitate a thorough evaluation of toxicological parameters related to their interaction with the skin.

EXERCISES

True or False

1. Skin barrier allows the permeation of water.

2. Skin barrier provides protection against ultraviolet radiation.

3. The skin barrier is composed of five different layers.

4. Molecules and macromolecules can enter into the skin through only two different pathways.

5. Dermatology is the area of medicine that deals with therapeutic or surgical intervention to treat a skin disease, disorder, or irritation.

Multiple Choice

6. Which of the following formulations are well known as delivery agents for nanodermatology? (you can choose more than one answer)

(a) Polymeric nanoparticles
(b) Titania nanoparticles
(c) Quantum dots
(d) Magnetic nanoparticles

7. Which of the following applications are *not* used in nanodermatology and nanocosmetics?
   (a) Photodynamic therapy
   (b) Nutritious crème formulation
   (c) Anti-inflammatory drug delivery
   (d) Nutritious crème formulation

Answers: 1. T; 2. T; 3. F; 4. F; 5. T; 6. a,b,d; 7. b

REFERENCES

Abramovits, W., Granowski, P., Arrazola, P., Applications of nanomedicine in dermatology: Use of nanoparticles in various therapies and imaging, *J. Cosmet. Dermatol.* **9**, 154–159 (2010).

Atiyeh, B. S., Costagliola, M., Hayek, S. N., Dibo, S. A., Effect of silver on burn wound infection control and healing, review of the literature, *Burns* **33**, 139–148 (2007).

Bartoli, B., Ennas, M. G., Loffredo, F., Isola, M., Pinna, R., López-Quintela, M. A., Penetration of metallic nanoparticles in human full-thickness skin, *J. Invest. Dermatol.* **127**, 1701–1712 (2007).

Battah, S., Balaratnam, S., Casas, A., O'Neill, S., Edwards, C., Batlle, A., Dobbin, P., MacRobert, A. J., Macromolecular delivery of 5-aminolaevulinic acid for photodynamic therapy using dendrimer conjugates, *Mol. Cancer Ther.* **6**, 876–885 (2007).

Bos, J. D., Meinardi, M. M. H. M., The 500 dalton rule for the skin penetration of chemical compounds and drugs, *Exp. Dermatol.* **9**, 165–169 (2000).

Camerin, M., Magaraggia, M., Soncin, M., Jori, G., Moreno, M., Chambrier, I., Cook, M. J., Russell, D. A., The *in-vivo* efficacy of phthalocyanine-nanoparticle conjugates for the photodynamic therapy of amelanotic melanoma, *Eur. J. Cancer* **46**, 1910–1918 (2010).

Castro, G. A., Orefice, R. L., Vilela, J. M. C., Andrade, M. S., Ferreira, L. A. M., Development of a new solid lipid nanoparticle formulation containing retinoic acid for topical treatment of acne, *J. Microencapsul.* **24**, 395–407 (2007).

Chaloupka, K., Malam, Y., Seifalian, A. M., Nanosilver as a new generation of nanoproduct in biomedical applications, *Trends Biotechnol.* **28**, 580–588 (2010).

Desai, P., Patlolla, R. R., Singh, M., Interaction of nanoparticles and cell-penetrating peptides with skin for transdermal drug delivery, *Mol. Membr. Biol.* **27**, 247–259 (2010).

Fong, J., Wood, F., nanocrystalline silver dressing in wound management: A review, *Int. J. Nanomed.* **1**, 441–449 (2006).

Fritsch, C., Goerz, G., Ruzicka, T., Photodynamic therapy in dermatology, *Arch. Dermatol.* **134**, 207–214 (1998).

Gamer, A. O., Leibold, E., van Ravenzwaay, B., The *in vitro* absorption of microfine zinc oxide and titanium dioxide through porcine skin, *Toxicol. In Vitro* **20**, 301–307 (2006).

Hagedorn-Leweke, U., Lippold, B. C., Accumulation of sunscreens and other compounds in keratinous substrates, *Eur. J. Pharm. Biopharm.* **46**, 215–221 (1998).

Hommoss, A., Souto, E. B., Müller, R. H., Assessment of the release profiles of a perfume incorporated into NLC dispersions in comparison to reference nanoemulsions, in *Annual Meeting and Exposition (AAPS)*, Nashville, TN, November 8–10, #M1238 (2005).

Ibbotson, S. H., Topical 5-aminolaevulinic acid photodynamic therapy for the treatment of skin conditions other than non-melanoma skin cancer, *Br. J. Dermatol.* **146**, 178–188 (2002).

Jensen, L. B., Magnussson, E., Gunnarsson, L., Vermehren, C., Nielsen, H. M., Petersson, K., Corticosteroid solubility and lipid polarity control release from solid lipid nanoparticles, *Int. J. Pharm.* **390**, 53–60 (2010).

Kanikkannan, N., Kandimalla, K., Lamba, S. S., Singh, M., Structure–activity relationship of chemical penetration enhancers in transdermal drug delivery, *Curr. Med. Chem.* **7**, 593–608 (2000).

Kim, S., Lim, Y. T., Soltesz, E. G., Soltesz, E. G., De Grand, A. M., Lee, J., Nakayama, A., Parker, J. A., Mihaljevic, T., Laurence, R. G., Dor, D. M., Cohn, L. H., Bawendi, M. G., Frangioni, J. V., Near-infrared fluorescent Type II quantum dots for sentinel lymph node mapping, *Nat. Biotechnol.* **22**, 93–97 (2004).

Korani, M., Rezayat, S., Gilani, K., Bidgoli, S. A., Adeli, S. Acute and subchronic dermal toxicity of nanosilver in guinea pig, *Int J Nanomedicine.* **6**, 855–862 (2011).

Lens, M., Use of fullerenes in cosmetics, *Recent Pat. Biotechnol.* **3**, 118–123 (2009).

Lowes, M. A., Bowcock, A. M., Krueger, J. G., Pathogenesis and therapy of psoriasis, *Nature* **445**, 866–873 (2007).

Luengo, J., Weiss, B., Schneider, M., Ehlers, A., Stracke, F., Konig, K., Kostka, K. H., Lehr, C. M., Schaefer, U. F., Influence of nanoencapsulation on human skin transport of flufenamic acid, *Skin Pharmacol. Physiol.* **19**, 190–197 (2006).

Martinez, L. R., Han, G., Chacko, M., Mihu, M. R., Jacobson, M., Gialanella, P., Friedman, A. J., Nosanchuk, J. D., Friedman, J. M., Antimicrobial and healing efficacy of sustained release nitric oxide nanoparticles against *staphylococcus aureus* skin infection, *J. Invest. Dermatol.* **129**, 2463–2469 (2009).

Mu, L., Sprando, R. L., Application of nanotechnology in cosmetics, *Pharm. Res.* **27**, 1746–1749 (2010).

Müller, R. H., Radtke, M., Wissing, S. A., Solid lipid nanoparticles (SLN) and nanostructured lipid carriers (NLC) in cosmetic and dermatological preparations, *Adv. Drug Deliv. Rev.* **54**, Suppl. 1, S131–S155 (2002).

Pople, P. V., Singh, K. K., Development and evaluation of topical formulation containing solid lipid nanoparticles of vitamin A, *AAPS Pharm. Sci. Tech.* **7**, 91–97 (2006).

Prow, T. W., Grice, J. E., Lin, L. L., Faye, R., Butler, M., Becker, W., Wurm, E. M. T., Yoong, C., Robertson, T. A., Soyer, H. P., Roberts, M. S., Nanoparticles and microparticles for skin drug delivery, *Adv. Drug Deliv. Rev.* **63**, 470–491 (2011).

Rao, J., Bissonnette, R., Taylor, C. R., Photodynamic therapy for the dermatologist (updated Jan. 11, 2010). Retrieved from the http://emedicine.medscape.com/article/1121517-overview on July 26, 2011.

Schäfer-Korting, M., Mehnert, W., Korting, H-C., Lipid nanoparticles for improved topical application of drugs for skin diseases, *Adv. Drug Deliv. Rev.* **59**, 427–443 (2007).

Smijs, T. G., Bouwstra, J. A., Focus on skin as a possible port of entry for solid nanoparticles and the toxicological impact, *J. Biomed. Nanotechnol.* **6**, 469–484 (2010).

Souto, E. B., Müller, R. H., Cosmetic features and applications of lipid nanoparticles (SLN®, NLC®), *Int. J. Cosmet. Sci.* **30**, 157–165 (2008).

Tandon, Y. K., Yang, M. F., Baron, E. D., Role of photodynamic therapy in psoriasis: A brief review, *Photodermatol. Photoimmunol. Photomed.* **24**, 222–230 (2008).

Touitou, E., Junginger, H. E., Weiner, N. D., Nagai, T., Mezei, M., Liposomes as carriers for topical and transdermal delivery, *J. Pharm. Sci.* **83**, 1189–1203 (1994).

Wakefield, G., Lipscomb, S., Holland, E., Knowland, J., The effects of manganese doping on UVA absorption and free radical generation of micronized titanium dioxide and its consequences for the photostability of UVA absorbing organic sunscreen components, *Photochem. Photobiol. Sci.* **3**, 648–652 (2004).

Wiechers, J. W., Musee, N., Engineered inorganic nanoparticles and cosmetics: Facts, issues, knowledge gaps and challenges, *J. Biomed. Nanotechnol.* **6**, 408–431 (2010).

Wissing, S. A., Mäder, K., Müller, R. H., Solid lipid nanoparticles (SLN™) as a novel carrier system offering prolonged release of the perfume allure (Chanel), *Int. Symp. Control. Release Bioact. Mater.* **27**, 311–312 (2000).

Wissing, S. A., Müller, R. H., The influence of solid lipid nanoparticles on skin hydration and viscoelasticity—*in vivo* study, *Eur. J. Pharm. Biopharm.* **56**, 67–72 (2003).

CHAPTER 20

NANODENTISTRY

This chapter deals with the application of nanotechnology to dental care. Nanotechnology can be expected to make a major impact on dentistry, the applications ranging from preventive care, to dental restoration, to delivery of local anesthesia and/or antibiotics, to treatment of various dental diseases, and finally to regenerative dentistry using tissue engineering to regenerate replacement for damaged dental structures [Subramani and Ahmed, 2012]. In addition, a nanoparticle platform also provides the ability to noninvasively detect and diagnose a dental problem. For the benefit of nonmedical readers, this chapter starts with a description of the structure and composition of a tooth in Section 20.1, along with an overview of nanotechnology application for dental care. The subsequent sections provide an account of some specific applications. Section 20.2 describes the applications of nanoparticles for preventive care, caries inhibition, and repair of earlier caries lesions. Section 20.3 covers the development on new formulations of dental adhesives and fillings using nanomaterials and nanocomposites. These nanocomposites can also provide antimicrobial benefits. Section 20.4 describes regenerative dentistry. Section 20.5 covers the utilization of nanoparticles to diagnose a dental problem using sensing and dental imaging. Finally, Section 20.6 presents a more futuristic approach of using nanorobots for dentistry, giving examples of local delivery of anesthesia and treatment of dental hypersensitivity.

Introduction to Nanomedicine and Nanobioengineering, First Edition. Paras N. Prasad.

20.1. NANOTECHNOLOGY FOR DENTAL CARE

In order to fully realize the impact of nanotechnology on dental care, it may be useful for the sake of nonmedical readers to briefly review the dental structure and composition first. Figure 20.1 provides a schematic of the tooth structure. The exterior coating of a tooth is the dental enamel, a highly mineralized structure that provides remarkable hardness and resistance to the tooth. The composition of the enamel is mainly inorganic, being 86% inorganic (primarily the hydroxy apatite), 2% organic, and 12% water. It consists of stacks of nanosize needles of calcium orthophosphate (labeled as apatites) in a nonstoichiometry form called hydroxyapatite, which are bundled together in a parallel-ordered prismatic configuration. Because of its primarily nonreplicating nature, tooth enamel can only scarcely self-repair. Thus artificial restoration is often necessary to repair damaged enamel for restoration of the tooth.

The layer underneath the enamel is the dentin, a porous biologic composite made up of apatite filler particles in a collagen matrix, which contains a tubular structure called dentinal tubules. Tubules are largest in diameter near the pulp (2.5 μm) and narrow at the dentin–enamel junction (0.8 μm). The composition

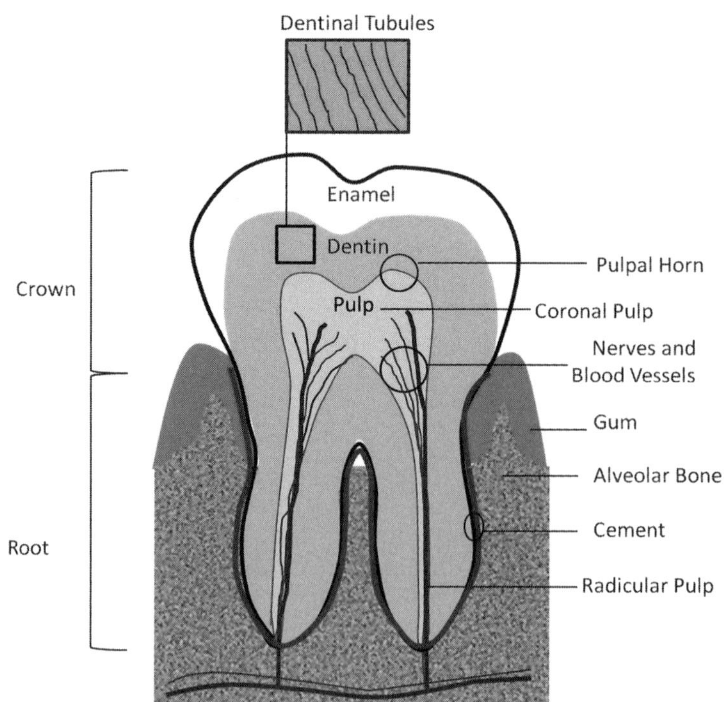

Figure 20.1. Schematic of the tooth structure (courtesy of Dr. Ana K. S. Braz).

of the dentin is mixed, with 60% being the inorganic block of apatite and the remaining 40% being organic biomaterials and water. The principal component of the organic matrix is the collagen type I. Beneath the dentin structure is the pulp, the soft tissue of the tooth. The pulp is made of a connective tissue that contains blood vessels and nerves that enter through the root. The dentin structure and the pulp together form an interactive network, often labeled as the dentin–pulp complex [Braz et al., 2009a]. A tooth is held in place by the gum, which provides a socket to hold and protect the lower tooth structure (cervical region). Decay or damage of the tooth enamel, excessive erosion of the gum, or bacterial infection may produce exposure of the dentin, pulp, or nerves in the root, leading to hypersensitivity, further decay, nerve damage, and even to the undesirable fate of tooth extraction.

Thus, dental care requires addressing care/treatment involving enamel, dentin, pulp, and the tooth/gum interface. In addition to restoration of the tooth enamel or treating tooth decay by caries, the final tooth structure must also have aesthetic appeal—that is, the appearance of close to that of a natural tooth. For determining the extent of decay and guiding the nature of restoration and precise filling of a cavity, dental imaging plays an important role. X-ray imaging is routinely utilized in dentistry. However, there is always a concern about the possibility of excessive exposure to the harmful X-ray radiation. Thus, there is a growing need for using noninvasive and safe imaging methods, such as those provided by optical imaging. Also, more futuristically, preventive care, based on risk care assessment and early detection of dental diseases, requires saliva-based tests that may identify specific biomarker expressions.

Nanotechnology provides solutions and approaches to all the issues of dental care discussed in the preceding paragraph. A multidisciplinary approach is required involving materials chemists, biologists, physicists, engineers, and practicing dentists to come up with innovative nanotechnology approaches for next-generation dental care. In this chapter, we will cover some important directions for application of nanotechnology for dental care. These selected areas of dentistry, where nanotechnology is expected to make major impacts, are listed in Figure 20.2. These areas are briefly outlined here and discussed in subsequent sections.

Nanoparticles for Preventive Dentistry. This is the area receiving a great deal of attention. The development of nanomaterials that mimic the enamel structure, bond to the enamel surface, exhibit long durability, and blend with the natural appearance of a tooth is being vigorously pursued [Saunders, 2009]. In the case of early caries lesions, nanomaterials are being developed with compositions and surface functionalities to adhere strongly to the enamel and restore the decay to provide protection. These developments are covered in Section 20.2.

Nanomaterials for Restorative Dentistry. Here the focus is on development of dental adhesive nanocomposites that adhere to dentin and have great

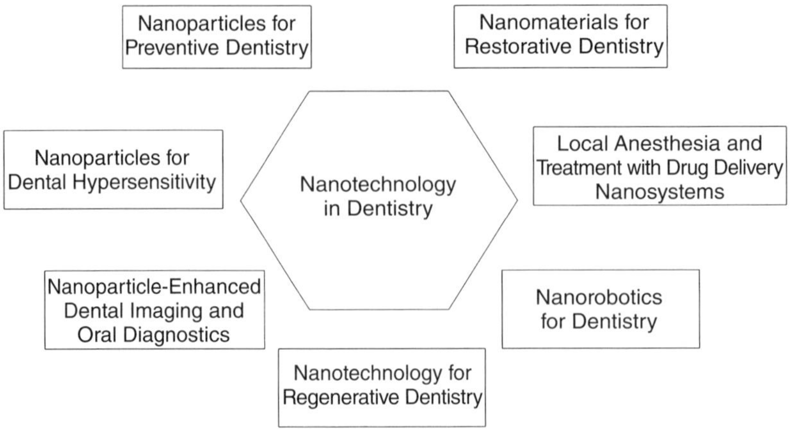

Figure 20.2. Major applications of nanotechnology in dentistry.

penetration into dentinal tubules. New resin-based composites for dental filling are being developed which exhibit small shrinkage upon photocuring, are wear-resistant, have strong long-term durability, and bond strongly to the adhesive layer. In addition, the adhesive layer and the nanocomposites should have antimicrobial properties to inhibit growth of any microbes leading to secondary caries, and thus overcome infections. These adhesives and nanocomposites contain nanofillers and nanoparticles and are discussed in Section 20.3.

Nanoparticle Enhanced Dental Imaging and Oral Diagnostics. Section 20.5 provides examples of the application of two optical imaging techniques, namely two-photon microscopy and optical coherence tomography, to dentistry. These two techniques have already been discussed in Chapter 8 ("Multimodal Biomedical Imaging"). They are noninvasive and do not utilize harmful radiation such as X ray. The emphasis of this section is to illustrate the utility of nanoparticles as emitters or contrast enhancers to enable dental imaging with improved resolution and/or sensitivity.

Also discussed in this section is the application of nanoparticles to identify patients with high risk for a dental disease such as caries or to detect a disease at a very early stage, using saliva. Saliva analysis can also provide an understanding of various oral diseases at molecular and cellular levels, which may ultimately lead to preventive care.

Nanomaterials for Dental Hypersensitivity. Nanotechnology can be employed to treat dental hypersensitivity. Dental hypersensitivity is pathological and is caused by pressure transmitted hydrodynamically to the pulp. This may be a result of dentinal tubules with surface densities significantly higher than those of nonsensitive tubules. The hypersensitivity is a result of dentinal tubules being exposed by the process of gingival recession and subsequent loss of cementum on root surfaces. Nanoparticles with tailored surfaces can be effectively used to provide appropriate protective coatings to cover the exposed dentin/pulp and hence reduce the hypersensitivity of the tooth. A more futuristic approach using nanorobots is presented in Section 20.6.

Nanoparticle Delivery of Anesthesia and Antibiotics. Using targeted nanoparticles that are external stimuli responsive to produce controlled and sustained release of anesthesia or antibiotics upon demand, more effective and highly localized effects can be produced. This will avoid the painful injection of local anesthesia which makes large areas of the mouth near the injection site numb. In the case of antibiotics, direct administration of topical nanoformulation of antibiotics at the infection site provides the benefit of producing high local concentration for a sustainable period to yield maximum benefit while minimizing the adverse effect of antibiotics. A futuristic approach is to use nanorobots, covered in Section 20.6, to deliver anesthetic materials.

Regenerative Dentistry. Regenerative dentistry employs tissue engineering to regenerate dental tissues for replacing diseased, missing, and traumatized dental structures. The strategy used for regenerative dentistry is the same as described in Chapter 18 ("Tissue engineering"). Regenerative dentistry differentiate itself from restorative dentistry in that it uses regeneration rather than restoration of a dental function.

Nanorobotics for Dentistry. There appears to be a great deal of enthusiasm among the dental research community to develop nanorobots that can navigate through various tissue structures to identify and localize at the biological site needing intervention. Then from an external command, the nanorobots can perform diagnostics, then provide controlled and sustained release of therapeutic agents. This concept in a sense is an extension of our nanoclinic concept introduced in Chapter 3 ("Nanomaterials"). Some perspectives on nanorobotics in relation to dentistry are presented in Section 20.6.

20.2. NANOPARTICLES FOR PREVENTIVE DENTISTRY

A major direction of modern dentistry is to provide for early prevention of decay of tooth enamel before a cavity forms needing to be filled [Hannig and Hannig, 2010]. This section describes the utilization of nanoparticles for preventive care, carries inhibition, and repair of tooth enamel. The next section then provides nanotechnology approaches for improving the treatment and filling of caries decay that lead to cavity formation.

Enamel is the exterior of a tooth and is exposed to the oral fluids. Therefore, it is subject to the hygienic condition of the mouth and susceptible to bacterial insults such as those producing caries lesions. It can also be eroded by dietary-related factors. To reiterate, the building blocks of enamel are fibrous apatite crystals, about 20 nm in diameter. Dental caries is a major cause of damage of dental enamel. Caries is produced by a number of streptococci and lactobacilli bacteria which produce excessive acids as by-products from the metabolism of fermentable carbohydrates (dietary sugar) that lead to demineralization of enamel, thus producing decay. The initial stage is the formation of an acquired pellicle on the enamel surface to which bacteria adhere. White spot lesions, appearing as a result of a noncontinuous demineralization process, are considered as a first evidence of tooth decay. In this section, we will describe the preventive methods using nanotechnology that deal with the control of the dental biofilms and the earlier stages of enamel caries lesions.

A preventive care can be provided by the application of an anti-adhesive surface coating that can control the formation of dental biofilms. Wear-resistant nanocomposite surface coatings have been developed, which involve nanosize inorganic particles in a fluoropolymer matrix [Hannig et al., 2007]. These biocompatible surface coatings facilitate the detachment of bacteria to produce preventive care.

Another approach used casein-phosphopeptide (CPP)-stabilized amorphous calcium phosphate (ACP) nanocomplexes to show *in vivo* reduction of bacterial adhesion [Reynolds, 2008]. In yet another approach tried *in vitro*, nonaggregated and clustered hydroxyapatite nanocrystalline particles of average size $100 \times 10 \times 5$ nm were shown to interact with bacterial adhesins and interfere with their binding to the tooth surface [Venegas et al., 2006].

The early stage of the decay, which is an area of demineralization restricted to the enamel, does not require filling. As the decay progresses, a cavity is formed which needs to be treated by filling the cavity. This stage is discussed in Section 20.3.

Application of bactericide and fluoride solution has been commonly used to aid remineralization of the lesion, if it is uncavitated and confined to enamel. Bactericides reduce the number of viable microorganisms, while fluoride inhibits demineralization and increases the remineralization efficacy, activating the deposition of minerals and causing redistribution of the mineral content within the lesion. A fluoride-rich acid-resistant mineral is formed on the surface of the white-spot lesion, reducing its porosity and rendering it more resistant to subsequent acid challenges [Lijima and Takagi, 2000]. However, this highly mineralized surface was found to be a serious obstacle to caries remineralization, restricting the penetration of fluoride, calcium, and phosphate into the body of the lesion [Larsen and Fejerskov, 1989]. Nanoparticles can have the same dimension as those of the hydroxyapatite and have the potential to infiltrate deeper into demineralized areas. By an optical coherence tomography characterization, which will be further described in Section 20.5, our group observed a high distribution of gold nanoparticles into a demineralized area surrounding a cavity (Figure 20.3). Thus, the use of nanoparticles with the capability to deliver fluoride can promote the deposition of minerals content not only at the surface but into the body of all demineralized white spot lesions.

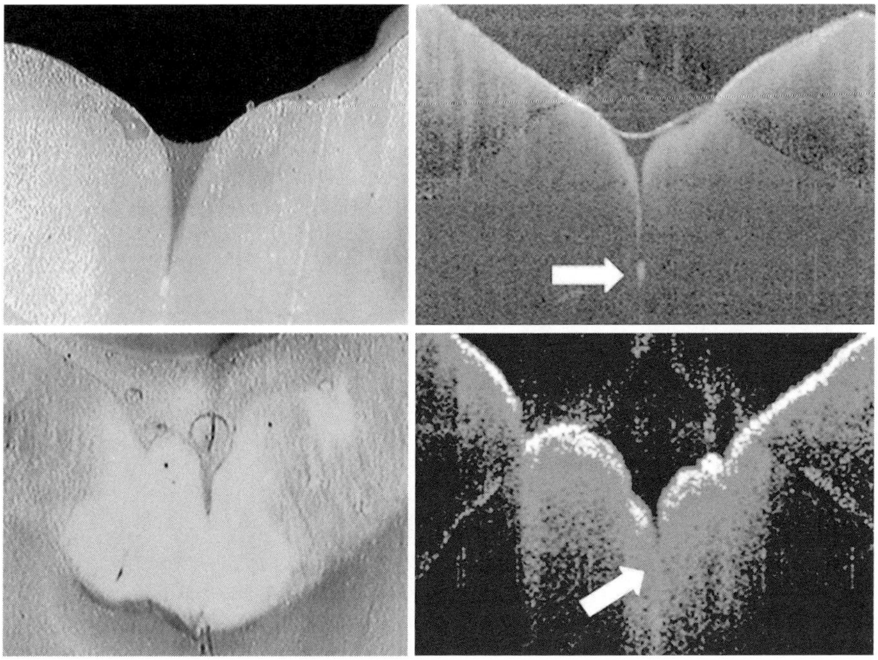

Figure 20.3. Optical microscopy of white spot enamel lesion (left). Noninvasive visualization by OCT (right). (Courtesy of Dr. Ana K. S. Braz.)

A number of approaches have been investigated *in vitro* for local repair of small decays in tooth enamel. Fan et al. [2009] used an amelogenin–hydroxyapatite composite with a full-length recombinant amelogenin (rP172) and showed that this nanocomposite in the presence of fluoride ions at a concentration of 1 mg/L produced restorative action. Amelogenin is the main protein involved in the morphogenesis of enamel. The key to this restoration approach is the use of nanosize (~20 nm) hydroxyapatite, which mimics the apatite structure in the natural enamel. Another approach utilized carbonate hydroxyapatite nanoparticles, clusters of which were incorporated into toothpaste [Lv et al., 2009] or mouth rinsing solutions. They are available as prophylactic products (BioRepair).

Another approach to repair tooth decay is tissue regeneration, the subject under regenerative dentistry covered in Section 20.4. *In situ* regeneration of enamel tissue can be used to repair the decay.

20.3. NANOMATERIALS FOR RESTORATIVE DENTISTRY

This section deals with a more advanced stage of decay where white spot lesions transform into a cavity that deepens with the progression of caries.

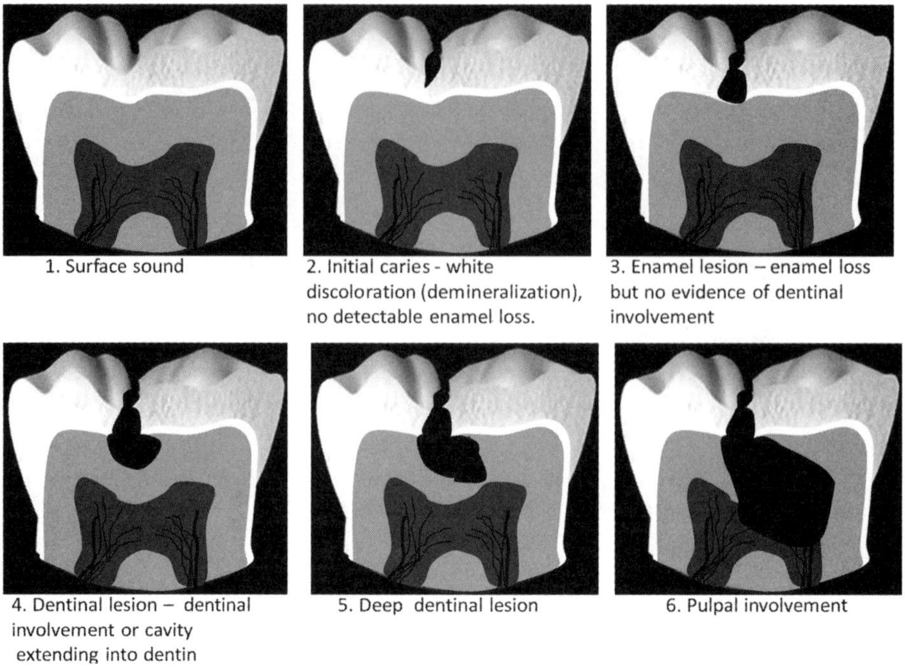

Figure 20.4. Different stages of progression of caries initiated tooth decay. (Courtesy of Dr. Ana K. S. Braz.)

Figures 20.4 and 20.5 illustrate the progression of caries and the nature of intervention it requires. Stages 1 and 2 require only prevention without the need of filling the decay, because no cavity pocket is yet formed. The nanotechnology-based approaches for stages 1 and 2 have already been discussed in section 20.2.

The present section discusses the approaches taken for filling the dental cavities formed by the progression of cavities and the implementation of nanotechnology to dental filling. Specifically stages 3, 4, and 5 are covered in this section.

Contemporary restorative techniques are based on the adhesive properties of tooth-colored resin-based materials. Filling of a cavity requires a number of considerations. Some of the major considerations are as follows:

- The filling must adhere to the walls of the cavity.
- The filling should be mechanically strong and wear-resistant.
- It should have small thermal expansion to be able to withstand temperature changes.
- It should have small polymerization shrinkage to avoid marginal leakage.
- In the case of a deeper cavity affecting the pulp areas, such as in case 5, a protective coating with antimicrobial properties may be desirable.

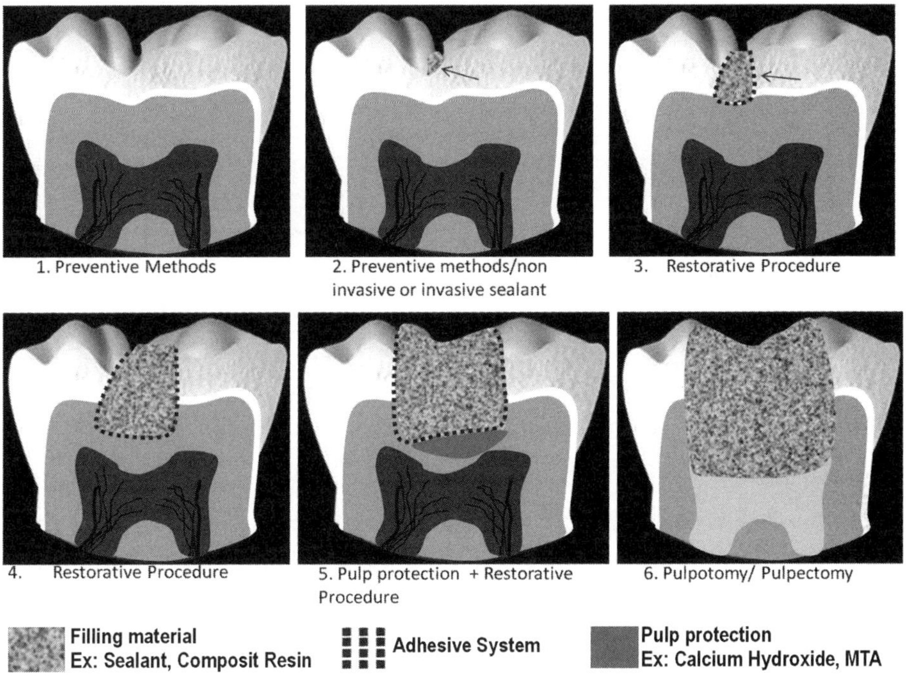

Figure 20.5. Schematic representation of the processes used for dental filling at various stages of caries-induced decay. (Courtesy of Dr. Ana K. S. Braz.)

In cases 3 and 4, a two-step process is used for filling. As depicted in Figure 20.5, an adhesive layer is first deposited to coat the cavity lining. Then the cavity is filled with a resin composite. Described below are the dental adhesives and the filling resin-based composites.

Dental Adhesives. The adhesive bonds on one side are to the enamel and the dentin, which are primarily inorganic (hydroxyapatite), and the bonds on the other side are to a resin composite, which is a primarily mineral-filled polymeric material that fills the cavity pocket. It therefore is a crucial layer, and its composition as well as processing to form within the enamel and dentin matrix are of prime significance.

The basic mechanism of adhesion involves the replacement of minerals (from hard tissue) by resin monomers (from adhesive system), which leads to micromechanical interlocking. Adhesion to enamel is more predictable, because this is a more homogeneous structure with high inorganic content. Adhesion to dentin is less predictable due to the higher variability and higher organic content of this structure. Therefore, efforts have been concentrated to improve the bonding properties to the dentin structure.

Preparation of tooth dentin with dental burs creates a smear layer of debris and bacteria. The smear layer is burnished into the underlying dentin and

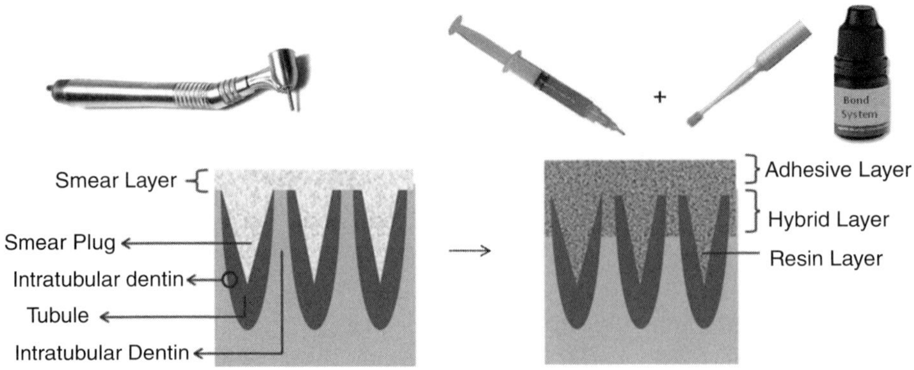

Figure 20.6. Adhesion process for dentin. (Courtesy of Dr. Ana K. S. Braz.)

dentin tubules. Dentin bonding removes or treats the smear layer, creating a hybrid layer of resin, collagen fibers, dentin surface structure, and intertubular structure. Over the hybrid layer forms an adhesive layer; and below the hybrid layer, into the tubules, forms the resin tags (Figure 20.6). An adhesive often involves a complex formulation containing many components. In contrast to the resin-based composite (RBC) filling materials, which are sufficiently described in detail in the literature, the compositions of adhesives often are kept as trade secrets. An excellent review on the various adhesive formulations and the processing used to produce adhesive layers is by Van Landuyt et al. [2007]. Generally, an adhesive contains a free-radical-initiated polymerizable monomer, an organic solvent, an initiator, and an inhibitor. In addition, recent trends have also been to introduce filler nanoparticles. The polymerizable monomers used are acrylates, methylacrylates, and methacrylamides (more resistant to water), which contain primarily hydrophobic polymerizable groups and can copolymerize with the filling RBC to provide good bonding with it. It also contains a hydrophilic functional group to provide good adhesion with the inorganic building block (hydroxyapatite) of the enamel and dentin. Also, tailored functional groups can be introduced to enhance wetting and demineralization of dentin for improved adherence as well as with the capability to release fluoride or impart antibacterial properties.

The curing processing involving polymerization is often initiated by a photoinitiator (e.g. camphorquinone, CQ) which, together with a co-initiator, can produce free radicals by absorption of blue light from a convenient and readily available blue LED. Often, a small amount of inhibitors that are free radical scavenging antioxidants is added to limit free radicals originating from prematurely reacted initiators; it has negligible effect on the actual polymerization of the monomer in the adhesive.

The adhesive layer, being in direct contact with the enamel surface and the dentin matrix, should play two important functions: (i) provide a strong interface between the tooth and the filling RBC and (ii) prevent the exposure of the enamel/dentin due to resin shrinkage, mechanical stress, or water leakage, so that secondary caries do not develop. Despite significant improvements of adhesive systems, the durability and stability of resin-bonded interfaces on dentin still remain questionable. The stability of the bonded interface relies on the creation of a compact and homogeneous hybrid layer (Figure 20.5). Among the different aging phenomena occurring at the dentin bonded interfaces, some are considered pivotal in degrading the hybrid layer: Insufficient resin impregnation of dentin, high permeability of the bonded interface, suboptimal polymerization, phase separation, and activation of endogenous collagenolytic enzymes are some of the recently reported factors that reduce the longevity of the bonded interface. In order to overcome these problems, recent studies indicated that resin impregnation techniques should be improved by using a hydrophobic last coating of the nonsolvated resin, extending curing time considered to reduce permeability and allowing a better polymerization of the adhesive film. Furthermore, protease inhibitors as additional primers should be used to increase the stability of the collagen fibrils within the hybrid layer inhibiting the intrinsic collagenolytic activity [Breschi et al., 2008].

In the case of stage 4 of caries progression, where the dentin matrix is affected, the dental adhesive should penetrate to maximum depth in all demineralized dentin to provide maximum bond strength, further protect the dentin–pulp complex, and even promote remineralization for regrowth of the dentin barrier. Nanoparticle fillers added to a dental adhesive can provide these abilities, in addition to enhancing mechanical strength, and increase the stability of the collagens fibrils [Haidekker et al., 2006]. Influence of gold nanoparticles on collagen fibril morphology has been quantified using transmission electron microscopy and image analysis [Haidekker et al., 2006; Castaneda et al., 2008]. Furthermore, as discussed in Section 20.5, the nanoparticle additives can also lend the ability to utilize advanced microscopic techniques for noninvasive evaluation of the adhesive layer, hybrid layer, and resin tags of the bonded dentin.

Nanoparticles, as opposed to micro-sized particles, produce the maximum benefit as an additive to the adhesive layer for the following reasons:

(i) The nanoparticles have a large surface-to-volume ratio, providing greater crosssurface interactions for improved bonding and also ease of surface functionalization to bring in multifunctionality.

(ii) Some of the functionalities, such as antibacterial activities, are significantly enhanced on nanosize scale. An example is the enhanced antibacterial activity of nanosize silver, which has already been discussed in Chapter 19.

(iii) These nanoparticles are of the same dimension as those of the hydroxyapatite and the interfibrillar spaces of the demineralized collagen network. Thus they can be expected to more readily infiltrate the demineralized collagen.

(iv) Surface functionalization, size, and surface charge control can prevent agglomeration of particles in the adhesive layer which often leads to a reduction in mechanical strength.

(v) The filler nanoparticles can also produce the ability to form a thicker adhesive layer that may provide a good relief of contraction stresses produced by the filling RBC material.

Colloidal silica nanoparticles of diameters <40 nm have been extensively utilized as nanofillers both in adhesive materials and in RBC composites (discussed below). They are suggested to enhance adhesion to the tooth [Manhart et al., 2000]. Often the surface of these nanofillers is silanized to allow them to form bonds with the resin matrix [Chen, 2010]. An example of silanization is coating the silica nanoparticles with 3-methacryloxypropyltrimethoxysilane (MPTS), which deters aggregation as well as promotes interfacial adhesion, because this coating allows the particle surface to copolymerize with the polymer matrix through covalent and hydrogen bonding.

Another variation of the nanosize filler is POSS (polyhedral oligomeric silsesquioxane) nanoparticles which have a unique –Si–O– cage structure that can covalently bond to surrounding functional groups (e.g. methacrylate groups). A new approach being investigated by our group is the application of ORMOSIL nanoparticles (see Chapters 3 and 4 for ORMOSIL) that contain both inorganic (–Si–O–) and organic groups, thus with an ability to bond both with the enamel and the filler resin. These ORMOSIL nanoparticles also provide the ability to bind with other functional units such as antibacterial nanoparticles, be conjugated with antibiotics, and contain units for sustained release of fluoride ions to promote remineralization. A recent emphasis has been on incorporating particles with antibacterial properties. Examples are silver nanoparticles [Kassaee et al., 2008] and quaternary ammonium polyethyleneimine (PEI) nanoparticles [Beyth et al., 2006].

Another direction for utilizing nanoparticles for advancing dental adhesive technology is to introduce responsiveness to external stimuli such as heat, light, magnetic field, or mechanical pressure [Jandt and Sigusch, 2009]. This external stimuli responsiveness can be utilized to release, on command, an antimicrobial compound or fluoride ions to fight infection or secondary caries formation. A direction pursued in our laboratory is to use near-IR light responsive nanoparticles that uncage upon excitation to release entrapped species. Another direction pursued by us is *in situ* generation of UV light, by action of external stimulus, to kill bacterial infection.

Another exciting direction is the development of self-heating materials. For example, nanoparticles filled with epoxy resins can be released under cracks and can react with an appropriate catalyst dispersed in the adhesive material to repair the cracks [White et al., 2001].

For case 5 in Figure 20.4, where the cavity is deep to affect pulp, a protective layer over the pulp is often used before coating the adhesive layer (Figure 20.5). This protective layer, in addition to protecting the pulp, also helps in avoiding the sensitivity to variation of temperature and pressure on the tooth. The protective layer usually consists of calcium hydroxide. If the pulp is not exposed, a calcium hydroxide paste can be used as the protective layer. If the pulp is exposed, calcium hydroxide powder or MTA is used. This protective layer can also include antibacterial nanoparticles to provide protection against any bacterial infection.

Resin-Based Composite Filling. This composite filling forms the bulk of the filling that is applied after the adhesive coating is placed. The main ingredient of the filling composite, again, is a mineral-filled polymerizable resin monomer that is usually photopolymerized using photoinitiators and co-initiators, as described above, to form a cross-linked polymer network that has strong mechanical strength to provide a highly desirable filling [Chen, 2010; Mitra et al., 2003]. In addition, inorganic nanofillers are used as additives to enhance the durability and promote other functionalities, similar to what has been discussed earlier in the context of an adhesive layer. Also, because this resin-based composite (RBC) dental filling appears at the surface of the tooth, its color has to match with that of the tooth to provide aesthetic appeal.

The most common monomeric materials used in dental composite filling is bis-GMA(2,2-bis[4-(2-hydroxy-3-methacryloxypropoxy)phenyl]-propane). Chen [2010] provides a list of other monomers. Since bis-GMA has a very high viscosity due to hydrogen bonding interactions involving the hydroxyl groups, it is often diluted with other monomers that are more fluid. A frequently used less viscous monomer, used as a diluent for bis-GMA-based composites, is TEG DMA (triethylene glycol dimethacrylate), which provides a fluid resin in which inorganic nanoparticle inclusions can be readily dispersed. However, TEG DMA can only be used as a diluent, because it produces more shrinkage with polymerization compared to bis-GMA. The common inorganic fillers in the RBC dental filling are colloidal silica and silica glass containing barium, strontium, and zirconium. They are used to increase the strength and modulus of elasticity, at the same time reducing polymerization shrinkage, the coefficient of thermal expansion, and water absorption. As discussed previously, the surfaces of these nanoparticles are silanized to chemically link them with the RBC matrix. For photopolymerization initiated by a blue LED (410–500 nm), the commonly used photoinitiator, again as described earlier, is camphorquinone (CQ) together with ethyl-4-(*N*,*N*-dimethylamino)benzoate (4EDMAB) as the co-initiator. CQ absorbs in the 450- to 500-nm range to produce free-radicals that initiate cross-linking.

Several nanocomposite products for dental filling are already in the market. One example is Filtek Supreme from 3M ESPE, USA, which contains 20- to 75-nm silica nanoparticles that are nonagglomerated and nanoclusters which are loosely bound agglomerates [Chen, 2010]. The smaller size (20 nm) nanoparticles are zirconia-silica nanoparticles. Again these nanoparticles are coated with MPTS as silanizing agent to provide coupling to the inorganic surface at one end and the methacrylate group of the resin on the other end.

Another example is Ceram-X from Dentsply DeTrey, Germany, which is an ORMOSIL-based nanocomposite, containing methacrylate modified silica nanoparticles of 20-nm diameter. Another nanoparticle used for improving the mechanical properties of RBC is TiO_2. The surfaces of TiO_2 nanoparticles of size <20 nm were coated with organosilane allyltriethoxysilane (ATES) and blended with the resin monomer. They were then added into 2100 dental RBC from 3M, ESPE. This formulation was found to improve the microhardness and flexure strength of the composites [Xia et al., 2008].

Another direction of development of dental RBC is the incorporation of caries-prevention fillers. For this purpose, nanoparticles of dicalcium phosphate anhydrous (DCPA) [Xu et al., 2007] and tetracalcium phosphate (TTCP) [Xu et al., 2009] have been developed which release calcium and phosphate ions. In one formulation, Xu et al. [2007] used DCPA nanoparticles with nano-sized silica fused whiskers which exhibited good reinforcement from the whisker to yield high strength, as well as Ca and PO_4 release. Similar results were found with TTCP nanoparticles and nanosilica-fused whiskers. Calcium fluoride nanoparticles have also been investigated as additives for inhibiting tooth demineralization and caries development as well as for strengthening neighboring enamel and dentin [Xu et al., 2008a]. An interesting combination is a nanocomposite containing both CaF_2 and DCPA, which can release F, Ca, and PO_4 ions.

Because of their exceptional mechanical properties, single-walled carbon nanotubes have also been studied as nanofillers [Zhang et al., 2008]. However, they impart a blackish color to the filling which does not have aesthetic appeal.

20.4. REGENERATIVE DENTISTRY

Regenerative dentistry is tissue engineering applied to replace diseased, missing, and traumatized dental tissue structures. Thus the concepts and strategies described in Chapter 18 ("tissue engineering") are applicable to regenerative dentistry.

Regenerative dentistry offers a clear advantage over restorative dentistry in that it uses regeneration rather than restoration of a dental function, using strategies involving individually or combined cells, growth factors, and scaffolds for tissue regeneration as detailed in Chapter 18. It is an exciting new area receiving considerable attention. This section provides a brief coverage of this topic. For a detailed exposition, the reader is referred to a good comprehensive coverage provided in the book *Regenerative Dentistry* edited by Professor Marei [Marei, 2010].

Tissue engineering has been applied for regeneration of the pulp-tissue complex and periodontal hard tissues of the tooth structure [Marei, 2010; Gu et al., 2011]. The pulp–dentin complex has already been described earlier in this chapter. Periodontal tissues, also labeled as periodontium, are the supporting structures of a tooth involving a complex of soft and hard tissues. The major components are: (1) the gum that covers the jaw bones, supporting the tooth structure, (2) the periodontal ligaments consisting of tough little connective tissue and elastic fibers that keep the tooth attached to the jaw, and (3) the alveolar bones consisting mainly of hydroxyapatite mineral-embedded protein (collagen) fiber, which is the jaw bone that keeps the tooth in place, feeding and protecting it.

The example of regenerative dentistry discussed here is regeneration of the pulp–dentin complex that constitutes the field of regenerative endodontics providing an alternative to root canal therapy. Regenerative endodontics enables removal and replacement of diseased/necrotic pulp tissues with regenerated pulp tissue to revitalize teeth [Sun et al., 2011; Tziafas, 2004; Nakashima and Akamine, 2005; Huang, 2009a]. Figure 20.7 illustrates the scheme of cell delivery-based strategy for dental pulp regeneration that involves harvesting tissue biopsies, expanding cell populations and subsequent transplantation back into the disinfected pulp space to regenerate lost pulp tissues. Incorporation of endogenous stem/progenitor cells into root canals to regenerate pulp tissues offers a major opportunity for dental pulp regeneration [Mao et al., 2010].

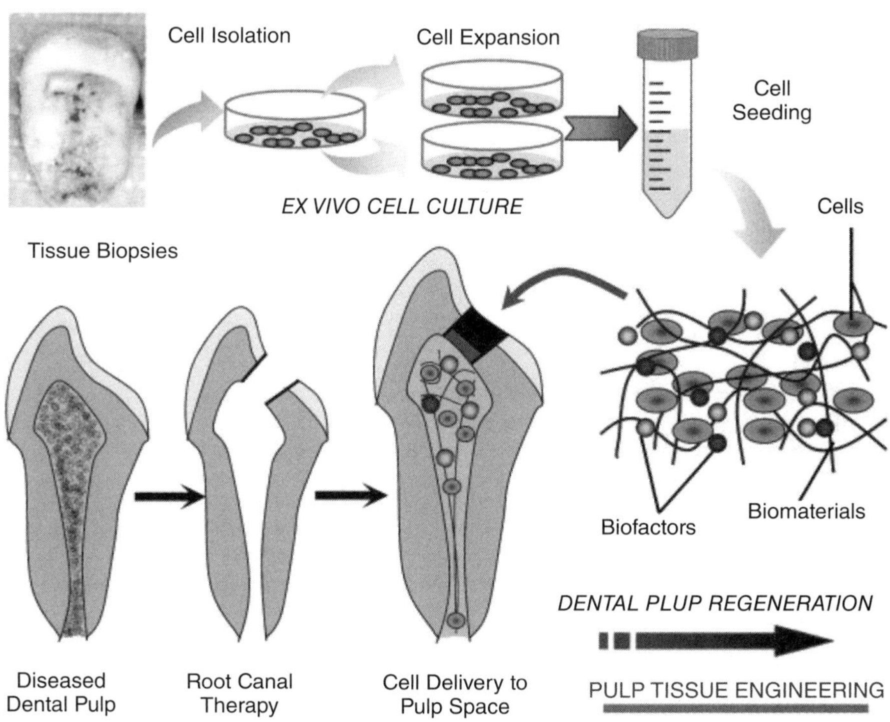

Figure 20.7. Scheme of cell delivery-based strategy for dental pulp regeneration (Reproduced with permission from Sun et al. [2011].)

Vital pulp therapy utilizes differentiation of a highly proliferative and clonogenic population of progenitor/stem cells that resides within the postnatal dental pulp, into hard tissue-forming cells that replace damaged odontoblasts by newly regenerated populations of odontoblast-like cells [Discher et al., 2009; Mao et al., 2010]. Cell homing is another approach which involves chemotaxis of the host's endogenous cells for *in situ* pulp and periodontal tissue regeneration [Sun et al., 2011]. Pulp tissue may be regenerated by an *in vivo* approach, which involves (a) the delivery, into root canals, of a combination of autologous or allogenic stem cells and appropriate scaffolds incorporated with specific growth factors or (b) pulp implantation (an *ex vivo* approach) involving the surgical implantation of engineered pulp tissue grown in the laboratory. Formation of pulp tissue was demonstrated *in vivo* by Cordeiro et al. [2008]. They seeded stem cells on to PLA scaffolds and inserted them into the pulp cavity of tooth slices after removing the original tissue and transplanted these constructs subcutaneously into immunodeficient mice. The dental stem cells differentiated into odontoblast-like cells.

Dental pulp regeneration requires scaffolds capable of slow release of relevant growth and differentiation factors used in combination with autologous dental pulp stem cells, or as cell-free systems, which should attract stem cells residing in the adjacent tissues and promote their proliferation and differentiation. Particularly challenging is the provision of sufficient vascularization over the length of the root canal which requires sprouting from vessels in the periapical region [Huang, 2009a, 2009b].

Tooth regeneration has utilized the various biodegradable polymer scaffolds that include poly-L-lactate-co-glycolate (PLGA) [Young et al., 2002], silk [Xu et al., 2008b], gelatin–chondroitinhyaluronan [Kuo et al., 2008], collagen gels or sponges [Honda et al., 2007; Ikeda et al., 2009], and self-assembled peptides [Kirkham et al., 2007]. Many of these nanostructured scaffolds have been discussed in Chapter 18.

Young et al. [2002] used PLGA, shaped into crowns of human incisors and molars, and coated with collagen type I gel. They seeded the total cell population of porcine third molar tooth buds into the PLGA and transplanted the constructs into the omentum of athymic rats. Another approach utilized a silk scaffold fabricated into porous sponges via salt porogen methods [Xu et al., 2008b]. Rat tooth bud cells, harvested and mixed with Matrigel, were seeded onto the porous silk scaffold, and subsequently transplanted into syngeneic Lewis rats, to form small mineralized particles distributed in the region of the scaffold.

Formation of small but well-organized enamel and dentin tissues were reported by Kuo et al., who used a gelatin–chondroitin–hyaluronan tri-copolymer scaffold and transplanted cultured porcine molar bud cells with the scaffold into the original alveolar sockets of donor swine [Kuo et al., 2008]. Another example is the use of beta-sheet-forming peptides, designed as a biomimetic scaffold for noncellular tissue engineering of enamel remineralization [Kirkham et al., 2007].

Honda et al. [2007] used a collagen sponge consisting of 75% (dry weight) type I atelocollagen and 25% type III atelocollagen from porcine skin, transplanted into porcine mandible with odontogenic mesenchymal and epithelial cells derived from porcine third molar buds. Enamel and dentin-like tissues formed in the transplanted groups where mesenchymal cells were plated within the collagen sponge and epithelial cells were plated on top.

Formation of dentin and pulp-like structures has been reported using ceramic scaffolds made from hydroxyapatite/tricalcium phosphate (HA/TCP) ceramic powder and with adult human dental pulp stem cells transplanted subcutaneously into immunocompromised mice [Gronthos et al., 2000]. Ikeda et al. [2009] reported the generation of whole functional teeth through the use of collagen gel, reconstituted *ex vivo* with highly concentrated mesenchymal and epithelial cells of mouse tooth bud. The reconstituted tooth germ was subsequently transplanted into an adult mouse socket from which a tooth had been extracted. These reports clearly show the promise of regenerative dentistry for tooth replacement therapy. A major impetus needed is availability of adult stem cells for the reconstitution of a bioengineered tooth germ and the regulation of stem cell differentiation into odontogenic cell lineage [Ikeda et al., 2009].

Regeneration and post-transplant healing of periodontal tissues can benefit from utilization of platelet-rich plasma (PRP), which is blood plasma enriched in platelets [Everts et al., 2006]. The PRP contains various growth factors, and cytokines which stimulate healing of bones and soft tissues, and mediate the immune response. These factors include but are not limited to platelet-derived growth factor, fibroblast growth factor, connective tissue growth factor, TGF-β, vascular endothelial growth factor, and interleukin-8. The PRP can be obtained from the same patient that undergoes the dental transplantation. The blood is usually drawn from the vein of a donor or patient in the presence of anticoagulants and enriched in platelets by differential centrifugation. The PRP is then instilled into the transplantation site, which makes healing more rapid.

20.5. NANOPARTICLE-ENHANCED DENTAL IMAGING AND ORAL DIAGNOSTICS

This section describes application of nanotechnology to dental imaging. The two optical imaging techniques which hold considerable promise for dentistry are optical coherence tomography (OCT) and multiphoton microscopy. Both these techniques have already been described in Chapter 8 on Bioimaging.

Braz, Gomes, and co-workers [Braz et al., 2009a] have demonstrated the application of optical coherence tomography (OCT) in imaging of the pulp–dentin complex as well as for evaluation of dental materials. OCT is a highly suitable method for imaging through strongly scattering media such as teeth. Braz et al. [2009a] showed that OCT could noninvasively generate images of the boundaries of the pulp and its relation to dentin. Thus OCT can be used to prevent iatrogenic exposures of the pulp and can be useful in predicting prognosis of dental treatment such as pulp capping therapy. Braz et al. have also successfully demonstrated the application of OCT to characterize slow crack propagation inside composite resin [Braz et al., 2009b] and to assure the integrity of dental sealant [Braz et al., 2011]. OCT provided a clear contrast between the sealant and the enamel, enabling identification of internal structures of both. They were able to detect failures at the interior as well as at the surface. Thus OCT can provide detailed information on features of fissures, bubbles, and failures to determine their retention in short and long terms.

Our group has collaborated with Braz and Gomes in introducing gold nanoparticles as contrast enhancers for dental OCT imaging. We introduced gold nanoparticles in commercial dental bonding adhesives by generating them *in situ* using photothermal reduction of a gold precursor. Then by OCT we monitored the presence of gold nanoparticles inside the hybrid layer and dentinal tubules. Future work will be performed to functionalize the structure of these nanoparticles to cross-link with dentinal collagen. The cross-linking is a way to improve the mechanical stability and slow down the biodegradation rate of collagen, which is considered one of the pivotal problems that affect the stability of the bonded interfaces as reported before [Breschi et al., 2008]. Figure 20.8 shows the OCT images, the bright spots indicating the presence of gold nanoparticles in the dentinal tubules.

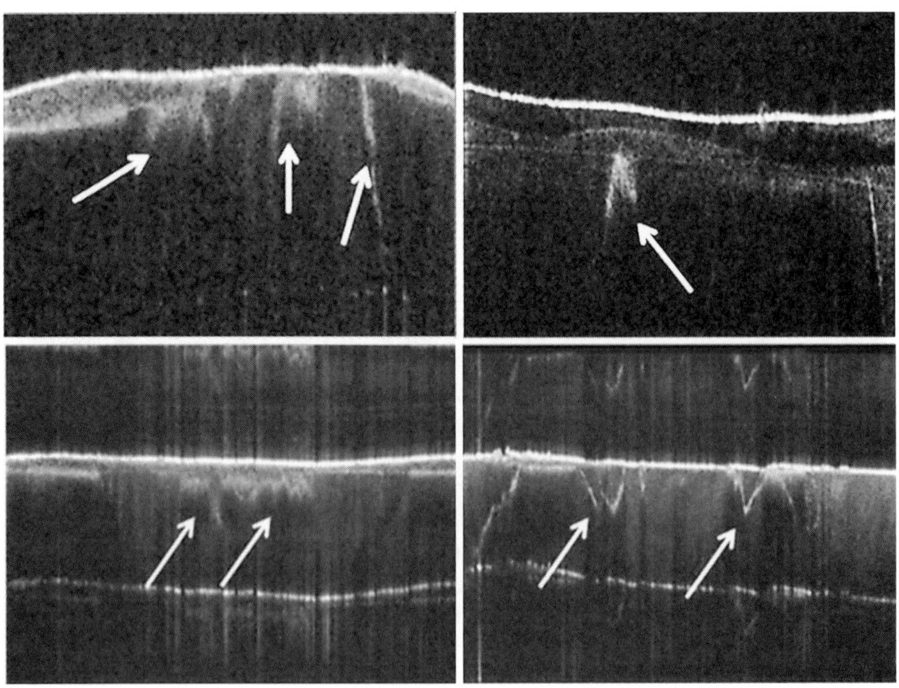

Figure 20.8. OCT images of gold nanoparticles in the dentinal tubules. (Courtesy of Dr. Ana K. S. Braz.)

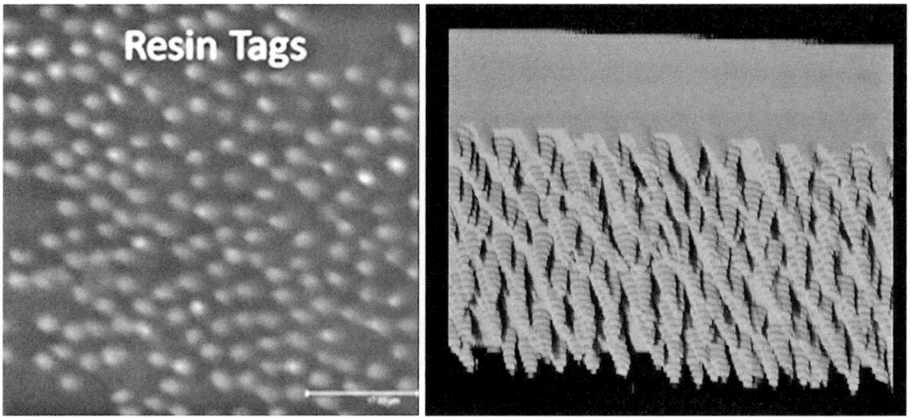

Figure 20.9. Two photon microscopic images showing penetration of dye-doped resin in dentinal tubules. Left panel: Plane (XY) image; Right panel: 3D reconstruction.

We have also collaborated with Braz and Gomes in application of two-photon microscopy, discussed in Chapter 8, for noninvasive evaluation of dental adhesives [Braz et al., 2010]. In this case, a two-photon absorbing fluorescent dye was doped in a commercial dental adhesive that was applied to dentin and photo-cured. Two-photon microscopic images (Figure 20.9) clearly reveal the dentinal tubule structures because of the penetration of the dye-doped adhesive. Two-photon microscopy permitted the noninvasive evaluation of the three structurally distinct zones of the dentin adhesion: adhesive layer, hybrid layer, and resin tags as well as also observed failures of adhesion in some samples as the presence of blisters and no micromechanical interlocking.

For next-generation, multiphoton microscopy, we plan to utilize an upconversion process in rare-earth-doped fluoride nanoparticles, which have also been discussed in Chapter 5 and again in Chapter 8 ("Multimodal Biomedical Imaging"). The appealing features offered by these nanoparticle upconverters are twofold:

(i) As the upconverted emission is a two-step process (Chapter 5), low-cost CW laser source emitting at 980 nm can be utilized, which makes it possible to use such an imaging system even in a dentist's office.

(ii) By utilizing the 980 nm to 800 nm upconversion [Chen et al., 2011], we can use maximum transmission both at the excitation (980 nm) and the emission (800 nm) wavelengths to enhance signal-to-noise ratio at obtained high contrast 3D images.

For sensing applied to dentistry, a saliva test using nanosensing can detect biomarkers present in the saliva who are early indicators of a gum disease. It may also be possible to determine risk factors by detecting patients who are predisposed to caries [Giannobile et al., 2011; Vieira et al., 2011]. The saliva test can be applied *in vitro* utilizing any of the methods already discussed in the Chapter 9 ("Biosensing").

20.6. NANOROBOTICS FOR DENTISTRY

There appears to be a considerable degree of excitement among the dental research community for the futuristic approach of using nanorobots for dental care [Gupta, 2011; Sahoo et al., 2007]. A strong appeal is that these nanorobots can use specific mobility mechanisms to crawl or swim through human tissue with great navigational precision [Patil et al., 2008]. This can use various sensor mechanisms to assess the oral hygiene, diagnose a disease, decay or bacterial infection and then treat appropriately on external command from a dentist.

An example is the use of nanorobots to get through the crown and mucosa and reach the dentin by migrating into the gingival sulcus, where they can enter the dentinal tubules and proceed to the pulp. Their navigation guidance can be provided externally by applying an external stimulus such as an electric, rf, or magnetic field controlled by the dentist. These nanorobots can provide a more effective treatment for hypersensitivity. A major clinical cause of hypersensitivity is the exposure of dentinal tubules because of gingival recession and subsequent loss of cementum on root surfaces. Nanorobots can selectively occlude at specific tubules within minutes and with great precision to offer a patient a quick and permanent cure [Freitas, 2000].

Another example is local delivery of anesthesia. The current practice of local delivery by injection is painful and creates a prolonged oral discomfort. The use of nanorobots to navigate through different dentinal tubular branching pattern to reach the pulp chamber and establish control over the nerve impulse traffic, then will allow a dentist to command release of anesthesia. Consequently, all sensitivity in a specific tooth needing treatment can be shut down on a highly selective basis.

As has been discussed at various places throughout this book, active and field directed targeting can be built on a nanoparticle platform to specifically localize them at a specific biological site. Magnetophoretic control using a magnet to guide a magnetic nanoparticle through a complex tissue structure can be effected. Discussed earlier in the chapter also is our approach to produce nanoparticles which upon external activation can release trapped chemicals. The concept of self-repair using a nanoparticle platform has also been introduced in Section 20.5. Thus the concept of using nanorobots with all these functions incorporated is not far from being realized.

HIGHLIGHTS OF THE CHAPTER

- The exterior coating of a tooth is the dental enamel, a mainly inorganic, highly mineralized structure that provides remarkable hardness and resistance to the tooth.

- The layer underneath the enamel is the dentin, a porous biologic composite made up of apatite filler particles in a collagen matrix that contains a tubular structure called dentinal tubules. The composition of dentin is mixed, containing inorganic block of apatite, organic biomaterials (collagen type I) and water.
- Beneath the dentin structure is the pulp, which is made of connective tissues that contain blood vessels and nerves that enter through the root.
- Dental care requires addressing care/treatment involving enamel, dentin, pulp, and the tooth–gum interface.
- Nanotechnology provides solutions and approaches to the issues of dental care which include using (i) *nanoparticles* for preventive dentistry, for enhanced dental imaging and oral diagnostics and for delivery of anesthesia and antibiotics, (ii) *nanomaterials* for restorative dentistry and for dental hypersensitivity, (iii) nanotechnology for regenerative dentistry involving tissue engineering to regenerate replacement for damaged tooth structures, and (iv) *nanorobots* for diagnostics and for controlled and sustained release of therapeutic agents.
- The preventive methods to control of the dental biofilms and the earlier stages of enamel caries lesions involve (i) wear-resistant nanocomposite surface coatings containing nanosize inorganic particles in a fluoropolymer matrix or casein-phosphopeptide (CPP)-stabilized amorphous calcium phosphate (ACP) nanocomplexes and (ii) local repair of small decays in tooth enamel using nanosize hydroxyapatite, which mimics the apatite structure in the natural enamel.
- In *restorative dentistry*, nanoparticles, as opposed to micro-sized nanoparticles, produce the maximum following benefits as an additive to the adhesive layer and filler: (i) a large surface-to-volume ratio, thus greater cross-surface interactions for improved bonding and also ease of surface functionalization to bring in multifunctionality; (ii) enhanced antibacterial activities; (iii) the same dimension as those of the hydroxyapatite and the interfibrillar spaces of the demineralized collagen network, thus readily infiltrating the demineralized collagen; and (iv) surface functionalization, size, and surface charge control to prevent agglomeration of particles in the adhesive layer, leading to a reduction in mechanical strength.
- The filler nanoparticles can also produce the ability to form a thicker adhesive layer that may provide a good relief of contraction stresses produced by the filling RBC material.
- Dental resin-based composite filling can also incorporate caries-prevention fillers. Nanoparticles of dicalcium phosphate anhydrous (DCPA) and tetracalcium phosphate (TTCP) have been developed which release calcium and phosphate ions, known to effect carries prevention.
- Optical coherence tomography (OCT) have demonstrated the application for imaging of the pulp–dentin complex as well as for evaluation of

dental materials. Gold nanoparticles can be introduced as contrast enhancers for dental OCT imaging.

- Two-photon microscopic imaging can be used for non-invasive evaluation of dental adhesives. It clearly revealed the dentinal tubule structures because of the penetration of the dye-doped adhesives.
- Next-generation multiphoton microscopy can utilize multiphoton upconversion in rare-earth-doped fluoride nanoparticles, which offers the following merits: (i) Because the upconverted emission is a two-step one-photon absorption, low-cost CW laser source emitting at 980 nm can be utilized, making it possible to apply it even in a dentist's office, and (ii) by utilizing the 980 nm to 800 nm upconversion, one can achieve maximum transmission both at the excitation (980 nm) and the emission (800 nm) wavelengths obtained high contrast 3D images.
- Regenerative dentistry is tissue engineering applied to replace diseased, missing, and traumatized dental tissue structures. It has been applied for regeneration of the pulp–tissue complex as well as of the periodontal tissues in the tooth structure.
- Nanorobots possessing the specific mobility mechanisms to crawl or swim through human tissue with great navigational precision can be used for diagnostics and for controlled and sustained release of therapeutic agents in specific sites of dental structure.

EXERCISES

True or False

1. The principal component of the dentin organic matrix is pulp.

2. Eighty-six percent of the enamel composition is inorganic structure.

3. Because of its primarily nonreplicating nature, tooth enamel can only scarcely self-repair.

4. Dental care requires addressing care/treatment involving only enamel, dentin, and pulp.

5. Two-photon microscopy and optical coherence tomography are noninvasive, do not utilize harmful radiation, such as X-ray, and can be used in dentistry.

Multiple choice

6. A preventive dentistry care can be provided by:

(a) Wear-resistant nanocomposite surface coatings to facilitate the detachment of bacteria

(b) X-ray imaging
(c) OCT imaging.
(d) Two-photon microscopy

7. Dental adhesives and fillers, which incorporates nanomaterials, possess the following features:
(a) Color close to natural
(b) Better infiltration and multifunctionality
(c) Antibiotics delivery ability
(d) Enhanced contrast for OCT imaging

Answers: 1. F; 2. T; 3. T; 4. F; 5. T; 6. a; 7. b, d

REFERENCES

Beyth, N., Yudovin-Farber, I., Bahir, R., Domb, A. J., Weiss, E. I., Antibacterial activity of dental composites containing quaternary ammonium polyethylenimine nanoparticles against *Streptococcus mutans*, *Biomaterials* **21**, 3995–4002 (2006).

Braz, A. K. S., Kyotoku, B. B. C., Gomes, A. S. L., *In vitro* tomographic image of human pulp-dentin complex: Optical coherence tomography and histology, *J. Endod.* **35**, 1218–1221 (2009a).

Braz, A. K. S., Kyotoku, B. B. C., Braz, R., Gomes, A. S. L., Evaluation of crack propagation in dental composites by optical coherence tomography, *Dent. Mater.* **25**, 74–79 (2009b).

Braz, A. K. S., Gomes, A. S. L., Ohulchanskyy, T. Y., Prasad, P. N., Failures of adhesion: Noninvasive and three-dimensional images using two-photon microscopy, *Dent. Mater* **26S**, e3 (2010).

Braz, A. K. S., Aguiara, C. M., Gomes, A. S. L. Evaluation of the integrity of dental sealants by optical coherence tomography, *Dent. Mater.* **27**, e60–e64 (2011).

Breschi, L., Mazzoni, A., Ruggeri, A., Cadenaro, M., Di Lenarda, R., De Stefano Dorigo, E., Dental adhesion review: Aging and stability of the bonded interface, *Dent. Mater.* **24**, 90–101 (2008).

Castaneda, L., Valle, J., Yang, N., Pluskat, S., Slowinska, K., Collagen cross-linking with Au nanoparticles, *Biomacromolecules* **9**, 3383–3388 (2008).

Chen, M.-H., Update on dental nanocomposites, *J. Dent. Res.* **89**, 549–560 (2010).

Chen G., Ohulchanskyy T. Y., Law W. C., Ågren H., Prasad P. N., Monodisperse $NaYbF_4$: Tm^{3+}/$NaGdF_4$ core/shell nanocrystals with near-infrared to near-infrared upconversion photoluminescence and magnetic resonance properties, *Nanoscale* **3**, 2003–2008 (2011).

Cordeiro, M. M., Dong, Z., Kaneko, T., Zhang, Z., Miyazawa, M., Shi, S., Smith, A. J., Nör, J. E., Dental pulp tissue engineering with stem cells from exfoliated deciduous teeth, *J. Endod.* **34**, 962–969 (2008).

Discher, D. E., Mooney,.DJ., Zandstra, P. W., Growth factors, matrices, and forces combine and control stem cells, *Science* **324**, 1673–1677 (2009).

Everts, P. A., Knape, J. T., Weibrich, G., Schonberger, J. P., Hoffmann, J., Overdevest, E. P., Box, H. A., van Zundert, A., Platelet-rich plasma and platelet gel: A review, *J. Extra Corpor. Technol.* **38**, 174–187 (2006).

Fan, Y., Sun, Z., Moradian-Oldak, J., Controlled remineralization of enamel in the presence of amelogenin and fluoride, *Biomaterials* **30**, 478–483 (2009).

Freitas, R. A. Jr., Nanodentistry, *J. Am. Dent. Assoc.* **131**, 1559–1566 (2000).

Giannobile, W. V,, Wong, D. T., Salivary diagnostics: Oral health and beyond! *J Dent. Res.* **90**, 1153–1154 (2011).

Gronthos, S., Mankani, M., Brahim, J., Robey, P. G., Shi, S., Postnatal human dental pulp stem cells (DPSCs) *in vitro* and *in vivo*, *Proc. Natl. Acad. Sci. USA* **97**, 13625–13630 (2000).

Gu, L., Kim, Y. K., Liu, Y., Ryou, H., Wimmer, C. E., Dai, L., Arola, D. D., Looney, S. W., Pashley, D. H., Tay, F. R., Biomimetic analogs for collagen biomineralization, *J. Dent. Res.* **90**, 82–87 (2011).

Gupta, J. Nanotechnology applications in medicine and dentistry, *J. Invest. Clin. Dent.* **2**, 81–88 (2011).

Haidekker, M. A., Boettcher, L. W., Suter, J. D., Rone, R., Grant, S. A., Influence of gold nanoparticles on collagen fibril morphology quantified using transmission electron microscopy and image analysis, *BMC Med. Imag.* **6**, 4 (2006).

Hannig, M., Hannig, C., Nanomaterials in preventive dentistry, *Nat. Nanotechnol.* **5**, 565–569 (2010).

Hannig, M., Kriener, L., Hoth-Hannig, W., Becker-Willinger, C., Schmidt, H., Influence of nanocomposite surface coating on biofilm formation *in situ*, *J. Nanosci. Nanotechnol.* **7**, 4642–4648 (2007).

Honda, M. J., Tsuchiya, S., Sumita, Y., Sagara, H., Ueda, The sequential seeding of epithelial and mesenchymal cells for tissue-engineered tooth regeneration, *Biomaterials* **28**, 680–689 (2007).

Huang, G. T., Apexification: The beginning of its end, *Int. Endod. J.* **42**, 855–866 (2009a).

Huang, G. T., Pulp and dentine tissue engineering and regeneration: Current progress, *Regen. Med.* **4**, 697–707 (2009b).

Ikeda, E., Morita, R., Nakao, K., Ishida, K., Nakamura, T., Takano-Yamamoto, T., Ogawa, M., Mizuno, M., Kasugai, S., Tsujia, T., Fully functional bioengineered tooth replacement as an organ replacement therapy, *Proc. Natl. Acad. Sci. USA.* **106**, 13475–13480 (2009).

Jandt, K. D., Sigusch, B. W. Future perspectives of resin-based dental materials, *Dent. Mater.* **25**, 1001–1006 (2009).

Kassaee, M. Z., Akhavan, A., Sheikh, N., Sodagar, A., Antibacterial effects of a new dental acrylic resin containing silver nanoparticles, *J. Appl. Pol. Sci.* **110**, 1699–1703 (2008).

Kirkham, J., Firth, A., Vernals, D., Boden, N., Robinson, C., Shore, R. C., Brookes, S. J., Aggeli, A., Self-assembling peptide scaffolds promote enamel remineralization, *J. Dent. Res.* **86**, 426–430 (2007).

Kuo, T. F., Huang, A. T., Chang, H. H., Lin, F. H., Chen, S. T., Chen, R. S., Chou, C. H., Lin, H. C., Chiang, H., Chen, M. H., Regeneration of dentin–pulp complex with cementum and periodontal ligament formation using dental bud cells in gelatin–

chondroitin–hyaluronan tricopolymer scaffold in swine, *J. Biomed. Mater. Res. A* **86**, 1062–1068 (2008).

Larsen, M. J., Fejerskov, O., Chemical and structural challenges in remineralization of dental enamel lesions, *Scand. J. Dent. Res.* **97**, 285–296 (1989).

Lijima, Y., Takagi, O., *In situ* acid resistance of *in vivo* formed white spot lesions, *Caries Res.* **34**, 388–394 (2000).

Lv, K., Zhang, J., Meng, X., Li, X. F., Remineralization effect of the nano-HA toothpaste on artificial caries, *Key Eng. Mater.* **330–332**, 267–270 (2009).

Manhart, J., Kunzelmann, K. H., Chen, H. Y., Hickel, R., Mechanical properties and wear behavior of light-cured packable composite resins, *Dent. Mater.* **16**, 33–40 (2000).

Marei, M. K. (ed.), *Regenerative Dentistry*, Morgan & Claypool Publishers, San Rafael, CA (2010).

Mao, J. J., Stosich, M. S., Moioli, E. K., Lee, C. H., Fu, S. Y., Bastian, B., Eisig, S. B., Zemnick, C., Ascherman, J., Wu, J., Rohde, C., Ahn, J., Facial reconstruction by biosurgery: Cell transplantation versus cell homing, *Tissue Eng. B Rev.* **16**, 257–262 (2010).

Mitra, S. B., Wu, D., Holmes, B. N., An application of nanotechnology in advanced dental materials, *J. Am. Dental Assoc.* **134**, 1382–1390 (2003).

Nakashima, M., Akamine, A., The application of tissue engineering to regeneration of pulp and dentine in endodontics, *J. Endod.* **31**, 711–718 (2005).

Patil, M., Mehta, D. S., Guwa, S., Future impact of nanotechnology on medicine and dentistry, *J. Indian Soc. Periodontol.* **12**, 34–40 (2008).

Reynolds, E. C., Calcium phosphate-based remineralization systems: Scientific evidence? *Aust. Dent. J.* **53**, 268–273 (2008).

Sahoo, S. K., Parveen, S., Panda, J. J. The present and future of nanotechnology in human health care, *Nanomedi. Nanotechnol. Biol. Medi.* **3**, 20–31 (2007).

Saunders, S. A., Current practicality of nanotechnology in dentistry, part 1: Focus on nanocomposite restoratives and biomimetics, *Clint. Cosmetic Invest. Dent.* **1**, 47–61 (2009).

Subramani, K., Ahmed, W., *Emerging Nanotechnologies in Dentistry Processes, Materials and Applications*, William Andrew, Oxford, UK (2012).

Sun, H. H., Jin, T., Yu, Q., Chen, F. M., Biological approaches toward dental pulp regeneration by tissue engineering, *J. Tissue Eng. Regen. Med.* **5**, e1–e16 (2011).

Tziafas, D., The future role of a molecular approach to pulp–dentinal regeneration, *Caries Res.* **38**, 314–320 (2004).

Van Landuyt, K. L., Snauwaert, J., De Munck, J., Peumans, M., Yoshida, Y., Poitevin, A., Coutinho, E., Suzuki, K., Lambrechts, P., Van Meerbeek, B., Systematic review of the chemical composition of contemporary dental adhesives, *Biomaterials* **28**, 3757–3785 (2007).

Venegas, S. C., Palacios, J. M., Apella, M. C., Morando, P. J., Blesa, M. A., Calcium modulates interactions between bacteria and hydroxyapatite, *J. Dent. Res.* **85**, 1124–1128 (2006).

Vieira, A. R., Deeley, K. B., Callahan, N. F., Noel, J. B., Anjomshoaa, I., Carricato, W. M., Schulhof, L. P., Desensi, R. S., Gandhi, P., Resick, J. M., Brandon, C. A., Rozhon, C., Patir, A., Yildirim, M., Poletta, F. A., Mereb, J. C., Letra, A., Menezes,

R., Wendell, S., Lopez-Camelo, J. S., Castilla, E. E.,, Orioli, I. M., Seymen, F., Weyant, R. J., Crout, R., McNeil, D. W., Modesto, A., Marazita, M. L., Detection of *Streptococcus mutans* genomic DNA in human DNA samples extracted from saliva and blood, *ISRN Dent.* **2011**, 543–561 (2011).

White, S. R., Sottos, N. R., Geubelle, P. H., Moore, J. S., Kessler, M. R., Sriram, S. R., Brown, E. R., Viswanathan, S, Autonomic healing of polymer composites, *Nature* **409**, 794–797 (2001).

Xia, Y., Zhang, F., Xie, H., Gu, N., Nanoparticle-reinforced resin-based dental nanocomposites, *J. Dent.* **36**, 450–455 (2008).

Xu, H. H., Weir, M. D., Sun, L., Takagi, S., Chow, L. C., Effect of calcium phosphate nanoparticles on Ca–PO_4 composite, *J. Dent. Res.* **86**, 378–383 (2007).

Xu, H. H., Moreau, J. L., Sun, L., Chow, L. C., Strength and fluoride release characteristics of a calcium fluoride based dental nanocomposite, *Biomaterials* **29**, 4261–4267 (2008a).

Xu, H. H., Weir, M. D., Sun, L., Calcium and phosphate ion releasing composite: Effect of pH release and mechanical properties, *Dent. Mater.* **25**, 535–542 (2009).

Xu, W. P., Zhang, W., Asrican, R., Kim, H. J., Kaplan, D. L., Yelick, P. C.. Accurately shaped tooth bud cell-derived mineralized tissue formation on silk scaffolds, *Tissue Eng. Part A* **14**, 549–57 (2008b).

Young, C. S., Terada, S., Vacanti J. P., Honda, M., Bartlett, J. D., Yelick, P. C., Tissue engineering of complex tooth structures on biodegradable polymer scaffolds, *J. Dent. Res.* **81**, 695–700 (2002).

Zhang, F. M., Xia, Y., Xu, L., Gu, N., Surface modification and microstructure of single-walled carbon nanotubes for dental resin-based composites, *J. Biomed. Mater. Res. B* **86B**, 90–97 (2008).

CHAPTER 21

NANOTOXICITY

Toxicity of nanomaterials is a very important issue to address in order for nanomedicine to have a major impact on our healthcare. Recently, a great deal of concern has been raised about the possible healthcare risk of nanomaterials, both by public press and by practicing physicians. Thus a critical assessment of risk/benefit of use of engineered nanoparticles for theranostics is extremely vital for the advancement of nanomedicine. This chapter focuses on nanotoxicity, bringing in focus the complexity of this subject and various methods of assessments of nanotoxicity, providing some examples of nanotoxicity studies. Section 21.1 presents an introductory discussion of nanotoxicity, along with the relevant issues and challenges to address. Section 21.2 describes the various mechanisms by which a specific nanoparticle can interact with a cell to produce cytotoxicity. This section is followed by Section 21.3, which lists *in vitro* assays to assess cytotoxicity. Section 21.4 discusses *in vivo* toxicity, together with the methods of studying them. Section 21.5 discusses some principal methods of evaluating overall toxicity *in vivo*. Section 21.6 provides selected examples of toxicity studies on some important classes of nanoparticles.

21.1. TOXICITY OF NANOPARTICLES

Toxicity of nanoparticles is a materials property that is dependent on the physicochemical nature of the particles. It determines how the nanoparticle

Introduction to Nanomedicine and Nanobioengineering, First Edition. Paras N. Prasad.

will react with its biological environment. Thus, nanotoxicology studies how the nanoparticles interact in the body to produce undesirable toxic effects. In the context of this monograph, the relevancy is to the toxicity of the nanoparticles used in nanomedicine.

The nanotoxicology deals with the determination of risk factors of the biomedical applications of nanomaterials. However, there is also considerable interest in study of the risk factors derived from unintentional environmental exposure to nanoparticles. This unintentional exposure could result from handling of nanoparticles or products containing nanoparticles, as well as from inhalation of airborne nanoparticles. Mechanistically, the adverse effects produced by both intentional and unintentional exposures are of same types, as discussed in the following sections.

Considering the importance of nanotoxicity, a major effort is being devoted worldwide to this subject. Consequently, a great deal of literature exists on this subject; it is not possible to enlist all of them here, considering the introductory nature of this monograph. Some comprehensive reviews on this subject are by Lewinski et al. [2008]; Oberdörster et al. [2005]; Aillon et al. [2009]; Dhawan and Sharma [2010]; Wani et al. [2011].

The toxicity of nanoparticles can be derived from the following two sources.

Nanoparticle Induced Toxicity. Here the physical and surface characteristics of the nanoparticle play the major role in producing toxicity. These characteristics are size, shape, surface charge, and surface coverage (coatings, chemically conjugated molecules). Because the surface-to-volume ratio in nanoparticles is very large compared to the corresponding bulk or even microscopic form of the same material, surface initiated processes are significantly enhanced. Thus, the toxicity studies performed in larger-size particles cannot be utilized to assess the toxicity of nanoparticles. Thus, size and shape have a clear role in determining the toxicity. Similarly, the surface charge on the nanoparticles has a pronounced impact on the toxicity produced.

Chemical Constituent Induced Toxicity. This type of toxicity is produced by the chemical constituents of the nanoparticle. The toxic effect may be produced by chemical interactions with the constituent materials on the surface. Even the surfactant such as CTAB, which has been extensively used to prepare and disperse gold nanorods by coating the surface, exhibits considerable toxicity. Another source of toxicity can be the release of toxic units from the nanoparticles to the surrounding, an example being the release of highly toxic Cd^{2+} ions from CdSe or CdTe quantum dots, which has been of great concern to the medical community. In biodegradable polymers, the fragments generated by degradation of nanoparticles may exhibit toxicity.

It is apparent from the above discussions that the nanotoxicity of specific nanoparticles is very much dependent on their physicochemical characteristics. This makes a consistent and reproducible toxicological study of nanoparticles much more complicated, compared to that of a conventional chemical. In fact,

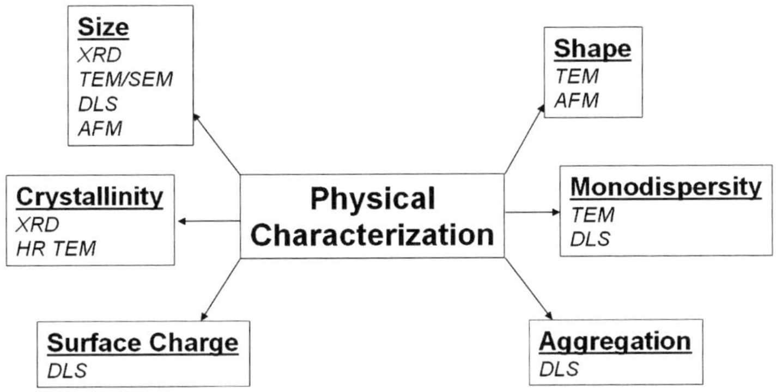

Figure 21.1. Various methods to characterize nanoparticles.

many conflicting results on the same type of nanoparticles have been reported; very often these discrepancies are derived from even subtle differences in their physicochemical characteristics. Thus, there is a need to formulate a set of characterization protocols to standardize the relevant tests and assays for an accurate characterization of nanotoxicity. Recognizing this need, a Nanotechnology Characterization Laboratory (ncl.cancer.gov/) has been established in the United States, which has standardized characterization protocols for nanotoxicity and has an extensive facility to provide nanotoxicology evaluation for a therapeutically promising nanoparticle formulation.

The key elements in evaluation of nanotoxicity are described below.

Physicochemical Characterization. A thorough characterization of the physical and chemical characteristics of the nanoparticles is the first step toward establishing the toxicity behavior. Figure 21.1 lists some important methods for the characterization of nanoparticles. One important physical characteristic is the size and its distribution. As described earlier in this book, the methods generally utilized for this purpose are X-ray diffraction, transmission electron microscopy, and dynamic light scattering. Another method not so commonly used is atomic force microscopy (AFM). The X-ray diffraction (XRD) is the simplest; and it provides not only a rough estimate of the size, but also information on the crystallinity and the crystalline phase of the nanoparticles. It is well known that the surface activity of materials is highly dependent on the type of crystalline phase. Transmission electron microscopy (TEM) provides the highest resolution and thus the most accurate information on the size, shape, and size dispersity of nanoparticles. High-resolution transmission electron microscopy (HRTEM) provides also information on the crystal structure. However, the electron microscopies are more time-consuming and expensive and require special sample preparation. Dynamic light scattering (DLS) is another method that provides the unique capability to measure the effective size (hydrodynamic radius) of the nanoparticles, when dispersed in a fluid. The DLS equipment also provides a measurement of the charge (zeta potential) on the surface of the nanoparticles.

The charge is another very important property that influences the toxicity behavior of the nanoparticles. Clear indications are that the cytotoxicity increases in the order neutral < anionic < cationic. The major reason for the increased toxicity of the positively charged nanoparticles is their affinity for the negatively charged cell membranes. Surface chemistry with the increased surface reactivity can be characterized by using various spectroscopic techniques such as X-ray photoelectron spectroscopy (XPS) or secondary ion mass spectrometry (SIMS), which provide information on surface chemical composition. Other techniques used for elemental analysis are inductively coupled plasma atomic emission spectroscopy (ICP-AES) [Szpunar, 2000], electron-dispersive X-ray analysis (EDS) [Asharani et al. 2008], and electron energy loss spectroscopy (EELS). The state of aggregation of nanoparticles, when dispersed in biological fluids, is another important factor in producing potential cytotoxicity. Aggregation or agglomeration is the phenomenon describing the nanoparticles coming together by attractive focus, such as van der Waals interaction, and adhering together to produce larger clusters that may affect intracellular uptake or may interfere with blood circulation. DLS measurements can also provide information on this characteristic.

Another recently introduced method utilizes the analysis of Brownian motions of nanoparticles. This method, known as nanoparticle tracking and analysis (NTA), involves particle-by-particle visualization and tracking, thus providing higher resolution than DLS and a better understanding of aggregation.

In Vitro Studies. The *in vitro* studies provide detailed information on cytotoxicity of nanoparticles. It is a controlled experiment where a systematic study can be used to determine not only the extent of toxicity, but also the cellular mechanism of toxicity. The *in vitro* studies are faster and less expensive and allow greater control for a systematic study compared to *in vivo* studies. However, the results of *in vitro* analysis are not directly applicable to fully predict *in vivo* results and, therefore, can only be used as initial indicators of potential toxicity. The various mechanisms of *in vitro* toxicity and the various assays used to determine them are described in detail in sections 21.2 and 21.3.

In Vivo Studies. The *in vivo* studies using animal models take into account an entire spectrum of biolmolecular interactions in real living organisms and are the next step of assessment of toxicity of nanomaterials. These toxicities are a measure of overall effect, including cytotoxicity and organotoxicity, and ultimately include entire organism. The interactions of nanoparticles here may be organ-specific, interacting with the tissue structures of the organ and interfering with its function. The *in vivo* toxicity and methods to determine them are described in detail in Section 21.4.

21.2. CYTOTOXICITY

Cytotoxicity describes toxicity at cellular level introduced by the interaction of nanoparticles with either cell membrane or the intracellular constituents and organelles. It can readily be characterized by *in vitro* studies under controlled experiments. There are various mechanisms by which a specific nanoparticle can interact with a cell and produce potential toxicity. Some of the main mechanisms are listed in Figure 21.2, although the distinctions between various mechanisms are not always clear. One mechanism, such as oxidative stress, may induce other mechanisms of cytotoxicity as discussed below. However, it is still useful to conceptually describe these processes separately as done here.

Oxidative Stress. This mechanism of cytotoxicity describes the disturbance in the oxidative balance of the cell caused by introduction of the nanoparticles. The oxidative stress, also discussed in Chapter 16 in a different context, is defined by the overexpression (increased production) of reactive oxygen species (ROS) such as superoxide (O_2^-), hydroxy radicals ($HO^•$), peroxide radicals ($ROO^•$), hydrogen peroxide (H_2O_2), and singlet oxygen as well as reactive nitrogen species (RNS) such as nitric oxide ($NO^•$), peroxinitrile anion (ONO^-), peroxynitrous acid (ONOOH), and nitrosoperoxy carbonate anion ($ONOO\ CO_2^-$). The abnormally large concentrations of these species induced by the interaction of nanoparticles with cells produces many manifestations by interactions with proteins, lipids, or nucleic acids to adversely affect cellular functions.

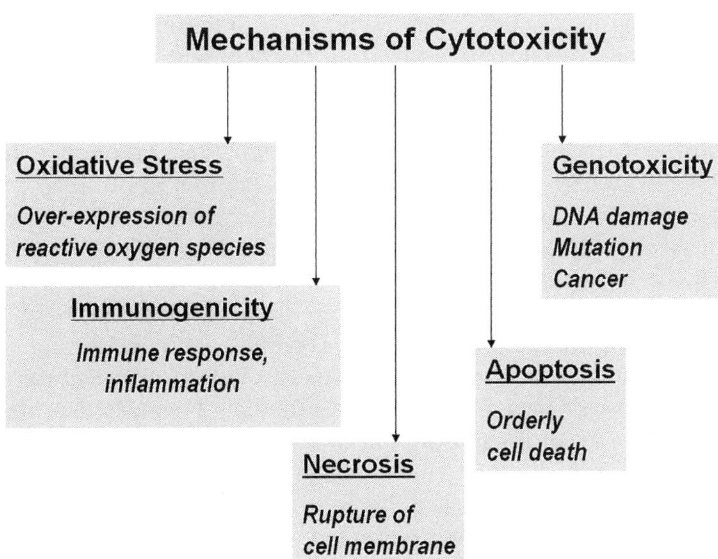

Figure 21.2. The various mechanisms of cytotoxicity.

At low level of oxidative stress, various gene transcription factors respond to increase the production of antioxidants and detoxifying enzymes, which destroy ROS and RNS, thus minimizing toxicity effect. At a higher level of oxidative stress, the protective antioxidants are not produced in sufficient quantities to adequately deal with the overexpression of ROS and RNS, which results in inflammatory response. Even higher level of oxidative stress truly overwhelms the antioxidant defense, primarily affecting mitochondrial function to lead eventually to apoptosis, the ordered cell death. The oxidative stress may also affect nuclear function by damaging DNA and by DNA mutation, producing cancer.

Genotoxicity. Genotoxicity describes the damage produced in DNA. Genotoxicity can be caused by either direct physical (e.g., intercalation into the double helix of DNA) or chemical interactions of nanoparticles with the DNA. Furthermore, the complexation of nanoparticles with DNA can induce oxidative stress which produces ROS and RNS that subsequently can damage the DNA. The effect can be DNA fragmentation, breakage of DNA double strand, suppression of DNA functions such as replication and transcription, and DNA mutations (erroneous base sequence) that may ultimately lead to malignant transformations.

Immunogenicity. This describes the immunogenic response generated by the nanoparticles as a result of modulation of cell-mediated immune responses. This is a result of the cell treating the nanoparticle entry as foreign object invasion and counteracting with immune response. This leads to increased expression of cytokines or protein signals of inflammatory response (e.g., TNF-α and chemokine IL-8), thus leading to inflammation. The inflammatory response *in vivo* may cause oxyradical release by the macrophage.

Apoptosis. Intracellular interaction of nanoparticles may induce apoptosis, which involves a series of orderly molecular processes to produce cell death. It may be a result of any of the mechanisms discussed here such as oxidative stress and genotoxicity. It is recognition by the cell of the damage and thus initiating self-destruction.

Necrosis. The interaction of nanoparticles with the cell membrane may compromise the integrity of the membrane. Ultimately, the cell membrane may be ruptured, leading to necrosis that destroys the cell. Necrosis may be a result of oxidative stress or direct interaction of nanoparticles with the lipid bilayer of the cell membrane.

Biofouling. This term has been used in the past in relation to protein absorption on a surface of nanoparticles to create undesirable effects. The association of proteins with nanoparticles can lead to the formation of a protein corona that can alter the properties of nanoparticles, affecting its biodistribution and interactions with cells and biostructures. It can trigger conformational changes in protein folding to lead to changes in the protein's biological function, and it can also affect the cell signaling pathways. Protein absorption can also lead to promote aggregation of nanoparticles, producing many adverse effects induced by aggregation.

21.3. *IN VITRO* CYTOTOXICITY ASSAYS

A number of cytotoxicity arrays have been used for *in vitro* studies to determine the mechanisms of toxicity of a specific type of nanoparticle. In addition, cell viability assays have been used to determine the eventual cytotoxicity leading to cell death. This so-called live/dead assay, also called the viability test, is an important aspect of assessing cytotoxicity.

Some of the principal *in vitro* methods to determine different types of cytotoxicity are described here. They can be broadly classified in two categories: (a) mechanistic assays, which deal with oxidative stress and DNA damage, and (b) viability assays, which investigate cell proliferation, apoptosis, necrosis, and live/dead cells relative populations. The assays commonly used generally employ optical detection, which could be colorimetric (color change), absorbance measurement at a specific wavelength, or a fluorescence detection.

Oxidative Stress. A variety of probes exhibiting a change in the fluorescence properties have been utilized [Marquis et al., 2009]. A widely used probe for determining oxidative stress is 2′,7′-dichlorofluorescein diacetate (DCFDA), which is nonfluorescent in its original form. When it crosses the cell membrane, it is hydrolyzed by esterases within the cell, to convert to a fluorescent form, 2′,7′-dichlorofluorescein, that can react with ROS. Thus, the presence of ROS can be quantified by monitoring the fluorescein (excitation at 512 nm, emission at 530 nm). However, this test may not always yield a reliable quantitative estimate of ROS, particularly in the case where apoptosis occurs which may trigger oxidation of DCFDA.

Another probe molecule for oxidative stress is dihydrorhodamine 123, which is oxidized by ROS and RNS to rhodamine 123, a highly fluorescent dye with the excitation maximum at 507 nm and emission maximum at 529 nm. Another probe molecule, particularly sensitive to superoxide, is hydroethidine, commercial name MitoSOX Red, which is used to detect superoxide production in mitochrondria where it forms an oxidative fluorescent product.

Yet another approach to detect oxidative stress is based on detecting the depletion of antioxidants caused by the presence of ROS and RNS. An example is the glutathione (GSH) test. It utilizes an essential antioxidant, GSH, which oxidizes during oxidative stress to form a GSM–GSH dimer (GSSH) involving a disulfide linkage. Thus the GSH/GSSG ratio is indicative of the oxidative stress. A simple method of detecting this ratio utilizes Ellman's reagent, 5,5′-dithio-bis-2-nitrobenzoic acid (DTNB), which reacts with the sulfhydryl group of GSH to form a yellow-colored produced, 5-thio-2-nitrobenzoic acid (TNB). TNB has a maximal absorbance at 412 nm; however, most plate readers are equipped with a 405-nm filter, and measurements are usually performed at this wavelength. Thus measuring the absorbance of TNB at 405 nm or 412 nm provides information on the concentration level of GSH.

Genotoxicity. The assessment of genotoxicity involves detection of DNA fragmentation, breakage of DNA double strand, and DNA mutation, as well as changes in regulation of DNA replication and gene expression. A variety of techniques can be utilized for a comprehensive characterization of genotoxicity.

The technique of flow cytometry, discussed in Chapter 10, can be utilized to quantify the DNA content of a cell, which may change due to an impaired DNA replication or due to aneuploidy (abnormal number of chromosomes) caused by erroneous cell cycle progression. For this purpose, a dye that intercalates into the double helix structure, thus selectively staining a double-stranded DNA, is used with the *in vitro* cellular specimen, which is then characterized as a flowing suspension in the flow cytometer. The flow cytometry steps are described in detail in Chapter 10.

A method commonly used to detect DNA fragmentation, called COMET, identifies the damaged DNA by using agarose gel electrophoreses of cell lysates. On the gel, damaged DNA appears like a comet, with the intact DNA portion as the head and the broken DNA pieces as migrating away like tails. A DNA-specific dye, propidium iodide, is utilized in the gel electrophoresis analysis to determine the amount of DNA in the tail portion, which is proportional to the amount of DNA damage. The gel electrophoresis separates DNA on the basis of molecular weight, thus producing a separate band for the DNA fragments which is trailing after electrophoresis.

A method called TUNEL (terminal deoxynucleotidyl transferase dUTP nick end labeling) can be used for the detection of DNA double-strand breakage. This method utilizes exogenous DNA polymerase I to repair the breakage in the presence of 5-bromo-2-deoxyuridine (BrdU), a synthetic analogue of the DNA base, deoxythymidine, incorporating BrdU into the repaired double-strand breaks. Subsequently a fluorescent dye conjugated to anti-BrdU antibody is used to label BrdU containing DNA, providing an estimate of the amount of DNA breakage.

DNA mutations can be investigated by using the DNA microarray technology discussed in Chapter 10, because it allows identification of all genomic sequences with a resolution of a single base mismatch. The DNA microarray technology also allows for studying the effect of nanoparticles on the expression of individual genes.

Immunogenicity. Nanoparticle interaction with the immune system and cell signaling pathways have been known to increase immunotoxicity to produce inflammatory response. The inflammation results in the up-regulation of cytokines and chemokines, which are proinflammatory mediators. The inflammatory response can be determined by ELISA or the microbead assay technologies discussed in Chapter 10, which provide quantitative assessments of cytokine and chemokine expressions.

Cell Proliferation and Viability. The cell proliferation is described by evaluation of the subpopulation of cells in the cell culture that undergoes the cell cycle progression. The cell cycle progression is very sensitive to the cell environment and can be easily interrupted by diverse factors. Thus a determination of cell proliferation is very commonly used to assess the nanotoxicity manifesting in adversely affecting cell metabolism. Apparently the easiest way to determine the population of proliferating cells is by using a mitotic index (ratio between the number of cells in mitosis and the total number of cells); this assay is usually performed by identification of mitotic cells in the cell culture using a transmission light microscope. Other approaches involve fluorescence staining of specific markers of the various cell cycle stages, an example being proliferating cell nuclear antigen (PCNA), a marker of the S phase. In a latter assay, the number of cycling cells can be derived from the ratio between the cells labeled with PCNA and total number of cells and taking into account the length of the cell cycle stages (Figure 2.8).

For the estimation of cells viability and indirect evaluation of proliferation activity, a commonly used method is the MTT colorimetric assay that measures the reduction of yellow 3-(4,5-dimethythiazol-2-yl)-2,5-diphenyl tetrazolium bromide (MTT) by mitochondrial enzyme. In a live cell, mitochondrial reductase in metabolically active mitochondria reduces this tetrazolium salt to produce insoluble purple color formazan, which can be readily detected colorimetrically. The updated version of this method, called MTS assay, utilizes the MTS (3-(4,5-dimethylthiazol-2-yl)-5-(3-carboxymethoxyphenyl)-2-(4-sulfophenyl)-2H-tetrazolium) dye with improved solubility.

Necrosis. Necrosis is identified by determining the integrity of the cell membrane and is assessed by monitoring the cellular uptake of a supravital drug, such as Trypan Blue (TB) or propidium iodide (PI). TB and PI being charged molecules can only penetrate cells with ruptured membranes. The blue staining of the cells with ruptured membranes using TB can be monitored by determining the strong absorbance of TB at ~605 nm. The PI, when being able to enter a cell with ruptured membrane, intercalates into DNA to produce a red fluorescence at 617 nm which can be used to identify necrosis.

Apoptosis. The most commonly used assay to determine nanoparticles induced apoptosis is the Annexin assay. It utilizes the feature of the early stages of apoptosis in which a phosphatidylserine in the inner layer of cell membrane is translocated to the outer layer of the membrane. Annexin V is a phosphatidylserine-specific binding molecule which then recognizes and binds to the phosphatidylserine in the outer layer of the apoptotic cell. This

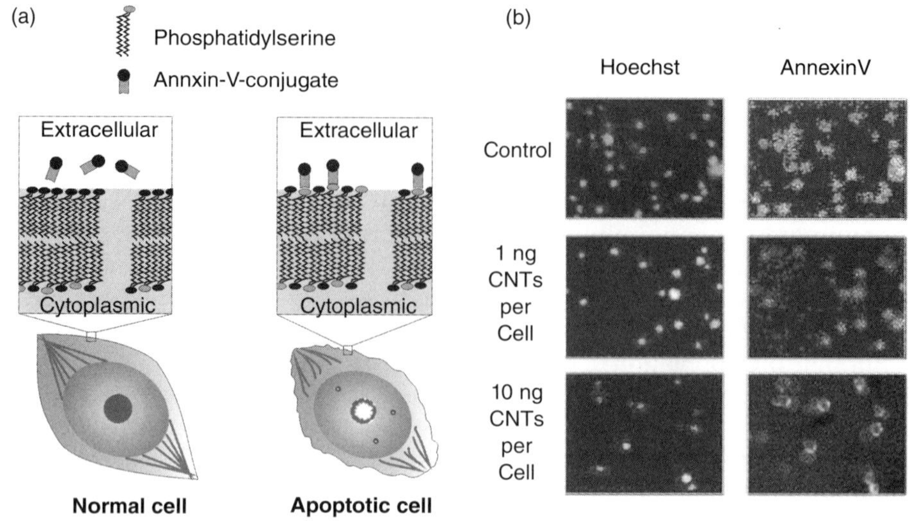

Figure 21.3. (a) Schematics of a cell undergoing apoptosis. (b) Annexin V assay for different loading of carbon nanotubes. Hoechst dye is used to stain the nucleus of the cells. (Reproduced with permission from Marquis et al. [2009].)

binding is then identified by labeling with a fluorescent dye such as FITC. Thus cells undergoing apoptosis can easily be identified by fluorescence detection, using a fluorescence microscopy.

Figure 21.3 illustrates the apoptotic behavior of the cells, cultured with the carbon nanotubes (CNT) by using a Hoechst DNA dye and Annexin-V-FITC. Apoptosis induction at a higher loading of CNT is evident.

Live/Dead Assay. This assay utilizes two dyes. The calcein acetoxymethyl (calcein AM) dye stains live cells by producing calcein with green fluorescence. Then another dye, ethidium homodimer, which is membrane impermiable, is used to stain dead cells in which the membranes are damaged and thus permeable. This dye thus stains dead cells, binding to the nucleic acids, to produce red fluorescence. A common excitation wavelength of 495 nm can be used to detect fluorescence from both calcin (green at 515 nm) and ethidium homodimer (red at 635 nm), making it quite easy to obtain a relative count of the live and red cells by fluorescence detection. An example provided is its application to the study of carbon nanoparticles, C_{60} and its derivatives, shown in Figure 21.4.

Biofouling. The protein absorption of the nanoparticles can be determined either spectroscopically or by measuring the flow characteristics/motions of the nanoparticles. Thus methods (e.g., NTA) described above to determine the

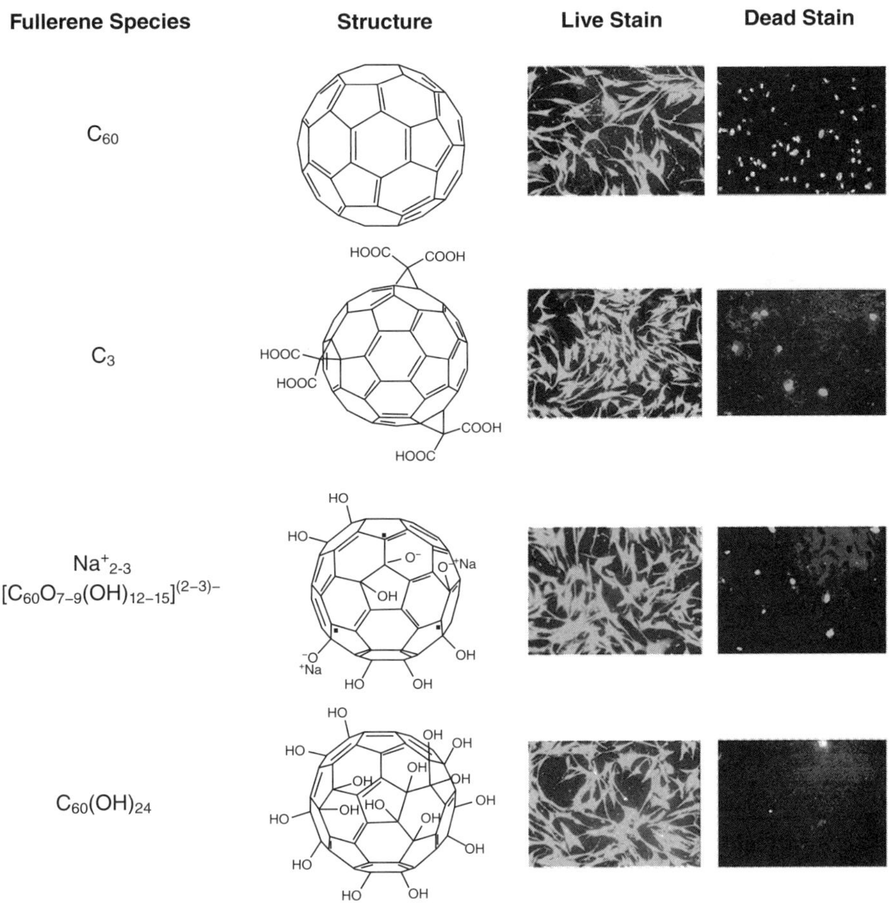

Figure 21.4. Live/dead assay for C_{60} and its derivatives. It is clear that C_{60} loading produces more dead cells, the least being exhibited by the hydroxy derivative. (Reproduced with permission from Sayes et al. [2004]).

hydrodynamic radius of the nanoparticles can be used to assess the degree of biofouling.

21.4. *IN VIVO* TOXICITY

The *in vivo* toxicity can be derived from interactions of nanoparticles with different cell constituents of the tissue structure in a given organ, as well as from interactions with the extracellular matrix (ECM). In addition, the nanoparticles can affect the cell–cell and cell–ECM interactions, which can be derived from the oxidative stress and inflammatory response. This *in vivo* toxicity can be organ-specific and manifested in a specific way dependent on

Figure 21.5. The various major types of *in vivo* toxicity of nanomaterials.

the mode of introduction of nanoparticles, or the specific organ being targeted (primary target organ). However, in a systemic delivery, where nanoparticles are directly introduced in the bloodstream, the high blood flow organs such as liver and spleen are the main target. In addition, translocation of nanoparticles to secondary organs and their tissues which are not targeted can also produce toxicity. These aspects are discussed in this section.

Figure 21.5 lists the various major types of *in vivo* toxicity of nanomaterials. They are briefly introduced here.

Pulmonary Toxicity. This organ toxicity refers to the side effect on the lungs created by interaction with nanoparticles. These effects are more pronounced when nanoparticles are introduced by instillation or inhalation within the respiratory tract.

The pulmonary deposition of the nanoparticles is segregated, according to their size, to the three different regions of the respiratory system [Simko and Mattsson, 2010]. Ninety percent of the of smaller-sized nanoparticles (~1 nm) are deposited in the nasopharyngeal, the rest being in the trachobronchial region. The particles in the size range 1–5 nm deposit in nasopharyngeal, tracheobronchial, and alveolar regions. Fifty percent of the larger-sized (~20 nm) nanoparticles deposit in the alveolar region; thus, the alveolar region is the major region for accumulation of nanoparticles of the size range typically used for nanomedicine, where the nanoparticles interact with macrophages. The adverse effects can be created by the ROS and RNS (see Section 21.2) to produce lung damage or by the pulmonary inflammatory response. In addition, translocation of nanoparticles, discussed below, can lead to their adverse interactions, with the interstitial alveolar epithelium or even transmigration across the blood–brain barriers to neurons. Specially nanoparticles of sizes <2.5 nm can reach deep lung, interacting with the alveolar epithelium and more readily translocating to other organs. The positively charged nanoparticles permeate the epithelium more effectively than the negatively charged ones.

The nanoparticle interaction with alveolar macrophages and the epithelial cells promote the release of cytokines involved in airway inflammation to generate acute inflammatory response in the lung.

Hepatotoxicity. This is the toxicity in the liver, which is a main organ known to capture nanoparticles, since it is a high blood flow system. Because it is the organ of high metabolic activity, the nanoparticles can adversely affect the metabolic activities by the oxidative stress or immunotoxicity mechanisms discussed in Section 21.2. The oxidative stress can inhibit the hepatocyte function and bile formation, while proinflammatory cytokines can lead to the pathology of liver disease.

Nephrotoxicity. This describes the toxicity in kidney, which is the body's filtration system, thus also a place of accumulation of nanoparticles. Again, the sources of nanotoxicity are mainly oxidative stress producing ROS and RNS as well as immunotoxicity generating inflammatory response which can cause kidney injury and kidney failure. Another source may be the reabsorption and accumulation of heavy metals released from nanoparticles containing them. An example is gadolinium-containing nanoparticles used for MRI contrast agent (discussed in Chapter 8, "Multimodal Biomedical Imaging"), which has been reported to create acute renal failure [Akgun et al., 2006].

Splenic Toxicity. As mentioned earlier, spleen is another organ of high blood flow and is a major target of oxidative stress due to the availability of oxygen.

Hematological Toxicity. This toxicity is mostly attributed to the red blood cells and characterized by morphological changes in these cells. The cationic nanoparticles have been reported to cause hemolysis and blood clotting. The hemolysis can also produce cardiovascular problems.

Translocation Induced Toxicity. As pointed out earlier, the nanoparticles can also reach secondary organs that are not the primary target. This process of translocation can occur through various mechanisms. An example is the translocation of inhaled nanoparticles into the brain. The nanoparticles introduced into blood circulation either by systemic injection or indirectly through epithelium absorption can cross the blood–brain barrier (BBB) to enter the central nervous system (CNS) of the brain. The BBB crossing can be due to size effect and pegylation as discussed in Chapter 6, or due to the damage of BBB by nanoparticle interaction. Neuronal cells in the CNS are very sensitive to stresses, because of their extensive, very thin, and fragile extensions (dendrites and axons), and they are also very sensitive to the aerobic metabolic activities. Also, these neurons are not renewable, making CNS very sensitive to any foreign object, causing cell death. Any inflammatory response created by the nanoparticles can cause many types of neurological disorder.

21.5. *IN VIVO* TOXICITY EVALUATION

A number of methods can be utilized to assess the *in vivo* toxicity in animal models. Most of the *in vivo* studies have been done utilizing mice. This section describes some principal methods utilized for the assessment of *in vivo* toxicity.

Behavioral Changes. This study consists of monitoring of behavior changes in an animal (e.g., mice) after injection of nanoparticles. Examples are loss of appetite, loss of weight, increased inactivity, and sleep disorder. These behavioral changes, monitored over a period of time, together with a control experiment, are indicative of toxicity effects produced by nanoparticles.

Hematology/Serum Analysis. Nanoparticle toxicity can also produce changes in populations of red blood cells, white blood cells, and platelets. These counts can readily be obtained by using flow cytometry.

Histology. An important *in vivo* assay for toxicity is examination of histological changes in the tissue structures of different organs, after introduction of nanoparticles. This requires sacrificing the animal, taking tissue specimen from different organs, fixing and sectioning them on a slide, and examining them using optical microscopy.

Excretion. An examination of renal excretion and feces provides information on the removal of nanoparticles from the body. It is not an indication of toxicity assessment of nanoparticles. However, nanoparticle excretion is definitely an indication of eliminating the possibility of any long-term health risk. Thus in the overall strategy of evaluating the benefit/risk ratio, study of excretion of nanoparticles, whether intact or biodegraded, is valuable. This can be accomplished by labeling nanoparticles by suitable contrast agents, whether optical, magnetic, or radioactive, and tracking their pathway through the animal body. Finally, one examines the excreted substances.

21.6. NANOTOXICITY STUDIES ON SELECTED NANOPARTICLES

A vast literature already exists on the toxicity studies of a wide variety of nanoparticles. Both *in vitro* and *in vivo* studies have been conducted. It is not possible to list all of them within the scope of this monograph. Suggested reviews listing some of these studies are by Lewinski et al. [2008] and Dhawan and Sharma [2010]. Despite these efforts, often conflicting and unclear results exist in the literature. Many of these are derived from the variations of experimental conditions and the materials properties of the nanoparticles. This section presents some selected types of nanoparticles, the nanotoxicity of which has been extensively investigated.

Quantum Dots. As discussed in various places in this book, quantum dots (QDs) have been extensively utilized for bioimaging and flow cytometric analysis. They have also been investigated for drug and gene delivery. In addition, they have shown promises for nonbiological applications, such as in solar cells. For this reason, a great deal of effort has been placed on evaluating their nanotoxicity, both at *in vitro* and *in vivo* levels [Lewinski et al., 2008; Hauck et al., 2010; Rzigalinski and Strobl, 2009].

The most widely used quantum dots, and hence extensively investigated for toxicity, are CdSe and CdTe [Rzigalinski and Strobl, 2009]. They have mostly been used within a core–shell geometry, with the shell being that of a wide bandgap semiconductor, usually ZnS. This coating enhances their luminescence properties as well as provides a barrier to the potential leakage of toxic Cd and Se (or Te) ions. In addition, there is a surfactant coating to make them water-dispersible and also biocompatible. Naturally the toxicity is also very much determined by the surface coatings. In addition, the size, shape, charge, and the dose (concentration) play important roles in determining toxicity.

Many reports have established that in the uncoated CdSe and CdTe-QDs, the main source of toxicity is the release of Cd^{2+} ions [Rzigalinski and Strobl, 2009; Derfus et al., 2004]. The cytotoxicity is created mainly by generation of free radicals (increased oxidative stress). Coating by a ZnS shell has been shown to reduce toxicity [Derfus et al., 2004]. However, the long-term protection provided by the ZnS is still in question, because it has been reported that the fluorescence intensity of CdSe/ZnS core–shell decreases and its peak shows a blue shift with time, suggesting that the ZnS shell degrades intracellularly [Zhang et al., 2006]. It is suggested that ZnS slowly oxidizes in the presence of air or water, generating the SO^{2-} radical.

As is suggested above, these nanoparticles have an additional layer of capping to make them dispersible. Tri-*N*-octylphosphine oxide (TOPO) is a capping ligand that has frequently been used for the Cd-based quantum dots. It has been shown that this capping ligand actually increases toxicity [Smith et al., 2008]. Thus, the choice of biocompatible and nontoxic capping ligand is very important.

In our own laboratory, we have used the quantum rods instead of quantum dots. We have performed the cytotoxicity study of silica-coated quantum rods [Kumar et al., 2010a].

We have carried out a series of MTS assay study over a period of 24–48 h with Panc1 cell line for three formulations of variable SiO_2 shell thickness using variable nanoparticles dose over a range of 0.25–20 μg/mL. Figure 21.6 shows the *in vitro* cytotoxicity effects of SiO_2/QR particles on the Panc1 and the RAW 264.7 cell lines. Both cell lines retained more than 80% viability even in the presence of a high SiO_2/QR nanoparticle loading of 20 μg/mL for 24 and 48 h. The low *in vitro* cytotoxicity of SiO_2-coated QRs can be attributed to a thick layer of ZnS and an outer layer of biocompatible silica to prevent the breakdown of the CdSe core.

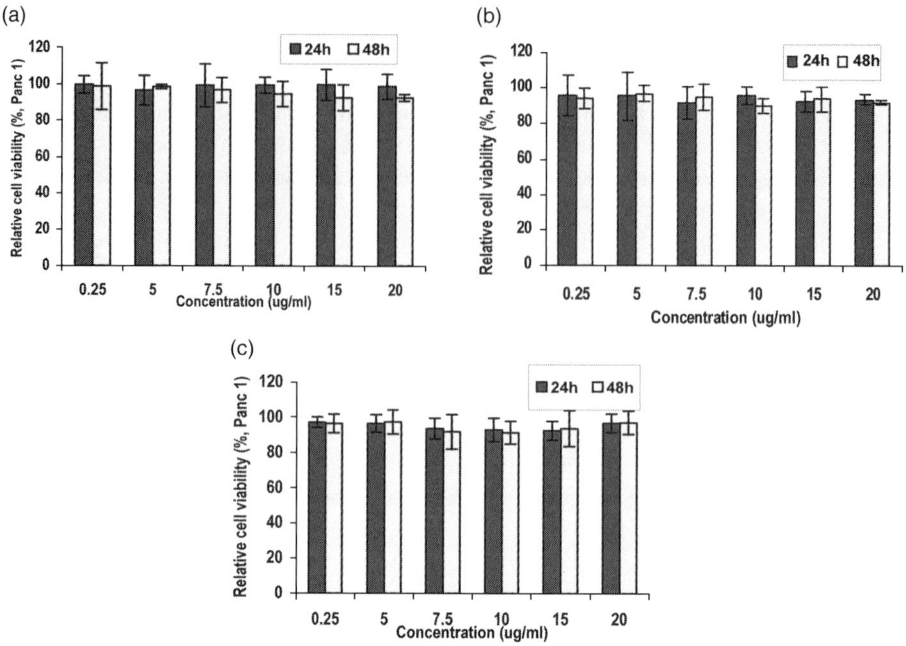

Figure 21.6. Cytotoxicity studies done by the MTS assay method for the three different formulations of SiO_2-coated QRs with different shell thickness: (a) 3 nm, (b) 10 nm, and (c) 21.3 nm. (Reproduced with permission from Kumar et al. [2010a].)

In vivo studies of toxicity of quantum dots have also been studied in various animal models. Even though the nanostructure for CdSe (or CdTe) quantum dots/rods used has the core–shell–ligand coating, which can prevent the leakage of Cd and Se ions in organs, significant concern still exists about (a) the long-term stability of the shell and (b) the coating disintegration of which can expose the CdSe(CdTe) core to the body. Cd^{2+} is known to produce ROS and oxidative stress, and its reabsorption and accumulation in kidneys can produce nephrotoxicity. Mice and rats have been used in the animal study. Rzigalinski and Strobl [2009] provide some accounts of these studies. They all point to absence of any significant toxicity at low dosage. The surface coating with a biocompatible ligand is also emerging as a strong requirement to minimize the toxicity. There also appears to be a size dependence: The smaller size, ~2.2-nm CdTe, which was localized in and around the nucleus, was found to exhibit greater toxicity than the larger, 5.2-nm size [Lovric et al., 2005], distributed within the cytoplasm.

We conducted the first toxicity study of semiconductor quantum dots in nonhuman primates, Rhesus macaque monkeys, which have a close genetic and physiological relation to humans [Ye et al., 2012]. They were intravenously injected with ~25 mg/kg of QDs and survived without any evident (observational or histological) toxic effects. These quantum dots were CdSe/CdS/ZnS core–shell–shell structures, 4–5 nm in diameter and coated with the phospholipid micelle, having a hydrodynamic radius of ~35 nm as determined by dynamic light scattering. All the measured biochemical markers were found to be in the normal range, and they showed no ill effects for six additional months. Body–weight fluctuations were insignificant compared to the period prior to treatment, suggesting that the formulation has minimal systemic effects. Also, no major fluctuation in the body temperature was observed.

Standard hematological and biochemical tests showed no sign of liver injury. Kidney function was also normal. The urine was generally transparent; occasional turbidity did not last for more than a day.

No pathological changes were found from the heart, lung, kidney, liver, and spleen. This result of absence of any toxicity is encouraging. However, the absence of any evidence for excretion of these heavy-metal-based QDs still poses concerns about the long-term effect. It thus is highly doubtful that the medical community will use them in the human body without further exhaustive studies on QDs nanotoxicity. On the other hand, these QDs are well-suited for *in vitro* diagnostics, as discussed in Chapters 8–10.

Recently, we have examined the suitability of a new-generation non-heavy-metal-based QDs. A very promising group is silicon quantum dots, which is biocompatible and also appears to be biodegradable. We have conducted similar Rhesus marquee monkey studies with silicon quantum dots and found no evidence of any toxicity.

Gold Nanoparticles and Nanorods. Gold nanostructures, in various forms such as spherical nanoparticles, nanorods, and nanoshells, have been used for a variety of applications such as imaging, sensing, photothermal therapy, and drugs, as well as gene delivery. These applications have been described at various places in this book. Considering the wide range of applications offered by nanosize gold, their toxicity is of significant interest and have been extensively investigated, both *in vitro* and *in vivo* [Lewinski et al., 2008; Aillon et al., 2009].

The literature on gold nanotixicity, again, is full of contradictions. A general consensus is that the cytotoxicity of gold nanoparticles is size-dependent, the smaller size (<5 nm) being more toxic than the larger size (≥15 nm) [Pan et al., 2007]. The smaller-sized gold nanoparticles can irreversibly bind to the major grooves of DNA to create genotoxicity, while the large-sized nanoparticles cannot because of the steric effect. A study of the effect of charge showed that cationic (amine-carrying) gold nanoparticles were moderately toxic, whereas the carboxyl-group-carrying anionic particles were non-toxic [Goodman et al., 2004].

In the case of gold nanorods, the primary cytotoxicity has been created by the cationic surfactant, CTAB, which is commonly used to produce and stabilize the nanorods [Wang et al., 2008]. Our own cell viability study of gold nanorods (GNRs) conducted using PANC1 pancreatic cancer cells showed that for both positively and negatively charged GNRs, greater than 80% cell viability was maintained even for particle concentrations as high as ~1000 μg/mL [Hu et al., 2009]. For 50% cell viability, also termed IC50, concentration values were found to be 9600 μg/mL for the gold nanorods. This high concentration value for IC50 indicates practically negligible *in vitro* toxicity for these bioconjugated GNRs which were coated with polyphospholipids.

We also conducted *in vivo* toxicity studies for GNRs [Kopwitthaya et al., 2010]. Histological studies were conducted on the major organs of mice injected with PEGylated GNR. Only minimal damage to the tissues was observed. To further investigate the toxicity of GNRs *in vivo*, hematoxylin and eosin (H&E) staining of the liver, spleen, heart, kidney, and lung tissue sections was performed. No pathological changes were observed upon comparing the tissue sections of treated mice with that of the control ones. Moreover, these anisotropic particles did not cause any significant damage to the spleen and liver, even at a concentration of 118 mg/ kg, as observed by the absence of any pathological injury in the structure of the spleen and liver. These results provide direct evidence that the prepared formulation has negligible tissue toxicity.

ORMOSILS. Our program in nanomedicine has extensively utilized organically modified silica (ORMOSIL) nanoparticles as presented in various chapters. We have carried out the histological analysis of the major organs (liver, spleen, kidney, and lung) at 360 h post injection of the DY776 conjugated ORMOSIL nanoparticles [Kumar et al., 2010b]. Tissues were harvested and fixed in 10% neutral buffered formalin and processed routinely into paraffin; the 4-μm sections were cut and stained with hematoxylin and eosin (H&E) and examined using light microscopy. No apparent tissue/cellular damages were observed in the mice injected with the DY776-conjugated nanoparticles, when compared to that obtained from mice injected with free DY776. Normal liver macrophages showed no evidence of cell injury. No pathological changes were observed in the spleen, kidney, liver, and lungs. Tissue sections were also studied for fluorescence; absence of any fluorescence from the tissues confirmed the clearance of the nanoparticles from these organs at 15 days post injection. Furthermore, mice injected with the ORMOSIL nanoparticles underwent physical and neurological evaluations for more than one month. No changes in eating, drinking, grooming, exploratory behavior, activity, physical features (e.g., weight and skin color), and neurological status were observed in these.

Carbon Nanostructures. Carbon nanoparticles, fullerene (C_{60}), and carbon nanotubes (single-walled, CWNT, as well as multiwalled, MWNT) have also been extensively investigated for biomedical applications. Because of the very large mechanical strengths of carbon nanotubes, they have also been of considerable interest for tissue engineering as discussed in Chapter 18. For this reason, the toxicological studies of these nanostructures have also received considerable attention [Lewinski et al., 2008; Aillon et al., 2009].

The cytotoxicity of C_{60} (fullerenes) have been found to be dependent on the surface functionalization as already discussed in Section 21.3. Furthermore, it also appears to be dependent on the cell type used [Lewinski et al., 2008]. While the results of *in vitro* toxicity of C_{60} are still mixed, SWNTs have been labeled to produce cytotoxicity at high concentrations, being more toxic than C_{60}. Manna et al. [2005] found an increase in oxidative stress and inhibition of cell proliferation in a dose- and time-dependent manner [Manna et al., 2005]. As in the case of C_{60}, the toxicity is also found to be highly dependent on the functionalization of the carbon nanotube surface. Some studies suggest that the cytotoxicity of the SWNTs may also be derived from the use of metal catalyst contaminants used in their preparation. Another source of toxicity, appearing in a time-dependent manner, is derived from the carbon nanotube's tendency to aggregate. The results for the MWNTs are similar to those for SWNTs [Lewinski et al., 2008].

In vivo studies show that pristine SWNTs remained stable in lungs, liver, and spleen up to 2 days [Yang et al., 2008]. This study reveals that the propensity of the carbon nanotube, to aggregate and clear slowly, can lead to oxidative stress in liver, lungs, and spleen and produce inflammation. Use of shorter length SWNT and functionalization to reduce aggregation appear to reduce toxicity [Aillon et al., 2009].

Dendrimers. Dendrimers have also been applied as major nanocarriers for drug delivery, as well as multifunctional units for theranostics. The toxicity of dendrimers has also been extensively investigated [Aillon et al., 2009]. Positive charge carrying dendrimers have been found to interact with blood compartments, destabilize cell membranes, and subsequently cause cell lysis [Stasko et al., 2007]. Hematological toxicity appears to be a major concern for dendrimers administered by i.v. injection. In the case of the G3 PAMAM dendrimers, high accumulation in kidney and liver has been reported to produce nephrotoxicity and hepatotoxicity [Roberts et al., 1996]. In general, the *in vivo* toxicity of dendrimers has been found to depend on many factors: chemical structure, size, and generation; exposure duration, nature of terminal (end) groups; and the charge. High-generation dendrimers and high doses lead to increased toxicity.

HIGHLIGHTS OF THE CHAPTER

- The toxicity of nanoparticles can be derived from two sources: nanoparticle-induced toxicity and chemical-constituent-induced toxicity.
- Nanoparticle-induced toxicity is acompanyed by the physical and surface characteristics of the nanoparticle—for example, size, shape, surface charge, and surface coverage (coatings, chemically conjugated molecules).

- The chemical-constituent-induced toxicity is produced by chemical interaction with the constituent materials on the surface. The source of toxicity can be the release of toxic units from the nanoparticles to the surrounding.
- A systematic evaluation of nanotoxicity involves (i) physicochemical characterization, (ii) *in vitro* studies, and (iii) *in vivo* studies.
- An understanding of the toxicity behavior of nanoparticles require characterizations of their chemical and physical properties such as the size, shape, crystallinity, surface charge, monodispersity, and aggregation.
- The charge is an important property that influences the toxicity behavior of the nanoparticles. Indications are that the cytotoxicity increases in the order neutral < anionic < cationic.
- The state of aggregation of nanoparticles is another important factor in producing potential cytotoxicity. Aggregation or agglomeration is the phenomenon describing the nanoparticles coming together by attractive forces, such as van der Waals interaction, and adhering together to produce larger clusters that may affect intracellular uptake or may interfere in blood circulation.
- *In vitro* studies are broadly classified in two categories: (a) mechanistic assays that deal with oxidative stress and DNA damage and (b) viability assays that investigate cell proliferation, apoptosis, necrosis, and live/dead cells relative populations.
- Cytotoxicity describes toxicity at cellular level introduced by the interaction of nanoparticles with either cell membrane or the intracellular constituents and organelles. Some of the main mechanisms are oxidative stress, immunogenicity, necrosis, apoptosis, and genotoxicity.
- At a higher level of oxidative stress, the protective antioxidants are not produced in sufficient quantities to adequately deal with the overexpression of ROS and RNS, which results in an inflammatory response.
- Genotoxicity describes the damage in DNA produced by interaction with the nanoparticles accumulated in the nucleus which creates DNA fragmentation, breakage of DNA double strand, and inhibition of DNA functions such as replication and transcription, as well as DNA mutation to change the base sequence leading to tumor growth.
- Immunogenicity describes the immunogenic response generated by the nanoparticles as a result of modulation of cell-mediated immune responses. The cells treat the nanoparticles entry as a foreign object invasion, counteracting with immune response.
- Apoptosis is a series of orderly step sequences to produce cell death. It can be a result of oxidative stress, genotoxicity, and immunogenicity.
- Necrosis describes the interaction of nanoparticles with the cell membrane that leads to compromising the integrity of the membrane.

- Biofouling involves protein absorption on a surface to create undesirable effects. The association of protein with nanoparticles can trigger conformational changes in protein folding to lead to changes in the protein's biological function and can also affect the cell signaling pathways.
- A widely used probe for determining oxidative stress is 2′,7′-dichlorofluorescein diacetate (DCFDA). When it crosses the cell membrane, it is hydrolyzed by esterases within the cell, to convert to a fluorescent form, 2′,7′-dichlorofluorescein, that can react with ROS. The presence of ROS can be quantified by monitoring the fluorescein.
- The assessment of genotoxicity involves detection of DNA fragmentation, breakage of DNA double strand, and DNA mutation, as well as changes in DNA replication and transcription to form m-RNA.
- The cell viability is described by the percentage of metabolically active cells. The most commonly used assay is the MTT assay, which utilizes a tetrazolium salt, 3-(4,5-dimethylthiazole-2-yl)-2,5-diphenyltetrazolium bromide. In a live cell, mitochondrial reductase in the metabolically active mitochondria reduces this tetrazolium salt to purple color formazan, which can be readily detected colorimetrically.
- Necrosis is assessed by monitoring the cellular uptake of a supravital drug, such as Trypan Blue (TB) or propidium iodide (PI). TB and PI, being charged molecules, can only penetrate cells with ruptured membranes and produce strong absorbance and fluorescence.
- The most commonly used assay to determine nanoparticles induced apoptosis is the Annexin assay. It utilizes the feature of the early stages of apoptosis in which a phosphatidylserine in the inner layer of cell membrane is translocated to the outer layer of the membrane.
- Live/dead assay utilizes two dyes. The calcein acetoxymethyl (calcein AM) dye stains live cells by producing calcein with green fluorescence. Then another dye, ethidium homodimer, which is membrane-impermiable, is used to stain dead cells in which the membranes are damaged and thus permeable.
- *In vivo* toxicity can be organ-specific and manifested in a specific way dependent on the mode of introduction of nanoparticles. For example, when the nanoparticles are directly introduced in the bloodstream, the high-blood-flow organs such as liver and spleen are the main target.
- Pulmonary toxicity refers to the side effect on the lungs created by interaction with nanoparticles. These effects are more pronounced when nanoparticles are introduced by instillation or inhalation within the respiratory tract.
- The nanoparticles, depending on the size, can deposit in areas of the three different regions of the respiratory system: nasopharyngeal, tracheobronchial, and in the alveolar regions.

- The nanoparticle interaction with alveolar macrophages and the epithelial cells promote the release of cytokines involved in airway inflammation to generate acute inflammatory response in the lung.
- Hepatotoxicity is the toxicity in the liver, the main organ known to capture nanoparticles. Because it is the organ of high metabolic activity, the nanoparticles can adversely affect the metabolic activities, by the oxidative stress or immunotoxicity mechanisms.
- Nephrotoxicity describes the toxicity in kidney which is the body's filtration system.
- Hematological toxicity is caused by changes in the morphology and hemology of red blood cells.
- *In vivo* toxicity can be evaluated by monitoring the behavioral changes in animal, hematology/serum analysis, histology, and excretion.
- Monitoring the behavioral changes of animals over a period of time, together with a control experiment, can provide indications of toxicity effects produced by nanoparticles.
- Hematology/serum analysis involves the monitoring of changes in the populations of red blood cells, white blood cells, and platelets.
- Histological changes in the tissue structures of different organs, after introduction of nanoparticles, are useful in assessing damage to an organ.
- An examination of renal excretion and feces provides information on the removal of nanoparticles through the body. Nanoparticle excretion is an indication of eliminating the possibility of any long-term health risk.

EXERCISES

True and False

1. Anionic nanoparticle is usually more toxic than the cationic one.
2. The alveolar region is the major region for accumulation of nanoparticles of the size range typically used for nanomedicine.
3. Genotoxicity is a result of the cell treating the nanoparticles entry as foreign object invasion and counteracting with immune response.
4. The mechanism of oxidative stress-induced toxicity describes the insufficient quantities of antioxidants to deal with the overexpression of ROS and RNS, which results in inflammatory response.
5. The proliferation of cell can be determined by a MTT assay.

Multiple Choice

6. Which of the following technique cannot be used to characterize the surface chemistry of nanoparticles?

(a) XPS
(b) TEM
(c) DLS
(d) SYMS

7. Which of the following assay can be used to determine the oxidative stress?
(a) MTT
(b) ELISA
(c) Trypan Blue
(d) DCFDA

8. How small of the nanoparticles can reach deep lung, interacting with the alveolar epithelium?
(a) <0.25 nm
(b) <2.5 nm
(c) < 25 nm
(d) <50 nm

9. In an MTT assay, a live cell will change the color of tetrazolium salt from (i) to (ii):
(a) (i) blue (ii) brown
(b) (i) brown (ii) blue
(c) (i) yellow (ii) purple
(d) (i) purple (ii) yellow

10. Which of the following is not a genotoxicity assessment?
(a) Ellman's reagent
(b) COMET
(c) Flow cytometry
(d) TUNEL

Answers: 1. F; 2. F; 3. F; 4. T; 5. T; 6. b; 7. d; 8. b; 9. c; 10. a

REFERENCES

Aillon, K. L., Xie, Y., El-Gendy, N., Berkland, C. J., Forrest, M. L., Effects of nanomaterials physicochemical properties on *in vivo* toxicity, *Advanced Drug Delivery Rev.* **61**, 457–466 (2009).

Akgun, H., Gonlusen, G., Cartwright, J., Jr., Suki, W. N., Truong, L. D., Are gadolinium-based contrast media nephrotoxic? *Arch. Pathol. Lab. Med.* **130**, 1354–1357 (2006).

Asharani, P. V., Wu, Y. L., Gong, Z. Y., Valiyaveettil, S., Toxicity of silver nanoparticles in zebrafish models, *Nanotechnology* **19**, 255102 (8 pp) (2008).

Derfus, A. M., Chan, W. C., Bhatia, S., Probing the cytotoxicity of semiconductor quantum dots, *Nano Lett.* **4**, 11–18 (2004).

Dhawan, A., Sharma, V., Toxicity assessment of nanomaterials: methods and challenges, *Anal. Bioanal. Chem.* **398**, 589–605 (2010).

Goodman, C. M., McCusker, C. D., Yilmaz, T., Rotello, V. M., Toxicity of gold nanoparticles functionalized with cationic and anionic side chains, *Bioconjug. Chem.* **15**, 897–900 (2004).

Hauck, T. S., Anderson, R. E., Fischer, H. C., Newbiggin, S., Chan, W. C. W., *In vivo* quantum-dot toxicity assessment, *Small* **6**, 138–144 (2010).

Hu, R., Yong, K. T., Ding, H., He, S., Prasad, P. N., Metallic nanostructures as localized plasmon resonance enhanced scattering probes for multiplex dark-field targeted imaging of cancer cells, *J. Phys. Chem. C* **113**, 2676–2684 (2009).

Kopwitthaya, A., Yong, K.-T., Hu, R., Roy, I., Ding, H., Vathy, L. A., Bergey, E. J., Prasad, P. N., Biocompatible PEGylated gold nanorods as colored contrast agents for targeted *In vivo* cancer applications, *Nanotechnology* **21**, 315101 (2010).

Kumar, R., Ding, H., Hu, R., Yong, K.-T., Roy, I., Bergey, E. J., Prasad, P. N., *In vitro* and *in vivo* optical imaging using water-dispersible, noncytotoxic, luminescent, silica-coated quantum rods, *Chem. Mater.* **22**, 2261–2267 (2010a).

Kumar, R., Roy, I., Ohulchanskky, T. Y., Vathy, L. A., Bergey, E. J., Sajjad, M., Prasad, P. N., *In vivo* biodistribution and clearance studies using multimodal organically modified silica nanoparticles, *ACS Nano* **4**, 699–708 (2010b).

Lewinski, N., Colvin, V., Drezek, R., Cytotoxicity of nanoparticles, *Small* **4**, 26–49 (2008).

Lovric, J., Bazzi, H. S., Cui, Y., Fortin, G. R. A., Winnik, F. M., Maysinger, D., Differences in subcellular distribution and toxicity of green and red Emitting CdTe quantum dots, *J. Mol. Med.* **83**, 377–385 (2005).

Manna, S., Sarkar, S., Barr, J., Wise, K., Barrera, E., Jejelowo, O., Rice-Ficht, A., Ramesh, G., Single-walled carbon nanotube induces oxidative stress and activates nuclear transcription factor-κB in human keratinocytes, *Nano Lett.* **5**, 1676–1684 (2005).

Marquis, B. J., Love, S. A., Braun, K. L., Haynes, C. L., Analytical methods to assess nanoparticle toxicity, *Analyst* **134** 425–439 (2009).

Oberdörster, G., Oberdörster, E., Oberdörster, J., Nanotoxicology: An emerging discipline evolving from studies of ultrafine particles, *Environ. Health Perspect.* **113**, 823–839 (2005).

Pan, Y., Neuss, S., Leifert, A., Fischler, M., Wen, F., Simon, U., Schmid, G., Brandau, W., Jahnen-Dechent, W., Size-dependent cytotoxicity of gold nanoparticles, *Small* **3**, 1941–1949 (2007).

Roberts, J. C., Bhalgat, M. K., Zera, R. T., Preliminary biological evaluation of polyamidoamine (PAMAM) starburst(TM) dendrimers, *J. Biomed. Mater. Res.* **30**, 53–65 (1996).

Rzigalinski, B. A., Strobl, J. S., Cadmium-containing nanoparticles: Perspective on pharmacology and toxicity of quantum dots, *Toxicol. Appl. Pharmacol.* **238**, 280–288 (2009).

Sayes, C., Fortner, J., Guo, W., Lyon, D., Boyd, A., Ausman, K., Tao, Y., Sitharaman, B., Wilson, L., Hughes, J., West, J., Colvin, V., The differential cytotoxicity of water-soluble fullerenes, *Nano Lett.* **4**, 1881–1887 (2004).

Simko, M., Mattsson, M-O., Risks from accidental exposures to engineering nanoparticles and neurological health effects: A critical review, *Particle Fibre Toxicol.* **7**, 1–15 (2010).

Smith, A. M., Duan, H., Mohs, A. M., Nie, S., Bioconjugated quantum dots for *in vivo* molecular and cellular imaging, *Adv. Drug Delivery Rev.* **60**, 1226–1240 (2008).

Stasko, N. A., Johnson, C. B., Schroenfisch, M. H., Johnson, T. A., Holmuhamedov, E. L., Cytotoxicity of polypropyleneimine dendrimer conjugates on cultured endothelial cells, *Biomacromolecules* **8**, 3853–3859 (2007).

Szpunar, J., Bio-inorganic speciation analysis by hyphenated techniques, *Analyst* **125**, 963–988 (2000).

Wang, S., Lu, W., Tovmachenko, O., Rai, U. S., Yu, H., Ray, P. C., Challenge in understanding size and shape dependent toxicity of gold nanomaterials in human skin keratinocytes, *Chem. Phys. Lett.* **463**, 145–149 (2008).

Wani, M. Y., Hashim, M. A., Nabi, F., Malik, M. A., Nanotoxicity: dimensional and morphological concerns, *Adv. Phys. Chem.* **2011**, 450912–450927 (2011).

Yang, S., Wang, X., Jia, G., Gu, Y., Wang, T., Nie, H., Ge, C., Wang, H., Liu, Y., Long-term accumulation and low toxicity of single-walled carbon nanotubes in intravenously exposed mice, *Toxicol. Lett.* **181**, 182–189 (2008).

Ye, L., Yong, K.-T., Liud, L., Roy, I., Hu, R., Zhu, J., Cai, H., Law, W.-C., Liu, J., Wang, K., Liu, J., Liu, Y., Hu, Y., Zhang, X., Swihart, M. T., and Prasad, P. N., A pilot study in nonhuman primates shows no adverse response to intravenous injection of quantum dots, *Nature Nanotechnol.* **7**, (2012).

Zhang, Y., He, J., Wang, P.-N., Chen, J.-Y., Lu, Z.-J., Da-Ru, L., Guo, J., Wang, C. C., Yang, W.-L., Time dependent photoluminescence blue shift of the quantum dots in living cells: Effects of oxidation by singlet oxygen, *J. Am. Chem. Soc.* **128**, 13396–13401 (2006).

INDEX

Note: Page numbers in *italics* indicate figures; tables are noted with *t*.